Lecture Notes in Mathematics

Volume 2275

More information about this series at http://www.springer.com/series/304

Tạ Quang Bửu Library, Hanoi University of Science and Technology. Location of VIASM

VIASM – The Vietnam Institute of Advanced Study in Mathematics, Hanoi

Mathematics in Vietnam goes back to ancient times. Over five hundred years ago in Hanoi the name of Lương Thế Vinh, an expert in geometry, was inscribed on a stele of honor in Văn Miếu.

Over sixty years ago, the Việt Minh published a geometry textbook written by Hoàng Tụy for schools in the liberated zones, a rare case of a guerrilla press publishing a mathematics book!

Founded in 2010 after the award of the Fields Medal to Ngô Bảo Châu, the Vietnam Institute for Advanced Study in Mathematics VIASM officially opened in Hanoi in 2011, aiming to become a leading research center where Vietnamese mathematicians can develop projects and nurture young talent. Ngô Bảo Châu, one of the initiators, became the scientific director in 2011.

VIASM engages in traditional research areas of pure and applied mathematics, as well as applying mathematics in other fields such as physics, computer science, biology and economics. The main activity of the Institute is the organization of research groups to conduct high quality research programs and projects. International and Vietnamese scientists in the same field gather and work together at the Institute. VIASM organizes conferences, workshops, seminars on topics associated with research groups working at the Institute, special schools for mathematics students, short-term training courses for mathematics teachers and common activities to disseminate scientific knowledge to the public and support the application of mathematics in socio-economic development.

The VIASM subseries of the Lecture Notes in Mathematics publishes high quality original articles or survey papers on topics of current interest. They are based on lectures delivered in special periods organized at the Vietnam Institute for Advanced Study in Mathematics (VIASM). With the agreement of the Editors of the LNM Series, and as a temporary arrangement, the first volumes are not subjected to the strict LNM rules of coherency for multi-author volumes.

Bruno Anglès • Tuan Ngo Dac

Editors

Arithmetic and Geometry over Local Fields

VIASM 2018

Editors

Bruno Anglès
Laboratoire de Mathématiques Nicolas
Oresme, CNRS UMR 6139
Université de Caen Basse-Normandie
Caen, France

Tuan Ngo Dac
Institut Camille Jordan, UMR 5208
Claude Bernard University Lyon 1
Villeurbanne, France

ISSN 0075-8434   ISSN 1617-9692   (electronic)
Lecture Notes in Mathematics
ISBN 978-3-030-66248-6   ISBN 978-3-030-66249-3   (eBook)
https://doi.org/10.1007/978-3-030-66249-3

Mathematics Subject Classification: Primary: 11Fxx, 11Gxx, 11Lxx, 14Fxx, 14Gxx; Secondary: 12Hxx

This Springer imprint is published by the registered company Springer Nature Switzerland AG.
The registered company address is: Gewerbestrasse 11, 6330 Cham, Switzerland

Preface

Arithmetic geometry is a very active branch of mathematics, with important and deep connections to various areas such as algebraic geometry, number theory, and Lie theory.

The goal of this volume is to introduce graduate students and young researchers to some recent research topics in arithmetic geometry over local fields. The lectures are centered around two common themes: the study of Drinfeld modules and non-Archimedean analytic geometry.

The notes of this volume grew out from the lectures given during the research program "Arithmetic and geometry of local and global fields," which took place at the Vietnam Institute of Advanced Study in Mathematics (VIASM) from June to August 2018. Two of them were given at the VIASM School on Number Theory (see https://hanoi-nt18.sciencesconf.org/). The others were presented as advanced courses during the research seminar.

The authors, all leading experts in the subject, have made a great effort to make the notes as self-contained as possible. In addition to introducing the basic tools, the lectures aim to present an overview of recent developments in the arithmetic and geometry of local fields and related topics. The included examples and suggested concrete research problems will enable young researchers to quickly reach the frontiers of this fascinating branch of mathematics.

Contents of This Volume

The volume consists of seven lectures:

3. *Igusa's Conjecture on Exponential Sums Modulo p^m and Local-Global Principle* by **Nguyen Huu Kien** (KU Leuven, Belgium),
4. *From the Carlitz Exponential to Drinfeld Modular Forms* by **Federico Pellarin** (Institute Camille Jordan and University of Saint-Etienne, France),
5. *Berkovich Curves and Schottky Uniformization I: The Berkovich Affine Line* by **Jérôme Poineau** (University of Caen Normandy, France) and **Daniele Turchetti** (Dalhousie University, Canada),
6. *Berkovich Curves and Schottky Uniformization II: Analytic Uniformization of Mumford Curves* by **Jérôme Poineau** (University of Caen Normandy, France) and **Daniele Turchetti** (Dalhousie University, Canada),
7. *On the Stark Units of Drinfeld Modules* by **Floric Tavares Ribeiro** (University of Caen Normandy, France).

The first lecture by D. Caro offers an introduction to p-differential methods in arithmetic geometry. First, he reviews Berthelot's ring of p-adic differential operators, which plays an important role in the theory of arithmetic $\mathscr{D}$-modules. Next, he extends it to some finite level on p-adic formal affine smooth schemes. Finally, concrete examples and a guide to further reading are also provided. The material assumes a basic knowledge of ring theory and algebraic geometry.

The second lecture by L. Di Vizio gives an overview of the Galois theory of difference equations. The first part presents a guide to the key definitions and results of difference Galois theory. In the second part, interesting applications to transcendence and differential transcendence are treated in detail. Note that the framework is the same as that of Papanikolas' theory in the setting of Drinfeld modules. The curious reader may wish to refer to the lectures of F. Pellarin and F. Tavares Ribeiro for more details.

The third lecture by K. Nguyen is a survey on Igusa's conjecture around exponential sums motivated by the study of local-global principles for forms of higher degree. After introducing the notion of exponential sums and those modulo p^m with some examples, he formulates a coarse form of Igusa's conjecture on a uniform bound of those exponential sums and explains its relations with Igusa's local zeta functions, the monodromy conjecture, and fiber integrals. He then states Igusa's conjecture on exponential sums and gives an overview of recent progress on this conjecture, in particular the most recent breakthrough of Cluckers, Mustaţă, and the author. The lecture ends with a general picture of the local-global principle for forms and the contribution of the aforementioned conjecture in this direction.

The fourth lecture by F. Pellarin presents a friendly introduction to the theory of Drinfeld modular forms attached to the affine line over a finite field. Drinfeld modular forms in positive characteristic are defined as analogues of classical modular forms by the pioneering works of Goss and Gekeler. The notes gradually introduce the very first basic elements of the arithmetic theory of Drinfeld modules, then the Drinfeld upper-half plane and its topology, and end with Drinfeld modular forms. The key notions are illustrated with many examples. The lecture also contains several advanced parts such as Drinfeld modular forms with values in Banach algebras. These course notes will enable the reader to gently enter into this rich and still developing theory.

The self-contained survey by J. Poineau and D. Turchetti consists of two lectures on non-Archimedean curves and Schottky uniformization from the point of view of Berkovich geometry. The first part, the fifth lecture, could be read as an elementary course on the theory of Berkovich spaces with emphasis on the affine line. The authors introduce basic definitions and properties with full details and proofs. The second part, the sixth lecture, is more advanced and deals with the theory of uniformization of curves under Berkovich's theory. After introducing the notion of Mumford curves and Schottky groups, the authors present an analytic proof of Schottky uniformization. Many examples, explicit research problems, and a guide for further reading are also provided. The reader who is interested in Schottky groups in the language of rigid analytic spaces is invited to read the relevant parts of the lecture of F. Pellarin.

The last lecture by F. Tavares Ribeiro presents some recent developments in the arithmetic theory of the special values of Goss zeta functions. This lecture is an exposition on Stark units of Drinfeld modules over the ring of polynomials over a finite field. The notes are also the occasion to introduce basic definitions of Drinfeld modules and more recent fundamental objects linked to an analytic class number formula obtained by L. Taelman: L-values, unit modules, and class modules attached to a Drinfeld module. The author then presents the notion of Stark units and gives their basic properties. Finally, he gives several applications of Stark units, in particular, to the study of congruence properties of Bernoulli–Carlitz numbers. He also gives hints for a general base ring.

Caen, France   Bruno Anglès
Lyon, France   Tuan Ngo Dac
October 2020

Acknowledgments

We would like to acknowledge the many people whose help was essential for this volume.

We would like to thank the speakers and the authors, Daniel Caro, Lucia Di Vizio, Nguyen Huu Kien, Federico Pellarin, Jérôme Poineau, Floric Tavares Ribeiro, Lenny Taelman, and Daniele Turchetti, for their enthusiastic investment to this semester and this volume. We also thank all the referees for many valuable comments and suggestions that helped in improving the exposition of each lecture as well as the whole volume.

We would like to express our gratitude to the directors, the researchers, and administrative staff of the VIASM for the hospitality and excellent working conditions during the special semester. Special thanks are due to Lê Minh Ha, Ngô Bao-Châu, Phung Hô Hai, Lê Thi Lan Anh, and Nguyên Hoang Anh.

Our semester was partially supported by the VIASM, the Institute of Mathematics in Hanoi (IMH), the European Research Council (ERC Starting Grant TOSSIBERG), the ANR Grant PerCoLaTor (ANR-14-CE25-0002-01), the Institut Henri Poincaré (IHP), the Foundation Compositio Mathematica, and the GDR Structuration de la théorie des nombres (CNRS).

Contents

Contributors

Daniel Caro Université de Caen Normandie, Normandie Univ, Laboratoire de Mathématiques Nicolas Oresme, CNRS UMR 6139 Caen, France

Lucia Di Vizio Laboratoire de Mathématiques UMR 8100, CNRS, Université de Versailles-St Quentin Versailles Cedex, France

Kien Huu Nguyen KU Leuven, Department of Mathematics Leuven, Belgium

Federico Pellarin Institut Camille Jordan, UMR 5208, Site de Saint-Etienne Saint-Etienne, France

Jérôme Poineau Laboratoire de mathématiques Nicolas Oresme, Université de Caen-Normandie, Normandie University Caen, France

Floric Tavares Ribeiro Laboratoire de Mathématiques Nicolas Oresme, Université de Caen Normandie, Normandie University Caen, France

Daniele Turchetti Department of Mathematics and Statistics, Dalhousie University Halifax, NS, Canada

Chapter 1
Some Elements on Berthelot's Arithmetic $\mathcal{D}$-Modules

Daniel Caro

Abstract This text is an introduction to Berthelot's theory of arithmetic $\mathcal{D}$-modules. We first review Berthelot's ring of differential operators of finite level on affine smooth p-adic formal schemes over a complete discrete valuation ring of mixed characteristic $(0, p)$ with perfect residue field. Berthelot's ring is a kind of weak p-adic completion of the usual ring of differential operators as defined by Grothendieck. We finish with the description and some finiteness properties of the constant coefficient which is constructed by adding overconvergent singularities. This lecture is suitable for graduate students and requires only basic knowledge of ring theory and algebraic geometry.

1.1 Introduction

These notes correspond to a course given at Vietnam Institute for Advanced Study in Mathematics (VIASM) in July 2018. The purpose was mainly to present an introduction on Berthelot's theory of arithmetic $\mathcal{D}$-modules for Ph.D. students. Except at the last page, we only work with affine (formal) schemes, i.e. with (p-adically complete) rings. The course until the very end does not require some background on Grothendieck's algebraic geometry. Beyond this paper, let us explain a bit the interest of such recent research topic. First, we have to underline that Berthelot's theory of arithmetic $\mathcal{D}$-modules gives a p-adic cohomology closed under Grothendieck six operations (see for instance, [CT12] with Frobenius structures or much more recently [Car20] for a context without Frobenius structures). Moreover, the theory of arithmetic $\mathcal{D}$-modules contain in some sense that of overconvergent F-isocrystals (more precisely, we can construct a canonical fully faithful functor). These latter coefficients lives over Tate rigid analytic spaces or more recently can be viewed as objects over Berkovich spaces. Berkovich spaces are very useful, as the

D. Caro (✉)
Université de Caen Normandie, Normandie Univ, Laboratoire de Mathématiques Nicolas Oresme, CNRS UMR 6139, Caen, France
e-mail: daniel.caro@unicaen.fr

© The Author(s), under exclusive license to Springer Nature Switzerland AG 2021
B. Anglès, T. Ngo Dac (eds.), *Arithmetic and Geometry over Local Fields*,
Lecture Notes in Mathematics 2275, https://doi.org/10.1007/978-3-030-66249-3_1

reader can see below via the lecture of Jérôme Poineau and Daniele Turchetti (see [PT20a, PT20b])

In the second chapter, we review some elements on the standard theory of the ring of differential operators. In the third chapter, we explain how to extend it to some finite level m in the sense of Berthelot. We also introduce the constant coefficient denoted by $\mathcal{O}_{\mathfrak{P}}(^{\dagger}T)_{\mathbb{Q}}$ and explain why we have to work with such a weak p-adic completion and not with the naive coefficient.

1.2 Ring of Differential Operators

1.2.1 Kähler Differential

Definition 1.2.1 Let R be a commutative ring, A be a commutative R-algebra, M be an A-module. An R-derivation of A into M is an R-linear map $d : A \to M$ such that

1. $d(a_1 a_2) = a_1 d(a_2) + a_2 d(a_1)$, for any $a_1, a_2 \in A$;

The set of R-derivation of A into M is denoted by $\mathrm{Der}_R(A, M)$. We check that $\mathrm{Der}_R(A, M)$ is an A-submodule of $\mathrm{Hom}_{\mathrm{Set}}(A, M)$, where the A-module structure is given by that of M.

Definition 1.2.2 A module of relative differential forms of A over R is an A-module $\Omega^1_{A/R}$ endowed with an R-derivation $d : A \to \Omega^1_{A/R}$ having the following universal property: for any A-module M, for any $D \in \mathrm{Der}_R(A, M)$, there exists a unique $\phi \in \mathrm{Hom}_A(\Omega^1_{A/R}, M)$ such that $D = \phi \circ d$.

Proposition 1.2.3 *A module of relative differential forms of A over R exists and is unique up to unique isomorphism. It will be denoted* $(\Omega^1_{A/R}, d)$.

Proof The uniqueness is standard. Let us check the existence. Let F be the free A-module generated by the symbols da, $a \in A$. Let E be the A-submodule of F generated by

  (i) dr, $r \in R$;
 (ii) $d(a_1 + a_2) - da_1 - da_2$, for any $a_1, a_2 \in A$.
(iii) $d(a_1 a_2) - a_1 da_2 - a_2 da_1$, for any $a_1, a_2 \in A$.

We put $\Omega^1_{A/R} := F/E$ and $d : A \to \Omega^1_{A/R}$ sends an element $a \in A$ to the class of da. We check that such $(\Omega^1_{A/R}, d)$ satisfies the universal property.   □

Remark 1.2.4 In other words, the proposition 1.2.3 means that the functor $M \mapsto \mathrm{Der}_R(A, M)$ from the category of A-modules to itself is representable. Moreover, following the proof of 1.2.3, the A-module $\Omega^1_{A/R}$ is generated by $\{da, a \in A\}$.

1.2.5 Let $\mu : A \otimes_R A \to A$ be the morphism of R-algebras given by $a \otimes a' \mapsto aa'$. Let $I := \ker \mu$. Since μ is surjective, we get the factorisation of R-algebras

$A \otimes_R A/I \xrightarrow{\sim} A$. The R-algebra $A \otimes_R A$ is endowed with two A-algebra structures, the left one and the right one. We remark that for both structures, the isomorphism $A \otimes_R A/I \xrightarrow{\sim} A$ is A-linear. Hence, if M is an $A \otimes_R A$-module, then the three way to endow M/IM with an A-module structure coincide. For instance, I/I^2 is endowed with a canonical structure of A-module.

For both A-algebra structure on $A \otimes_R A$, I is generated as A-submodule by the elements of the form $1 \otimes a - a \otimes 1$, with $a \in A$. Indeed, for the left case, if $\sum a_l \otimes a'_l \in I$, then $\sum_l a_l \otimes a'_l = \sum_l a_l (1 \otimes a'_l - a'_l \otimes 1)$.

Proposition 1.2.6 *Let* $d \colon A \to I/I^2$ *given by* $a \mapsto 1 \otimes a - a \otimes 1 \mod I^2$. *Then* $(I/I^2, d)$ *is a module of relative differential forms of A over R is an A-module, i.e. in particular* $I/I^2 \xrightarrow{\sim} \Omega^1_{A/R}$.

Proof The R-linearity of d and the property $d(r) = 0$, for any $r \in R$ are obvious. Let $a, a' \in A$. To fix ideas, we use by default the left A-algebra structure the $A \otimes_R A$. Then I^n can be viewed as an A-submodule of $A \otimes_R A$. We remark that the induced A-module structure on I/I^2 is the same than that given by its canonical $A \otimes_R A/I$-module structure. We get in $A \otimes_R A$ the equality $a(1 \otimes a' - a' \otimes 1) + a'(1 \otimes a - a \otimes 1) = a \otimes a' - 2aa' \otimes 1 + a' \otimes a$. Since $1 \otimes aa' - a \otimes a' - a' \otimes a + aa' \otimes 1 = (1 \otimes a - a \otimes 1)(1 \otimes a' - a' \otimes 1) \in I^2$, this yields $a(1 \otimes a' - a' \otimes 1) + a'(1 \otimes a - a \otimes 1) \equiv 1 \otimes aa' - aa' \otimes 1 \mod I^2$. Hence, $a\,da' + a'\,da = d(aa')$.

It remains to check that d satisfies the universal property. Let M be an A-module and $D \in \mathrm{Der}_R(A, M)$. Let $A * M$ be the trivial extension ring, i.e. this is an A-algebra such that $A * M = A \oplus M$ as A-module, and the multiplication is given by $(a, m)(a', m') = (aa', am' + a'm)$. We remark that M is an ideal of $A * M$ such that $M^2 = 0$. Let $\theta \colon A \otimes_R A \to A * M$ be the A-linear map (for the left structure of $A \otimes_R A$) such that $a_1 \otimes a_2 \mapsto (a_1 a_2, a_1 Da_2)$. We check that θ is in fact a homomorphism of A-algebras. Indeed, we compute $\theta((a_1 \otimes a_2)(a'_1 \otimes a'_2)) = (a_1 a_2 a'_1 a'_2, a_1 a'_1 Da_2 a'_2) = \theta(a_1 \otimes a_2)\theta(a'_1 \otimes a'_2)$. Let $i = \sum_l a_l \otimes a'_l \in I$. Since $\sum a_l a'_l = 0$, we get $\theta(i) \in M$. Hence $\theta(I^2) \subset M^2 = 0$. Hence, θ induces $\phi \colon I/I^2 \to M$ given by $\sum_l a_l \otimes a'_l \mod I^2 \mapsto \sum_l a_l Da'_l$. Since $D = \phi \circ d$, we are done. $\square$

Examples 1.2.7 Suppose $A = R[T_1, \ldots, T_d]$. Then $A \otimes_R A = R[T_1 \otimes 1, \ldots, T_d \otimes 1, 1 \otimes T_1, \ldots, 1 \otimes T_d]$. The diagonal morphism $\mu \colon A \otimes_R A \to A$ is given by $T_i \otimes 1 \mapsto T_i$ and $1 \otimes T_i \mapsto T_i$. Let $\tau_i := 1 \otimes T_i - T_i \otimes 1$. Let $P \in I$. By doing successive Euclidian divisions, we get $P = \tau_1 Q_1 + \cdots + \tau_d Q_d + S$, where $S \in R[1 \otimes T_1, \ldots, 1 \otimes T_d]$ and $Q_i \in A \otimes_R A$. Since $\mu(P) = 0$, this yields $S = 0$. Hence, I is generated by $\tau_1, \ldots, \tau_d$. For any $\underline{\alpha} \in \mathbb{N}$, we set $\underline{\tau}^{\underline{\alpha}} := \tau_1^{\alpha_1} \cdots \tau_d^{\alpha_d}$, and $|\underline{\alpha}| := \alpha_1 + \cdots + \alpha_d$. For any $n \in \mathbb{N}$, this implies that I^n is generated by $\{\underline{\tau}^{\underline{\alpha}}, \sum_i \alpha_i = n\}$.

The A-module I^n/I^{n+1} is generated by $\{\underline{\tau}^{\underline{\alpha}} \mod I^{n+1}, \sum_i \alpha_i = n\}$. Indeed, let $P = \sum_{|\underline{\alpha}|=n} P_{\underline{\alpha}} \underline{\tau}^{\underline{\alpha}} \in I^n$, with $P_{\underline{\alpha}} \in A$. If $P \in I^{n+1}$, then either the total degree in $1 \otimes T_1, \ldots, 1 \otimes T_d$ of P is $\geq n + 1$ or $P = 0$. Hence, we are done.

1.2.2 Ring of Differential Operators

Let A be a commutative R-algebra.

Definition 1.2.8 For any $n \in \mathbb{N}$, we set $P_{A/R}^n := (A \otimes_R A)/I^{n+1}$. The ring $P_{A/R}^n$ has two A-algebra structures: the left one induced by the left structure of $A \otimes_R A$, and the right one. We denote by $d_0^n : A \to P_{A/R}^n$ (or simply d_0) the homomorphism given by the left structure, and $d_1^n : A \to P_{A/R}^n$ (or simply d_1) that given by the right one.

Definition 1.2.9 An R-linear map $h : A \to A$ is a "differential operator of A/R of order $\leq n$" if and only if there exists an A-linear map $\widetilde{h} : d_{0*}(P_{A/R}^n) \to A$ making commutative the diagram

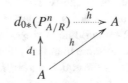

We denote by $\mathrm{Diff}_{A/R}^n$ the set of differential operators of A/R of order $\leq n$. Since $d_{0*}(P_{A/R}^n)$ is generated as A-module by the image of d_1, then we get the uniqueness of such $\widetilde{h}$ if it exists. Hence we get the bijection $\mathrm{Diff}_{A/R}^n \xrightarrow{\sim} \mathrm{Hom}_A(d_{0*}(P_{A/R}^n), A)$ given by $h \mapsto \widetilde{h}$. We set $D_{A/R,n} := \mathrm{Hom}_A(d_{0*}(P_{A/R}^n), A)$.

Since $P_{A/R}^n$ has two structures of A-algebra inducing a structure of A-bimodules, then $D_{A/R,n}$ is endowed with a structure of A-bimodule. The left (resp. right) structure of A-module of $D_{A/R,n}$ is by definition that coming from the left (resp. right) structure of A-algebra of $P_{A/R}^n$.

Notation 1.2.10 Let $n, n' \in \mathbb{N}$. When we write $P_{A/R}^n \otimes_A P_{A/R}^{n'}$, we use the right A-algebra structure of $P_{A/R}^n$ and the left A-algebra structure of $P_{A/R}^{n'}$ to define the tensor product. Hence, we get a structure of the left (resp. right) A-algebra structure on $P_{A/R}^n \otimes_A P_{A/R}^{n'}$ coming from that of $P_{A/R}^n$ (resp. $P_{A/R}^{n'}$).

Lemma 1.2.11 *For any integers* $n, n' \in \mathbb{N}$, *there exists a unique homomorphism* $\delta^{n,n'} : P_{A/R}^{n+n'} \to P_{A/R}^n \otimes_A P_{A/R}^{n'}$ *of A-algebras for both left and right structures of A-algebras making commutative the following diagram*

$$
\begin{CD}
A @>{d_1}>> P_{A/R}^{n+n'} \\
@V{d_1}VV @VV{\delta^{n,n'}}V \\
P_{A/R}^{n'} @>{d_1}>> P_{A/R}^n \otimes_A P_{A/R}^{n'}.
\end{CD}
\tag{1.1}
$$

Proof Let us check the unicity. Computing the image of 1 in the diagram 1.1, we get $\delta^{n,n'}(\overline{1 \otimes 1}) = \overline{1 \otimes 1} \otimes \overline{1 \otimes 1}$. Using the fact that $\delta^{n,n'}$ is an homomorphism of A-algebras for both left and right structures of A-algebras, this yields $\delta^{n,n'}(\overline{a \otimes a'}) = \overline{a \otimes 1} \otimes \overline{1 \otimes a'}$. It remains to prove the existence. Let $\delta \colon A \otimes_R A \to (A \otimes_R A) \otimes_A (A \otimes_R A)$ defined by $a \otimes a' \mapsto (a \otimes 1) \otimes (1 \otimes a')$. Set $1\!\!\!/ := 1 \otimes 1$. We compute $\delta(1 \otimes a - a \otimes 1) = 1\!\!\!/ \otimes (1 \otimes a - a \otimes 1) + (1 \otimes a - a \otimes 1) \otimes 1\!\!\!/$. Since I is generated as A-module (for both structures) by $1 \otimes a - a \otimes 1$ with $a \in A$, then $\delta(I) \subset I \otimes_A (A \otimes_R A) + (A \otimes_R A) \otimes_A I$. Hence $\delta(I^{n+n'+1}) \subset I^{n+1} \otimes_A (A \otimes_R A) + (A \otimes_R A) \otimes_A I^{n'+1}$, i.e. δ induces the morphism of A-algebras $\delta^{n,n'}$. $\qquad\square$

Proposition 1.2.12 *Let $h \in \mathrm{Diff}^n_{A/R}$, $h' \in \mathrm{Diff}^{n'}_{A/R}$. Then $h \circ h' \in \mathrm{Diff}^{n+n'}_{A/R}$.*

Proof We check by definition, functoriality or by using 1.1 the commutativity of the following diagram:

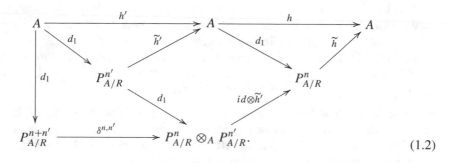

$$\tag{1.2}$$

$\qquad\square$

Definition 1.2.13 By convention, R-algebras are always unital and associative. Following 1.2.12, $\mathrm{Diff}_{A/R} := \cup_{n \in \mathbb{N}} \mathrm{Diff}^n_{A/R}$ is an R-subalgebra of $\mathrm{Hom}_R(A, A)$. We get a natural action of $\mathrm{Diff}_{A/R}$ on A as follows: $\forall P \in \mathrm{Diff}_{A/R}$, $\forall f \in A$, $P \cdot f := P(f)$.

Definition 1.2.14 Via the bijection $\mathrm{Diff}^n_{A/R} \xrightarrow{\sim} D_{A/R,n}$, we get an R-algebra structure on $D_{A/R} := \cup_{n \in \mathbb{N}} D_{A/R,n}$ making the bijection $\mathrm{Diff}_{A/R} \xrightarrow{\sim} D_{A/R}$ an isomorphism of R-algebras and of A-bimodules. The R-algebra $D_{A/R}$ is called the ring of differential operators on A/R. The multiplication can be described via the diagram 1.2. We also get a natural action of $D_{A/R}$ on A. Finally, the mapping $A \to D_{A/R}$ given by $a \mapsto a \cdot 1$ is the same for the left or right structure of A-module. Moreover, the mapping $A \to D_{A/R}$ is an homomorphism of R-algebras, and we can view A as an R-subalgebra of $D_{A/R}$.

Notation 1.2.15

1. If $P \in D_{A/R}$ and $f \in A$, we set $P(f) := P(d_1(f))$, which is consistant with the identification of P with an element of $\mathrm{Diff}_{A/R}$.
2. For any $P \in D_{A/R}$, we set $\mathrm{ord}(P) := \min\{n \; ; \; P \in D_{A/R,n}\}$.

1.2.3 Smooth Differential Case

From now, we suppose R Noetherian, A/R of finite type. Let $t_1, \ldots, t_d \in A$, and $\tau_i := 1 \otimes t_i - t_i \otimes 1 \in A \otimes_R A$. We suppose that $\tau_1, \ldots, \tau_d \in I$ is quasi-regular, i.e.

(a) $\overline{\tau}_1, \ldots, \overline{\tau}_d$ is an A-basis of $I/I^2 = \Omega^1_{A/R}$.
(b) the morphism $S^\bullet_A(I/I^2) \to \mathrm{gr}^\bullet_I(A \otimes_R A)$ induced by $\overline{\tau}_1, \ldots, \overline{\tau}_d$ (see notation 1.2.17 below) is an isomorphism of graded A-algebras.

This also means that A/R is differentially smooth in the sense of Grothendieck.

Examples 1.2.16

1. Suppose A is equal to the polynomial algebra $R[T_1, \ldots, T_d]$. Set $\tau_i := 1 \otimes T_i - T_i \otimes 1 \in A \otimes_R A$ for any $i = 1, \ldots, d$ Then $\tau_1, \ldots, \tau_d \in I$ is quasi-regular.
2. We have the following example which is fundamental but can be skipped if the reader is not familiar with Grothendieck's notion of étaleness. Let $f \colon R[T_1, \ldots, T_d] \to A$ be an étale morphism. Let $t_1, \ldots, t_d \in A$ be the image of $T_1, \ldots, T_d$ via f. Set $\tau_i := 1 \otimes t_i - t_i \otimes 1 \in A \otimes_R A$ for any $i = 1, \ldots, d$. Then $\tau_1, \ldots, \tau_d \in I$ is quasi-regular (see [EGA IV, 0.15.1.7]).

Notation 1.2.17 If M is an A-module, we denote by $S^\bullet_A(M)$, or simply $S_A(M)$ if we forget the filtration, the symmetric algebra of M over A endowed with its canonical filtration. Recall $M \mapsto S_A(M)$ is the left adjoint functor of the inclusion from the category of commutative A-algebras to that of A-modules.

If B is an algebra and J is an ideal of B, we denote by $\mathrm{gr}^\bullet_J(B)$ the graded ring $\mathrm{gr}^\bullet_J(B) := \oplus_{d \geq 0} J^d / J^{d+1}$.

Notation 1.2.18 For any $\underline{\alpha} \in \mathbb{N}^d$, we set $\underline{T}^{\underline{\alpha}} := T_1^{\alpha_1} \cdots T_d^{\alpha_d}$ and $\underline{\tau}^{\underline{\alpha}} := \tau_1^{\alpha_1} \cdots \tau_d^{\alpha_d}$. By hypothesis, we have the isomorphism of graded A-algebras of the form

$$A[T_1, \ldots, T_d] \xrightarrow{\;\sim\;} \mathrm{gr}^\bullet_I(A)$$

given by $\underline{T}^{\underline{\alpha}} \mapsto \underline{\tau}^{\underline{\alpha}} \bmod I^{n+1}$, for any $\underline{\alpha} \in \mathbb{N}^d$ such that $|\underline{\alpha}| = n$. Moreover, we can check $\underline{\tau}^{\underline{\alpha}} \bmod I^{n+1}$ is an A-basis of $P^n_{A/R}$ for both A-algebra structures. We denote by $\{\partial^{[\underline{k}]}, |\underline{k}| \leq n\}$ be the dual basis of $D_{A/R,n} = \mathrm{Hom}_A(d_{0*}(P^n_{A/R}), A)$ of $\{\underline{\tau}^{\underline{k}}, |\underline{k}| \leq n\}$. Since the morphism $P^{n+1}_{A/R} \twoheadrightarrow P^n_{A/R}$ sends $\underline{\tau}^{\underline{\alpha}} \bmod I^{n+2}$ to $\underline{\tau}^{\underline{\alpha}}$ $\bmod I^{n+1}$, then the monomorphism $D_{A/R,n} \hookrightarrow D_{A/R,n+1}$ sends $\partial^{[\underline{k}]}$ to $\partial^{[\underline{k}]}$ for $|\underline{k}| \leq n$. Hence, we get on $D_{A/R}$ the basis $\{\partial^{[\underline{k}]}, \underline{k} \in \mathbb{N}^d\}$ as A-module.

Let $\underline{\epsilon}_i = (0, \ldots, 0, 1, 0, \ldots, 0)$ where 1 is at the ith place. Set $\partial_i := \partial^{[\underline{\epsilon}_i]}$.

Proposition 1.2.19 *We have the following relations:*

1. $\forall f \in A$, $\forall n \geq 0$, *we have in* $P^n_{A/R}$ *the formula* $d_1(f) = \sum_{|\underline{k}| \leq n} \partial^{[\underline{k}]}(f) \underline{\tau}^{\underline{k}}$.
2. $\forall \underline{k} \leq \underline{i}$, $\partial^{[\underline{k}]}(\underline{t}^{\underline{i}}) = \binom{\underline{i}}{\underline{k}} \underline{t}^{\underline{i} - \underline{k}}$.

3. $\forall \underline{k}', \underline{k}'' \in \mathbb{N}^d$, $\underline{\partial}^{[\underline{k}']} \underline{\partial}^{[\underline{k}'']} = \binom{\underline{k}'+\underline{k}''}{\underline{k}'} \underline{\partial}^{[\underline{k}'+\underline{k}'']}$.

4. $\forall \underline{k} \in \mathbb{N}^d$, $\forall f \in A$, $\underline{\partial}^{[\underline{k}]} f = \sum_{\underline{k}'+\underline{k}''=\underline{k}} \underline{\partial}^{[\underline{k}']}(f) \underline{\partial}^{[\underline{k}'']}$.

Proof With notation 1.2.15, the part 1) is obvious from the fact that $\{\underline{\partial}^{[\underline{k}]}, |\underline{k}| \leq n\}$ is the dual basis of $\{\underline{\tau}^{\underline{k}}, |\underline{k}| \leq n\}$. Using $d_1(\underline{t}^{\underline{i}}) = (1 \otimes \underline{t})^{\underline{i}} = (\underline{\tau} + \underline{t} \otimes 1)^{\underline{i}} = \sum_{\underline{k} \leq \underline{i}} \binom{\underline{i}}{\underline{k}} \underline{t}^{\underline{i}-\underline{k}} \underline{\tau}^{\underline{k}}$, this yields the second formula. From $\delta^{n,n'}(\tau_i) = 1\!\!\!\!\!/ \otimes \tau_i + \tau_i \otimes 1\!\!\!\!\!/$, we get $\delta^{n,n'}(\underline{\tau}^{\underline{l}}) = \sum_{\underline{l}'+\underline{l}''=\underline{l}} \underline{\tau}^{\underline{l}'} \otimes \underline{\tau}^{\underline{l}''}$. Hence, $(id \circ \underline{\partial}^{[\underline{k}'']}) \circ \delta^{n,n'}(\underline{\tau}^{\underline{l}}) = \binom{\underline{l}}{\underline{l}-\underline{k}''} \underline{\tau}^{\underline{l}-\underline{k}''}$. This yields $\underline{\partial}^{[\underline{k}']} \underline{\partial}^{[\underline{k}'']}(\underline{\tau}^{\underline{l}}) = \underline{\partial}^{[\underline{k}']} \circ (id \circ \underline{\partial}^{[\underline{k}'']}) \circ \delta^{n,n'}(\underline{\tau}^{\underline{l}}) = \delta_{\underline{l},\underline{k}'+\underline{k}''} \binom{\underline{k}'+\underline{k}''}{\underline{k}'}$, where $\delta_{\underline{l},\underline{k}'+\underline{k}''}$ is the Kronecker symbol. Hence, we get the third formula. By using 1.2, we get the commutative diagram

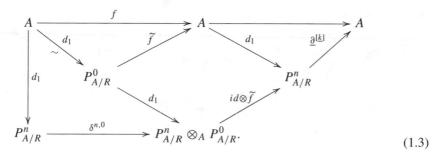

$$(1.3)$$

Hence, by identifying A with a R-subalgebra of $D_{A/R}$ and A with $P^0_{A/R}$, this yields by definition of the product in $D_{A/R}$ the formula $\underline{\partial}^{[\underline{k}]} f(\underline{\tau}^{\underline{i}}) = \underline{\partial}^{[\underline{k}]}((1 \otimes f)\underline{\tau}^{\underline{i}})$. Hence, using the formula 1), we get $\underline{\partial}^{[\underline{k}]} f(\underline{\tau}^{\underline{i}}) = \underline{\partial}^{[\underline{k}]} \left(\sum_{|\underline{k}'| \leq n} \underline{\partial}^{[\underline{k}']}(f) \underline{\tau}^{\underline{k}'+\underline{i}} \right) = \underline{\partial}^{[\underline{k}-\underline{i}]}(f)$. $\qquad\square$

1.2.20

1. If R is an $\mathbb{F}_p$-algebra, then using 1.2.19. 3, we get $\underline{\partial}^{\underline{k}} = \underline{\partial}^{[\underline{k}]} \underline{k}!$. Hence, $\partial_i^p = p \underline{\partial}^{[p\epsilon_i]} = 0$.

2. Suppose R is a $\mathbb{Q}$-algebra. Then $\underline{\partial}^{[\underline{k}]} = \underline{\partial}^{\underline{k}}/\underline{k}!$. Hence $D_{A/R}$ is generated by $\partial_1, \dots, \partial_d$ and A as R-algebra.

Remark 1.2.21

1. Using 1.2.19.4, we can check $\mathrm{gr} D_{A/R} := \oplus_{n \geq 0} D_{A/R,n}/D_{A/R,n+1}$ is a commutative A-algebra.

2. Suppose R is a $\mathbb{Q}$-algebra and $A = R[T_1, \dots, T_d]$. Then $D_{A/R}$ is the non commutative R-algebra generated by $T_1, \dots, T_d, \partial_1, \dots, \partial_d$ subject to the relations $[T_i, T_j] = 0$, $[\partial_i, \partial_j] = 0$, $[\partial_i, T_j] = \partial_i T_j - T_j \partial_i = \delta_{ij}$, $\forall i, j$.

1.2.4 Left $D_{A/R}$-Modules

We keep notation and hypotheses of the Sect. 1.2.3.

Definition 1.2.22 Let E be an A-module. A stratification on E is a collection of $P^n_{A/R}$-linear isomorphisms

$$\epsilon_n \colon P^n_{A/R} \otimes_A E \xrightarrow{\sim} E \otimes_A P^n_{A/R}$$

such that

1. $\epsilon_0 = id$ and the family $(\epsilon_n)_n$ is compatible with the restrictions $P^n_{A/R} \to P^{n'}_{A/R}$, for any $n \leq n'$;
2. for any integers n, n', the diagram

$$(1.4)$$

where $q^{n,n'}_0 \colon P^{n+n'}_{A/R} \twoheadrightarrow P^n_{A/R} \xrightarrow{id \otimes d_0} P^n_{A/R} \otimes_A P^{n'}_{A/R}$, and $q^{n,n'}_1 \colon P^{n+n'}_{A/R} \twoheadrightarrow P^{n'}_{A/R} \xrightarrow{d_1 \otimes id} P^n_{A/R} \otimes_A P^{n'}_{A/R}$, is commutative.

Remark 1.2.23 Let $\pi_{ij} \colon A \otimes_R A \to A \otimes_R A \otimes_R A$ be the morphism corresponding to the projection at the i and j places. Let $r_i \colon A \to A \otimes_R A \otimes_R A$ be the morphism corresponding to the projection at the ith place. Let $\mu(2) \colon A \otimes_R A \otimes_R A \to A$ be the morphism corresponding to the diagonal morphism, and et $I(2) := \ker \mu(2)$. We denote by $P^n_{A/R}(2) := A \otimes_R A \otimes_R A / I(2)^{n+1}$ for any $n \in \mathbb{N}$ and by $r^n_i \colon A \to P^n_{A/R}(2)$ the homomorphism induced by r_i.

Modulo the canonical isomorphism $A \otimes_R A \otimes_R A \xrightarrow{\sim} (A \otimes_R A) \otimes_A (A \otimes_R A)$, we have $I(2) = I \otimes_A (A \otimes_R A) + (A \otimes_R A) \otimes_A I$. Indeed, let $x = \sum_{i,j,k} a_i \otimes a'_j \otimes a''_k \in A \otimes_R A \otimes_R A$. Then, $\sum_{i,j,k} a_i \otimes a'_j \otimes a''_k = \sum_{i,j,k} (a_i \otimes a'_j) \otimes (1 \otimes a''_k) \equiv \sum_{i,j,k} (1 \otimes a_i a'_j) \otimes (1 \otimes a''_k) \mod I \otimes_A (A \otimes_R A) = \sum_{i,j,k} (1\!\!\!/ \otimes (a_i a'_j \otimes a''_k) \equiv \sum_{i,j,k} (1\!\!\!/ \otimes (1 \otimes a_i a'_j a''_k) \mod (A \otimes_R A) \otimes_A I$. Hence, $I(2) \subset I \otimes_A (A \otimes_R A) + (A \otimes_R A) \otimes_A I$. The converse inclusion is obvious.

Hence $I(2)^{n+n'+1} \subset I^{n+1} \otimes_A (A \otimes_R A) + (A \otimes_R A) \otimes_A I^{n'+1} \subset I(2)^{\min\{n,n'\}+1}$. This yields the commutative diagram

$$
\begin{array}{ccc}
A \otimes_R A \xrightarrow[\substack{\pi_{01} \\ \pi_{12} \\ \pi_{02}}]{} A \otimes_R A \otimes_R A \xrightarrow{\sim} (A \otimes_R A) \otimes_A (A \otimes_R A) \\
\downarrow \qquad\qquad \downarrow \qquad\qquad \downarrow \\
P_{A/R}^{n+n'} \xrightarrow[\substack{\pi_{01}^{n+n'} \\ \pi_{12}^{n+n'} \\ \pi_{02}^{n+n'}}]{} P_{A/R}^{n+n'}(2) \longrightarrow P_{A/R}^{n} \otimes_A P_{A/R}^{n'} \\
\| \qquad\qquad\qquad\qquad \| \\
P_{A/R}^{n+n'} \xrightarrow[\substack{q_0^{n,n'} \\ q_1^{n,n'} \\ \delta^{n,n'}}]{} P_{A/R}^{n} \otimes_A P_{A/R}^{n'},
\end{array}
\tag{1.5}
$$

where $\pi_{ij}^{n+n'} : P_{A/R}^{n+n'} \to P_{A/R}^{n+n'}(2)$ denotes the unique homomorphism making commutative the left top square. Hence, the cocycle condition is equivalent to the following condition:

⋆ for any $n \in \mathbb{N}$, we have the equality $\pi_{02}^{n*}(\epsilon_n) = \pi_{01}^{n*}(\epsilon_n) \circ \pi_{12}^{n*}(\epsilon_n)$, i.e. the following diagram commutes

$$
\begin{array}{ccccc}
& \pi_{12}^{n*} d_1^{n*}(E) \xrightarrow[\pi_{12}^{n*}(\epsilon_n)]{\sim} \pi_{12}^{n*} d_0^{n*}(E) & \\
r_2^{n*}(E) == \pi_{02}^{n*} d_1^{n*}(E) \qquad\qquad \pi_{01}^{n*} d_1^{n*}(E) == r_1^{n*}(E) \\
\pi_{02}^{n*}(\epsilon_n) \downarrow \sim \qquad\qquad \sim \downarrow \pi_{01}^{n*}(\epsilon_n) \\
r_0^{n*}(E) == \pi_{02}^{n*} d_0^{n*}(E) == \pi_{01}^{n*} d_0^{n*}(E) == r_0^{n*}(E).
\end{array}
\tag{1.6}
$$

Proposition 1.2.24 *Let E be an A-module.*

1. The following datum are equivalent:

   (a) A structure of left $D_{A/R}$-module on E extending its structure of A-module via the homomorphism of R-algebras $A \to D_{A/R}$.
   (b) A compatible family of A-linear map $\theta_n : E \to E \otimes_A P_{A/R}^n$ such that $\theta_0 = id$ and such that for any $n, n' \in \mathbb{N}$ the diagram

$$
\begin{array}{ccc}
E \otimes_A P_{A/R}^{n+n'} & \xrightarrow{id \otimes \delta^{n,n'}} & E \otimes_A P_{A/R}^n \otimes_A P_{A/R}^{n'} \\
\theta_{n+n'} \uparrow & & \uparrow \theta_n \otimes id \\
E & \xrightarrow{\theta_{n'}} & E \otimes_A P_{A/R}^{n'}
\end{array}
\tag{1.7}
$$

   is commutative.
   (c) A stratification $(\epsilon_n)_{n \in \mathbb{N}}$ on E.

2. *An A-linear homomorphism* $\phi \colon E \to F$ *between two left* $D_{A/R}$-*module is* $D_{A/R}$-*linear if and only if* ϕ *is horizontal, i.e., the following diagram is commutative for any* $n \in \mathbb{N}$

$$
\begin{array}{ccc}
P^n_{A/R} \otimes_A E & \xrightarrow{\underset{\sim}{\epsilon_n}} & E \otimes_A P^n_{A/R} \\
\downarrow{\scriptstyle id \otimes \phi} & & \downarrow{\scriptstyle \phi \otimes id} \\
P^n_{A/R} \otimes_A F & \xrightarrow{\underset{\sim}{\epsilon_n}} & F \otimes_A P^n_{A/R}.
\end{array}
\tag{1.8}
$$

Proof A structure of left $D_{A/R}$-module on E extending its A-module structure is equivalent to a family of A-linear maps $\mu_n \colon D_{A/R,n} \otimes_A E \to E$ such that $\mu_0 = id$ and such that, $\forall n, n'$, the following diagrams

$$
\begin{array}{ccc}
D_{A/R,n} \otimes_A E & \xrightarrow{\mu_n} & E, \\
\Big\uparrow & \nearrow{\scriptstyle \mu_{n'}} & \\
D_{A/R,n'} \otimes_A E & &
\end{array}
\qquad
\begin{array}{ccc}
D_{A/R,n'} \otimes_A D_{A/R,n} \otimes_A E & \xrightarrow{\mu_n} & D_{A/R,n'} \otimes_A E \\
\downarrow & & \downarrow{\scriptstyle \mu_{n'}} \\
D_{A/R,n+n'} \otimes_A E & \xrightarrow[\mu_{n+n'}]{} & E
\end{array}
\tag{1.9}
$$

are commutative. By adjunction, the morphism μ_n is equivalent to a morphism of the form $\theta_n \colon E \to \operatorname{Hom}_A(D_{A/R,n}, E) \xrightarrow{\sim} E \otimes_A P^n_{A/R}$. Since $D_{A/R,n'} \otimes_A D_{A/R,n} \to D_{A/R,n+n'}$ is given by $P' \otimes P \mapsto P' \circ (id \otimes P) \circ \delta^{n,n'}$, we can check that the commutativity of the diagram 1.7 is equivalent to the right one of 1.9. Hence, we get the equivalence between (a) and (b). We get the equivalence between the datum θ_n and ϵ_n via the commutative diagram

$$
\begin{array}{ccc}
E & \xrightarrow{\theta_n} & E \otimes_A P^n_{A/R} \\
& \searrow{\scriptstyle d_1 \otimes id} & \uparrow{\scriptstyle \epsilon_n} \\
& & P^n_{A/R} \otimes_A E.
\end{array}
\tag{1.10}
$$

Finally, the fact that ϵ_n is necessarily an isomorphism comes from Lemma 1.2.25 below which gives an explicit formula of the inverse function. $\square$

Lemma 1.2.25 *Let* E *be a left* $D_{A/R}$-*module. Let* $x \in E$. *We have the formulas*

$$
\theta_n(x) = \epsilon_n(1 \otimes x) = \sum_{|\underline{k}| \leq n} \underline{\partial}^{[\underline{k}]}(x) \otimes \underline{\tau}^{\underline{k}},
\tag{1.11}
$$

$$
\epsilon_n^{-1}(x \otimes 1) = \sum_{|\underline{k}| \leq n} (-1)^{|\underline{k}|} \underline{\tau}^{\underline{k}} \otimes \underline{\partial}^{[\underline{k}]}(x).
\tag{1.12}
$$

Proof Let us check the formula 1.11. Since $P^n_{A/R}$ is a free A-module (for its left structure of A-module, by default), the canonical morphism ev: $P^n_{A/R} \to \mathrm{Hom}_A(D_{A/R,n}, A)$ is an isomorphism. Let $\{\underline{\partial}^{[\underline{k}]*}, |\underline{k}| \leq n\}$ be the dual basis of $\{\underline{\partial}^{[\underline{k}]}, |\underline{k}| \leq n\}$. Via this isomorphism, $\underline{\tau}^{\underline{k}}$ is sent to $\underline{\partial}^{[\underline{k}]*}$. Hence, the composition $E \otimes_A P^n_{A/R} \xrightarrow{\sim} E \otimes_A \mathrm{Hom}_A(D_{A/R,n}, A) \xrightarrow{\sim} \mathrm{Hom}_A(D_{A/R,n}, E)$ sends $\sum x_{\underline{k}} \otimes \underline{\tau}^{\underline{k}}$ to $(\underline{\partial}^{[\underline{k}]} \mapsto x_{\underline{k}})$. Since $\mu_n \colon D_{A/R,n} \otimes_A E \to E$, is given by $\underline{\partial}^{[\underline{k}]} \otimes x \mapsto \underline{\partial}^{[\underline{k}]}(x)$, then $E \to \mathrm{Hom}_A(D_{A/R,n}, E)$ is given by $x \mapsto (\underline{\partial}^{[\underline{k}]} \mapsto \underline{\partial}^{[\underline{k}]}(x))$, and we are done.

By using 1.2.19, we can check that the $P^n_{A/R}$-linear morphism defined via the right term of 1.12 is the inverse function of ϵ_n. $\square$

Corollary 1.2.26 *Let E, F be two left $D_{A/R}$-modules.*

1. *There exists a unique structure of $D_{A/R}$-module on $E \otimes_A F$ extending its canonical structure of A-module such that, for any $\underline{k} \in \mathbb{N}^d$, $x \in E$, $y \in F$,*

$$\underline{\partial}^{[\underline{k}]}(x \otimes y) = \sum_{\underline{i} \leq \underline{k}} \underline{\partial}^{[\underline{i}]}(x) \otimes \underline{\partial}^{[\underline{k}-\underline{i}]}(y). \tag{1.13}$$

2. *There exists a unique structure of $D_{A/R}$-module on $\mathrm{Hom}_A(E, F)$ extending its canonical structure of A-module such that, $\forall \underline{k} \in \mathbb{N}^d$, for any $\phi \in \mathrm{Hom}_A(E, F)$*

$$(\underline{\partial}^{[\underline{k}]}\phi)(x) = \sum_{\underline{i} \leq \underline{k}} (-1)^{|\underline{i}|} \underline{\partial}^{[\underline{k}-\underline{i}]} \left(\phi(\underline{\partial}^{[\underline{i}]}x) \right). \tag{1.14}$$

Proof The unicity is obvious. Let us check the existence. The stratification of $E \otimes_A F$ is given by the composition

$$\epsilon_n^{E \otimes F} \colon P^n_{A/R} \otimes_A E \otimes_A F \xrightarrow[\epsilon_n^E \otimes id]{\sim} E \otimes_A P^n_{A/R} \otimes_A F \xrightarrow[id \otimes \epsilon_n^F]{\sim} E \otimes_A F \otimes_A P^n_{A/R}.$$

We compute $1 \otimes x \otimes y \mapsto \sum_{\underline{i}} \underline{\partial}^{[\underline{i}]}(x) \otimes \underline{\tau}^{\underline{i}} \otimes y \mapsto \sum_{\underline{i}} \sum_{\underline{j}} \underline{\partial}^{[\underline{i}]}(x) \otimes \underline{\partial}^{[\underline{j}]}(y) \otimes \underline{\tau}^{\underline{j}} \underline{\tau}^{\underline{i}}$. We proceed similarly for the second part. $\square$

Definition 1.2.27 Let E be an A-module. A connection on E is an additive map $\nabla \colon E \to E \otimes_A \Omega^1_{A/R}$ such that $\nabla(ax) = a\nabla(x) + x \otimes da$, for any $a \in A$, $x \in E$. We denote by $\Omega^i_{A/R} := \wedge^i \Omega^1_{A/R}$. We get the map $\nabla \colon E \otimes_A \Omega^i_{A/R} \to E \otimes_A \Omega^{i+1}_{A/R}$, given by $x \otimes \omega \mapsto \nabla(x) \wedge \omega + (-1)^i x \wedge d\omega$ for any $x \in E$ and $\omega \in \Omega^i_{A/R}$. We say that the connection is integral if $\nabla^2 = 0$.

Theorem 1.2.28 *Suppose R is a $\mathbb{Q}$-algebra. There is a bijection between the data of an integrable connection on E and that of $D_{A/R}$-module extending its structure of A-module.*

Proof Left to the reader. $\square$

1.3   Berthelot's Ring of Differential Operators of Finite Level and Infinite Order

Let R be a complete discrete valuation ring of mixed characteristic $(0, p)$ with perfect residue field k, uniformizer π, ramification index e. Let K be the fraction field of R. For example, R might be the Witt vectors of a perfect field of characteristic p.

We focus in the affine case as follows (except for the last subsection). Let A be a commutative flat R-algebra separated and complete for the p-adic topology. For any $n \in \mathbb{N}$, we set $A_n := A/\pi^{n+1}A$, $R_n := R/\pi^{n+1}R$. We denote by I the kernel of the canonical morphism $\mu\colon A\widehat{\otimes}_R A \to A$, where $A\widehat{\otimes}_R A$ is the p-adic completion of $A \otimes_R A$. Let $t_1, \ldots, t_d \in A$, $\tau_i := 1 \otimes t_i - t_i \otimes 1 \in A\widehat{\otimes}_R A$ for any $i = 1, \ldots, d$. For any $n \in \mathbb{N}$, for any $i = 1, \ldots, d$, let τ_{in} be the image of τ_i in $A\widehat{\otimes}_R A/\pi^{n+1}(A\widehat{\otimes}_R A) \xrightarrow{\sim} A_n \otimes_{R_n} A_n$.

(*)   We suppose in this chapter that for any $n \in \mathbb{N}$ the sequence $\tau_{1n}, \ldots, \tau_{dn}$ is a quasi-regular sequence of $A_n \otimes_{R_n} A_n$.

Examples Let $R\{T_1, \ldots, T_d\}$ be the p-adic completion of the polynomial algebra $R[T_1, \ldots, T_d]$. Then $\tau_i := 1 \otimes T_i - T_i \otimes 1$ for any $i = 1, \ldots, d$ satisfy the condition (*).

More generally, let $f\colon R\{T_1, \ldots, T_d\} \to A$ be an étale morphism. Let $t_1, \ldots, t_d \in A$ be the image of $T_1, \ldots, T_d$ via f. Then $\tau_i := 1 \otimes t_i - t_i \otimes 1 \in A\widehat{\otimes}_R A$ for any $i = 1, \ldots, d$ satisfy the condition (*).

1.3.1   Formal Affine Case, Standard Ring of Differential Operators

Definition 1.3.1 For any $n \in \mathbb{N}$, we set $P^n_{A/R} := (A\widehat{\otimes}_R A)/I^{n+1}$. The ring $P^n_{A/R}$ has two A-algebra structures: the left one induced by the left structure of $A\widehat{\otimes}_R A$, and the right one. We denote by $d_0\colon A \to P^n_{A/R}$ the homomorphism given by the left structure, and $d_1\colon A \to P^n_{A/R}$ that given by the right one.

Definition 1.3.2 An R-linear map $h\colon A \to A$ is a "differential operator of A/R of order $\leq n$" if and only if there exists an A-linear map $\widetilde{h}\colon d_{0*}(P^n_{A/R}) \to A$ making commutative the diagram

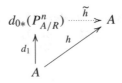

We denote by $\mathrm{Diff}^n_{A/R}$ the set of differential operators of A/R of order $\leq n$. Since $d_{0*}(P^n_{A/R})$ is generated as A-module by the image of d_1, then we get the uniqueness of such $\tilde{h}$ if it exists. Hence we get the bijection $\mathrm{Diff}^n_{A/R} \to \mathrm{Hom}_A(d_{0*}(P^n_{A/R}), A)$ given by $h \mapsto \tilde{h}$. We set $D_{A/R,n} := \mathrm{Hom}_A(d_{0*}(P^n_{A/R}), A)$.

Since $P^n_{A/R}$ has two structures of A-algebra inducing a structure of A-bimodules, then $D_{A/R,n}$ is endowed with a structure of A-bimodule. The left (resp. right) structure of A-module of $D_{A/R,n}$ is by definition that coming from the left (resp. right) structure of A-algebra of $P^n_{A/R}$.

Lemma 1.3.3 *There exists a unique homomorphism* $\delta^{n,n'} : P^{n+n'}_{A/R} \to P^n_{A/R} \otimes_A P^{n'}_{A/R}$ *of A-algebras for both left and right structures of A-algebras making commutative the following diagram*

$$
\begin{array}{ccc}
A & \xrightarrow{\ d_1\ } & P^{n+n'}_{A/R} \\
{\scriptstyle d_1}\big\downarrow & & \big\downarrow{\scriptstyle \delta^{n,n'}} \\
P^{n'}_{A/R} & \xrightarrow{\ d_1\ } & P^n_{A/R} \otimes_A P^{n'}_{A/R}.
\end{array}
\tag{1.15}
$$

Proposition 1.3.4 *Let* $h \in \mathrm{Diff}^n_{A/R}$, $h' \in \mathrm{Diff}^{n'}_{A/R}$. *Then* $h \circ h' \in \mathrm{Diff}^{n+n'}_{A/R}$.

Proof We copy the proof of 1.2.12. $\qquad\qquad\square$

Definition 1.3.5 Via the bijection, $\mathrm{Diff}^n_{A/R} \xrightarrow{\sim} D_{A/R,n}$, we get an R-algebra structure on $D_{A/R} := \cup_{n \in \mathbb{N}} D_{A/R,n}$ making the bijection $\mathrm{Diff}_{A/R} \xrightarrow{\sim} D_{A/R}$ an isomorphism of R-algebras and of A-bimodule. The multiplication can be described via the diagram 1.2. We also get a natural action of $D_{A/R}$ on A. This is the ring of differential operators on A/R. Finally, the mapping $A \to D_{A/R}$ given by $a \mapsto a \cdot 1$ is the same for the left or right structure of A-module. Moreover, the mapping $A \to D_{A/R}$ is an homomorphism of R-algebras, and we can view A as an R-subalgebra of $D_{A/R}$.

Notation 1.3.6 We check that $\underline{\tau}^{\underline{\alpha}}$ mod I^{n+1} is a basis of $P^n_{A/R}$ for both A-algebra structures. We denote by $\{\underline{\partial}^{[\underline{k}]},\ |\underline{k}| \leq n\}$ be the dual basis of $D_{A/R,n} = \mathrm{Hom}_A(d_{0*}(P^n_{A/R}), A)$ of $\{\underline{\tau}^{\underline{k}},\ |\underline{k}| \leq n\}$. Since the morphism $P^{n+1}_{A/R} \twoheadrightarrow P^n_{A/R}$ sends $\underline{\tau}^{\underline{\alpha}}$ mod I^{n+2} to $\underline{\tau}^{\underline{\alpha}}$ mod I^{n+1}, then the monomorphism $D_{A/R,n} \hookrightarrow D_{A/R,n+1}$ send $\underline{\partial}^{[\underline{k}]}$ to $\underline{\partial}^{[\underline{k}]}$ for $|\underline{k}| \leq n$. Hence, we get on $D_{A/R}$ the basis $\{\underline{\partial}^{[\underline{k}]},\ \underline{k} \in \mathbb{N}^d\}$ as A-module. Let $\underline{\epsilon}_i = (0, \ldots, 0, 1, 0, \ldots, 0)$ where 1 is at the ith place. Set $\partial_i := \underline{\partial}^{[\underline{\epsilon}_i]}$.

Proposition 1.3.7 *We have the following relations:*

1. $\forall f \in A$, $\forall n \geq 0$, *we have in* $P^n_{A/R}$ *the formula* $d_1(f) = \sum_{|\underline{k}| \leq n} \underline{\partial}^{[\underline{k}]}(f)\underline{\tau}^{\underline{k}}$.
2. $\forall \underline{k} \leq \underline{i}$, $\underline{\partial}^{[\underline{k}]}(\underline{t}^{\underline{i}}) = \binom{\underline{i}}{\underline{k}} \underline{t}^{\underline{i}-\underline{k}}$.

3. $\forall \underline{k}', \underline{k}'' \in \mathbb{N}^d,\ \partial^{[\underline{k}']}\partial^{[\underline{k}'']} = \left(\dfrac{\underline{k}'+\underline{k}''}{\underline{k}'} \right) \partial^{[\underline{k}'+\underline{k}'']}.$

4. $\forall \underline{k} \in \mathbb{N}^d,\ \forall f \in A,\ \partial^{[\underline{k}]} f = \sum_{\underline{k}'+\underline{k}''=\underline{k}} \partial^{[\underline{k}']}(f)\partial^{[\underline{k}'']}.$

Proof We copy the proof of 1.2.19.   □

1.3.2 Formal Affine Case, Berthelot's Ring of Differential Operators of Level m

Fix a level $m \in \mathbb{N}$.

1.3.8 For any $k \in \mathbb{N}$, we set $k = q_k^{(m)} p^m + r_k^{(m)}$, where $q_k^{(m)} \in \mathbb{N}$ and $0 \le r_k^{(m)} < p^m$. For any $k, k' \in \mathbb{N},\ k' \le k$, set

$$\left\{ \begin{matrix} k \\ k' \end{matrix} \right\}_{(m)} := \frac{q_k^{(m)}!}{q_{k'}^{(m)}! q_{k-k'}^{(m)}!},\quad \left\{ \begin{matrix} k \\ k' \end{matrix} \right\}_{(\infty)} := 1.$$

For any $m \in \mathbb{N} \cup \{\infty\}$, we set $\left\langle \begin{matrix} k \\ k' \end{matrix} \right\rangle_{(m)} := \dfrac{\binom{k}{k'}}{\left\{ \begin{matrix} k \\ k' \end{matrix} \right\}_{(m)}}.$

Notation 1.3.9 Let $k \in \mathbb{N}$. We write uniquely $k = \sum_i a_i p^i$, such that $a_i \in \{0, \ldots, p-1\}$. We set $\sigma(k) := \sum_i a_i$. Then we compute that $v_p(k!) = \frac{k-\sigma(k)}{p-1}$.

Lemma 1.3.10 Let $k', k'' \in \mathbb{N},\ k := k' + k''$. We write $k = \sum_i a_i p^i,\ k' = \sum_i a_i' p^i,\ k'' = \sum_i a_i'' p^i$, with $a_i, a_i', a_i'' \in \{0, \ldots, p-1\}$.

1. $\sigma(k' + k'') \le \sigma(k') + \sigma(k'')$.
2. We have $\sigma(k' + k'') = \sigma(k') + \sigma(k'')$ if and only if $a_i = a_i' + a_i''$ for any $i \in \mathbb{N}$.

Proof Since

$$0 \le v_p \left(\binom{k}{k'} \right) = \frac{\sigma(k') + \sigma(k'') - \sigma(k)}{p-1},$$

then we get the first statement.

If $\{i \mid a_i' + a_i'' \ge p\}$ is empty then we get $\sigma(k' + k'') = \sigma(k') + \sigma(k'')$. Conversely, suppose $\{i \mid a_i' + a_i'' \ge p\}$ is not empty. Let i_0 be its smallest element. Then, for any $i < i_0$, we get $a_i = a_i' + a_i''$. Hence, we can suppose, for any $i < i_0,\ 0 = a_i = a_i' = a_i''$. Let $i_1 = \min\{i > i_0;\ a_i' + a_i'' < p - 1\}$. Since $a_{i_1}' + a_{i_1}'' < p - 1$, then $\sum_{i \le i_1} a_i' p^i + \sum_{i \le i_1} a_i'' p^i = \sum_{i \le i_1} a_i p^i$. Hence, $\sum_{i > i_1} a_i' p^i + \sum_{i > i_1} a_i'' p^i = \sum_{i > i_1} a_i p^i$. Following the first part, this yields the formula $\sum_{i > i_1} a_i' + \sum_{i > i_1} a_i'' \ge \sum_{i > i_1} a_i$.

On the other hand, we compute more precisely $\sum_{i \leq i_1} a_i' p^i + \sum_{i \leq i_1} a_i'' p^i = (a_{i_0}' + a_{i_0}'' - p) p^{i_0} + \sum_{i_0 < i < i_1} (a_i' + a_i'' + 1 - p) p^i + (a_{i_1}' + a_{i_1}'' + 1) p^{i_1}$, and then $a_{i_0} = a_{i_0}' + a_{i_0}'' - p$, $a_i = a_i' + a_i'' + 1 - p$ for $i_0 < i < i_1$, and $a_{i_1} = a_{i_1}' + a_{i_1}'' + 1$. Hence, $\sum_{i \leq i_1} a_i = (1-p)(i_1 - i_0) + \sum_{i \leq i_1} a_i' + a_i'' < \sum_{i \leq i_1} a_i' + a_i''$. This yields, $\sigma(k' + k'') < \sigma(k') + \sigma(k'')$. $\qquad \square$

Lemma 1.3.11

(a) For any $k, k' \in \mathbb{N}$, $\left\{ {k \atop k'} \right\}_{(m)} \in \mathbb{Z}$, and $\left({k \atop k'} \right)_{(m)} \in \mathbb{Z}_{(p)}$.

(b) For any $j \in \mathbb{N}$, for any $q' \leq q$, suppose either $j \geq m$ or $q < p$. Then $\left({p^j q \atop p^j q'} \right)_{(m)} \in \mathbb{Z}_{(p)}^*$.

(c) For any $j, r, q \in \mathbb{N}$, such that $r < p^j$, we have $\left({p^j q + r \atop p^j q} \right)_{(m)} \in \mathbb{Z}_{(p)}^*$.

Proof

(1) Since $k = (q_{k'}^{(m)} + q_{k-k'}^{(m)}) p^m + (r_{k'}^{(m)} + r_{k-k'}^{(m)})$, then $q_{k'}^{(m)} + q_{k-k'}^{(m)} \leq q_k^{(m)}$. Hence, $\left\{ {k \atop k'} \right\}_{(m)} \in \mathbb{Z}$. Set $k'' := k - k'$. We compute $v_p(\left\{ {k \atop k'} \right\}_{(m)}) = \frac{(k - \sigma(k)) - (k' - \sigma(k')) - (k'' - \sigma(k'')) - (q_k^{(m)} - \sigma(q_k^{(m)})) + (q_{k'}^{(m)} - \sigma(q_{k'}^{(m)})) + (q_{k''}^{(m)} - \sigma(q_{k''}^{(m)}))}{p-1}$. Since $\sigma(k) = \sigma(q_k^{(m)}) + \sigma(r_k^{(m)})$ (and with some primes), we get

$$(p-1) v_p(\left\langle {k \atop k'} \right\rangle_{(m)}) = q_{k'}^{(m)} + q_{k''}^{(m)} - q_k^{(m)} + \sigma(r_{k'}^{(m)}) + \sigma(r_{k''}^{(m)}) - \sigma(r_k^{(m)}). \tag{1.16}$$

We have two case: either $q_k^{(m)} = q_{k'}^{(m)} + q_{k''}^{(m)}$ and $r_k^{(m)} = r_{k'}^{(m)} + r_{k''}^{(m)}$ or $q_k^{(m)} = q_{k'}^{(m)} + q_{k''}^{(m)} + 1$ and $p^m + r_k^{(m)} = r_{k'}^{(m)} + r_{k''}^{(m)}$. Using the first part of Lemma 1.3.10, in the first case we get $\sigma(r_{k'}^{(m)}) + \sigma(r_{k''}^{(m)}) - \sigma(r_k^{(m)}) \geq 0$. Using again Lemma 1.3.10, in the second case we get $1 + \sigma(r_k^{(m)}) = \sigma(r_{k'}^{(m)} + r_{k''}^{(m)}) \leq \sigma(r_{k'}^{(m)}) + \sigma(r_{k''}^{(m)})$, and we are done.

(2) Set $k := p^j q$, $k' := p^j q'$, $k'' := k - k'$. a) If $j \geq m$, then $k = p^m(p^{j-m} q)$, $k' = p^m(p^{j-m} q')$, $k'' := k - k' = p^m(p^{j-m} q - p^{j-m} q')$. Hence, $0 = r_k^{(m)} = r_{k'}^{(m)} = r_{k''}^{(m)}$, and $q_k^{(m)} = q_{k'}^{(m)} + q_{k''}^{(m)}$. By using the formula 1.16, this yields $v_p(\left\langle {p^j q \atop p^j q'} \right\rangle_{(m)}) = 0$.

(b) Now suppose $q < p$ and $j < m$. In that case $p^j q < p^m$, hence $q_k^{(m)} = 0$ and $r_k^{(m)} = k$; and similarly with some primes. Hence, we are done.

(3) Since $r < p^j$, there exists $a_i \in \{0, \ldots, p-1\}$ such that we can write $k' := p^j q = \sum_{i \geq j} a_i p^i$ and $k'' := r = \sum_{i < j} a_i p^i$. Hence, $k := p^j q + r = \sum_{i \geq 0} a_i p^i$. This yields $q_k^{(m)} = q_{k'}^{(m)} + q_{k''}^{(m)}$ and $\sigma(r_k^{(m)}) = \sigma(r_{k'}^{(m)}) + \sigma(r_{k''}^{(m)})$. We conclude by using the formula 1.16.

$\qquad \square$

Lemma 1.3.12 *For any* $k', k'' \in \mathbb{N}$, *we set*

$$C_{k'',k'}^{(m)} := \frac{q_{k'k''}^{(m)}!}{\left(q_{k'}^{(m)}!\right)^{k''} q_{k''}^{(m)}!}.$$

Then $C_{k'',k'}^{(m)} \in \mathbb{N}$.

Proof We have $C_{k'',k'}^{(0)} = \frac{(k'k'')!}{(k'!)^{k''} k''!} = \prod_{i=1}^{k''-1} \binom{(i+1)k'-1}{k'-1} \in \mathbb{N}$.

Set $q = q_{k'k''}^{(m)}$, $q' = q_{k'}^{(m)}$, $q'' = q_{k''}^{(m)}$, $r = r_{k'k''}^{(m)}$, $r' = r_{k'}^{(m)}$, $r'' = r_{k''}^{(m)}$, $s = q_{rr'}^{(m)}$, $t = r_{rr'}^{(m)}$. We get $q = p^m q'q'' + q'r'' + q''r' + s$. We compute

$$C_{k'',k'}^{(m)} = r''!(q''r'+s)!(p^m!)^{q''} C_{r'',q'}^{(0)} C_{q'',p^m q'}^{(0)} (C_{p^m,q'}^{(0)})^{q''} \frac{q!}{(p^m q'q'')!(q'r'')!(q''r'+s)!} \in \mathbb{N}.$$

$\square$

1.3.13 Let B be a flat commutative R-algebra and J be an ideal of B. For any $x \in J$, any $k \in \mathbb{N}$, we set

$$x^{\{k\}(m)} := \frac{x^k}{q_k^{(m)}!} \in B \otimes_R K.$$

The operations $x \mapsto x^{\{k\}(m)}$ satisfy the following properties:

(i) $\forall x \in J$, $x^{\{0\}(m)} = 1$, $x^{\{1\}(m)} = x$, $\forall k \geq 1$, $x^{\{k\}(m)} \in J \otimes_R K$;

(ii) $\forall x \in J$, $\forall b \in B$, $\forall k \in \mathbb{N}$, $(bx)^{\{k\}(m)} = b^k x^{\{k\}(m)}$.

(iii) $\forall x, y \in J$, $\forall k \in \mathbb{N}$, $(x+y)^{\{k\}(m)} = \sum_{k'+k''=k} \left\{ {k \atop k'} \right\}_{(m)} x^{\{k'\}(m)} y^{\{k''\}(m)}$;

(iv) $\forall x \in J$, $\forall k', k'' \in \mathbb{N}$, $x^{\{k'\}(m)} x^{\{k''\}(m)} = \left\{ {k \atop k'} \right\}_{(m)} x^{\{k'+k''\}(m)}$;

(v) $\forall x \in J$, $\forall k', k'' \in \mathbb{N}$, $\left(x^{\{k'\}(m)} \right)^{\{k''\}(m)} = C_{k'',k'}^{(m)} x^{\{k'k''\}(m)}$

The m-PD envelop $P_{(m)}(B, J)$ of J as an ideal of B is the B-subalgebra of $B \otimes_R K$ generated by the elements $x^{\{k\}(m)}$ for any $x \in J$ and any $k \in \mathbb{N}$. We denote by $\overline{J}$ the ideal of $P_{(m)}(B, J)$ generated by $x^{\{k\}(m)}$ for any $x \in J$ and any integer $k \geq 1$. Using the above properties of the operations $x \mapsto x^{\{k\}(m)}$, we can check that $\overline{J}$ is stable under these operations and that $(P_{(m)}(B, J), \overline{J})$ satisfies the corresponding universal property. Berthelot defined in a more general context the notion of m-PD envelop but our particular case is sufficient for our purpose. Using Lemma 1.3.12 and the above properties of the operations $x \mapsto x^{\{k\}(m)}$, we also check that if J is generated by $(x_\alpha)_{\alpha \in L}$, then $P_{(m)}(B, J)$ generated as B-algebra by the elements $x_\alpha^{\{k\}(m)}$ for any $\alpha \in L$ and any $k \in \mathbb{N}$, and the ideal $\overline{J}$ of $P_{(m)}(B, J)$ is generated by $x_\alpha^{\{k\}(m)}$ for any $\alpha \in L$ and any integer $k \geq 1$.

Notation 1.3.14 We denote by $P_{A/R,(m)}^n$ the m-PD envelop of $(P_{A/R}^n, I/I^{n+1})$. In other words, $P_{A/R,(m)}^n$ is the subring of $P_{A/R}^n \otimes_R K$ generated by $\{x^{\{k\}(m)}, \forall k \in \mathbb{N}, \forall x \in I/I^{n+1}\}$.

For any $\underline{k} \in \mathbb{N}^d$, we set $\underline{\tau}^{\{\underline{k}\}(m)} := \tau_1^{\{k_1\}(m)} \cdots \tau_d^{\{k_d\}(m)}$, where $\tau_1, \ldots, \tau_d$ are by abuse of notation the image of $\tau_1, \ldots, \tau_d$ in $P_{A/R}^n \subset P_{A/R}^n \otimes_R K$. We check that $P_{A/R,(m)}^n$ is an A-subalgebra for both structures of $P_{A/R}^n \otimes_R K$ and that $P_{A/R,(m)}^n$ is a free A-module for both structure with the basis $\{\underline{\tau}^{\{\underline{k}\}(m)}, \forall \underline{k} \in \mathbb{N}^d, |\underline{k}| \leq n\}$.

Notation 1.3.15 Let $\underline{k} = (k_1, \ldots, k_d) \in \mathbb{N}^d$. We set $q_{\underline{k}}^{(m)}! := q_{k_1}^{(m)}! \cdots q_{k_d}^{(m)}!$. For any $\underline{k}, \underline{k}' \in \mathbb{N}^d$, we set $\left\{ \frac{\underline{k}}{\underline{k}'} \right\}_{(m)} := \left\{ \frac{k_1}{k_1'} \right\}_{(m)} \cdots \left\{ \frac{k_d}{k_d'} \right\}_{(m)}$ and $\left\langle \frac{\underline{k}}{\underline{k}'} \right\rangle_{(m)} := \left\langle \frac{k_1}{k_1'} \right\rangle_{(m)} \cdots \left\langle \frac{k_d}{k_d'} \right\rangle_{(m)}$.

Since $q_{\underline{k}}^{(m+1)}!$ divides $q_{\underline{k}}^{(m)}!$ for any $\underline{k} \in \mathbb{N}^d$, then we have the inclusions $P_{A/R,(m+1)}^n \subset P_{A/R,(m)}^n$ and $\cap_{m \in \mathbb{N}} P_{A/R,(m)}^n = P_{A/R}^n$. We denote by $D_{A/R,n}^{(m)}$ the dual of $P_{A/R,(m)}^n$ for the left A-algebra structure. We denote by $\{\partial^{<\underline{k}>(m)}, \underline{k} \in \mathbb{N}^d$ such that $|\underline{k}| \leq n\}$ the dual basis of $\{\underline{\tau}^{\{\underline{k}\}(m)}, \underline{k} \in \mathbb{N}^d$ such that $|\underline{k}| \leq n\}$. The inclusions $P_{A/R}^n \subset P_{A/R,(m+1)}^n \subset P_{A/R,(m)}^n$ induce by duality $D_{A/R,n}^{(m)} \to D_{A/R,n}^{(m+1)} \to D_{A/R,n}$. We compute that $\partial^{<\underline{k}>(m)}$ is sent to $\frac{q_{\underline{k}}^{(m)}!}{q_{\underline{k}}^{(m+1)}!} \partial^{<\underline{k}>(m+1)}$ in $D_{A/R,n}^{(m+1)}$ and is sent to $q_{\underline{k}}^{(m)}! \partial^{[\underline{k}]}$ in $D_{A/R,n}$. In particular, they are injective. By identifying $D_{A/R,n}^{(m)}$ as an A-subalgebra of $D_{A/R,n}$, we get the equality $\partial^{<\underline{k}>(m)} = q_{\underline{k}}^{(m)}! \partial^{[\underline{k}]}$.

We set $D_{A/R}^{(m)} := \cup_{n \in \mathbb{N}} D_{A/R,n}^{(m)} \subset \cup_{n \in \mathbb{N}} D_{A/R,n} = D_{A/R}$. Then $D_{A/R}^{(m)}$ is equal to the free A-submodule of $D_{A/R}$ (for the right or left structure) whose basis consists in $\{\partial^{<\underline{k}>(m)}, \underline{k} \in \mathbb{N}^d\}$. This yields the equality $D_{A/R,n}^{(m)} = D_{A/R}^{(m)} \cap D_{A/R,n}$.

Remark 1.3.16 We have the inclusions $D_{A/R}^{(0)} \subset D_{A/R}^{(1)} \subset \cdots \subset D_{A/R}^{(m)} \subset D_{A/R}$. Moreover, since $q_{\underline{k}}^{(m)} = 0$ for m large enough, we get $\cup_{m \in \mathbb{N}} D_{A/R}^{(m)} = D_{A/R}$. Hence, we might write $D_{A/R}^{(\infty)} := D_{A/R}$.

Proposition 1.3.17

1. Then $D_{A/R}^{(0)}$ is equal to the R-subalgebra of $D_{A/R}$ generated by $\partial_1, \ldots, \partial_d$. This is the ring of differential operators of level 0 of A/R.
2. The ring $D_{A/R}^{(0)}$ is left and right Noetherian.

Proof

(1) We have to check that $\oplus_{\underline{k} \in \mathbb{N}^d} A \partial^{\underline{k}}$ is an R-subalgebra of $D_{A/R}$. Using 1.3.7. 4, we compute $\forall \underline{k} \in \mathbb{N}^d, \forall f \in A, \partial^{\underline{k}} f = \sum_{\underline{k}' + \underline{k}'' = \underline{k}} \partial^{[\underline{k}']}(f) \frac{\underline{k}!}{\underline{k}''!} \partial^{\underline{k}''}$, which proves the first part.

(2) Let ξ_i be the image of ∂_i in $\mathrm{gr}D_{A/R}^{(0)} = \oplus_{n\in\mathbb{N}} D_{A/R,n}^{(0)}/D_{A/R,n-1}^{(0)}$. Using 1.3.7. 4, we can check that $R[T_1,\ldots,T_d] \to \mathrm{gr}D_{A/R}^{(0)}$ given by $T_i \mapsto \xi_i$ is an isomorphism of graded rings. This yields that $\mathrm{gr}D_{A/R}^{(0)}$ is Noetherian and then so is $D_{A/R}^{(0)}$.

$\square$

To extend the above proposition to the level m, we will need the following propositions.

Proposition 1.3.18

1. *For any $\underline{k}', \underline{k}'' \in \mathbb{N}^d$, we have* $\underline{\partial}^{<\underline{k}'>(m)}\underline{\partial}^{<\underline{k}''>(m)} = \left\{ \dfrac{\underline{k}+\underline{k}'}{\underline{k}'} \right\}_{(m)} \underline{\partial}^{<\underline{k}+\underline{k}'>(m)}.$

2. *For any $\underline{k} \in \mathbb{N}^d$, $f \in A$,*

$$\underline{\partial}^{<\underline{k}>(m)} f = \sum_{\underline{k}'+\underline{k}''=\underline{k}} \left\{ \frac{\underline{k}}{\underline{k}'} \right\}_{(m)} \underline{\partial}^{<\underline{k}'>(m)}(f)\underline{\partial}^{<\underline{k}''>(m)}.$$

Proof This is straightforward from 1.3.7. $\square$

Proposition 1.3.19 *Let $k = \sum_{j=0}^m a_j p^j$, such that $a_j \in \mathbb{N}$ for any j and $0 \le a_j \le p-1$ for any $0 \le j \le m-1$. There exists $u \in \mathbb{Z}_{(p)}^*$ such that*

$$\partial_i^{<k>(m)} = u \prod_{j=0}^m \left(\partial_i^{[p^j]} \right)^{a_j}.$$

Proof For any $j \in \mathbb{N}$, by induction in n we compute $\left(\partial_i^{<p^j>(m)} \right)^n = \partial_i^{<np^j>(m)} \prod_{l=2}^n \left\langle \dfrac{lp^j}{p^j} \right\rangle_{(m)}$. Moreover,

$$\prod_{j=0}^m \partial_i^{<a_j p^j>(m)} = \partial_i^{<k>(m)} \prod_{j=0}^{m-1} \left\langle \frac{\sum_{l=0}^{j+1} a_l p^l}{a_j p^j} \right\rangle_{(m)}.$$

Hence, $\prod_{j=0}^m \left(\partial_i^{<p^j>(m)} \right)^{a_j} = \prod_{j=0}^m \partial_i^{<a_j p^j>(m)} \prod_{l=2}^{a_j} \left\langle \frac{lp^j}{p^j} \right\rangle_{(m)} = \partial_i^{<k>(m)}$ $\prod_{j=0}^{m-1} \prod_{l=2}^{a_j} \left\langle \frac{lp^j}{p^j} \right\rangle_{(m)} \left\langle \frac{\sum_{l=0}^{j+1} a_l p^l}{a_j p^j} \right\rangle_{(m)}$. Using 1.3.11, we can check $\prod_{j=0}^{m-1} \prod_{l=2}^{a_j} \left\langle \frac{lp^j}{p^j} \right\rangle_{(m)}$ $\left\langle \frac{\sum_{l=0}^{j+1} a_l p^l}{a_j p^j} \right\rangle_{(m)} \in \mathbb{Z}_{(p)}^*$. Since for any $j \le m$, we have $\partial_i^{<p^j>(m)} = \partial_i^{[p^j]}$, then we are done.

$\square$

Corollary 1.3.20 *The module $D_{A/R}^{(m)}$ is the R-subalgebra of $D_{A/R}$ generated by $D_{A/R,p^m}$. This is the ring of differential operators of level m of A/R.*

Proof This is a straightforward consequence of Propositions 1.3.11, 1.3.18 and 1.3.19. □

Corollary 1.3.21 *The R-algebra $D_{A/R}^{(m)}$ is twosided Noetherian.*

Proof We have $D_{A/R,n}^{(m)} = \oplus_{|\underline{k}|\leq n} A\underline{\partial}^{<\underline{k}>(m)}$. Using 1.3.18, we can check that $\mathrm{gr}D_{A/R}^{(m)}$ is a commutative A-algebra. We get from 1.3.19, that $\mathrm{gr}D_{A/R}^{(m)}$ is a commutative A-algebra generated by the image of $\partial_i^{<p^j>(m)}$, for $j = 0, \ldots, m$ and $i = 1, \ldots, d$ via the map $D_{A/R}^{(m)} \to \mathrm{gr}D_{A/R}^{(m)}$. Hence, $\mathrm{gr}D_{A/R}^{(m)}$ is Noetherian, and we are done.

□

Remark 1.3.22 This is false that $D_{A/R}$ is twosided Noetherian.

1.3.3 Left $D_{A/R}^{(m)}$-modules

Let $m \in \mathbb{N}$. We have seen in 1.3.20 that $D_{A/R}^{(m)}$ is the R-subalgebra of $D_{A/R}$ generated by $D_{A/R,p^m}$. In fact we can endow canonically $D_{A/R}^{(m)}$ with an R-algebra structure such that $D_{A/R}^{(m)} \to D_{A/R}$ is an homomorphism of R-algebras as follows.

Lemma 1.3.23 *There exists a unique homomorphism $\delta_{(m)}^{n,n'}: P_{A/R,(m)}^{n+n'} \to P_{A/R,(m)}^n \otimes_A P_{A/R,(m)}^{n'}$ of A-algebras for both left and right structures of A-algebras making commutative the following diagrams*

$$
\begin{array}{ccc}
A & \xrightarrow{d_1} & P_{A/R,(m)}^{n+n'} \\
\downarrow{d_1} & & \downarrow{\delta_{(m)}^{n,n'}} \\
P_{A/R,(m)}^{n'} & \xrightarrow{d_1} & P_{A/R,(m)}^n \otimes_A P_{A/R,(m)}^{n'},
\end{array}
\qquad
\begin{array}{ccc}
P_{A/R}^{n+n'} \otimes_R K & \hookrightarrow & P_{A/R,(m)}^{n+n'} & \hookrightarrow & P_{A/R}^{n+n'} \otimes_R K \\
\downarrow{\delta^{n,n'}} & & \downarrow{\delta_{(m)}^{n,n'}} & & \downarrow{\delta^{n,n'}\otimes id} \\
P_{A/R}^n \otimes_A P_{A/R}^{n'} & \hookrightarrow & P_{A/R,(m)}^n \otimes_A P_{A/R,(m)}^{n'} & \hookrightarrow & P_{A/R}^n \otimes_A P_{A/R}^{n'} \otimes_R K.
\end{array}
$$

(1.17)

Proof We check by a straightforward computation that the homomorphism $\delta^{n,n'} \otimes id: P_{A/R}^{n+n'} \otimes_R K \to P_{A/R}^n \otimes_A P_{A/R}^{n'} \otimes_R K$ of 1.3.3 sends $P_{A/R,(m)}^{n+n'}$ to $P_{A/R,(m)}^n \otimes_A P_{A/R,(m)}^{n'}$. This yields the homomorphism $\delta_{(m)}^{n,n'}: P_{A/R,(m)}^{n+n'} \to P_{A/R,(m)}^n \otimes_A P_{A/R,(m)}^{n'}$. We conclude using 1.15. □

1.3.24 Let $P \in D_{A/R,n}^{(m)}$, $P' \in D_{A/R,n'}^{(m)}$. We define PP' to be the composite morphism

$$P_{A/R,(m)}^{n+n'} \xrightarrow{\delta_{(m)}^{n,n'}} P_{A/R,(m)}^{n} \otimes_A P_{A/R,(m)}^{n'} \xrightarrow{id \otimes P'} P_{A/R,(m)}^{n} \xrightarrow{P} A.$$

We check that this is well defined, i.e. this is independent on the choice of such n and n'. Moreover, we compute this yields an R-algebra structure on $D_{A/R}^{(m)}$. Following the formal analogue of 1.2, we can check that $D_{A/R}^{(m)} \to D_{A/R}$ is an homomorphism of R-algebras.

Definition 1.3.25 Let E be an A-module. An m-PD stratification on E is a collection of $P_{A/R,(m)}^{n}$-linear isomorphisms

$$\epsilon_n : P_{A/R,(m)}^{n} \otimes_A E \xrightarrow{\sim} E \otimes_A P_{A/R,(m)}^{n}$$

such that

1. $\epsilon_0 = id$ and the family $(\epsilon_n)_n$ is compatible with the restrictions $P_{A/R,(m)}^{n} \to P_{A/R,(m)}^{n'}$, for any $n \leq n'$;
2. for any integers n, n', the diagram

$$(1.18)$$

where $q_0^{n,n'} : P_{A/R,(m)}^{n+n'} \twoheadrightarrow P_{A/R,(m)}^{n} \xrightarrow{id \otimes d_0} P_{A/R,(m)}^{n} \otimes_A P_{A/R,(m)}^{n'}$, and $q_1^{n,n'} : P_{A/R,(m)}^{n+n'} \twoheadrightarrow P_{A/R,(m)}^{n'} \xrightarrow{d_1 \otimes id} P_{A/R,(m)}^{n} \otimes_A P_{A/R,(m)}^{n'}$, is commutative.

Remark 1.3.26 Let $\pi_{ij} : A \widehat{\otimes}_R A \to A \widehat{\otimes}_R A \widehat{\otimes}_R A$ be the morphism corresponding to the projection at the i and j places. Let $r_i : A \to A \widehat{\otimes}_R A \widehat{\otimes}_R A$ be the morphism corresponding to the projection at the ith place. Let $\mu(2) : A \widehat{\otimes}_R A \widehat{\otimes}_R A \to A$ be the morphism corresponding to the diagonal morphism, and et $I(2) := \ker \mu(2)$. We denote by $P_{A/R}^{n}(2) := A \widehat{\otimes}_R A \widehat{\otimes}_R A / I(2)^{n+1}$ for any $n \in \mathbb{N}$. For any $n \in \mathbb{N}$, let $P_{A/R,(m)}^{n}(2)$ be the m-PD envelop of $(P_{A/R}^{n}(2), I(2)/I(2)^{n+1})$, and $r_i^n : A \to P_{A/R}^{n}(2)$ be the homomorphism induced by r_i.

Using the universal property of the m-PD envelop (or by an easy computation), we get the homomorphism $\pi_{ij}^n : P_{A/R,(m)}^n \to P_{A/R,(m)}^n(2)$ induced by π_{ij} (see also 1.2.23). We get the commutative diagram

$$
\begin{array}{ccc}
P_{A/R,(m)}^{n+n'} \xrightarrow[\;\;\;\;\;]{\substack{\pi_{01}^{n+n'} \\ \pi_{12}^{n+n'} \\ \pi_{02}^{n+n'}}} P_{A/R,(m)}^{n+n'}(2) & \longrightarrow & P_{A/R,(m)}^n \otimes_A P_{A/R,(m)}^{n'} \\
\Big\| & & \Big\| \\
P_{A/R,(m)}^{n+n'} \xrightarrow[\;\;\;\;\;]{\substack{q_0^{n,n'} \\ q_1^{n,n'} \\ \delta_{(m)}^{n,n'}}} & & P_{A/R,(m)}^n \otimes_A P_{A/R,(m)}^{n'}.
\end{array}
\tag{1.19}
$$

Hence, the cocycle condition is equivalent to the following condition:

$\star$ for any $n \in \mathbb{N}$, we have the equality $\pi_{02}^{n*}(\epsilon_n) = \pi_{01}^{n*}(\epsilon_n) \circ \pi_{12}^{n*}(\epsilon_n)$, i.e. the following diagram commutes

$$
\begin{array}{ccccccc}
& & \pi_{12}^{n*} d_1^{n*}(E) & \xrightarrow[\pi_{12}^{n*}(\epsilon_n)]{\sim} & \pi_{12}^{n*} d_0^{n*}(E) & & \\
& \nearrow & \Big\| & & \Big\| & \searrow & \\
r_2^{n*}(E) \!=\!\!=\! \pi_{02}^{n*} d_1^{n*}(E) & & & & \pi_{01}^{n*} d_1^{n*}(E) \!=\!\!=\! r_1^{n*}(E) & & \\
\pi_{02}^{n*}(\epsilon_n) \Big\downarrow \sim & & & & \sim \Big\downarrow \pi_{01}^{n*}(\epsilon_n) & & \\
r_0^{n*}(E) \!=\!\!=\! \pi_{02}^{n*} d_0^{n*}(E) & \!=\!\!=\! & \pi_{01}^{n*} d_0^{n*}(E) & \!=\!\!=\! & r_0^{n*}(E). & &
\end{array}
\tag{1.20}
$$

Proposition 1.3.27 *Let E be an A-module.*

1. The following datum are equivalent:

(a) A structure of left $D_{A/R}^{(m)}$-module on E extending its structure of A-module via the homomorphism of R-algebras $A \to D_{A/R}^{(m)}$.

(b) A compatible family of A-linear map $\theta_n : E \to E \otimes_A P_{A/R,(m)}^n$ such that $\theta_0 = id$ and such that for any $n, n' \in \mathbb{N}$ the diagram

$$
\begin{array}{ccc}
E \otimes_A P_{A/R,(m)}^{n+n'} & \xrightarrow{\;id \otimes \delta_{(m)}^{n,n'}\;} & E \otimes_A P_{A/R,(m)}^n \otimes_A P_{A/R,(m)}^{n'} \\
\theta_{n+n'} \Big\uparrow & & \theta_n \otimes id \Big\uparrow \\
E & \xrightarrow[\;\;\;\;\theta_{n'}\;\;\;\;]{} & E \otimes_A P_{A/R,(m)}^{n'}
\end{array}
\tag{1.21}
$$

is commutative.

(c) An m-PD stratification $(\epsilon_n)_{n \in \mathbb{N}}$ on E.

2. *An A-linear homomorphism* $\phi \colon E \to F$ *between two left* $D_{A/R}^{(m)}$-*module is* $D_{A/R}^{(m)}$-*linear if and only if* ϕ *is horizontal, i.e., the following diagram is commutative for any* $n \in \mathbb{N}$

$$
\begin{array}{ccc}
P_{A/R,(m)}^n \otimes_A E & \xrightarrow[\sim]{\epsilon_n} & E \otimes_A P_{A/R,(m)}^n \\
\downarrow{\scriptstyle id \otimes \phi} & & \downarrow{\scriptstyle \phi \otimes id} \\
P_{A/R,(m)}^n \otimes_A F & \xrightarrow[\sim]{\epsilon_n} & F \otimes_A P_{A/R,(m)}^n.
\end{array}
\tag{1.22}
$$

Proof By adding some (m) at the write place, we can copy word by word the proof of 1.2.24.

$\square$

Lemma 1.3.28 *We have the formula* $\forall x \in E$, $\theta_n(x) = \epsilon_n(1 \otimes x) = \sum_{|\underline{k}| \leq n} \underline{\partial}^{<\underline{k}>(m)}(x) \otimes \underline{\tau}^{\{\underline{k}\}(m)}$.

Proof Since $P_{A/R,(m)}^n$ is a free A-module, the canonical morphism ev: $P_{A/R,(m)}^n \to \mathrm{Hom}_A(D_{A/R,n}^{(m)}, A)$ is an isomorphism. Let $\{\underline{\partial}^{<\underline{k}>(m)*}, |\underline{k}| \leq n\}$ be the dual basis of $\{\underline{\partial}^{<\underline{k}>(m)}, |\underline{k}| \leq n\}$. Via this isomorphism, $\underline{\tau}^{\{\underline{k}\}(m)}$ is sent to $\underline{\partial}^{<\underline{k}>(m)*}$. Hence, the composition $E \otimes_A P_{A/R,(m)}^n \xrightarrow{\sim} E \otimes_A \mathrm{Hom}_A(D_{A/R,n}^{(m)}, A) \xrightarrow{\sim} \mathrm{Hom}_A(D_{A/R,n}^{(m)}, E)$ sends $\sum x_{\underline{k}} \otimes \underline{\tau}^{\{\underline{k}\}(m)}$ to $(\underline{\partial}^{<\underline{k}>(m)} \mapsto x_{\underline{k}})$. Since $\mu_n \colon D_{A/R,n}^{(m)} \otimes_A E \to E$, is given by $\underline{\partial}^{<\underline{k}>(m)} \otimes x \mapsto \underline{\partial}^{<\underline{k}>(m)}(x)$, then $E \to \mathrm{Hom}_A(D_{A/R,n}^{(m)}, E)$ is given by $x \mapsto (\underline{\partial}^{<\underline{k}>(m)} \mapsto \underline{\partial}^{<\underline{k}>(m)}(x))$, and we are done. $\square$

Corollary 1.3.29 *Let* E, F *be two left* $D_{A/R}^{(m)}$-*modules.*

1. *There exists a unique structure of* $D_{A/R}^{(m)}$-*module on* $E \otimes_A F$ *extending its canonical structure of A-module such that, for any* $\underline{k} \in \mathbb{N}^d$, $x \in E$, $y \in F$,

$$
\underline{\partial}^{<\underline{k}>(m)}(x \otimes y) = \sum_{\underline{i} \leq \underline{k}} \left\{ \begin{matrix} \underline{k} \\ \underline{i} \end{matrix} \right\}_{(m)} \underline{\partial}^{<\underline{i}>(m)}(x) \otimes \underline{\partial}^{<\underline{k}-\underline{i}>(m)}(y).
\tag{1.23}
$$

2. *There exists a unique structure of* $D_{A/R}^{(m)}$-*module on* $\mathrm{Hom}_A(E, F)$ *extending its canonical structure of A-module such that,* $\forall \underline{k} \in \mathbb{N}^d$, *for any* $\phi \in \mathrm{Hom}_A(E, F)$

$$
(\underline{\partial}^{<\underline{k}>(m)} \phi)(x) = \sum_{\underline{i} \leq \underline{k}} (-1)^{|\underline{i}|} \left\{ \begin{matrix} \underline{k} \\ \underline{i} \end{matrix} \right\}_{(m)} \underline{\partial}^{<\underline{k}-\underline{i}>(m)} \left(\phi(\underline{\partial}^{<\underline{i}>(m)} x) \right).
\tag{1.24}
$$

Proof The m-PD stratification of $E \otimes_A F$ is the composition $\epsilon_n^{E \otimes F} \colon P_{A/R,(m)}^n \otimes_A E \otimes_A F \xrightarrow[\epsilon_n^E \otimes id]{\sim} E \otimes_A P_{A/R,(m)}^n \otimes_A F \xrightarrow[id \otimes \epsilon_n^F]{\sim} E \otimes_A F \otimes_A P_{A/R,(m)}^n$. We get $1 \otimes x \otimes y \mapsto$

$$\sum_{\underline{i}} \partial^{<\underline{i}>(m)}(x) \otimes \underline{\tau}^{\{\underline{i}\}(m)} \otimes y \mapsto \sum_{\underline{i}} \sum_{\underline{j}} \partial^{<\underline{i}>(m)}(x) \otimes \partial^{<\underline{j}>(m)}(y) \otimes \underline{\tau}^{\{\underline{j}\}(m)} \underline{\tau}^{\{\underline{i}\}(m)} =$$

$\sum_{\underline{i}} \sum_{\underline{j}} \left\{ \begin{matrix} \underline{i}+\underline{j} \\ \underline{i} \end{matrix} \right\}_{(m)} \partial^{<\underline{i}>(m)}(x) \otimes \partial^{<\underline{j}>(m)}(y) \otimes \underline{\tau}^{\{\underline{i}+\underline{j}\}(m)}$. We proceed similarly for the second part. $\square$

1.3.4 Weak p-Adic Completion

1.3.30 Let $m \in \mathbb{N}$. Let $\widehat{D}_{A/R}^{(m)} := \varprojlim_n D_{A/R}^{(m)}/p^n D_{A/R}^{(m)}$ be the p-adic completion of $D_{A/R}^{(m)}$. Let $P \in \widehat{D}_{A/R}^{(m)}$. Then there exists a unique sequence $(a_{\underline{k}})_{\underline{k} \in \mathbb{N}^d}$ (resp. $(b_{\underline{k}})_{\underline{k} \in \mathbb{N}^d}$) of elements of A such that $a_{\underline{k}} \to 0$ when $|\underline{k}| \to \infty$ (resp. $b_{\underline{k}} \to 0$ when $|\underline{k}| \to \infty$) and $P = \sum_{\underline{k} \in \mathbb{N}^d} a_{\underline{k}} \partial^{<\underline{k}>(m)}$ (resp. $P = \sum_{\underline{k} \in \mathbb{N}^d} \partial^{<\underline{k}>(m)} b_{\underline{k}}$).

Berthelot has proved that the homomorphisms $\widehat{D}_{A/R}^{(m)} \otimes_R K \to \widehat{D}_{A/R}^{(m+1)} \otimes_R K$ are flat (see [Ber96, 3.5.4]). We sketch the proof as follows. Let D' be the subset of $\widehat{D}_{A/R}^{(m+1)}$ of elements which can be written in the form $P + Q$ with $P \in \widehat{D}_{A/R}^{(m)}$ and $Q \in D_{A/R}^{(m+1)}$. Using the fact that $D_{A/R}^{(m+1)}$ and $\widehat{D}_{A/R}^{(m)}$ are Noetherian, with some technical computations we can prove D' is a Noetherian ring. Hence, the extension $D' \to \widehat{D}'$ is flat. Since $D_{A/R}^{(m+1)} \otimes_R K = D_{A/R}^{(m)} \otimes_R K = D_{A/R} \otimes_R K$, we get $D' \otimes_R K = \widehat{D}_{A/R}^{(m)}$. Moreover, we compute $\widehat{D}' = \widehat{D}_{A/R}^{(m+1)} \otimes_R K$. Hence, we are done.

Notation 1.3.31 $D_{A/R}^{\dagger} := \cup_{m \in \mathbb{N}} \widehat{D}_{A/R}^{(m)}$. This is Berthelot's ring of differential operators of finite level and infinite order. Following proposition 1.3.35 below, we can view $D_{A/R}^{\dagger}$ as the p-adic weak completion of $D_{A/R}$, where the weakness means that the majoration appearing in 1.3.35 is satisfied.

The ring $\widehat{D}_{A/R}^{(m)}$ is Noetherian, but this is not the case of $D_{A/R}^{\dagger}$ or $D_{A/R}^{\dagger} \otimes_R K$. Fortunately, since the homomorphisms $\widehat{D}_{A/R}^{(m)} \otimes_R K \to \widehat{D}_{A/R}^{(m+1)} \otimes_R K$ are flat, then $D_{A/R}^{\dagger} \otimes_R K$ is coherent. Hence, the notions of (left or right) $D_{A/R}^{\dagger} \otimes_R K$-coherent module and of (left or right) $D_{A/R}^{\dagger} \otimes_R K$-module of finite presentation are then identical. It is likely that $D_{A/R}^{\dagger}$ is not coherent but it is an open question.

Lemma 1.3.32 *We denote by* $|.|$ *the norm on* $\mathbb{Z}$ *defined by* $|n| = p^{-v_p(n)}$ *for any* $n \in \mathbb{Z}$.

1. $\forall m \in \mathbb{N},\ \exists \eta < 1,\ \exists c \in \mathbb{R}$ *such that* $|q_k^{(m)}!| \leq c\eta^k$ *for any* $k \in \mathbb{N}$.
2. $\forall \eta < 1,\ \exists m \in \mathbb{N},$ *such that* $\eta^k \leq |q_k^{(m)}!|$ *for any* $k \in \mathbb{N}$.

Proof

(1) Put $q := q_k^{(m)}$. If $q = \sum_{i=0}^{r} a_i p^i$ with $a_i \in \{0, \dots, p-1\}$ for $i = 0, \dots, r-1$ and $a_r \in \{1, \dots, p-1\}$, then $0 \leq \sigma(q) \leq (p-1)(r+1) < (p-1)(\log_p(q+1)+1) \leq (p-1)(\log_p(k+1)+1)$. Moreover $k/p^m - 1 <$

$q \leq k/p^m$. Since $v_p(q!) = (q - \sigma(q))/(p-1)$, we get

$$k/p^m(p-1) - 1/(p-1) - \log_p(k+1) - 1 < v_p(q!) \leq k/p^m(p-1). \quad (1.25)$$

Set $f(k) := 1/(p-1) + \log_p(k+1) + 1 = p/(p-1) + \log_p(k+1)$. There exists k_0 large enough such that for any $k \geq k_0$, we get $0 \leq k/2p^m(p-1) - f(k)$. This yields, for any $k \in \mathbb{N}$, the inequality

$$k/2p^m(p-1) - f(k_0) < v_p(q!). \quad (1.26)$$

Indeed, on one hand, if $k \leq k_0$, then $f(k) \leq k_0$ and then $k/2p^m(p-1) - f(k_0) \leq k/p^m(p-1) - f(k) < v_p(q!)$. On the other hand, if $k \geq k_0$, then $-f(k_0) \leq 0 \leq k/2p^m(p-1) - f(k)$. Hence, $k/2p^m(p-1) - f(k_0) \leq k/p^m(p-1) - f(k) < v_p(q!)$.

From the inequality 1.26, we get $|q!| < p^{f(k_0)}(1/p^{1/2p^m(p-1)})^k$. We can take $\eta = 1/p^{1/2p^m(p-1)} < 1$, $c = p^{f(k_0)}$.

(2) Fix $\eta < 1$. Using 1.25, we get $|q_k^{(m)}!| \geq (1/p)^{k/p^m(p-1)} = (1/p^{1/p^m(p-1)})^k$. For m large enough, $1/p^{1/p^m(p-1)} \geq \eta$. Hence, we are done.

$\square$

1.3.33 For any $i \in \mathbb{N}$, the canonical morphism $D_{A/R}/\pi^{i+1}D_{A/R} \to D_{A_i/R_i}$ is an isomorphism. This yields the isomorphism $\widehat{D}_{A/R}/\pi^{i+1}\widehat{D}_{A/R} \xrightarrow{\sim} D_{A_i/R_i}$.

Remark 1.3.34 The ring $A \otimes_R K$ is a Tate K-algebra. For any isomorphism of K-algebras of the form $K\{T_1, \ldots, T_r\}/I \xrightarrow{\sim} A \otimes_R K$, we get a quotient norm which endows $A \otimes_R K$ with a structure of Banach K-algebra. Such norms are all equivalents (see [BGR84, 3.7.3, prop. 3]).

Proposition 1.3.35 *Choose* $\| \cdot \|$ *a Banach norm on* $A \otimes_R K$. *Let* $P = \sum_{k \in \mathbb{N}^d} a_k \underline{\partial}^{[k]} \in \widehat{D}_{A/R}$. *For any* $i \in \mathbb{N}$, *let* $P_i \in D_{A_i/R_i}$ *be the image of* P. *The following conditions are equivalent:*

1. *$P \in D^\dagger_{A/R}$;*
2. *$\exists \alpha, \beta \in \mathbb{R}$ such that $\mathrm{ord}(P_i) \leq \alpha i + \beta$ for any $i \in \mathbb{N}$;*
3. *$\exists c, \eta \in \mathbb{R}_+$ such that $\eta < 1$ and $\| a_k \| \leq c\eta^{|k|}$, for any $\underline{k} \in \mathbb{N}^d$.*

Proof 1 $\Rightarrow$ 2. Suppose $P \in D^\dagger_{A/R}$. Then there exists $m \in \mathbb{N}$ large enough such that $P \in \widehat{D}_{A/R}^{(m)}$. Let us fix such m. Following 1.3.30, there exists a unique sequence $(b_k)_{k \in \mathbb{N}^d}$ of elements of A such that $b_k \to 0$ when $|k| \to \infty$ and $P = \sum_{k \in \mathbb{N}^d} b_k \underline{\partial}^{<k>(m)}$. Since $\underline{\partial}^{<\underline{k}>(m)} = q_k^{(m)}! \underline{\partial}^{[k]}$, we get $P = \sum_{k \in \mathbb{N}^d} b_k q_k^{(m)}! \underline{\partial}^{[k]}$, i.e. $a_k = b_k q_k^{(m)}!$. Using 1.3.32. 1, $\exists \eta < 1$, $\exists c \in \mathbb{R}$ such that $|q_k^{(m)}!| \leq c\eta^{|k|}$ for any $\underline{k} \in \mathbb{N}^d$. Setting $\alpha := e \log_p(1/c)$ and $\beta := e \log_p(1/\eta)$, this yields $v_\pi(a_k) \geq v_\pi(q_k^{(m)}!) = e v_p(q_k^{(m)}!) \geq \alpha|k| + \beta$, where v_p is the p-adic valuation of A (i.e. for any $a \in A$, $v_p(a) := \sup\{n \in \mathbb{N} \; ; \; a \in p^n A\}$) and v_π is the π-adic valuation of A. This is equivalent to saying that $\mathrm{ord}(P_i) \leq \alpha i + \beta$ for any $i \in \mathbb{N}$.

$2 \Rightarrow 3$. Since Banach norms on $A \otimes_R K$ are equivalent, we can choose $\| \cdot \|$ defined by $\| c \| := p^{-v_p(c)/e}$ where the v_p is defined by the p-adic topology $(p^n A)_{n \in \mathbb{N}}$ on $A \otimes_R K$ (i.e for any $b \in A \otimes_R K$, $v_p(b) := \sup\{n \in \mathbb{Z} \; ; \; b \in p^n A\}$). Suppose $\exists \alpha, \beta \in \mathbb{R}$ such that $\mathrm{ord}(P_i) \leq \alpha i + \beta$ for any $i \in \mathbb{N}$. Let $\underline{k} \in \mathbb{N}^d$. Set $i_{\underline{k}} := v_\pi(a_{\underline{k}})$. Since the image of $a_{\underline{k}}$ in $A_{i_{\underline{k}}}$ is not null, then $|\underline{k}| \leq \alpha i_{\underline{k}} + \beta$. Hence, $\| a_{\underline{k}} \| = p^{-i_{\underline{k}}/e} \leq p^{-(|\underline{k}|-\beta)/\alpha e} = p^{\beta/\alpha e} \left(p^{-1/\alpha e} \right)^{|\underline{k}|}$. Hence, we can choose $c = p^{\beta/\alpha e}$ and $\eta = p^{-1/\alpha e}$.

$3 \Rightarrow 1$. Suppose $\exists c, \eta \in \mathbb{R}_{>0}$ such that $\eta < 1$ and $\| a_{\underline{k}} \| \leq c\eta^{|\underline{k}|}$, for any $\underline{k} \in \mathbb{N}^d$. We have to prove that for m large enough, $a_{\underline{k}}/q_{\underline{k}}^{(m)}! \in A$. The inequality $\| a_{\underline{k}} \| \leq c\eta^{|\underline{k}|}$ is equivalent to $v_p(a_{\underline{k}}) \geq \lambda |\underline{k}| + \mu$, with $\mu = -\log_p(c)$ and $\lambda = -\log_p(\eta) > 0$. Using 1.25, this yields

$$v_p(a_{\underline{k}}/q_{\underline{k}}^{(m)}!) \geq \lambda |\underline{k}| + \mu - |\underline{k}|/p^m (p-1) = \left(\lambda - 1/p^m(p-1) \right) |\underline{k}| + \mu. \quad (1.27)$$

Suppose m large enough such that $\lambda - 1/p^m(p-1) > 0$. Hence, if $\mu \geq 0$, then $v_p(a_{\underline{k}}/q_{\underline{k}}^{(m)}!) \geq 0$, i.e. $a_{\underline{k}}/q_{\underline{k}}^{(m)}! \in A$ and we are done. Suppose now $\mu < 0$ and m large enough such that the inequalities hold

$$p^m > -\mu/\left(\lambda - 1/p^m(p-1) \right) \Leftrightarrow p^m \lambda - 1/(p-1) \geq -\mu \Leftrightarrow p^m \geq (-\mu + 1/(p-1))/\lambda. \quad (1.28)$$

Let $\underline{k} \in \mathbb{N}^d$. If $|\underline{k}| \geq -\mu/\left(\lambda - 1/p^m(p-1) \right)$, then $(\lambda - 1/p^m(p-1)) |\underline{k}| + \mu \geq 0$ and we are done thanks to 1.27. On the other hand, if $|\underline{k}| \leq -\mu/\left(\lambda - 1/p^m(p-1) \right)$, then from 1.28 we get $|\underline{k}| \leq p^m$. Hence, $q_{\underline{k}}^{(m)}! = 0$ and then $a_{\underline{k}}/q_{\underline{k}}^{(m)}! \in A$. $\qquad \square$

Examples 1.3.36 We give here a fundamental example of a left $D_{A/R}^\dagger$-module. Let $f \in A$. We denote by $A_{\{f\}}$ the p-adic completion of A_f. We denote by $A_f^\dagger$ the weak p-adic completion of A_f as A-algebra. More precisely, the element x of $A_f^\dagger$ are the subset of $A_{\{f\}}$ whose elements can be written in the form $x = \sum_{n \geq 0} \frac{a_n}{f^n}$ such that $\exists c, \eta \in \mathbb{R}_+$ such that $\eta < 1$ and $\| a_n \| \leq c\eta^n$, for any $n \in \mathbb{N}$.

The ring $A_{\{f\}}$ has a canonical structure of left $\widehat{D}_{A/R}$-module. With the description of 1.3.35, we compute that $A_f^\dagger$ is a $D_{A/R}^\dagger$-submodule of $A_{\{f\}}$.

1.3.5 Sheafification, Coherence

We denote by $\mathfrak{S} := \mathrm{Spf}\, R$ the formal scheme associated to R endowed with the p-adic topology. Let $\mathfrak{P}$ be a smooth formal scheme over $\mathfrak{S}$, and $f : \mathfrak{P} \to \mathfrak{S}$ be the structural morphism. We will keep the following notation.

- We denote by $R\{T_1, \ldots, T_d\}$ the p-adic completion of $R[T_1, \ldots, T_d]$ and $\widehat{\mathbb{A}}_R^d :=$ Spf $R\{T_1, \ldots, T_d\}$ is the p-adic completion of $\mathbb{A}_R^d$. For any point $x \in \mathfrak{P}$, there exists an open affine formal subscheme $\mathfrak{U}_x$ of $\mathfrak{P}$ containing x together with an étale morphism $\mathfrak{U}_x \to \widehat{\mathbb{A}}_R^d$. We denote by $\mathfrak{B}$ the set of open affine formal subschemes $\mathfrak{U}$ of $\mathfrak{P}$ such that there exists an étale morphism of the form $\mathfrak{U} \to \widehat{\mathbb{A}}_R^d$. We get presheaf $\mathcal{D}_{\mathfrak{P}/\mathfrak{S}}^{(m)}$ on $\mathfrak{B}$ defined by $\mathfrak{U} \in \mathfrak{B} \mapsto D_{\Gamma(\mathfrak{U},\mathfrak{P})/R}^{(m)}$. If $\mathfrak{V}$ is a standard open subset of $\mathfrak{U}$, then the canonical morphism $D_{\Gamma(\mathfrak{U},\mathfrak{P})/R}^{(m)} \otimes_{\Gamma(\mathfrak{U},\mathfrak{P})} \Gamma(\mathfrak{V}, \mathfrak{P}) \to D_{\Gamma(\mathfrak{V},\mathfrak{P})/R}^{(m)}$ is an isomorphism. Hence, $\mathcal{D}_{\mathfrak{P}/\mathfrak{S}}^{(m)}$ is a sheaf on $\mathfrak{B}$. We still denote by $\mathcal{D}_{\mathfrak{P}/\mathfrak{S}}^{(m)}$ the corresponding sheaf on $\mathfrak{P}$.

- Let $\widehat{\mathcal{D}}_{\mathfrak{P}/\mathfrak{S}}^{(m)} := \varprojlim_n \mathcal{D}_{\mathfrak{P}/\mathfrak{S}}^{(m)} / p^n \mathcal{D}_{\mathfrak{P}/\mathfrak{S}}^{(m)}$ be the p-adic completion of $\mathcal{D}_{\mathfrak{P}/\mathfrak{S}}^{(m)}$. We put $\mathcal{D}_{\mathfrak{P}/\mathfrak{S}}^{\dagger} := \cup_{m \in \mathbb{N}} \widehat{\mathcal{D}}_{\mathfrak{P}/\mathfrak{S}}^{(m)}$, and $\mathcal{D}_{\mathfrak{P}/\mathfrak{S},\mathbb{Q}}^{\dagger} := \mathcal{D}_{\mathfrak{P}/\mathfrak{S}}^{\dagger} \otimes_{\mathbb{Z}} \mathbb{Q} = \mathcal{D}_{\mathfrak{P}/\mathfrak{S}}^{\dagger} \otimes_R K$. Using 1.3.31, we can check that $\mathcal{D}_{\mathfrak{P}/\mathfrak{S},\mathbb{Q}}^{\dagger}$ is coherent.

- Let k be the residue field of R. We denote by P the special fiber of $\mathfrak{P}$, i.e. P is the k-scheme equal to the reduction modulo π of $\mathfrak{P}$. Let T be a divisor of P. Let $\mathfrak{U}$ be an open affine formal subscheme of $\mathfrak{P}$ such that there exists a section $f \in \Gamma(\mathfrak{U}, \mathcal{O}_{\mathfrak{P}})$ whose reduction modulo π is an equation of T in P. Then $A_f^{\dagger} \otimes_R K$ does not depend on the choice of the lifting f of an equation of T. We denote it by $\mathcal{O}_{\mathfrak{U}}(^{\dagger}T)_{\mathbb{Q}}$. We get a presheaf of K-algebras $\mathcal{O}_{\mathfrak{P}}(^{\dagger}T)_{\mathbb{Q}}$ on such open formal subschemes of $\mathfrak{P}$ defined by $\mathfrak{U} \mapsto \mathcal{O}_{\mathfrak{U}}(^{\dagger}T)_{\mathbb{Q}}$. Similarly to $\mathcal{D}_{\mathfrak{P}/\mathfrak{S}}^{\dagger}$, we can check that $\mathcal{O}_{\mathfrak{P}}(^{\dagger}T)_{\mathbb{Q}}$ is in fact a sheaf. Using 1.3.36, the sheaf $\mathcal{O}_{\mathfrak{P}}(^{\dagger}T)_{\mathbb{Q}}$ is endowed with a canonical structure of left $\mathcal{D}_{\mathfrak{P}/\mathfrak{S},\mathbb{Q}}^{\dagger}$-module.

Proposition 1.3.37 (Berthelot) *Suppose T is a strict normal crossing divisor of P. Let $x \in P$. Choose an open affine formal subscheme $\mathfrak{U}$ of $\mathfrak{P}$ such that there exist local coordinates $t_1, \ldots, t_d$ satisfying $t_1, \ldots, t_r \in \Gamma(\mathfrak{U}, \mathcal{O}_{\mathfrak{U}})$ with $r \leq d$, and $T \cap U = V(\bar{t}_1 \cdots \bar{t}_r)$ where U is the special fiber of U and $\bar{t}_1, \ldots, \bar{t}_r$ are the images of respectively $t_1, \ldots, t_r$ in $\Gamma(U, \mathcal{O}_U)$. We have the exact sequence*

$$(\mathcal{D}_{\mathfrak{U}/\mathfrak{S},\mathbb{Q}}^{\dagger})^d \xrightarrow{\psi} \mathcal{D}_{\mathfrak{U}/\mathfrak{S},\mathbb{Q}}^{\dagger} \xrightarrow{\phi} \mathcal{O}_{\mathfrak{U}}(^{\dagger}T \cap U)_{\mathbb{Q}} \to 0, \tag{1.29}$$

where $\phi(P) = P \cdot (1/t_1 \cdots t_r)$, and ψ is defined by

$$\psi(P_1, \ldots, P_d) = \sum_{i=1}^{r} P_i \partial_i t_i + \sum_{i=r+1}^{d} P_i \partial_i. \tag{1.30}$$

Proof By devissage, we reduce to the case where $r = 1$. Then, this is a (technical) computation. The reader can find a proof at [Ber90, 4.3.2]. □

Theorem 1.3.38 (Berthelot) *The left $\mathcal{D}_{\mathfrak{P}/\mathfrak{S},\mathbb{Q}}^{\dagger}$-module $\mathcal{O}_{\mathfrak{P}}(^{\dagger}T)_{\mathbb{Q}}$ is coherent.*

Proof Sketch of the proof: using de Jong's desingularisation theorem, we reduce to the case where T is a strict normal crossing divisor of P, which has already been checked.   □

Corollary 1.3.39 *Suppose* $\mathfrak{P}$ *proper. The de Rham cohomological spaces* $H^n \left(\mathfrak{P}, \Omega^\bullet_{\mathfrak{P}/\mathfrak{S}} \otimes_{\mathcal{O}_{\mathfrak{P}}} \mathcal{O}_{\mathfrak{P}}(^\dagger T)_{\mathbb{Q}} \right)$ *of* $\mathcal{O}_{\mathfrak{P}}(^\dagger T)_{\mathbb{Q}}$ *are* K-*vector spaces of finite type.*

Remark 1.3.40 Suppose $\mathfrak{P}$ is the p-adic completion of $\mathbb{P}^1_R$, $\mathfrak{U}$ is the p-adic completion of $\mathbb{A}^1_R$, and $T = P \backslash U$ is the divisor corresponding to the point at infinity. Then $H^1 \left(\mathfrak{U}, \Omega^\bullet_{\mathfrak{U}/\mathfrak{S}} \otimes_{\mathbb{Z}} \mathbb{Q} \right)$ is an infinite K-vector space. Indeed, this corresponds to the cokernel of the composite $R\{t\} \otimes_R K \to \Gamma(\mathfrak{U}, \Omega^1_{\mathfrak{U}/\mathfrak{S}}) \otimes_{\mathbb{Z}} \mathbb{Q} \xrightarrow{\sim} R\{t\} \otimes_R K$ which is given by the derivation with respect to t. On the other hand, the cokernel of the map $R[t]^\dagger \otimes_R K \to R[t]^\dagger \otimes_R K$ given by the derivation with respect to t is zero as desired, i.e. $H^1 \left(\mathfrak{P}, \Omega^\bullet_{\mathfrak{P}/\mathfrak{S}} \otimes_{\mathcal{O}_{\mathfrak{P}}} \mathcal{O}_{\mathfrak{P}}(^\dagger T)_{\mathbb{Q}} \right) = 0$. This is the reason why we should replace the naive constant coefficient $\mathcal{O}_{\mathfrak{U},\mathbb{Q}}$ by $\mathcal{O}_{\mathfrak{P}}(^\dagger T)_{\mathbb{Q}}$. In other words, the analogue of $\mathcal{O}_{\mathbb{A}^1_{\mathbb{C}}}$ is $\mathcal{O}_{\mathfrak{P}}(^\dagger T)_{\mathbb{Q}}$, where $\mathbb{C}$ is the field of complex numbers, i.e. we can view $\mathcal{O}_{\mathfrak{P}}(^\dagger T)_{\mathbb{Q}}$ as the right object corresponding to the constant coefficient $\mathcal{O}_{\mathbb{A}^1_k}$.

Acknowledgments I thank the Vietnam Institute for Advanced Study in Mathematics (VIASM) for their invitation and the opportunity to expand the theory of Berthelot of arithmetic $\mathcal{D}$-modules.

Appendix: Further Reading

The context of 1.3.2 has allowed us to strongly simplify the notion of m-PD-envelopes. Indeed, in this context we almost work with $\mathbb{Q}$-rings which give us an obvious definition of partial of level m divided powers (see 1.3.13). In a wider context, e.g. for $\mathbb{F}_p$-rings, Berthelot has introduced a notion of m-envelope which behaves in the same way (see [Ber96, 1]). This makes it possible for any smooth morphism $X \to S$ of schemes to define the ring of operators $\mathcal{D}^{(m)}_{X/S}$ of level m of X/S. The reader can find a detailed construction in [Ber96, 2].

One key fundamental property in Berthelot's arithmetic $\mathcal{D}$-modules is his Frobenius descent theorem. Let us describe this theorem in the simpler context where S is a scheme over Spec $\mathbb{F}_p$. Let X' be the base change of X under the absolute Frobenius of S. We get the relative Frobenius $F \colon X \to X'$ which is an S-morphism. Then the inverse image functor F^* under F induces an equivalence of categories between the category of left $\mathcal{D}^{(m)}_{X'/S}$-modules and that of left $\mathcal{D}^{(m+1)}_{X/S}$-modules. The proof is highly technical and uses fundamental properties of m-PD-envelops (see [Ber00, 2] for a wider context and more details). One important consequence is the following. Suppose S regular. Then using standard techniques we can check that

$\mathcal{D}_{X/S}^{(0)}$ is of finite cohomological dimension. Using Frobenius descent theorem, this yields that $\mathcal{D}_{X/S}^{(m)}$ is of finite cohomological dimension for any non negative integer m.

We denote by $\mathfrak{S} := \mathrm{Spf}\, R$ the formal scheme associated to R endowed with the p-adic topology. Let $\mathfrak{P}$ be a smooth formal scheme over $\mathfrak{S}$, and $f : \mathfrak{P} \to \mathfrak{S}$ be the structural morphism. We have already explained why $\mathcal{D}_{\mathfrak{P}/\mathfrak{S},\mathbb{Q}}^{\dagger}$ is coherent. Using Frobenius descent as described previously, since $\mathcal{D}_{P/S}^{(0)}$ is of finite cohomological dimension, then so is $\mathcal{D}_{\mathfrak{P}/\mathfrak{S},\mathbb{Q}}^{\dagger}$. This is a key property which implies that the category of perfect complexes of $\mathcal{D}_{\mathfrak{P}/\mathfrak{S},\mathbb{Q}}^{\dagger}$-modules is the category of coherent complexes of $\mathcal{D}_{\mathfrak{P}/\mathfrak{S},\mathbb{Q}}^{\dagger}$-modules (with bounded cohomology). This yields that coherent complexes are stable under duality. The reader might find an introduction to the construction of the duality and other cohomological operations in [Ber02].

References

[Ber90] P. Berthelot, Cohomologie rigide et théorie des $\mathcal{D}$-modules, in *p-Adic Analysis (Trento, 1989)* (Springer, Berlin, 1990), pp. 80–124

[Ber96] P. Berthelot, $\mathcal{D}$-modules arithmétiques. I. Opérateurs différentiels de niveau fini. Ann. Sci. École Norm. Sup. (4) **29**(2), 185–272 (1996)

[Ber00] P. Berthelot, $\mathcal{D}$-modules arithmétiques. II. Descente par Frobenius. Mém. Soc. Math. Fr. (N.S.) (81), vi+136 (2000)

[Ber02] P. Berthelot, Introduction à la théorie arithmétique des $\mathcal{D}$-modules. Astérisque (279), 1–80 (2002). Cohomologies p-adiques et applications arithmétiques, II

[BGR84] S. Bosch, U. Güntzer, R. Remmert, *Non-Archimedean Analysis* (Springer, Berlin, 1984). A systematic approach to rigid analytic geometry

[Car20] D. Caro, Arithmetic $\mathcal{D}$-modules over algebraic varieties of characteristic $p > 0$. arXiv:1909.08432 (2020)

[CT12] D. Caro, N. Tsuzuki, Overholonomicity of overconvergent F-isocrystals over smooth varieties. Ann. Math. (2) **176**(2), 747–813 (2012)

[EGA IV] A. Grothendieck, Éléments de géométrie algébrique. IV. Étude locale des schémas et des morphismes de schémas. I. Inst. Hautes Études Sci. Publ. Math. (20), 259 (1964)

[PT20a] J. Poineau, D. Turchetti, Berkovich curves and Schottky uniformization I: The Berkovich affine line, in *Arithmetic and Geometry Over Local Fields*, ed. by B. Anglès, T. Ngo Dac (Springer International Publishing, Cham, 2020)

[PT20b] J. Poineau, D. Turchetti, Berkovich curves and Schottky uniformization II. Analytic uniformization of Mumford curves, in *Arithmetic and Geometry Over Local Fields*, ed. by B. Anglès, T. Ngo Dac (Springer International Publishing, Cham, 2020)

Chapter 2
Difference Galois Theory for the "Applied" Mathematician

Lucia Di Vizio

Abstract The lecture notes below correspond to the course given by the author in occasion of the VIASM school on Number Theory (18–24 June 2018, Hanoi). We have chosen to omit the proofs that are already presented in details in many references in the literature, although they were explained during the lectures, and we have devoted more space to statements useful in the applications. The applications concern many different mathematical settings, where linear difference equations naturally arise. We cite in particular the case of Drinfeld modules, which is considered in [Pel] and [TR].

2.1 Introduction

The initial data of classical Galois theory are a field K, let's say of characteristic 0, and an irreducible polynomial $P \in K[x]$, with coefficients in K. Then *the* minimal field extension L of K containing a *full set* of roots of P is constructed and one defines the Galois group G of L/K, namely the group of the field automorphisms of L that fix the elements of K. The group G is finite and acts on L by shuffling the roots of P. The idea behind this construction is that the structure of the group G should reveal hidden algebraic relations among the roots of P, other than the evident relations given by P itself.

The same kind of philosophy applied to functional equations has been the starting point of differential Galois theory, first, and difference Galois theory, later. The references are numerous and it is almost impossible to list them all. We refer to [vdPS03], for the differential case, and to [vdPS97] and [HSS16], for the difference case.

L. Di Vizio (✉)
Laboratoire de Mathématiques UMR 8100, CNRS, Université de Versailles-St Quentin, Versailles Cedex, France
e-mail: lucia.di.vizio@math.cnrs.fr

© The Author(s), under exclusive license to Springer Nature Switzerland AG 2021
B. Anglès, T. Ngo Dac (eds.), *Arithmetic and Geometry over Local Fields*,
Lecture Notes in Mathematics 2275, https://doi.org/10.1007/978-3-030-66249-3_2

Let K be a base field and τ be an endomorphism of K. We consider a linear difference equation $\tau(Y) = AY$, where A is an invertible square matrix with coefficients in K and Y is a square matrix of unknowns. The usual approach in Galois theory of difference equations is to construct an abstract K-algebra L/K containing the entries of a fundamental (i.e., invertible) matrix of solutions, under the assumption that the characteristic of K is zero, that the field of constants k is algebraically closed and that τ is surjective. The idea is that, in order to study the properties of the solutions a difference equation, it is not "important" to solve it, but only to understand the structure of its Galois group. However, in applications, it usually happens that a set of solutions is given in some specific K-algebra: When that's the case, it is not always easy to understand which properties transfer from the abstract solutions to the ones that we have found.

In this paper we suppose that we have a fundamental solution is some field L/K, equipped with an extension of τ. The divergence between the classical approach and the apparently more pedestrian approach that we are considering here, starts immediately: Indeed in general we cannot assume that the field L exists and only the existence of a pseudo field is ensured (see Remark 2.3.6 below). In [CHS08], the authors reconcile these two points of view and the point of view of model theory, in the special case of q-difference equations: They construct a group using given solutions in a specific algebra and compare it with the group constructed in [vdPS97], but they assume that τ is an automorphism and the statements on the comparison of the Galois groups are not easy to apply in other settings.

More recently, a very general abstract approach has been considered by A. Ovchinnikov and M. Wibmer in [OW15, §2.2], where, in contrast with the more classical references, the authors do not make any assumption on the characteristic of the field, they do not require that the endomorphism is surjective, and they do not assume that the constants are algebraically closed. They do not even assume that K is a field, but only that K is a pseudo field.

M. Papanikolas in [Pap08, §4] chooses a framework which is in between the two examples above: He works on a field K equipped with an automorphisms τ, but the characteristic can be positive and the field of constants is not necessarily algebraically closed. Moreover he supposes that he already has a fundamental solution in a field extension of K. We will consider the same setting as Papanikolas, apart from the fact that we only ask that τ is an endomorphism. This seems to be a reasonable framework for many applications. For the proofs, we usually refer to [OW15], which is the reference with more general assumptions. Notice that Papanikolas has a more geometric approach while Ovchinnikov and Wibmer prefer algebraic arguments.

Finally we point out that we assume that the characteristic is zero in Sects. 2.6.2 and 2.7 and that τ is an automorphism in Sect. 2.8.

Remarks on the Content and the Organization of the Text Below The text below is meant to be a guide to the existing literature. From this perspective, I will give references for the proofs, rather than writing a self-contained exposition. I'm addressing with particular attention readers that need to apply Galois theory of difference equations, therefore a large space is devoted to statements that may be useful in applications.

The exposition is divided in three parts. The first part quickly explains the fundamentals results and ideas of difference Galois theory. In order to be precise and correct I have been obliged to use some more sophisticated tools. The second part, devoted to applications, is more accessible and applies the statements of the first part as black boxes. We conclude with a last paragraph on the role of normal subgroups in the Galois correspondence.

2.2 Glossary of Difference Algebra

We give here a short glossary of terminology in difference algebra. Classical references are [Coh65] and [Lev08].

We consider a field F equipped with an endomorphism τ. We will call the pair (F, τ) a τ-field or a difference field, when the reference to τ is clear from the context. The set $k = \{f \in F : \tau(f) = f\}$, also denoted F^τ, is naturally a field and is called the field of constants of F. All along this exposition, we will assume that τ is non-periodic on F (i.e. there exists $f \in F$ such that for any $n \in \mathbb{Z}$ we have $\tau^n f \neq f$) and we won't assume that k is algebraically closed. The example below will be our playground until the end of the paper.

Example 2.2.1 We consider $k = \mathbb{C}$, $F = k(x)$ and τ a non-periodic homography acting on x, so that $\tau(f(x)) = f(\tau(x))$. Supposing that τ has one or two fixed points, we can assume without loss of generality that $\tau(x) = x + 1$ or that $\tau(x) = qx$, for some $q \in \mathbb{C} \smallsetminus \{0, 1\}$, not a root of unity.

We will add the prefix τ to the usual terminology in commutative algebra, to signify the invariance with respect to τ. For instance:

- A τ-subfield K of a τ-field F is a subfield of F such that τ induces an endomorphism of K, i.e., such that $\tau(K) \subset K$.
- A τ-K-algebra is a K-algebra equipped with an endomorphism extending τ (which we still call τ for simplicity).
- An ideal I of a τ-K-algebra R is a τ-ideal if $\tau(I) \subset I$.
- A τ-K-algebra R is τ-simple if its only τ-ideals are R and 0.

Example 2.2.2 For any $x \in F$ transcendental over k, such that $\tau(x) \in k(x)$ and that for any $n \in \mathbb{Z}$ we have $\tau^n(x) \neq x$, we can consider the field $K := k(x)$, on which τ induces a non-periodic endomorphism. Since we don't ask τ to be surjective, the situation is a little bit more general than Example 2.2.1. For instance τ can be the Mahler operator $\tau(x) = x^\kappa$, where $\kappa \geq 2$ is an integer.

For any positive integer n, we can consider $k_n = F^{\tau^n}$ and the difference fields $(K_n := k_n(x), \tau)$ and $(K_n := k_n(x), \tau^n)$.

Sometimes it is useful to consider (k, id), where id stands for the identity map, as a difference field, therefore we will call it a trivial τ-field. In general we will say that a k-algebra is equipped with a trivial action of τ, when τ acts on it as the identity. For further reference we state the following lemma:

Lemma 2.2.3 *Let B be a k-algebra endowed with a trivial action of τ, K a τ-subfield of F such that $K^\tau = k$ and let $R \subset F$ be a τ-K-algebra. Then $R \otimes_k B$ has a natural structure of τ-K-algebra defined by*

$$\tau(r \otimes b) = \tau(r) \otimes b, \text{ for any } r \in R \text{ and } b \in B. \tag{2.1}$$

Moreover $R \otimes_k B \hookrightarrow F \otimes_k B$ and $(R \otimes_k B)^\tau = k \otimes_k B \cong B$.

Remark 2.2.4 We have chosen to detail the proof, even if the lemma above is included in many references and could partially be proved invoking the flatness of F over k. The argument is indeed an instance of a classical way of reasoning in difference algebra and is useful in many situations.

Proof Clearly Eq. (2.1) defines a ring endomorphism, that is the tensor product of τ and the identity in the category of rings. Since $R \subset F$, we have a natural map of τ-K-algebras $R \otimes_k B \to F \otimes_k B$. We want to prove that this map is injective. By absurdum, we suppose that the kernel is non-trivial and therefore we choose a non-zero element $\sum_{i=1}^n r_i \otimes b_i \in R \otimes_k B$ in the kernel such that n is minimal, i.e., we suppose that there exists no element of $R \otimes_k B$ in the kernel, that can be written as a sum of less than n elements of the form $r \otimes b \in R \otimes_k B$. This implies in particular that the r_i's are linearly independent over k and that all the r_i's and b_i's are non-zero. Since F is a field, in $F \otimes_k B$ we can multiply by $r_n^{-1} \otimes 1$, hence we have: $1 \otimes b_n + \sum_{i=1}^{n-1} r_i r_n^{-1} \otimes b_i = 0$ in $F \otimes_k B$. We conclude that the image of

$$\sum_{i=1}^{n-1} \left(\tau(r_i r_n^{-1}) - r_i r_n^{-1} \right) \otimes b_i \in R \otimes_k B$$

in $F \otimes_k B$ is zero. The minimality of n, together with the fact that $b_i \neq 0$, implies that $\tau(r_i r_n^{-1}) - r_i r_n^{-1} = 0$ and hence that $r_i r_n^{-1} \in k$ for any $i = 1, \ldots, n-1$. The linear independence of the r_i's over k implies that $n = 1$ and hence that $r_1 \otimes b_1 = 0$ in $F \otimes_k B$, with $r_1 \neq 0$. Since F is a field, $r_1^{-1} \otimes 1 \in F \otimes_k B$ and hence:

$$1 \otimes b_1 = (r_1^{-1} \otimes 1) \cdot (r_1 \otimes b_1) = 0$$

in $F \otimes_k B$, and we obtain the contradiction $b_1 = 0$. This proves the injectivity.

To conclude it is enough to prove that $(R \otimes_k B)^\tau = k \otimes_k B$. First of all, notice that if $r \otimes b \in (R \otimes_k B)^\tau$, with $r \neq 0 \neq b$, then $r \in k$. By absurdum, let us suppose that there exists $\sum_{i=1}^n r_i \otimes b_i \in (R \otimes_k B)^\tau \smallsetminus k \otimes_k B$, such that no r_i belongs to k and n is minimal. Once again, the minimality of n implies that the r_i's and the

b_i's are linearly independent over k. Moreover we know that necessarily $n \geq 2$. We conclude that

$$\sum_{i=1}^{n} (\tau(r_i) - r_i) \otimes b_i = \tau \left(\sum_{i=1}^{n} r_i \otimes b_i \right) - \left(\sum_{i=1}^{n} r_i \otimes b_i \right) = 0.$$

There are two possibility: either the $(\tau(r_i) - r_i)$'s are linearly dependent over k or they are not. If they are linearly dependent over k, we can find $\lambda_1, \ldots, \lambda_n \in k$, not all zero, such that $\sum_{i=1}^{n} \lambda_i(\tau(r_i) - r_i) = 0$. We can suppose without loss of generality that $\lambda_n \neq 0$. It implies that

$$\sum_{i=1}^{n} \lambda_i r_i = \sum_{i=1}^{n} \lambda_i \tau(r_i) = \tau \left(\sum_{i=1}^{n} \lambda_i r_i \right),$$

and hence that $c := \sum_{i=1}^{n} \lambda_i r_i \in k$. Since the r_i's are linearly independent over k, c is not zero. The k-linearity of the tensor product implies:

$$\sum_{i=1}^{n} r_i \otimes b_i = \sum_{i=1}^{n-1} r_i \otimes b_i + \lambda_n^{-1} \left(c - \sum_{i=1}^{n-1} \lambda_i r_i \right) \otimes b_n$$

$$= \sum_{i=1}^{n-1} r_i \otimes (b_i - \lambda_n^{-1} \lambda_i b_n) + \lambda_n^{-1} c \otimes b_n \in (R \otimes_k B)^\tau.$$

Because $\lambda_n^{-1} c \otimes b_n \in k \otimes_k B \subset (R \otimes_k B)^\tau$, we conclude that $\sum_{i=1}^{n-1} r_i \otimes (b_i - \lambda_n^{-1} \lambda_i b_n) \in (R \otimes_k B)^\tau$. The minimality of n implies that $b_i = \lambda_n^{-1} \lambda_i b_n$, for any $i = 1, \ldots, n-1$. Finally we obtain:

$$\sum_{i=1}^{n} r_i \otimes b_i = \left(\sum_{i=1}^{n} \lambda_n^{-1} \lambda_i r_i \right) \otimes b_n \in (R \otimes_k B)^\tau,$$

and therefore that $\sum_{i=1}^{n} r_i \otimes b_i \in k \otimes_k B$, in contradiction with our choice of $\sum_{i=1}^{n} r_i \otimes b_i$. We still have to consider the case in which $(\tau(r_i) - r_i)$'s are linearly independent over k, with $\sum_{i=1}^{n} (\tau(r_i) - r_i) \otimes b_i = 0$ in $F \otimes_k B$, but we have seen in the first part of the proof that this cannot happen, unless $\tau(r_i) = r_i$, which is against our assumptions. This ends the proof of the whole lemma. □

2.3 Picard-Vessiot Rings

We now consider a field F with an endomorphism $\tau : F \to F$. Our base field will be a τ-subfield K of F, containing $k := F^\tau$, which implies that $K^\tau = k$. We do assume neither that k is algebraically closed, nor that τ induces an automorphism of K, but we assume that τ is non-periodic over K. Moreover we assume that F/k is a separable extension.

Remark 2.3.1 In the classical theory, one usually assumes that k is algebraically closed, which simplifies a little the theory, although not in a fundamental way. The main difference comes from the fact that the Galois group, that we will define in the following section, is an algebraic group over the field k. Therefore if k is algebraically closed one can avoid a more sophisticated geometric point of view and simply identify the Galois group to a group of invertible matrices with coefficients in k. This remark will become clearer in what follows. We will comment again on the consequences of the fact that k is not algebraically closed.

We also point out that the assumption $F^\tau = K^\tau = k$ is crucial, otherwise we may end up introducing new meaningless solutions in the theory. For more details, see Example 2.3.7 and the proof of Proposition 2.6.1 below.

We consider a linear difference system

$$\tau(Y) = AY, \text{ where } A \in \mathrm{GL}_d(K), \tag{2.2}$$

and we suppose that *there exists a fundamental solution matrix $U \in \mathrm{GL}_d(F)$ of* (2.2). Then the field $L := K(U) \subset F$ is obviously stable by τ. We have made an abuse of notation that we will repeat frequently: By $K(U)$ we mean the field generated over K by the entries of U.

There are two main situations that the readers, according to their background, could keep in mind as a guideline through the text below:

Example 2.3.2 One can consider the following two classical situations:

1. F is the field of meromorphic functions over $\mathbb{C}$ in the variable x and $\tau : f(x) \mapsto f(x + 1)$. Then k is the field of meromorphic 1-periodic functions over $\mathbb{C}$.
2. F is the field of meromorphic functions over $\mathbb{C}^*$ in the variable x and $\tau : f(x) \mapsto f(qx)$, for a fixed complex number $q \in \mathbb{C} \setminus \{0, 1, \text{ roots of unity}\}$. Here k is the field of meromorphic q-elliptic functions over $\mathbb{C}^*$.

In both cases, we can chose K to be any τ-subfield of F, containing k. A typical choice for K is $k(x)$, which is the point of view taken in [CHS08], as far as q-difference equations is concerned.

Let as consider a linear system $\tau(\vec{y}) = A\vec{y}$ with coefficients in K, of the form (2.2). In the settings above, plus the assumption $|q| \neq 1$ in the q-difference case, Praagman proves that $\tau(\vec{y}) = A\vec{y}$ has a fundamental matrix of solutions with coefficients in F. See [Pra86, Theorem 1 and Theorem 3]. This does not mean that all possible solutions are meromorphic. Indeed it is enough to multiply a solution matrix by a matrix whose entries are functions with some essential singularities and that are constant with respect to τ.

Example 2.3.3 In [Pap08, §4.1] the triple (k, K, F) is called a τ-admissible triple. He is specifically interested in the study of t-motifs and, hence, of the associated triple $(\mathbb{F}_q(t), \mathbb{K}, \mathbb{L})$, defined as follows [Pap08, §2.1]:

- $\mathbb{F}_q(t)$ is the field of rational functions in the variable t and with coefficients in the field $\mathbb{F}_q$ with q elements, where q is an integer power of a prime p;

- $\mathbb{K}$ is the smallest algebraically closed and complete extension of a field of rational functions $\mathbb{F}_q(\theta)$, with respect to the θ^{-1}-adic valuation.
- $\mathbb{L}$ is the field of fractions of the ring of all power series in $\mathbb{K}[[t]]$ convergent on the closed unit disk for the θ^{-1}-adic valuation.

The automorphism τ is the inverse of the Frobenius morphism on the algebraic closure of $\mathbb{F}_q$, sends θ to $\theta^{1/q}$, and is defined on $\mathbb{K}[[t]]$ as

$$\tau\left(\sum_{n\geq 0} a_n t^n\right) = \sum_{n\geq 0} a_n^{1/q} t^n.$$

Finally, it extends to $\mathbb{L}$ by multiplicativity. Then $\mathbb{L}^\tau = \mathbb{K}^\tau = \mathbb{F}_q(t)$, as proved in Lemma 3.3.2 in *loc.cit.*.

We start mentioning a technical but fundamental property of the τ-K-algebra generated by the entries of the solution matrices of (2.2), which allows to reconnect our point of view with the more classical approach of the Picard-Vessiot theory.

Proposition 2.3.4 *In the notation above, let* $U \in \mathrm{GL}_d(F)$ *verify* $\tau(U) = AU$. *Then* $R = K\left[U, \det U^{-1}\right] \subset F$ *is* τ-*simple.*

Proof The statement is a special case of [OW15, Proposition 2.14], as one can see from Definitions 2.2 and 2.4 in *loc.cit.*. □

At this point, the reader should pay attention to the terminology in the literature. The ring R in the proposition above and its field of fractions L are a *weak* Picard-Vessiot ring and a *weak* Picard-Vessiot field, respectively, according to [CHS08, Definition 2.1]. Indeed it follows immediately from their definition that $R^\tau = L^\tau = k$. Proposition 2.3.4 shows that R and L are respectively a Picard-Vessiot ring and a Picard-Vessiot field, following also the definition [OW15, Defintion 2.12]. For the purpose of this paper, we will use a terminology in between [CHS08] and [OW15], knowing that R and L satisfy the definitions in both the cited references, and that, therefore, the results in both references apply here.

Definition 2.3.5 A τ-simple τ-K-algebra R is called a Picard-Vessiot ring (for (2.2)) if there exists $V \in \mathrm{GL}_d(R)$ such that $\tau(V) = AV$ and $R = K\left[V, \det V^{-1}\right]$.

A difference field L is called a Picard-Vessiot field over K (for (2.2)) if $L^\tau = k$ and $L = K(V)$ for a $V \in \mathrm{GL}_d(L)$ such that $\tau(V) = AV$. We will call L/K a Picard-Vessiot extension.

Remark 2.3.6 If we do not have a field where to find enough solutions of our equation, we have to construct an abstract Picard-Vessiot extension. To do so, one considers the ring of polynomials in the d^2 variables $X = (x_{i,j})$ with coefficients in K. Inverting $\det X$ and setting $\tau(X) = AX$, we obtain a ring $K[X, \det X^{-1}]$ with an endomorphism τ. Any of its quotients by a maximal τ-invariant ideal is a Picard-Vessiot ring of $\tau(\vec{y}) = A\vec{y}$ over K. It is important to notice that the ring R does not need to be a domain. It can be written has a direct sum $R_1 \oplus \cdots \oplus R_r$, such that: $R_i = e_i R$, for some $e_i \in R$ such that $e_i^2 = e_i$; R_i is a domain; there exists a permutation σ of $\{1, \ldots, r\}$ such that $\tau(R_i) \subset R_{\sigma(i)}$. See [vdPS97, Cor. 1.16]. Its total field of fractions is a pseudo field, that is the tensor product of the fraction field of each R_i. For a precise definition of pseudo field, see for instance [OW15, Definition 2.2].

The following example shows the importance of the assumption $L^\tau = k$ in the definition of Picard-Vessiot field.

Example 2.3.7 Let us consider the equation $\tau(y) = -y$, with k, K and F as above. We suppose that there exists a two-dimensional k-vector space of solutions of $\tau(y) = -y$ in F, which coincides with $k_2 := F^{\tau^2}$. The Picard-Vessiot field (contained in F) of $\tau(y) = -y$ over K is $L := K(k_2)$.

We could also have considered a field of rational function $F(T)$ with coefficients in F and in the variable T. Since T is transcendental, we can set $\tau(T) = -T$ and obtain an endomorphism of $F(T)$. If we do not assume that the Picard-Vessiot field has the same field of constants than the base field K, we see that $K(k_2)(T)$ is a Picard-Vessiot field for $\tau(y) = -y$, whose field of constants is $k_2(T^2)$. Of course, the solution T is somehow artificial and the extension $K(k_2)(T)/K$ is much bigger (i.e. has many more automorphisms, see next section) than $K(k_2)/K$.

The expected relations between Picard-Vessiot rings and Picard-Vessiot fields are verified. If R, $L \subset F$, the proof is actually straightforward.

Corollary 2.3.8 ([OW15, Proposition 2.15])

1. *Let R be domain which is a Picard-Vessiot ring. Then its field of fractions is a Picard-Vessiot field.*
2. *Let L be a Picard-Vessiot field for (2.2) and let $U \in \mathrm{GL}_d(L)$ be a solution of (2.2). Then $K[U, \det U^{-1}]$ is a Picard-Vessiot ring.*

Notice that one can always compare two different Picard-Vessiot rings, up to an algebraic extension of the field of constants:

Proposition 2.3.9 *Picard-Vessiot rings have the following uniqueness properties:*

1. *Let R_1 and R_2 be two Picard-Vessiot rings for (2.2), both contained in F. Then $R_1 = R_2$.*
2. *Let $R \subset F$ and R' be two Picard-Vessiot rings for (2.2). (Notice that we do not suppose that $R' \subset F$!) Then there exists an algebraic field extension $\tilde{k}$ of k, containing a copy of $k' := (R')^\tau$, such that $R \otimes_k \tilde{k}$ is isomorphic to $R' \otimes_{k'} \tilde{k}$ as a $K \otimes_k \tilde{k}$-τ-algebra.*

Proof The first assertion follows from the fact that any pair of fundamental solutions $U, V \in \mathrm{GL}_d(F)$ verifies $\tau(U^{-1}V) = U^{-1}V$, i.e., $U^{-1}V \in \mathrm{GL}_d(k)$. In fact, this implies that $K[U, \det U^{-1}] = K[V, \det V^{-1}] \subset F$. For the second assertion, see [OW15, Theorem 2.16]. Notice that the key-point of its proof is Lemma 2.13 in *loc.cit.*, which ensure that k'/k must be an algebraic extension. □

We give an explicit example to explain the necessity of extending the constants to $\widetilde{k}$ in the statement above.

Example 2.3.10 Let k be the field of 1-periodic meromorphic functions over $\mathbb{C}$ and let $K = k(x)$. The Picard-Vessiot ring of the equation $\tau(y) = xy$ over K, contained in F, is $R := K[\Gamma(x), \Gamma(x)^{-1}]$, where $\Gamma(x)$ is the Euler Gamma function.

Now let $f(x)$ be a 2-periodic function, algebraic over k, but not meromorphic over $\mathbb{C}$. It means that $f(x)$ lives on an analytic two-fold covering of $\mathbb{C}$ and has some branching points. The function $c(x) := f(2x)$ is 1-periodic but does not belong to F, and hence not to k. The K-algebra $R' := K[\widetilde{\Gamma}(x), \widetilde{\Gamma}(x)^{-1}]$, where $\widetilde{\Gamma}(x) := c(x)\Gamma(x)$, is a Picard-Vessiot ring for $\tau(y) = xy$. In the notation of the proposition above, we have $k' = k$ and $\widetilde{k} := k(c(x))$. Indeed, $\Gamma(x) \mapsto c(x)\Gamma(x)$ defines an automorphism from $R \otimes_k \widetilde{k}$ to $R' \otimes_k \widetilde{k}$ as $K \otimes_k \widetilde{k}$-algebras.

We close the section with a couple of easy, yet crucial, examples:

Example 2.3.11 Let a be a non-zero element of K and let us consider the rank-one equation $\tau(y) = ay$. By assumption there exists a solution $z \in F$ verifying $\tau(z) = az$. Hence $K[z, z^{-1}]$ is a Picard-Vessiot ring for $\tau(y) = ay$ and $K(z)$ is a Picard-Vessiot field. Generically, z is transcendental over K, but not always. For instance, if F is the field of meromorphic functions over $\mathbb{C}$ in the variable x, $\tau(f(x)) = f(x + 1)$ for any $f \in F$, $K = k(x)$ and $a = -1$, we can take $z = \exp(\pi i x) \in F$. In this case $K(z)$ is a finite extension of degree 2, since $\exp(\pi i x)^2 = \exp(2\pi i x)$ is a 1-periodic function, belonging to k.

Example 2.3.12 Let $f \in K$ and let us consider the inhomogeneous equation $\tau(y) = y + f$. Such an equation is equivalent to the matrix system

$$\tau(Y) = \begin{pmatrix} 1 & f \\ 0 & 1 \end{pmatrix} Y,$$

whose fundamental solution is given by $Y = \begin{pmatrix} 1 & z \\ 0 & 1 \end{pmatrix}$, where $z \in F$ verifies $\tau(z) = z + f$. Since $Y^{-1} = \begin{pmatrix} 1 & -z \\ 0 & 1 \end{pmatrix}$, the Picard-Vessiot ring of $\tau(y) = y + f$ is $K[z]$ and the Picard-Vessiot field is $K(z)$.

2.4 The Galois Group

The Galois group of a difference system of the form (2.2) is a linear algebraic group defined over the field of constants k. As we have already pointed out, since we have chosen not to assume that k is algebraically closed, we cannot stick to a naive approach to linear algebraic groups as sets of matrices with entries in the base field, but we have to use the point of view of group schemes. For the reader's convenience we recall informally a minimal amount of definitions that are necessary in what follows. They are contained in any classical reference on group schemes, for instance [Wat79].

2.4.1 A Short Digression on Group Schemes

A group scheme G over the field k is a covariant functor from the category of k-algebras to the category of groups:

$$G : k\text{-algebras} \to \text{Groups}$$
$$B \mapsto G(B)$$

An affine group scheme is a group scheme which is representable, i.e., there exists a k-algebra $k[G]$ such that the functor G and $\text{Hom}_k(k[G], -)$ are naturally isomorphic. This implies, in particular, that $G(B)$ and $\text{Hom}_k(k[G], B)$ are isomorphic as groups, for any k-algebra B.

Example 2.4.1 For $k = \mathbb{Q}$, we can look at GL_n as an affine group scheme over $\mathbb{Q}$ in the following way:

$$\text{GL}_{n,\mathbb{Q}} : \mathbb{Q}\text{-algebras} \to \text{Groups}$$
$$B \mapsto \text{GL}_n(B)$$

We recall that, for a general $\mathbb{Q}$-algebra B, $GL_n(B)$ is the group of square matrices with coefficients in B, whose determinant is an invertible element of B. We have:

$$\mathbb{Q}[\text{GL}_{n,\mathbb{Q}}] = \frac{\mathbb{Q}[t, x_{i,j}, i, j = 1, \ldots, n]}{(t \det(x_{i,j}) - 1)}.$$

Of course an analogue definition holds for the affine group scheme $\text{GL}_{n,k}$, define over a generic field k. For $n = 1$, we obtain the multiplicative affine group scheme, for whom we will use the notation $\mathbb{G}_{m,k}$ rather then $\text{GL}_{1,k}$. The additive affine group scheme $\mathbb{G}_{a,k}$ is defined as follows:

$$\mathbb{G}_{a,k} : k\text{-algebras} \to \text{Groups}$$
$$B \mapsto \left\{ \begin{pmatrix} 1 & b \\ 0 & 1 \end{pmatrix} \mid b \in B \right\}$$

We have:

$$k[\mathbb{G}_{a,k}] = \frac{k[\text{GL}_{2,k}]}{(x_{1,1} - 1, x_{2,2} - 1, x_{2,1})}.$$

It follows that $k\left[\mathbb{G}_{a,k}\right]$ can be naturally identified with the algebra $k[x]$ of polynomial in the variable x and coefficients in k. One can define in an analogous way the affine group scheme $\mathrm{SL}_{n,k}$ over k, whose algebra is $k\left[\mathrm{SL}_{n,k}\right] = \frac{k[x_{i,j}, i, j=1,\ldots,n]}{(\det(x_{i,j})-1)}$.

The Yoneda Lemma ensures that $k[G]$ is unique up to isomorphism. An important property (that we won't use, because we are not getting into the details of the proofs) of $k[G]$ is that it has a natural structure of Hopf algebra.

The affine group scheme G is said to be an algebraic group if $k[G]$ is a finitely generated k-algebra. This means that $k[G]$ can be identified with a quotient $k[x_1, \ldots, x_n]/I$ of a ring of polynomials by a convenient (Hopf) ideal I. It allows to identify $G(B)$ with the set of zeros of I in B^n, for any k-algebra B. In other words, G can be identified with an affine variety defined over k. All the affine group schemes in Example 2.4.1, as well as all the affine group schemes appearing in this paper, are algebraic groups.

If G is an algebraic group and G' is another affine group scheme defined over k, we say that G' is an affine subgroup scheme of G if there exists a surjective morphism of Hopf algebras $k[G] \rightarrow k[G']$. This implies that $G'(B)$ can be identified naturally to a subgroup of $G(B)$, for any k-algebra B. We say that G' is a normal algebraic subgroup of G, if $G'(B)$ is a normal subgroup of $G(B)$, for any k-algebra B.

Example 2.4.2 In the notation of Example 2.4.1, we have:

1. The additive affine group scheme $\mathbb{G}_{a,k}$ is an algebraic subgroup of $\mathrm{GL}_{2,k}$.
2. We have a surjective morphism from $k\left[\mathrm{GL}_{n,k}\right]$ to $k\left[\mathrm{SL}_{n,k}\right]$ defined by $t \mapsto 1$, therefore $\mathrm{SL}_{n,k}$ is naturally an algebraic subgroup of $\mathrm{GL}_{n,k}$.

For further reference, we describe the algebraic subgroups of $\mathbb{G}_{m,k}$. As we have already pointed out, we have $k[\mathbb{G}_{m,k}] = \frac{k[x,t]}{(xt-1)}$, therefore we can write for short $k[\mathbb{G}_{m,k}] = k\left[x, \frac{1}{x}\right]$. The algebraic subgroups of $\mathbb{G}_{m,k}$ are the defined by equations of the form $x^n - 1$, for any non-negative integer n. They are represented by the quotients $\frac{k\left[x,\frac{1}{x}\right]}{(x^n-1)}$. If G is one of those subgroups, with $n \geq 1$, and B is a k-algebra, then $G(B)$ is nothing more that the group of n-th roots of unity contained in B. For $n = 1$, we obtain the trivial algebraic subgroup $\{1\}$ of $\mathbb{G}_{m,k}$, while for $n = 0$ we obtain the whole $\mathbb{G}_{m,k}$.

Let us consider the algebraic group $G_{m,k}^n$, for some positive integer n. We have $k[G_{m,k}^n] = k\left[x_1, \frac{1}{x_1}, \ldots, x_n, \frac{1}{x_n}\right]$. The algebraic subgroups of $G_{m,k}^n$ are defined by polynomials of the form $x_1^{\alpha_1} \ldots x_n^{\alpha_n} - 1$, where $\alpha_1, \ldots, \alpha_n \in \mathbb{Z}$.

We will also need the description of the algebraic subgroups of $\mathbb{G}_{a,k}^n$, where n is a positive integer. For any k-algebra B, $\mathbb{G}_{a,k}^n(B)$ can be naturally identified to B^n. Therefore, an algebraic subgroup G of $\mathbb{G}_{a,k}^n$ is defined by an ideal generated by at most n independent linear equations with coefficients in k. In particular, we will use the fact that a proper algebraic subgroup of $\mathbb{G}_{a,k}^n$ is always an algebraic subgroup of the group represented by the algebra $\frac{k[x_1,\ldots,x_n]}{(\alpha_1 x_1 + \cdots + \alpha_n x_n)}$, for some $\alpha_1, \ldots, \alpha_n \in k$, not all zero. Notice that, for $n = 1$, the algebraic group $G_{a,k}$ does not have any proper algebraic subgroup.

2.4.2 The Galois Group of a Linear Difference System

Let $K \subset F$ be our base field, with $k = K^\tau = F^\tau$, and let us consider a system of the form (2.2). From now on we will assume *implicitly* that all Picard-Vessiot rings and all Picard-Vessiot fields are contained in F. So let R ($\subset F$) be a Picard-Vessiot ring for (2.2) and let L be its field of fractions.

For more details on what follows, see [OW15, §2.7].[1]

Definition 2.4.3 ([OW15, 2.50]) We call the (difference) Galois group of (2.2) the following group scheme:

$$\mathrm{Gal}(L/K) : k\text{-Algebras} \to \qquad \text{Groups}$$
$$B \qquad \mapsto \mathrm{Aut}^\tau(R \otimes_k B/K \otimes_k B),$$

where:

1. the k-algebra B is endowed with a structure of trivial τ-k-algebra, so that $\tau(f \otimes b) = \tau(f) \otimes b$, for any $f \in R$ and any $b \in B$;
2. $\mathrm{Aut}^\tau(R \otimes_k B/K \otimes_k B)$ is the group of the ring automorphisms of $R \otimes_k B$, that fix $K \otimes_k B$ and commute with τ.

The functor $\mathrm{Gal}(L/K)$ acts on morphisms by extension of constants, namely, each morphism of k-algebras $\alpha : B_1 \to B_2$ defines a structure of B_1-algebra over B_2 and the definition of $\mathrm{Gal}(L/K)(\alpha) : \mathrm{Aut}^\tau(R \otimes_k B_1/K \otimes_k B_1) \to \mathrm{Aut}^\tau(R \otimes_k B_2/K \otimes_k B_2)$ relies on the fact that $R \otimes_k B_2 \cong R \otimes_k B_1 \otimes_{B_1,\alpha} B_2$.

For any choice of a fundamental solution matrix $U \in \mathrm{GL}_d(R)$ of (2.2), for any k-algebra B and any $\varphi \in \mathrm{Gal}(L/K)(B)$ we have that

$$\tau(U^{-1}\varphi(U)) = U^{-1}A^{-1}\varphi(A)\varphi(U) = U^{-1}\varphi(U) \in \mathrm{GL}_d(B),$$

where we have identified U and $U \otimes 1$ in $R \otimes_k B$, making an abuse of notation that we will repeat frequently. (Notice that we have used the fact that $(R \otimes_k B)^\tau = B$. See Lemma 2.2.3.) The maps $\varphi \mapsto U^{-1}\varphi(U)$ represents $\mathrm{Gal}(L/K)(B)$ as a subgroup of $\mathrm{GL}_d(B)$. The linearity of the difference system (2.2) immediately implies that another choice of the fundamental solution matrix leads to a conjugated representation, so that most of the times we can identify $\mathrm{Gal}(L/K)(B)$ with a subgroup scheme of $\mathrm{GL}_{d,k}(B)$, forgetting to mention the matrix U. The following

---

[1] In the notation of [OW15], one has to take $\Phi = \tau$ and σ to be the identity.

proposition says that such a representation is functorial in B, in the sense that $\mathrm{Gal}(L/K)$ is an algebraic subgroup of $\mathrm{GL}_{d,k}$, as in the next example.

Example 2.4.4 Let us consider the rank-one difference equation $\tau(y) = ay$, where $a \in K$, as in Example 2.3.11. By assumption, there exists $z \in F$ such that $\tau(z) = az$ and $R = K[z, z^{-1}]$ is its Picard-Vessiot ring. For any k-algebra B and any $\varphi \in \mathrm{Gal}(L/K)(B)$, the element $\varphi(z \otimes 1)$ of $R \otimes_k B$ must be a solution of $\tau(y) = ay$, and hence there exists $c_\varphi \in B^*$ such that $\varphi(z \otimes 1) = c_\varphi(z \otimes 1)$. This means that $\mathrm{Gal}(L/K)$ can be identified with a subgroup of the multiplicative group $\mathbb{G}_{m,k}$ defined over k, therefore it coincides either with $\mathbb{G}_{m,k}$ or with a cyclic group. If for instance $a = -1$, then we must have $z^2 \in k$, and therefore $c_\varphi^2 = 1$.

Proposition 2.4.5 ([OW15, Lemma 2.51]) *The Galois group* $\mathrm{Gal}(L/K)$ *is an algebraic group defined over k, represented by the k-algebra* $(R \otimes_K R)^\tau$.

Remark 2.4.6 We remind that the statement above means that $(R \otimes_K R)^\tau$ is a finitely generated k-algebra and that the functors $\mathrm{Gal}(L/K)$ and $\mathrm{Hom}_{k\text{-Algebra}}((R \otimes_K R)^\tau, -)$ are naturally isomorphic. In particular for any k-algebra B, we have can identify $\mathrm{Gal}(L/K)(B)$ with $\mathrm{Hom}_{k\text{-Algebra}}((R \otimes_K R)^\tau, B)$.

The proposition above says that there exists an ideal I of the ring of polynomials $k[X, \det X^{-1}]$, with $X = (x_{i,j})_{i,j=1,\dots,d}$, such that for any k-algebra B the image of the group morphism defined above

$$\mathrm{Gal}(L/K)(B) \to \quad \mathrm{GL}_d(B)$$
$$\varphi \qquad \mapsto [\varphi]_U := U^{-1}\tau(U)$$

is exactly the set of zeros of I in $\mathrm{GL}_d(B)$. The idea of the proof is to consider the k-algebra $k[Z, \det Z^{-1}] \hookrightarrow (R \otimes_k R)^\tau$, where $Z := (U^{-1} \otimes 1)(1 \otimes U)$. Then one can prove that we have the series of isomorphisms:

$$R \otimes_K R \cong R.k[Z, \det Z^{-1}] \cong R \otimes_k (R \otimes_K R)^\tau.$$

This allows to prove a series of group isomorphisms showing that for any k-algebra B we have:

$$\mathrm{Hom}_{\tau\text{-}(K \otimes_k B)\text{-alg}}(R \otimes_k B, R \otimes_k B) \cong \mathrm{Hom}_{k\text{-alg}}((R \otimes_K R)^\tau, B).$$

See [OW15] for details.

Example 2.4.7 Let F be the field of meromorphic functions over $\mathbb{C}^*$ and let τ be defined by $\tau : f(x) \mapsto f(qx)$, for $q \in \mathbb{C}$, such that $|q| > 1$. The field of constants k is the field of meromorphic functions over the torus $\mathbb{C}^*/q^{\mathbb{Z}}$ and we set $K = k(x)$. The Jacobi Theta function

$$\Theta(x) = \sum_{x \in \mathbb{Z}} q^{-n(n+1)/2} x^n \in F$$

is solution of the difference equation $\tau(y) = xy$. Its differential Galois group G is the multiplicative group. Indeed for any k-algebra B and for any $\varphi \in G(B)$, φ multiplies $\Theta(x)$ by an invertible element of B. On the other hand, since $\Theta(x)$ is a transcendental function, any invertible constant of B defines an automorphism of $K[\Theta(x), \Theta(x)^{-1}] \otimes_k B$.

Now let us consider an integer $r \geq 2$ and choose a r-th root $q^{1/r}$ of q. The meromorphic function $z(x) := \Theta(q^{1/r}x)/\Theta(x) \in F$ is a solution of the finite difference equation $\tau(y) = q^{1/r}y$ and its difference Galois group is the cyclic subgroup of $G_{m,k}$ of order r. To prove the last claim it is enough to notice that $z(qx)^r = qz(x)^r$, hence $z(x)^r$ is a meromorphic function of the form $xg(x)$, with $g(x) \in k \subset K$.

Example 2.4.8 Let us consider an element $f \in K$ and the inhomogeneous difference equation $\tau(y) = y + f$. By assumption, there exists a solution $z \in F$ and we have already noticed that $R = k[z]$. See Example 2.3.12. For any k-algebra B and any $\varphi \in \mathrm{Gal}(L/K)$, the element $\varphi(z)$ of $R \otimes_k B$ must be another solution of $\tau(y) = y + f$, hence there exists $c_\varphi \in B$ such that $\varphi(z) = z + c_\varphi$ (here we have identified z and $z \otimes 1$). It follows that $\mathrm{Gal}(L/K)$ is a algebraic subgroup of the additive group $G_{a,k}$. This means that either $\mathrm{Gal}(L/K) = \{1\}$ or $\mathrm{Gal}(L/K) = G_{a,k}$.

2.4.3 Transcendence Degree of the Picard-Vessiot Extension

We can now state the first important result from the point of view of applications to number theory and more specifically to transcendence. It compares the dimension of G over k as an algebraic variety and the transcendence degree of R over K and its proof is based on Proposition 2.4.5.

Theorem 2.4.9 ([OW15, Lemma 2.53]) *Let R, L and G be as above. Then the dimension of G as an algebraic variety over k is equal to the transcendence degree of R (or of L) over K:*

$$\dim_k G = \mathrm{trdeg}_K R = \mathrm{trdeg}_K L.$$

Example 2.4.10 Let us consider the case of finite difference equations, i.e, let F be the field of meromorphic functions over $\mathbb{C}$ equipped with the operator $\tau : f(x) \mapsto f(x + 1)$. Then k is the field of 1-periodic functions and we set $K = k(x)$. We consider the difference equations $\tau(y) = xy$, which is satisfied by the Euler Gamma function $\Gamma(x)$. Its Picard-Vessiot ring is $K[\Gamma(x), \Gamma(x)^{-1}]$, as discussed in Example 2.3.11. As in Example 2.4.4, its Galois group G is a subgroup of $G_{m,k}$. For any k-algebra B and any $\varphi \in G(B)$, there exists an invertible element c_φ of B such that $\varphi(\Gamma) = c_\varphi \Gamma$. As already explained, the proper subgroups of $G_{m,k}$ are the finite cyclic groups. If G was a the cyclic group, $\Gamma(x)$ would be an algebraic function, by the previous theorem. Proving that the functional equations of the Gamma function implies that it is not algebraic over $k(x)$ is an exercise that we leave to the reader. Therefore the difference Galois group G is the whole $\mathbb{G}_{m,k}$.

Example 2.4.11 Let us consider the system

$$
\tau(Y) = \begin{pmatrix} x & 1 & 0 \\ 0 & x & 1 \\ 0 & 0 & x \end{pmatrix} Y.
$$

We denote by $(\cdot)'$ the derivation $\frac{d}{dx}$, with respect to x. Since τ and $\frac{d}{dx}$ commute, we have:

$$
\tau\left(\Gamma'(x)\right) = \frac{d}{dx}\left(x\Gamma(x)\right) = x\Gamma'(x) + \Gamma(x)
$$

and

$$
\tau\left(\frac{\Gamma''(x)}{2}\right) = \frac{1}{2}\frac{d}{dx}\left(x\Gamma'(x) + \Gamma(x)\right) = x\frac{\Gamma''(x)}{2} + \Gamma'(x).
$$

Therefore a solution matrix is given by

$$
Y = \begin{pmatrix} \Gamma(x) & \Gamma'(x) & \Gamma''(x)/2 \\ 0 & \Gamma(x) & \Gamma'(x) \\ 0 & 0 & \Gamma(x) \end{pmatrix},
$$

so that the associated Picard-Vessiot ring is $R = K\left[\Gamma(x), \Gamma'(x), \Gamma''(x), \Gamma(x)^{-1}\right]$. Let G be its difference Galois group and let B be a k-algebra. For any $\varphi \in G(B)$, φ commutes to the action of τ over $R \otimes_k B$, therefore it must send any element of R, which is solution of a τ-difference equation, into a solution of the same equation. We know that Γ is solution of a homogenous order 1 equation, while $\frac{\Gamma'(x)}{\Gamma(x)}$ and $\frac{d}{dx}\left(\frac{\Gamma'(x)}{\Gamma(x)}\right)$ are solutions inhomogeneous order 1 equations. Therefore, as in Examples 2.4.4 and 2.4.8, there must exist $c_\varphi \in \mathbb{G}_{m,k}(B)$ and $(d_{1,\varphi}, d_{2,\varphi}) \in \mathbb{G}_{a,k}(B)^2$ such that

$$
\begin{cases}
\varphi\left(\Gamma(x)\right) = c_\varphi \Gamma(x), \\[2mm]
\varphi\left(\dfrac{\Gamma'(x)}{\Gamma(x)}\right) = \dfrac{\Gamma'(x)}{\Gamma(x)} + d_{1,\varphi}, \\[2mm]
\varphi\left(\dfrac{d}{dx}\left(\dfrac{\Gamma'(x)}{\Gamma(x)}\right)\right) = \dfrac{d}{dx}\left(\dfrac{\Gamma'(x)}{\Gamma(x)}\right) + d_{2,\varphi} = \dfrac{\Gamma''(x)}{\Gamma(x)} - \dfrac{\Gamma'(x)}{\Gamma^2(x)} + d_{2,\varphi}.
\end{cases}
$$

So we have represented G as a subgroup of $\mathbb{G}_{m,k} \times \mathbb{G}_{a,k}^2$. Since $\Gamma(x), \Gamma'(x), \Gamma''(x)$ are algebraically independent by Hölder theorem [Höl87] (see Proposition 2.7.3 below for a proof), it is actually an isomorphism, thanks to Theorem 2.4.9. The construction above can be easily generalized to system associated to a Jordan block of eigenvalue x and order higher than 3, using higher order derivatives of Γ.

Now let us consider the finite difference equation $(\tau - x)^n(y) = 0$, for some positive integer n. Such an equation occurs in generalized Carlitz modules studied in [HR97] and more extensively in [Pel13]. We have seen that Γ is a solution for $n = 1$. One can verify recursively that, for $n > 1$, a basis of solution over k is given by the higher order derivatives of Γ with respect to x, namely the set $\Gamma(x), \Gamma'(x), \ldots, \Gamma^{(n-1)}(x)$. It follows that the Picard-Vessiot ring is the same as the one of the system above, namely $R = \left[\Gamma(x), \Gamma'(x), \ldots, \Gamma^{(n-1)}(x), \Gamma(x)^{-1}\right]$. We conclude that the Galois group is $\mathbb{G}_{m,k} \times \mathbb{G}_{a,k}^{n-1}$.

Before being able to prove the most common results used in the applications, we need to state the Galois correspondence and its properties.

2.5 The Galois Correspondence (First Part)

We consider $R, L \subset F$ and $G := \mathrm{Gal}(L/K)$ as above.

Definition 2.5.1 Let $\frac{r}{s} \in L$, with $r, s \in R$ and $s \neq 0$, and let $\varphi \in G(B)$, for a k-algebra B. We say that $\frac{r}{s}$ is invariant under the action of φ if in $R \otimes_k B$ we have:

$$\varphi(r \otimes 1)(s \otimes 1) = (r \otimes 1)\varphi(s \otimes 1).$$

If H is an algebraic subgroup of G defined over k, then $\frac{r}{s}$ is invariant under the action of H if $\frac{r}{s}$ is invariant under the action of φ, for all k-algebras B and all $\varphi \in H(B)$.

We denote by L^H the set of elements of L invariant under the action of H.

Remark 2.5.2 Notice that if $\varphi \in G(k)$, then the condition above simply means that $\frac{\varphi(r)}{\varphi(s)} = \frac{r}{s}$ in L.

If M is an intermediate field of L/K, stable by τ, then we can consider (2.2) as a system defined over M. Indeed if L is a Picard-Vessiot field over K, it must be a Picard-Vessiot field over M and we can define $\mathrm{Gal}(L/M)$.

Theorem 2.5.3 ([OW15, 2.52]) *There exists a one-to-one correspondence between the algebraic subgroup of G defined over k and the intermediate fields of L/K stable by τ. In the notation above we have two maps that are one the inverse of the other and are defined by:*

$$H \mapsto L^H, \quad M \mapsto \mathrm{Gal}(L/M).$$

Moreover if H is an algebraic subgroup of G, then $H = G$ if and only if $L^H = K$.

As usual in Galois theories, the last statement is the key-point of the Galois correspondence.

2.6   Application to Transcendence and Differential Transcendence

2.6.1   General Statements

From the theorems above we deduce a first result on transcendency that is very useful in many settings. The criteria of transcendence below are contained in [HS08, §3], up to some reformulations. They are the key-point of several applications, for instance in [DHR18, DHR16, DHRS18, DHRS20, DH19].

In the proposition below the assumption $k = F^\tau = K^\tau$ is crucial as well as in all the subsequent results in this section.

Proposition 2.6.1 *Let $f_1, \ldots, f_d \in K^*$ and let $z_1, \ldots, z_d \in F$ be a solution of the following inhomogeneous difference system:*

$$\left\{ \tau(z_i) = z_i + f_i, \text{ for } i = 1, \ldots, d. \right. \tag{2.3}$$

The following assertions are equivalent:

1. *There exist $\lambda_1, \ldots, \lambda_d \in k$, not all zero, and $g \in K$ such that $\lambda_1 f_1 + \cdots + \lambda_d f_d = \tau(g) - g$.*
2. *There exist $\lambda_1, \ldots, \lambda_d \in k$, not all zero, such that $\lambda_1 z_1 + \cdots + \lambda_d z_d \in K$.*
3. *There exist $\lambda_1, \ldots, \lambda_d \in K$, not all zero, such that $\lambda_1 z_1 + \cdots + \lambda_d z_d \in K$.*
4. *$z_1, \ldots, z_d$ are algebraically dependent over K.*

Remark 2.6.2 Notice that the first statement above is about the f_i's, while the others are about the z_i's.

Proof Let us assume that we are in the situation of the first assertion. We have:

$$\tau \left(\sum_{i=1}^d \lambda_i z_i - g \right) = \sum_{i=1}^d \lambda_i z_i - \tau(g) + \sum_{i=1}^d \lambda_i f_i = \sum_{i=1}^d \lambda_i z_i - g,$$

hence $\sum_{i=1}^d \lambda_i z_i - g \in k$, which proves 2. Moreover, the implications $2 \Rightarrow 3 \Rightarrow 4$ are tautological.

We conclude by proving that $4 \Rightarrow 1$. As in Example 2.4.8, the system (2.3) is equivalent to the following linear system of order $2d$:

$$
\tau(Y) = \begin{pmatrix}
\boxed{\begin{matrix} 1 & f_1 \\ 0 & 1 \end{matrix}} & 0 & \cdots & 0 \\
0 & \boxed{\begin{matrix} 1 & f_2 \\ 0 & 1 \end{matrix}} & \ddots & \vdots \\
\vdots & \ddots & \ddots & 0 \\
0 & \cdots & 0 & \boxed{\begin{matrix} 1 & f_d \\ 0 & 1 \end{matrix}}
\end{pmatrix} Y,
$$

so that its Picard-Vessiot ring is $R = k[z_1, \ldots, z_d]$. It follows that $\mathrm{Gal}(L/K)$ is an algebraic subgroup of $\mathbb{G}_{a,k}^d$, defined over k, and that for any k-algebra B and any $\varphi \in \mathrm{Gal}(L/K)(B)$ there exists $c_{\varphi,1}, \ldots, c_{\varphi,d} \in B$ such that $\varphi(z_i) = z_i + c_{\varphi,i}$.

Since $z_1, \ldots, z_d$ are algebraically dependent over K, Theorem 2.4.9 implies that the dimension of $\mathrm{Gal}(L/K)$ is strictly smaller than d and hence $\mathrm{Gal}(L/K)$ is a proper subgroup scheme of $\mathbb{G}_{a,k}^d$. All the proper algebraic subgroup of $\mathbb{G}_{a,k}^d$ are contained in a hyperplane and $\mathrm{Gal}(L/K)$ is defined over k, therefore there exist $\lambda_1, \ldots, \lambda_d \in k$, not all zero, such that for any k-algebra B and any $\varphi \in \mathrm{Gal}(L/K)(B)$, we have $\sum_{i=1}^d \lambda_i c_{\varphi,i} = 0$. We conclude that $g := \sum_{i=1}^d \lambda_i z_i$ verifies

$$
\varphi(g) = \sum_{i=1}^d \lambda_i z_i + \sum_{i=1}^d \lambda_i c_{\varphi,i} = \sum_{i=1}^d \lambda_i z_i = g,
$$

and hence that $g \in K$, by the Galois correspondence. Finally we have:

$$
\tau(g) - g = \tau \left(\sum_{i=1}^d \lambda_i z_i \right) - \sum_{i=1}^d \lambda_i z_i = \sum_{i=1}^d \lambda_i f_i.
$$

This ends the proof. $\square$

In an analogous way, taking into account that the algebraic subgroup of $\mathbb{G}_{m,k}^d$ are defined by equations of the form $x_1^{\alpha_1} \cdots x_d^{\alpha_d} = 1$, for some $\alpha_1, \ldots \alpha_d \in \mathbb{Z}$, it is possible to prove the following proposition:

Proposition 2.6.3 Let $a_1, \ldots a_d \in K^*$ and let $z_1, \ldots, z_d \in F^*$ be a solution of the following difference system:

$$
\{ \tau(z_i) = a_i z_i, \text{ for } i = 1, \ldots, d. \tag{2.4}
$$

The following assertions are equivalent:

1. *There exist* $\lambda_1, \ldots, \lambda_d \in \mathbb{Z}$, *not all zero, and* $g \in K$ *such that* $a_1^{\lambda_1} \cdots a_d^{\lambda_d} = \tau(g)/g$.
2. *There exist* $\lambda_1, \ldots, \lambda_d \in \mathbb{Z}$, *not all zero, such that* $z_1^{\lambda_1} \cdots z_d^{\lambda_d} \in K$.
3. $z_1, \ldots, z_d$ *are algebraically dependent over* K.

Remark 2.6.4 Notice that Proposition 2.6.3 has one more characterization of the algebraic dependency of the solutions. Here we only have 3 assertions because of the multiplicative form of the algebraic subgroups of $\mathbb{G}_{m,k}^d$.

2.6.2 Differential Algebraicity and D-Finiteness

We now switch our attention to the characterization of differential algebraicity, hence in this subsection we assume that we are in characteristic zero. We assume that the field F comes equipped with a derivation ∂ that commutes with the endomorphism τ. This implies in particular that ∂ induces a derivation on both k and K.

Example 2.6.5 In the notation of Example 2.2.1, we can take $\partial = \frac{d}{dx}$ for $\tau : f(x) \mapsto f(x+1)$ and and $\partial = x\frac{d}{dx}$ for $\tau : f(x) \mapsto f(qx)$.

We recall the following definition:

Definition 2.6.6 We say that $f \in F$ is differentially algebraic (with respect to ∂) over K, if there exists an integer $n \geq 0$ such that $f, \partial(f), \ldots, \partial^n(f)$ are algebraically dependent over K, or, equivalently, if f is solution of an algebraic differential equation over K. We say that f is differentially transcendental over K if it is not differentially algebraic over K and that it is D-finite over K if it is differentially algebraic and, moreover, it is the solution of a linear differential equation with coefficients in K.

We say that F is differentially algebraic over K is all elements of F are differentially algebraic over k.

Once again, the assumption $F^\tau = K^\tau$ is crucial in the following corollaries:

Corollary 2.6.7 *Let* $f \in K^*$ *and* $z \in F$ *be a solution of* $\tau(y) = y + f$. *The following statements are equivalent:*

1. *There exist* $n \geq 0$, $\lambda_0, \ldots, \lambda_n \in k$, *not all zero, and* $g \in K$ *such that* $\lambda_0 f + \lambda_1 \partial(f) + \cdots + \lambda_n \partial^n(f) = \tau(g) - g$.
2. *There exist* $n \geq 0$, $\lambda_0, \ldots, \lambda_n \in k$, *not all zero, such that* $\lambda_0 z + \lambda_1 \partial(z) + \cdots + \lambda_n \partial^n(z) \in K$.
3. z *is D-finite over* K.
4. z *is differentially algebraic over* K.

In particular, the corollary above says that:

Corollary 2.6.8 *In the notation of Corollary 2.6.7, if z is not D-finite over K then z is differentially transcendental over K.*

Proof of Corollary 2.6.7 The assumption on the commutativity of ∂ and τ implies that z satisfies all the following difference equations:

$$\tau\left(\partial^i(z)\right) = \partial^i(z) + \partial^i(f), \text{ for all } i = 0, 1, 2, \ldots.$$

The statement follows from Proposition 2.6.1. $\square$

Corollary 2.6.9 *Let $a \in K^*$ and $z \in F^*$ be a solution of $\tau(y) = ay$. The following statement are equivalent:*

1. *There exist $n \geq 0$, $\lambda_0, \ldots, \lambda_n \in k$, not all zero, and $g \in K$ such that $\lambda_0 \frac{\partial(a)}{a} + \lambda_1 \partial\left(\frac{\partial(a)}{a}\right) + \cdots + \lambda_n \partial^n\left(\frac{\partial(a)}{a}\right) = \tau(g) - g$.*
2. *There exist $n \geq 0$, $\lambda_0, \ldots, \lambda_n \in k$, not all zero, such that $\lambda_0 \frac{\partial(z)}{z} + \lambda_1 \partial\left(\frac{\partial(z)}{z}\right) + \cdots + \lambda_n \partial^n\left(\frac{\partial(z)}{z}\right) \in K$.*
3. *$\frac{\partial(z)}{z}$ is D-finite over K.*
4. *z is differentially algebraic over K.*

Proof Notice that z is differentially algebraic over K if and only if $\partial(z)/z$ is differentially algebraic over K. In fact, $\partial^n\left(\frac{\partial(z)}{z}\right) \in \frac{1}{z^n} K[z, \partial(z), \ldots, \partial^n(x)]$, therefore an elementary algebraic manipulation allows to transform an algebraic differential equation satisfied by z into an algebraic differential equation satisfied by $\partial(z)/z$ and *vice versa*. Taking the logarithmic derivative of $\tau(z) = az$, we obtain:

$$\tau\left(\frac{\partial(z)}{z}\right) = \frac{\partial(z)}{z} + \frac{\partial(a)}{a}.$$

The statement follows from the Corollary 2.6.7. $\square$

Remark 2.6.10 The generalization of the last two corollaries to systems of order 1 equations and to an arbitrary set of commuting derivations is straightforward. For the generalization to the case of equation of the form $\tau(y) = ay + f$, we refer to [HS08, Propositions 3.8, 3.9, and 3.10].

2.7 Applications to Special Cases

In this section we suppose that F has characteristic zero. The results in Sects. 2.7.1 and 2.7.3 have been originally proven in [HS08, §3].

2.7.1 Finite Difference Equations and Hölder Theorem

We are in the situation of Example 2.4.10, i.e., let F be the field of meromorphic functions over $\mathbb{C}$ equipped with the operator $\tau : f(x) \mapsto f(x + 1)$. Then k is the field of meromorphic 1-periodic functions and we set $K = k(x)$. We set $\partial = \frac{d}{dx}$, which commutes with τ.

Corollary 2.7.1 *In the notation above, let $f \in K^*$ and $z \in F$ be such that $\tau(z) = z + f$. The following assertions are equivalent:*

1. *z is differentially algebraic over K.*
2. *z is D-finite over K.*
3. *There exist a positive integer n, $\lambda_0, \ldots, \lambda_n \in k$ and $g \in K$ such that*

$$\lambda_0 f + \lambda_1 \partial (f) + \cdots + \lambda_n \partial^n (f) = g(x + 1) - g(x).$$

If $f \in \mathbb{C}(x)$ (resp. $f \in \overline{\mathbb{Q}}(x)$), then they are also equivalent to:

4. *There exist a positive integer n, $\lambda_0, \ldots, \lambda_n \in \mathbb{C}$ (resp. $\in \overline{\mathbb{Q}}$) and $g \in \mathbb{C}(x)$ (resp. $\in \overline{\mathbb{Q}}(x)$) such that*

$$\lambda_0 f + \lambda_1 \partial (f) + \cdots + \lambda_n \partial^n (f) = g(x + 1) - g(x).$$

Proof Notice that $1 \Leftrightarrow 2 \Leftrightarrow 3$ follow from Corollary 2.6.7. Moreover $4 \Rightarrow 3$ is trivial. Let us prove that $3 \Rightarrow 4$, by a classical descent argument. Let $C = \mathbb{C}$ or $\overline{\mathbb{Q}}$, so that $f \in C(x)$. Let N be the degree of the denominator of g and M the degree of its denominator. We consider a ring of polynomials of the form $C[\Lambda_0, \ldots, \Lambda_n, A_0, \ldots, A_N, B_0 \ldots, B_M]$, so that we can write the equality

$$\Lambda_0 f + \Lambda_1 \partial (f) + \cdots + \Lambda_n \partial^n (f) = \frac{A_0 + A_1(x + 1) + \cdots + A_N(x + 1)^N}{B_0 + B_1(x + 1) + \cdots + B_M(x + 1)^M}$$
$$- \frac{A_0 + A_1 x + \cdots + A_N x^N}{B_0 + B_1 x + \cdots + B_M x^M}. \tag{2.5}$$

Equalizing the coefficients of each integer powers of x in (2.5), we obtain a system of polynomial equations with coefficients in C, that has a solution in k, by assumption. Since C is an algebraically closed field contained in k, it must also have a solution in C. This proves the corollary. □

Although the last assertion of Corollary 2.7.1 is stated over $\mathbb{C}$ (or over $\overline{\mathbb{Q}}$), we cannot conclude the differential algebraicity of z over $\mathbb{C}(x)$. See Example 2.7.4 that it is based on the fact that there are meromorphic 1-periodic functions that are differentially transcendental over $\mathbb{C}(x)$. For now, notice that the statement above only implies the following:

Corollary 2.7.2 *We consider the same notation as in the previous corollary, with* $C = \overline{\mathbb{Q}}$ *or* $\mathbb{C}$ *and* $f \in C(x)$. *Suppose that for any* $n \geq 0$, *any* $\lambda_0, \ldots, \lambda_n \in C$ *and any* $g \in C(x)$, *we have:* $\lambda_0 f + \lambda_1 \partial (f) + \cdots + \lambda_n \partial^n (f) \neq g(x+1) - g(x)$. *Then* z *is differentially transcendent over* K *and hence over* $C(x)$.

The Euler Gamma function, that we have already mentioned in some examples, is a meromorphic function over $\mathbb{C}$ satisfying the functional equation $\Gamma(x + 1) = x\Gamma(x)$. Hölder theorem [Höl87] says that the Gamma function is differentially transcendental over $\mathbb{C}(x)$ and we are now able to prove it, using a Galoisian argument that has first appeared in [Har08] and [HS08]. Notice that in [BK78] there is a similar proof of the differential transcendency of the Gamma function, which relies on a statement similar to Corollary 2.7.2 in the specific case of the Gamma function, proven by an elementary argument of complex analysis.

Proposition 2.7.3 *The Gamma function* Γ *is differentially transcendental over* $\mathbb{C}(x)$.

Proof As in the proof of Proposition 2.6.9, the Gamma function Γ is differentially transcendental over $\mathbb{C}(x)$ if and only if the function $\psi(x) := \partial(\Gamma)(x)/\Gamma(x)$, that verifies the functional equation

$$\tau(\psi(x)) = \psi(x) + \frac{1}{x},$$

is differentially transcendental over $\mathbb{C}(x)$. Suppose that there exist a positive integer n, $\lambda_0, \ldots, \lambda_n \in \mathbb{C}$ and $g \in \mathbb{C}(x)$ such that

$$\lambda_0 \frac{1}{x} + \lambda_1 \partial \left(\frac{1}{x}\right) + \cdots + \lambda_n \partial^n \left(\frac{1}{x}\right) = g(x+1) - g(x).$$

Since the left-hand side has all its poles at 0, while the right-hand side must have at least a non-zero pole, we find a contradiction, by Corollary 2.7.2. $\square$

The following is a counterexample, based on Hölder theorem, for the fact that we cannot conclude the differential algebraicity over $\mathbb{C}(x)$ in Corollary 2.7.1.

Example 2.7.4 The meromorphic function $\Gamma(\exp(2i\pi x))$ is 1-periodic, hence belongs to $k \subset K$, but is not differentially algebraic over $\mathbb{C}(x)$, since it is the composition of a differentially algebraic function and a differentially transcendental function. In other words, K itself is differentially transcendental over $\mathbb{C}(x)$.

Corollary 2.7.5 ([HS08, Corollary 3.4]) *Let* $a(x) \in \mathbb{C}(x)^*$ *and let* z *be a meromorphic function over* $\mathbb{C}$ *solution of* $z(x + 1) = a(x)z(x)$. *Then* $z(x)$ *is differentially algebraic over* $k(x)$ *if and only if* $a(x) = c\frac{g(x+1)}{g(x)}$, *for some* $g(x) \in \mathbb{C}(x)$ *and* $c \in \mathbb{C}$.

Proof First of all, replacing $z(x)$ with $z(x)g(x)^{-1}$, for a convenient $g(x) \in \mathbb{C}(x)$, and $a(x)$ with $a(x)\frac{g(x)}{g(x+1)}$, we can suppose that two distinguished poles of $a(x)$ do not differ by an integer.

It follows from Corollary 2.7.1, that $z(x)$ is differentially algebraic over $k(x)$ if and only if there exist a positive integer n, $\lambda_0, \ldots, \lambda_n \in \mathbb{C}$ and $g \in \mathbb{C}(x)$ such that

$$\lambda_0 \frac{\partial(a)}{a} + \lambda_1 \partial \left(\frac{\partial(a)}{a} \right) + \cdots + \lambda_n \partial^n \left(\frac{\partial(a)}{a} \right) = g(x+1) - g(x).$$

In the differential relation above, the right hand side must have at least two pole in any τ-orbit where it has a pole. while the left hand side has at worst one pole per τ-orbit. We conclude that $a(x)$ is constant.

On the other hand, if $a(x) = c\frac{g(x+1)}{g(x)}$ and we choose a logarithm $\log c$ of c, a general solution of $y(x+1) = a(x)y(x)$ has the form $z(x) = p(x)\exp(x\log c)$, with $p(x) \in k$. The latter is differentially algebraic over $k(x)$.   □

2.7.2 Linear Inhomogeneous q-Difference Equations of the First Order

We consider the setting of q-difference equations, i.e., F is the field of meromorphic functions over $\mathbb{C}^*$, q is a fixed complex number such that $|q| > 1$, $\tau : f(x) \mapsto f(qx)$, $K = k(x)$, with $k = F^\tau$. We consider the derivation $\partial = x\frac{d}{dx}$, that commutes with τ.

With respect to differential algebraicity, the case of q-difference equations deeply differs from the case of finite difference equation because of the following property (see Example 2.7.4):

Lemma 2.7.6 *The field of elliptic functions k is differentially algebraic over $\mathbb{C}$.*

To prove Lemma 2.7.6, it suffices to write the torus $\mathbb{C}^*/q^{\mathbb{Z}}$ in the form $\mathbb{C}/\mathbb{Z} + i\tau\mathbb{Z}$, where $q = \exp(2i\pi\tau)$, using the exponential function, and remember that the Weierstrass function $\wp$ is differentially algebraic over $\mathbb{C}(x)$, which is itself differentially algebraic over $\mathbb{C}$.

For further reference, we state the following corollary which is a consequence of the fact that, if we have a tower of differentially algebraic extensions $\widetilde{k}/k'$ and k'/k, then $\widetilde{k}/k$ is also differentially algebraic:

Corollary 2.7.7 *For a meromorphic function $f \in F$, it is equivalent to be differentially algebraic over the following fields: $k(x)$, k, $\mathbb{C}(x)$, $\mathbb{C}$.*

Taking into account the previous lemma, the proof of the corollary below follows word by word the proof of Corollary 2.7.1:

Corollary 2.7.8 *In the notation above, let $f \in K^*$ and $z \in F$ be such that $\tau(z) = z + f$. The following assertions are equivalent:*

1. *f is differentially algebraic over K.*
2. *f is differentially algebraic over $\mathbb{C}(x)$.*
3. *f is D-finite over K.*
4. *There exist a positive integer n, $\lambda_0, \ldots, \lambda_n \in k$ and $g \in K$ such that*

$$\lambda_0 f + \lambda_1 \partial(f) + \cdots + \lambda_n \partial^n(f) = g(qx) - g(x).$$

Moreover, if $f \in \overline{\mathbb{Q}}(x)$ (resp. $\mathbb{C}(x)$), they are also equivalent to:

1. *There exist a positive integer n, $\lambda_0, \ldots, \lambda_n \in \overline{\mathbb{Q}}$ (resp. $\mathbb{C}$) and $g \in \overline{\mathbb{Q}}(x)$ (resp. $\mathbb{C}(x)$) such that*

$$\lambda_0 f + \lambda_1 \partial(f) + \cdots + \lambda_n \partial^n(f) = g(qx) - g(x).$$

We finally conclude by proving a result for homogenous order 1 q-difference equations:

Corollary 2.7.9 ([HS08, Corollary 3.4]) *Let $a(x) \in \mathbb{C}(x)^*$ and let z be a meromorphic function over $\mathbb{C}^*$ (resp. $\mathbb{C}$) be a solution of $z(qx) = a(x)z(x)$. The $z(x)$ is differentially algebraic over $\mathbb{C}(x)$ (or equivalently over $k(x)$) if and only if $a(x) = cx^n \frac{g(qx)}{g(x)}$, for some $g(x) \in \mathbb{C}(x)$, $n \in \mathbb{Z}$ and $c \in \mathbb{C}$ (resp. $n = 0$ and $c \in q^{\mathbb{Z}}$).*

Proof If $a(x) = cx^n \frac{g(qx)}{g(x)}$, then a meromorphic solution in F is given by $z(x) = p(x) \frac{\Theta(cx)}{\Theta(x)} \Theta(x)^n g(x)$, where $p(x) \in k$ and $\Theta(x) = \sum_{n \in \mathbb{Z}} q^{-n(n+1)/2} x^n \in F$ is the Jacobi Theta function, which verifies the functional equation $\Theta(qx) = x\Theta(x)$. Notice that $\partial \left(\frac{\partial(\Theta(x))}{\Theta(x)} \right) \in k$, therefore z is differentially algebraic over $\mathbb{C}(x)$. In particular, if $n = 0$, and c is an integer power of q, the solution is also meromorphic at zero.

Let us prove the inverse. We assume that z is meromorphic over $\mathbb{C}^*$. First of all, replacing $z(x)$ with $z(x)g(x)^{-1}$, for a convenient $g(x) \in \mathbb{C}(x)$, and $a(x)$ with $a(x) \frac{g(x)}{g(qx)}$, we can suppose that two distinguished poles of $a(x)$ do not differ by an integer power of q.

It follows from Corollary 2.7.8, that $z(x)$ is differentially algebraic over $\mathbb{C}(x)$ if and only if there exist a positive integer n, $\lambda_0, \ldots, \lambda_n \in \mathbb{C}$ and $g \in \mathbb{C}(x)$ such that

$$\lambda_0 \frac{\partial(a)}{a} + \lambda_1 \partial \left(\frac{\partial(a)}{a} \right) + \cdots + \lambda_n \partial^n \left(\frac{\partial(a)}{a} \right) = g(qx) - g(x).$$

The differential relation above shows that $a(x)$ must have at least two poles in any non-zero τ-orbit, which is in contradiction with our assumptions, therefore we conclude that $a(x) = cx^n$, for some $c \in \mathbb{C}$ and $n \in \mathbb{Z}$.

If moreover z is has a pole at zero, rather than an essential singularity, we can take the expansion of z in $\mathbb{C}((x))$. Plugging it into the equation $z(qx) = cx^n z(x)$, we see that $n = 0$ and hence that c must be an integer power of q.   □

2.7.3 A Particular Case of the Ishizaki-Ogawara's Theorem

In the case of q-difference equations we give a Galoisian proof of the following statement, which is a particular case of Ogawara's theorem [Oga14, Theorem 2]. As already noted by Ogawara, Ishizaki's theorem [Ish98, Theorem 1.2] can be deduced from his formal result. The latter is proved using elementary complex analysis and it is a crucial ingredient of [DHRS20]. Both Ishizaki's and Ogawara's results are based on the idea that q-difference equations "do not have many solutions which are meromorphic in a neighborhood of 0". In this subsection we only need to assume that $q \neq 0$ is not a root of unity, hence we allow q to have norm equal to 1.

Proposition 2.7.10 ([Oga14, Theorem 2]) *Let $q \in \mathbb{C} \smallsetminus \{0, \text{roots of unity}\}$, $f \in \mathbb{C}(x)$, $f \neq 0$ and let $z \in \mathbb{C}((x))$ be a formal power series solution of $\tau(z) = z + f$. The following assertions are equivalent:*

1. *$z \in \mathbb{C}(x)$.*
2. *z is algebraic over $\mathbb{C}(x)$.*
3. *z is D-finite over $\mathbb{C}(x)$.*
4. *z is differentially algebraic over $\mathbb{C}(x)$.*

Proof The implications $1 \Rightarrow 2 \Rightarrow 3 \Rightarrow 4$ are trivial. We prove that $4 \Rightarrow 1$. We decompose $f(x)$ into elementary fractions and we take care of each part of the decomposition separately. We consider a pole $\alpha \in \mathbb{C}^*$ of $f(x)$ such that all the other poles of $f(x)$ in $\alpha q^{\mathbb{Z}}$ are of the form $q^{-n}\alpha$, with $n \geq 0$. Let N_α the largest integer such that $q^{-N_\alpha}\alpha$ is a pole of $f(x)$ and $\sum_i \frac{a_i}{(x-q^{-N_\alpha}\alpha)^i}$ be the polar part of $f(x)$ at $q^{-N_\alpha}\alpha$. We set $h_1(x) = \sum_i \frac{q^i a_i}{(x-q^{-N_\alpha}\alpha)^i}$. Replacing $z(x)$ with $z_1(x) = z(x) + h_1(x)$, we are reduced to consider a new functional equation $y(qx) = y(x) + f_1(x)$, with $f_1(x) = f(x) - h_1(qx) + h_1(x)$, which has a smaller N_α. Iterating the argument we obtain a q-difference equation with an inhomogeneous term $f(x)$ having at most a single pole in each q-orbit $\alpha q^{\mathbb{Z}}$. Corollary 2.6.7 (for $F = \mathbb{C}((x))$ and $K = \mathbb{C}(x)$) implies that there exist $n \geq 0$, $\lambda_0, \ldots, \lambda_n \in \mathbb{C}$, not all zero, and $g \in \mathbb{C}(x)$ such that $\lambda_0 f + \lambda_1 \partial(f) + \cdots + \lambda_n \partial^n(f) = \tau(g) - g$. Since $\tau(g) - g$ cannot have a single pole in $q^{\mathbb{Z}}\alpha$, for $\alpha \neq 0$, we conclude that the rational function $f(x)$ must have no pole at all in $q^{\mathbb{Z}}\alpha$. We are reduced to prove the claim in the case $f \in \mathbb{C}[x, x^{-1}]$, but this assumption obliges $z(x) \in \mathbb{C}((x))$ to be an element of $\mathbb{C}[x, x^{-1}]$, as one can see directly from the equation $f(x) = z(qx) - z(x)$, identifying the coefficients of x^n, for every integer n. This ends the proof.   □

The expansion at zero defines an injective morphism from the field of meromorphic functions over $\mathbb{C}$ to $\mathbb{C}((x))$, which commutes to the action of ∂, therefore we obtain:

Corollary 2.7.11 ([Ish98, Theorem 1.2]) *Let $f \in \mathbb{C}(x)$, $f \neq 0$ and let $z \in F$ be a meromorphic function over $\mathbb{C}$, solution of $\tau(z) = z + f$. Then the assertions of Proposition 2.7.10 are equivalent for z.*

Remark 2.7.12 The reader can find a Galoisian proof of the Ishizaki theorem in whole generality in [HS08, Proposition 3.5], i.e., for equations of the form $\tau(y) = ay + f$. The general statement can be proven using the parameterized Galois theory of difference equations.

Remark 2.7.13 We make some comments on the relation between convergent and meromorphic solutions, under the assumption that $|q| \neq 1$:

1. An important property of q-difference equations is the following:

   > For a solution of a linear q-difference equation with meromorphic coefficients over $\mathbb{C}^*$, it is equivalent to be meromorphic in a neighborhood of zero and to be meromorphic over $\mathbb{C}$.

   The proof is quite easy and relies on the fact that we have supposed that $|q| \neq 1$. In fact, this allows to consider a meromorphic continuation of the solution thanks to the fact that any point can be "brought next to zero" with a repeated application of τ or of τ^{-1}. It seems that this remark is originally due to H. Poincaré [Poi90, page 318].

2. Let us suppose that z is a meromorphic function over $\mathbb{C}$ and algebraic over $\mathbb{C}(x)$. Then z is a meromorphic function over $\mathbb{C}$, which has at worst a pole at ∞, hence it is rational. This proves that $2 \Rightarrow 1$ in Proposition 2.7.10 is true for all linear q-difference equations with rational coefficients, as soon as the solution is meromorphic at 0.

2.8 The Galois Correspondence (Second Part)

In this section we are going to focus on the role of normal subgroups in the Galois correspondence, under the following assumption.

Assumption 2.8.1 We suppose that τ is an automorphism of F and induces an automorphism of K. In difference algebra, when τ is an automorphism, i.e. admits an inverse, is usually called inversive.

The assumption above immediately implies that τ is also an automorphism of any Picard-Vessiot ring and any Picard-Vessiot field contained in F. In fact, if U is a fundamental solution of a system $\tau(Y) = AY$ as in (2.2), we also have $\tau^{-1}(U) = \tau^{-1}(A^{-1})U$. Notice that we continue to work under the assumption of Sect. 2.3 and in particular that F contains a fundamental solution of the linear system $\tau(Y) = AY$

with coefficients in K, and hence, that all our Picard-Vessiot ring and extensions are contained in F.

The main result of this section is the following:

Theorem 2.8.2 *In the notation of Theorem 2.5.3 above, let H be an algebraic subgroup of G defined over k and let $M = L^H$. The following assertions are equivalent:*

1. *H is a normal subgroup of G;*
2. *M is a Picard-Vessiot field over K (for a convenient linear difference equation).*

Assuming the equivalent conditions above, the algebraic group $\mathrm{Gal}(M/K)$ is naturally isomorphic to G/H.

In order to complete the proof of the Galois correspondence, we need to prove a quite classical proposition on the action of the Galois group on the elements of the Picard-Vessiot extension. The proof is not difficult and indeed it is quite similar to the differential case [vdPS03, Corollary 1.38], but, to the best of my knowledge, it is not detailed anywhere in the literature. Notice that the hypothesis that τ is inversive is a central ingredient.

Proposition 2.8.3 *Let $R \subset F$ be the Picard-Vessiot ring for a linear difference system of the form (2.2) over K and f an element of the field of fractions of R. The following statements are equivalent:*

1. *$f \in R$;*
2. *the K-vector space spanned by $\{\tau^n(f), n \geq 0\}$ has finite dimension.*

Proof Let us prove that (1) $\Rightarrow$ (2). We remind that there exists a fundamental solution matrix U of a difference system of the form (2.2), such that $R = K[U, \det U^{-1}]$. Let us denote by $t_1, \ldots, t_{d^2+1}$ the elements of the matrix U, plus $\det U^{-1}$. Since $\tau(U) = AU$ and $\tau(\det U^{-1}) = \det A^{-1} \cdot \det U^{-1}$, for any integer $r \geq 1$, the K-vector space generated by the monomials of degree r in the t_i's and their τ-iterated has finite dimension over K. This proves the statement, because any $f \in R$ can be written as a polynomial in the t_i's and hence the K-vector space spanned by $\{\tau^n(f), n \geq 0\}$ is contained in a finite dimensional K-vector space.

We now show that (2) $\Rightarrow$ (1). Let W be the K-vector space generated by $\{\tau^n(f), n \geq 0\}$. We consider the ideal of R defined by $I := \{a \in R | aW \subset R\}$. Since $f \in L$ and L is the field of fractions of R, the ideal I is non-zero. Moreover τ is inversive, hence $W \subset \tau^{-1}W$. Since W and $\tau^{-1}(W)$ are vector spaces of the same dimension, this implies that $\tau^{-1}(W) = W$. We conclude that $\tau(a)W \subset \tau(aW) \subset R$ for any $a \in I$ and therefore that I is τ-invariant. Finally, $1 \in I$, because R is τ-simple, and $f \in W \subset R$.   □

Corollary 2.8.4 *Let L/K be a Picard-Vessiot extension and R be the Picard-Vessiot ring of L.*

1. *Let M be an intermediate field which is itself a Picard-Vessiot field over K. Moreover let R_M be its Picard-Vessiot ring. Then $R_M = M \cap R$.*

2. We fix $f \in R$ and a linear difference equation $\mathcal{L}(y) = 0$ with coefficients in K such that $\mathcal{L}(f) = 0$. Furthermore, we suppose that the operator associated with the equation $\mathcal{L}(y) = 0$ has minimal order m in τ. Then the solutions of $\mathcal{L}(y) = 0$ in R form a k-vector space of solutions of maximal dimension m.

Proof

1. The statement follows from the previous proposition, since $f \in M \cap R$ if and only if $f \in M$ and f is a solution of a linear difference equation with coefficients in K.
2. Let W be the space of solution of $\mathcal{L}(y) = 0$ in R. Since $f \in W$, we know that $W \neq 0$ and we can consider a k-basis $w_1, \ldots, w_r$ of W. The following formula

$$\det \begin{pmatrix} w_1 & w_2 & \cdots & w_r & y \\ \tau(w_1) & \tau(w_2) & \cdots & \tau(w_r) & \tau(y) \\ \vdots & \vdots & \ddots & \vdots & \vdots \\ \tau^r(w_1) & \tau^r(w_2) & \cdots & \tau^r(w_r) & \tau^r(y) \end{pmatrix} = 0$$

gives a τ-difference equation $\widetilde{\mathcal{L}}(y) = 0$ with coefficients in L having W as space of solutions. By definition of the Galois group, $\varphi(W \otimes_k B) = W \otimes_k B$, for any $\varphi \in G(B)$ and any k-algebra B. The Galois correspondence and the invariance by the action of the Galois group show that the coefficients of $\widetilde{\mathcal{L}}(y) = 0$ are actually in K. Because of the minimality of the order of the operator associated with $\mathcal{L}(y) = 0$, we conclude that $\mathcal{L}$ and $\widetilde{\mathcal{L}}$ coincide up to the multiplication of a non-zero element of K. This implies that $W \subset R$ has maximal dimension m over k.

This proves the claims.   □

Remark 2.8.5 Notice that in the proof of the second statement above, we could replace R by any τ-K-algebra $\widetilde{R} \subset R$, such that for any k-algebra B and any $\psi \in G(B)$, we have $\psi(\widetilde{R} \otimes B) \subset (\widetilde{R} \otimes B)$.

Proof of Theorem 2.8.2 Let M be a Picard-Vessiot field. Then $R_M := R \cap M$ is a Picard-Vessiot ring, which is generated by the entries of a matrix U solution of a difference linear system with coefficients in K, and its inverse. By definition of the difference Galois group, for any k-algebra B and any $\psi \in G(B)$ we have $\psi(R_M \otimes B) \subset R_M \otimes B$. It implies that we have a natural group morphism $G(B) \to \mathrm{Gal}(M/K)(B)$, given by the restriction of the morphisms. The kernel coincides with $H(B)$, hence H is a normal subgroup of G.

Let us suppose that H is a normal subgroup of G. We set $M = L^H$ and $R_M = R \cap M$, so that any $\varphi \in H(B)$ induces the identity over $R_M \otimes B$. Because of the normality of $H(B)$ in $G(B)$, any $\psi \in G(B)$ verifies $\psi(R_M \otimes B) \subset R_M \otimes B$. Finally Remark 2.8.5 shows that R_M is generated by the solution of a linear difference equations, and hence that it is a Picard-Vessiot ring. We deduce that M is the field of fraction of R_M because they both coincide with L^H.   □

Remark 2.8.6 In the notation of the proof, for any k-algebra B, we have a functorial isomorphism

$$(G/H)(B) \cong \mathrm{Aut}^\tau (R_M \otimes_k B/K \otimes_k B),$$

which is actually an isomorphism of algebraic group.

We remind that $k_n = K^{\tau^n}$. See Example 2.2.2.

Corollary 2.8.7 *Let L and G be as above and let G° be the connected component of the identity of G. Then, L^{G° is the relative algebraic closure of K in L (and, hence in our framework coincides with $K(k_n)$, for a convenient positive integer n).*

Proof Since G° is a normal subgroup of G, the finite quotient G/G° is isomorphic to $\mathrm{Gal}(L^{G^\circ}/K)$. Theorem 2.4.9 implies that L^{G°/K is an algebraic extensions, which is also finitely generated.

Let $\tilde{L}$ be the relative algebraic closure of K in L. Then $L^{G^\circ} \subset \tilde{L}$. Since any algebraic element of L over K is solution of a differential equation over K, $\tilde{L}$ is also a Picard-Vessiot field, that therefore correspond to an algebraic subgroup H of G, in the sense that $\tilde{L} = L^H$. The inclusion $L^{G^\circ} \subset \tilde{L}$ implies that $H \subset G^\circ$. Moreover G/H is a finite group, because it must have dimension 0, after Theorem 2.4.9. Since G° is the smallest group such that the quotient G/G° is finite, we deduce that $H = G^\circ$ and therefore, from the Galois correspondence, that $\tilde{L} = L^{G^\circ}$. □

Appendix: Behavior of the Galois Group with Respect to the Iteration of τ

Let us consider the system (2.2) and its n-th iteration:

$$\tau^n y = A_n y, \quad \text{where } A_n := \tau^{n-1}(A) \cdots \tau(A)A. \tag{A.1}$$

We want to compare the Galois group of (2.2) with the Galois group of (A.1).

It follows from the Definition 2.3.5 of Picard-Vessiot ring and field that, if R (resp. L) is a Picard-Vessiot ring (resp. field) for (2.2) over K, $R(k_n)$ (resp. $L(k_n)$) is also a Picard-Vessiot ring (resp. field) for (A.1) over $K(k_n)$. Let $G_n := \mathrm{Gal}^{\tau^n}(L(k_n)/K(k_n))$, where we have add the superscript τ^n to the notation with the obvious meaning, to avoid any confusion.

Let G_1° be the identity component of G_1. By Corollary 2.8.7, there exists r, such that $k_r \subset L$ and that $\mathrm{Gal}^\tau(L/K(k_r)) = G_1^\circ$.

Lemma A.1 *In the notation above, we have $\mathrm{Gal}^{\tau^r}(L/K(k_r)) = G_1^\circ \otimes_k k_r$.*

Proof By definition, for any k_r—algebra B, we have an injective morphisms from $G_1^\circ(B) \to \mathrm{Gal}^{\tau^r}(L/K(k_r))(B)$, indeed if a morphisms commutes with τ, it commutes also with τ^n. The equality follows from the fact that the groups are connected and that they have the same dimension, by Theorem 2.4.9. $\qquad\square$

Proposition A.2 *In the notation introduced above, for any $n \geq r$, the Galois group of (2.2) over $K(k_n)$ is isomorphic to the Galois group of (A.1) over $K(k_n)$.*

Proof The Galois group of (2.2) over $K(k_n)$ is $\mathrm{Gal}^{\tau}(L(k_n)/K(k_n)) \cong G_1^\circ \otimes_k k_n$. It can be naturally seen as a subgroup of $\mathrm{Gal}^{\tau^n}(L(k_n)/K(k_n))$. Equality follows from Theorem 2.4.9 and the connectedness of the groups, as in the proof of the previous lemma. $\qquad\square$

Acknowledgments I'm indebted to the organizers of the VIASM school on Number Theory, in particular to Bruno Anglès and Tuan Ngo Dac for their invitation both to give the course and to write the this paper. I'd like to thank the participants of the *Groupe de travail autour des marches dans le quart de plan*, who have endured several talks on this topic, that have influenced the text below. In particular, I'm grateful to Alin Bostan, Frédéric Chyzac and Marni Mishna for their remarks and their attentive reading of various drafts of this survey and to the anonymous referees for the many useful and constructive comments.

This project has received funding from the ANR project DeRerumNatura, ANR-19-CE40-0018.

References

[BK78] S.B. Bank, R.P. Kaufman, A note on Hölder's theorem concerning the gamma function. Math. Annal. **232**(2), 115–120 (1978)

[CHS08] Z. Chatzidakis, C. Hardouin, M.F. Singer, On the definitions of difference Galois groups. *Model Theory with Applications to Algebra and Analysis*, vol. 1. London Mathematical Society Lecture Note series, vol. 349 (Cambridge University Press, Cambridge, 2008), pp. 73–109

[Coh65] R.M. Cohn, *Difference Algebra* (Interscience Publishers John Wiley & Sons, New York, 1965)

[DH19] T. Dreyfus, C. Hardouin, Length derivative of the generating series of walks confined in the quarter plane. arXiv:1902.10558 (2019)

[DHR16] T. Dreyfus, C. Hardouin, J. Roques, Functional relations of solutions of q-difference equations. arXiv:1603.06771 (2016)

[DHR18] T. Dreyfus, C. Hardouin, J. Roques, Hypertranscendence of solutions of Mahler equations. J. Eur. Math. Soc. **20**(9), 2209–2238 (2018)

[DHRS18] T. Dreyfus, C. Hardouin, J. Roques, M.F. Singer, On the nature of the generating series of walks in the quarter plane. Invent. Math. **213**(1), 139–203 (2018)

[DHRS20] T. Dreyfus, C. Hardouin, J. Roques, M.F. Singer, Walks in the quarter plane: genus zero case. J. Combin. Theory Ser. A **174**, 105251, 25 (2020)

[Har08] C. Hardouin, Hypertranscendance des systèmes aux différences diagonaux. Compos. Math. **144**(3), 565–581 (2008)

[HS08] C. Hardouin, M.F. Singer, Differential Galois theory of linear difference equations. Math. Ann. **342**(2), 333–377 (2008). MR MR2425146 (2009j:39001)

[HSS16] C. Hardouin, J. Sauloy, M.F. Singer, *Galois Theories of Linear Difference Equations: An Introduction*. Mathematical Surveys and Monographs, vol. 211 (American Mathematical Society, Providence, 2016), Papers from the courses held at the CIMPA Research School in Santa Marta, July 23–August 1, 2012

[HR97] Y. Hellegouarch, F. Recher, Generalized t-modules. J. Algebra **187**(2), 323–372 (1997)

[Höl87] O. Hölder, Ueber die Eigenschaft der Gammafunction keiner algebraischen Differentialgleichung zu genügen. Math. Ann. **28**, 1–13 (1887)

[Ish98] K. Ishizaki, Hypertranscendency of meromorphic solutions of a linear functional equation. Aequationes Math. 56(3), 271–283 (1998)

[Lev08] A. Levin, *Difference Algebra*. Algebra and Applications, vol. 8 (Springer, New York, 2008)

[Oga14] H. Ogawara, Differential transcendency of a formal Laurent series satisfying a rational linear q-difference equation. Funkcial. Ekvac. **57**(3), 477–488 (2014)

[OW15] A. Ovchinnikov, M. Wibmer, σ-Galois theory of linear difference equations. Int. Math. Res. Not. (12), 3962–4018 (2015)

[Pap08] M.A. Papanikolas, Tannakian duality for Anderson-Drinfeld motives and algebraic independence of Carlitz logarithms. Invent. Math. **171**(1), 123–174 (2008)

[Pel13] F. Pellarin, On the generalized Carlitz module. J. Number Theory **133**(5), 1663–1692 (2013)

[Pel] F. Pellarin, From the Carlitz exponential to Drinfeld modular forms, in *Arithmetic and Geometry over Local Fields*, ed. by B. Anglès, T. Ngo Dac. Lecture Notes in Mathematics, vol. 2275 (Springer, Cham, 2021). https://doi.org/10.1007/978-3-030-66249-3

[Poi90] H. Poincaré, Sur une classe nouvelle de transcendantes uniformes. Journal de Mathématiques pures et appliquées **6**, 313–365 (1890)

[Pra86] C. Praagman, Fundamental solutions for meromorphic linear difference equations in the complex plane, and related problems. J. Reine Angew. Math. **369**, 101–109 (1986). MR 850630 (88b:39004)

[TR] F. Tavares Ribeiro, On the Stark units of Drinfeld modules, in *Arithmetic and Geometry over Local Fields*, ed. by B. Anglès, T. Ngo Dac. Lecture Notes in Mathematics, vol. 2275 (Springer, Cham, 2021). https://doi.org/10.1007/978-3-030-66249-3

[vdPS97] M. van der Put, M.F. Singer, *Galois Theory of Difference Equations*. Lecture Notes in Mathematics, vol. 1666 (Springer, Berlin, 1997)

[vdPS03] M. van der Put, M.F. Singer, Galois theory of linear differential equations, in *Grundlehren der Mathematischen Wissenschaften [Fundamental Principles of Mathematical Sciences]*, vol. 328 (Springer, Berlin, 2003)

[Wat79] W.C. Waterhouse, Introduction to affine group schemes, in *Graduate Texts in Mathematics*, vol. 66 (Springer, New York, 1979)

Chapter 3
Igusa's Conjecture on Exponential Sums Modulo p^m and the Local-Global Principle

Kien Huu Nguyen

Abstract In this survey we discuss the conjecture of Igusa on exponential sums modulo p^m and some progress of this conjecture. We also present a connection between this conjecture and the local-global principle for forms of higher degree.

3.1 Introduction

Exponential sums play an important role in number theory with many deep applications. One of which is the use of the quadratic Gauss sums in Gauss's proof of the law of quadratic reciprocity that is the first example of reciprocity laws (see [Ire90, Chapters 5 and 6]). Exponential sums modulo p have a deep connection with the Riemann hypothesis over finite fields by the works of Weil, Deligne, Katz, Laumon among others (see for example [Del77, Del74, Del80, Kat85, Kat99, Kat89, Wei48]).

This survey aims to introduce Igusa's conjecture on exponential sums modulo p^m. We report the progress made towards its resolution and its connection with the local-global principle for forms which was indeed one of the initial goals of Igusa.

We begin with one important class of exponential sums depending on a non-constant polynomial f in n variables with integer coefficients. Let N be a positive integer. We define the exponential sum modulo N associated to f by

$$E_N(f) := \frac{1}{N^n} \sum_{\overline{x} \in (\mathbb{Z}/N\mathbb{Z})^n} \exp(\frac{2\pi i f(x)}{N}). \tag{3.1}$$

K. H. Nguyen (✉)
KU Leuven, Department of Mathematics, Leuven, Belgium
e-mail: kien.nguyenhuu@kuleuven.be

© The Author(s), under exclusive license to Springer Nature Switzerland AG 2021   61
B. Anglès, T. Ngo Dac (eds.), *Arithmetic and Geometry over Local Fields*,
Lecture Notes in Mathematics 2275, https://doi.org/10.1007/978-3-030-66249-3_3

Our goal is to look for good upper bounds of these sums. The Chinese remainder theorem allows us to simplify slightly the previous problem. In fact, we can express

$$\frac{1}{N} = \sum_{i=1}^{k} \frac{a_i}{p_i^{m_i}}$$

where $p_1, \ldots, p_k$ are distinct primes, $a_1, \ldots, a_k$ and $m_1, \ldots, m_k$ are integers such that $(a_i, p_i) = 1$ and $m_i \geq 1$ for all $1 \leq i \leq k$. It follows that

$$E_N(f) = \prod_{i=1}^{k} E_{p_i^{m_i}}(a_i f). \tag{3.2}$$

Thus it is sufficient to find good estimates of the exponential sums

$$E_{p,m}(f) := E_{p^m}(f) = \frac{1}{p^{mn}} \sum_{\overline{x} \in (\mathbb{Z}/p^m\mathbb{Z})^n} \exp(\frac{2\pi i f(x)}{p^m})$$

for all primes p and all $m \geq 1$.

Example 3.1.1 We consider the simplest example where $f(x) = x$. We see easily that for $N > 1$, we have

$$E_N(f) = 0.$$

Example 3.1.2 We now consider a more complicated polynomial by taking $f(x) = x^2$. Let p be a prime and m be a positive integer. We write $m = 2k + r$ where $k \geq 1$ and $r \in \{0, 1\}$. We calculate directly $E_{p^m}(f)$ by distinguishing two cases.
Case 1: p is an odd prime. We see that if $(a, p) = 1$ and $0 \leq \alpha \leq k - 1$, then

$$\sum_{b=1}^{p^{\alpha+1}} \exp(\frac{2\pi i (p^\alpha a + p^{m-1-\alpha} b)^2}{p^m}) = 0.$$

Thus we get

$$E_{p^{2k}}(f) = \frac{1}{p^{2k}} \sum_{a=1}^{p^k} \exp(\frac{2\pi i (p^k a)^2}{p^{2k}}) = \frac{1}{p^k}.$$

We also have

$$E_{p^{2k+1}}(f) = \frac{1}{p^{2k+1}} \sum_{a=1}^{p^{k+1}} \exp(\frac{2\pi i (p^k a)^2}{p^{2k+1}}) = \frac{1}{p^{k+1}} \sum_{a=1}^{p} \exp(\frac{2\pi i a^2}{p}).$$

Thus

$$|E_{p^{2k+1}}(f)| = \frac{1}{p^{k+\frac{1}{2}}}.$$

Here the above equality is a consequence of the following fact about quadratic Gauss sums (see for example [Ire90, Chapter 6])

$$(\sum_{a=1}^{p} \exp(\frac{2\pi i a^2}{p}))^2 = p.$$

Case 2: $p = 2$. It is still true that if $(a, 2) = 1$ and $0 \leq \alpha \leq k - 2$, then

$$\sum_{b=1}^{2^{\alpha+2}} \exp(\frac{2\pi i (2^\alpha a + 2^{m-2-\alpha} b)^2}{2^m}) = 0.$$

Thus

$$E_{2^{2k}}(f) = \frac{1}{2^{2k}} \sum_{a=1}^{2^{k+1}} \exp(\frac{2\pi i (2^{k-1} a)^2}{2^{2k}}) = \frac{1}{2^{k+1}} \sum_{a=1}^{4} \exp(\frac{2\pi i a^2}{4}) = \frac{1+i}{2^k}.$$

Further, we have

$$E_{2^{2k+1}}(f) = \frac{1}{2^{2k+1}} \sum_{a=1}^{2^{k+2}} \exp(\frac{2\pi i (2^{k-1} a)^2}{2^{2k+1}}) = \frac{1}{2^{k+2}} \sum_{a=1}^{8} \exp(\frac{2\pi i a^2}{8}) = \frac{1+i}{2^{\frac{2k+1}{2}}}.$$

By the same calculation, for all primes p, all positive integers m and all non-zero integers A such that $(A, p) = 1$, we have

$$|E_{p^m}(Ax^2)| \leq c_p \, p^{-\frac{m}{2}},$$

where

$$c_p = \begin{cases} 1 & \text{if } p \neq 2, \\ \sqrt{2} & \text{otherwise.} \end{cases} \tag{3.3}$$

Hence by (3.2) we conclude that for all non-zero integers N,

$$|E_N(f)| \leq \sqrt{2} N^{-\frac{1}{2}}.$$

The equality holds if $N > 1$ is a square number.

In the case where f is a polynomial in one variable, exponential sums modulo p^m have been studied by many mathematicians and we refer the reader to [Coc99] for more details.

For polynomials f in n variables, Igusa showed that for each prime p, there exist a constant $\sigma_p \le +\infty$ and a positive constant c_p such that for all $\sigma < \sigma_p$ and all $m \ge 1$, we have

$$|E_{p^m}(f)| \le c_p \, p^{-m\sigma}. \tag{3.4}$$

Furthermore, either $\sigma_p = +\infty$ or $-\sigma_p$ is the real part of a pole of the Igusa local zeta function associated to f. Thus we would like to know how to obtain a global information from the local information for each prime p, i.e. the dependence of c_p and σ_p in p.

Example 3.1.3 In Example 3.1.1, for each prime p we can take $\sigma_p = +\infty$ and an arbitrary positive constant $c_p > 0$.

In Example 3.1.2 we can take $\sigma_p = \frac{1}{2}$ for all primes p and

$$c_p = \begin{cases} 1 & \text{if } p \ne 2, \\ \sqrt{2} & \text{otherwise.} \end{cases} \tag{3.5}$$

In order to prove (3.4), Igusa found a way to understand exponential sums via singularity theory. In fact, exponential sums $E_{p^m}(f)$ modulo p^m can be computed by certain Igusa local zeta functions (see Sect. 3.2 for more details). As a consequence, the asymptotic expansion of $E_{p^m}(f)$ for $m > 1$ could be given in terms of poles of these Igusa local zeta functions.

We now give more details about the above discussion. First we recall some basic facts about p-adic fields and then express exponential sums modulo p^m as p-adic integrals. Letting p be a prime, we define the p-adic norm $|.|_p$ on the field of rational numbers $\mathbb{Q}$ as follows. We set $|0|_p := 0$ and for all integers a, b, k with $(a, p) = (b, p) = 1$,

$$|\frac{a}{b} p^k|_p := p^{-k}$$

We denote by $\mathbb{Q}_p$ the completion of $\mathbb{Q}$ with respect to this norm and by $\mathbb{Z}_p$ the closure of $\mathbb{Z}$ in $\mathbb{Q}_p$. Then $\mathbb{Q}_p$ is a locally compact field equipped with the norm $|.|_p$ which extends $|.|_p$ over $\mathbb{Q}$. Further, $\mathbb{Z}_p$ is a closed and open subring of $\mathbb{Q}_p$ and

$$\mathbb{Z}_p = \{x \in \mathbb{Q}_p \mid |x|_p \le 1\}.$$

It is a discrete valuation ring with the unique maximal ideal

$$\mathcal{M}_p = p\mathbb{Z}_p = \{x \in \mathbb{Q}_p \mid |x|_p < 1\}.$$

Let x be an element of $\mathbb{Q}_p$. We can write

$$x = \sum_{i \geq k} a_i p^i$$

for some integers k and a_i with $0 \leq a_i \leq p - 1$. If $x = 0$, then we set $\mathrm{ord}_p(x) :=$ $+\infty$. Otherwise, we can suppose that $a_k \neq 0$ and set $\mathrm{ord}_p(x) := k$. Then it is clear that

$$|x|_p = p^{-\mathrm{ord}_p(x)}.$$

Here we take the convention that $p^{-\infty} = 0$. We note that $x \in \mathbb{Z}_p$ if and only if $\mathrm{ord}_p(x) \geq 0$ and $x \in \mathcal{M}_p$ if and only if $\mathrm{ord}_p(x) > 0$.

The standard additive character of $\mathbb{Q}_p$ is the homomorphism of abelian groups

$$\psi_1 := \exp : (\mathbb{Q}_p, +) \to (\mathbb{C}^*, \times)$$

which sends x to $\exp(2\pi i x')$ with $x' \in \mathbb{Z}[\frac{1}{p}] \cap (x + \mathbb{Z}_p)$. It is well-defined since the value $\exp(2\pi i x')$ does not depend on the choice of $x' \in \mathbb{Z}[\frac{1}{p}] \cap (x + \mathbb{Z}_p)$. An additive character ψ of $\mathbb{Q}_p$ is defined to be a continuous homomorphism from $(\mathbb{Q}_p, +)$ to $(\mathbb{C}^*, \times)$ with compact image. For such an additive character ψ there exists a unique $z \in \mathbb{Q}_p$ such that

$$\psi_z(x) := \psi_1(xz) = \psi(x).$$

Since $\mathbb{Q}_p$ is locally compact, we can endow $\mathbb{Q}_p^n$ with the Haar measure $|dx|$ normalized such that $\mathbb{Z}_p^n$ has volume 1. It follows immediately that

$$E_{p^m}(f) = \int_{\mathbb{Z}_p^n} \psi_{p^{-m}}(f(x)) |dx|.$$

This suggests that to any additive character ψ of $\mathbb{Q}_p$ we can associate an exponential sum by

$$E_\psi(f) := \int_{\mathbb{Z}_p^n} \psi(f(x)) |dx|.$$

This integral is an example of Igusa local zeta functions.

More generally, letting L be a non-Archimedean local field which is a finite extension of either the p-adic field $\mathbb{Q}_p$ or the field of Laurent series $\mathbb{F}_q((t))$ with coefficients in a finite field $\mathbb{F}_q$, we can associate an exponential sum $E_\psi(f)$ to any polynomial $f \in L[x_1, \ldots, x_n]$ and any additive character ψ of L.

As mentioned earlier, the asymptotic expansion of $E_{p^m}(f)$ for $m > 1$ could be given in terms of poles of the associated Igusa local zeta function. To determine the poles of Igusa local zeta functions, Igusa formulated the so-called strong monodromy conjecture which relates these poles to eigenvalues of monodromy and roots of Bernstein-Sato polynomials (see Sect. 3.2.3). As a consequence, if the strong monodromy conjecture holds for f, then the size of $E_{p^m}(f)$ can be bounded in terms of the biggest non-trivial root of the Bernstein-Sato polynomial b_f of f.

We now state a coarse form of Igusa's conjecture for a uniform bound of exponential sums modulo p^m when p and m go to infinity.

Conjecture 3.1.4 *Let f be a non-constant polynomial in n variables with coefficients in $\mathbb{Z}$ and σ be a positive real number. Suppose that for all primes p large enough, there exists a constant $c_p > 0$ such that we have*

$$|E_{p^m}(f)| \le c_p \, p^{-m\sigma}$$

for all $m \ge 2$. Then there exists a constant $C > 0$ such that

$$|E_{p^m}(f)| \le C \, p^{-m\sigma}$$

for all primes p large enough and all $m \ge 2$.

Remark 3.1.5 We rediscover the original conjecture of Igusa for homogeneous polynomials f. We refer the reader to Sect. 3.4.2 for a discussion about this conjecture as well as a variant of this conjecture due to Cluckers [Clu08a] and Cluckers and Veys [Clu16].

Remark 3.1.6 The condition $m \ge 2$ in Conjecture 3.1.4 can be replaced by a weaker condition $m \ge 1$ in many cases (see Example 3.1.2). However, in general, we have to treat separately the case $m = 1$ as explained below.

Let us consider the polynomial $f = x_1 - x_1^2 x_2$. We show that $E_{p^m}(f) = 0$ for all primes p and all $m > 1$ (see Remark 3.3.3 for more details). But for all primes p we have

$$E_p(f) = \frac{1}{p^2} \Big(\sum_{x_1 \ne 0 \ \mathrm{mod}\ p} \sum_{x_2 \in \mathbb{Z}/p\mathbb{Z}} \exp(\frac{2\pi i (x_1 - x_1^2 x_2)}{p}) + \sum_{x_2 \in \mathbb{Z}/p\mathbb{Z}} 1 \Big) = \frac{1}{p}.$$

Let $\sigma > 1$ then

$$|E_{p^m}(f)| \le p^{\sigma-1} p^{-m\sigma}$$

for all primes p and all $m \ge 1$ but we cannot find a constant C such that

$$|E_{p^m}(f)| \le C \, p^{-m\sigma}$$

for all primes p large enough and all $m \ge 1$.

Remark 3.1.7 We keep the notation of Conjecture 3.1.4. Suppose that there exist a positive integer M and a constant $C \leq 1$ such that

$$|E_{p^m}(af)| \leq C\, p^{-m\sigma}$$

for all primes $p > M$, all integers a with $(a, p) = 1$ and all $m \geq 1$. Moreover, for each prime $p \leq M$ there exists a constant c_p such that

$$|E_{p^m}(af)| \leq c_p\, p^{-m\sigma}$$

for all integers a with $(a, p) = 1$ and all $m \geq 1$. Thus (3.2) implies immediately

$$|E_N(f)| \leq C'N^{-\sigma}$$

for some constant $C' > 0$ and all $N \geq 1$.

Remark 3.1.8 The statement of Conjecture 3.1.4 extends without difficulty to an arbitrary global field K (i.e a finite extension of $\mathbb{Q}$ or a function field of an algebraic curve over a finite field) and a non-constant polynomial $f \in \mathcal{O}_K[x_1, \ldots, x_n]$ where $\mathcal{O}_K$ is the ring of integers of K.

In fact, for any finite place v of K, we denote by K_v the completion of K at v equipped with the norm $|.| : K_v \to \mathbb{R}$ and by $\mathcal{O}_v$ the ring of integers of K_v. Let π_v be a uniformizer of $\mathcal{O}_v$. We fix an additive character ψ_1 of K_v such that $\psi_1|_{\mathcal{O}_v} = 1$ but $\psi_1|_{\pi_v^{-1}\mathcal{O}_v} \neq 1$ (see Sects. 3.2 and 3.3 for more details). Let σ be a positive real number such that for all but finitely many finite places v of K and all $z \in K_v \setminus \pi_v^{-1}\mathcal{O}_v$, we have

$$|E_{\psi_z}(f)| = |\int_{\mathcal{O}_v^n} \psi_1(zf(x))|dx|| \leq c_v|z|^{-\sigma}.$$

Then we can ask whether there exists a constant C such that $c_v \leq C$ for all but finitely many finite places v.

In Sect. 3.4 we give an overview of progress on this conjecture due to many mathematicians. We begin with the work of Igusa in the non-degenerate case and end with the most recent result of Cluckers, Mustaţă and the author in case of non-rational singularities.

We should mention that Igusa's work [Igu78] around exponential sums modulo p^m was motivated by his ultimate hope to extend the local-global principle to forms of higher degree (i.e. homogeneous polynomials of degree at least 3). Recall that for a form $f \in \mathbb{Z}[x_1, \ldots, x_n]$ of degree d, we say that the local-global principle holds for f if the following assertion is true: f represents zero in $\mathbb{Q}$ if and only if it represents zero in $\mathbb{R}$ and in all fields $\mathbb{Q}_p$. The Hasse-Minkowski theorem states that the local-global principle holds for quadratic forms. The idea of Igusa to generalize the Hasse-Minkowski theorem to forms of higher degree is divided into two steps. First, a good uniform bound of exponential sums modulo p^m in p

and m together with some extra conditions would imply the existence of a certain Poisson formula (see Sect. 3.3.2 and Proposition 3.3.7). Second, one derives the local-global principle from this Poisson formula (see Sect. 3.5).

The above discussion illustrates one of the common approaches of this volume which is to apply analytic techniques in the study of arithmetic geometry. The reader is strongly encouraged to read other chapters for "further examples" in different settings, in particular, the lecture of Poineau and Turchetti [Poi20a, Poi20b] and to discover possible connections among them.

We close this section by saying some words about function fields. In this lecture we only consider Conjecture 3.1.4 for number fields K but it is natural to ask whether one could extend the results in Sects. 3.2, 3.3 and 3.4 to the case where K is a function field which means the function field of an algebraic curve over a finite field $\mathbb{F}_q$. The answer is yes for non-constant polynomials $f \in K[x_1, \ldots, x_n]$ such that for all critical values a of f, $f^{-1}(a)$ admits an embedded resolution with good reduction at all but finitely many places v of K (see Sect. 3.2 for the definition of such a resolution). For number fields the existence of an embedded resolution for all polynomials f is guaranteed by Hironaka's theorem in [Hir64]. However, the resolution of singularities in positive characteristic is more complicated and the existence of such a resolution for general f is still unknown. Hence we hope that some young mathematicians could attack this challenging question in the future.

3.2 Igusa Local Zeta Functions and Exponential Sums Modulo p^m

In this section we review the notion of Igusa local zeta functions and exponential sums modulo p^m over an arbitrary non-Archimedean local field of characteristic 0. We refer the reader to the excellent survey of Denef [Den91] and the work of Igusa [Igu78] for more details.

3.2.1 Local Fields

For the rest of this paper we fix a positive integer $n \geq 1$.

In what follows, we consider a non-Archimedean local field L of characteristic 0. It means that L is a finite extension of $\mathbb{Q}_p$ defined as in Sect. 3.1 for some prime p. To simplify, we will say that L is a p-adic field and we set $p_L := p$.

We remark that the norm $|.|_p$ on $\mathbb{Q}_p$ extends uniquely to a norm $|.|_L$ in L. We will write $|.|$ instead of $|.|_L$ if no confusion results. Let $\mathcal{O}_L$ be the ring of integers in L. Then

$$\mathcal{O}_L = \{x \in L \mid |x| \leq 1\}.$$

It is a discrete valuation ring with the maximal ideal $\mathcal{M}_L$ given by

$$\mathcal{M}_L = \{x \in L \mid |x| < 1\}.$$

We denote by $k_L = \mathcal{O}_L/\mathcal{M}_L$ the residue field of L. This field is a finite extension of $\mathbb{F}_p$ and we denote by q_L the cardinality of k_L. Let ϖ be a uniformizer of L, i.e. ϖ is a generator of $\mathcal{M}_L$. For each non-zero element $x \in L$, we can write in a unique way $x = \lambda \varpi^\alpha$ where $\lambda \in \mathcal{O}_L^*$ and $\alpha \in \mathbb{Z}$. We set

$$\mathrm{ac}(x) := \lambda, \quad \mathrm{ord}(x) := \alpha,$$

and

$$\overline{\mathrm{ac}}(x) := \mathrm{ac}(x) \bmod \mathcal{M}_L.$$

We can extend the maps ac and ord to L by setting $\mathrm{ac}(0) = 0$ and $\mathrm{ord}(0) = +\infty$.

We introduce the following three functions which will play an important role in the sequel. First, the *standard additive character* of L is the homomorphism $\psi_1 : L \to \mathbb{C}^*$ given by

$$\psi_1 := \exp(\mathrm{Tr}_{L/\mathbb{Q}_p}(x))$$

where $\exp(.)$ is the map given in Sect. 3.1. Any *additive character* ψ of L can be written in the form $\psi(x) := \psi_z(x) = \psi(zx)$ for some element $z \in L$. We put

$$m(\psi) := -\mathrm{ord}(z).$$

Second, a multiplicative character χ of $\mathcal{O}_L^*$ is defined to be a continuous homomorphism from $(\mathcal{O}_L^*, \times)$ to $(\mathbb{C}^*, \times)$ with finite image. For a multiplicative character χ, let $c(\chi)$ be the smallest integer such that $\chi|_{1+\mathcal{M}_L^{c(\chi)}}$ is trivial. It is called the *conductor* of χ. We set $\chi(0) := 0$. It is clear that χ induces a character of $\mathcal{O}_L^*/(1 + \mathcal{M}_L^{c(\chi)})$. In particular, if $c(\chi) = 1$, then χ induces a character of k_L^* which is still denoted by χ and we extend χ to k_L by setting $\chi(0) = 0$.

Third, a *Schwartz-Bruhat function* $\Phi : L^n \to \mathbb{C}$ is a locally constant function with compact support, denoted by $\mathrm{Supp}(\Phi)$. We say that Φ is *residual* if $\mathrm{Supp}(\Phi) \subset \mathcal{O}_L^n$ and if $\Phi(x)$ only depends on $x \bmod \mathcal{M}_L$. If Φ is residual, then Φ induces a function $\overline{\Phi} : k_L^n \to \mathbb{C}$.

As in case of $\mathbb{Q}_p^n$, we will endow L^n with a Haar measure $|dx|$ such that the volume of $\mathcal{O}_L^n$ is 1.

3.2.2   Embedded Resolutions

Let K be a field of characteristic 0. Let $f \in K[x_1, \ldots, x_n]$ be a non-constant polynomial in n variables. We set

$$X = \mathbb{A}_K^n = \operatorname{Spec} K[x_1, \ldots, x_n],$$

and

$$D = f^{-1}(0) = \operatorname{Spec} K[x_1, \ldots, x_n]/(f).$$

An *embedded resolution* (Y, h) of D in X is a closed smooth subscheme Y of the projective space $\mathbb{P}_X^m$ over X for some m such that the restriction h to Y of the projection $\mathbb{P}_X^m \to X$ has the following properties:

(i) $h : Y \backslash h^{-1}(D) \to X \backslash D$ is an isomorphism,
(ii) the reduced scheme $(h^{-1}(D))_{\mathrm{red}}$ associated to $h^{-1}(D)$ has simple normal crossings as a subscheme of Y (i.e. its irreducible components are smooth and intersect transversally).

Let $E_i, i \in \mathcal{T}$, be the irreducible components of $(h^{-1}(D))_{\mathrm{red}}$. For each $i \in \mathcal{T}$, let N_i be the multiplicity of E_i in the divisor of $f \circ h$ on Y and let $v_i - 1$ be the multiplicity of E_i in the divisor of $h^*(dx_1 \wedge \ldots \wedge dx_n)$. The set $\{(N_i, v_i)_{i \in \mathcal{T}}\}$ are called the *numerical data* of the resolution.

Further, for each subset $I \subset \mathcal{T}$, we define

$$E_I := \cap_{i \in I} E_i \qquad \text{and} \qquad \overset{\circ}{E}_I := E_I \backslash \cup_{j \in \mathcal{T} \backslash I} E_j.$$

In particular, when $I = \emptyset$ we have $E_\emptyset = Y$.

We also denote by $C_f \subset X$ be the critical locus of $f : X \to \mathbb{A}_K^1$.

We remark that such a resolution exists by the seminal work of Hironaka [Hir64, Main Theorem II]. It can be obtained from a series of blow-ups with smooth centers.

Remark 3.2.1 Let K' be a field extension of K. By the functoriality of embedded resolutions, h induces an embedded resolution $h : Y_{K'} \to \mathbb{A}_{K'}^n = X_{K'}$ of $D_{K'}$ in $X_{K'}$. We remark that each blow-up center C of h may be written as a union of finitely many irreducible components C_i over K' and we can replace the blow-up with center C by the composition of blow-ups with center C_i. If K' is an algebraically closed field, then h induces an embedded resolution which can be obtained by successive blow-ups at irreducible smooth varieties. Similarly, each irreducible component E_i can be split into a disjoint union of finitely many irreducible components E_{ij} over K'. But we always have $N_i = N_{ij}, v_i = v_{ij}$.

In what follows, let K be a number field, $\mathcal{O}_K$ be its ring of integers and $f \in \mathcal{O}_K[x_1, \ldots, x_n]$ be a non-constant polynomial in n variables. Let (Y, h) be an embedded resolution of D in X. If Z is a closed subscheme of Y and $\mathfrak{p}$ is a prime ideal of $\mathcal{O}_K$, we denote by $\overline{Z}$ the reduction modulo $\mathfrak{p}$ of Z (see [Shi55]). We say that the embedded resolution (Y, h) of D in X has *good reduction modulo* $\mathfrak{p}$ if the following conditions are satisfied:

(i) $\overline{Y}$ and $\overline{E}_i$ are smooth for all $i \in \mathcal{T}$,
(ii) $\cup_{i \in \mathcal{T}} \overline{E}_i$ has simple normal crossings,
(iii) the schemes $\overline{E}_i$ and $\overline{E}_j$ have no common components for all $i, j \in \mathcal{T}$ with $i \neq j$.

One can show that there exists a finite subset S of $\operatorname{Spec} \mathcal{O}_K$, such that for all $\mathfrak{p} \notin S$, we have $f \in \mathcal{O}_{\mathfrak{p}}[x]$, $f \not\equiv 0 \bmod \mathfrak{p}$ and that the resolution (Y, h) for f has good reduction mod $\mathfrak{p}$ (see [Den87, Theorem 2.4]). Then for $\mathfrak{p} \notin S$ and $I \subset \mathcal{T}$, one can show that $\overline{E}_I = \cap_{i \in I} \overline{E}_i$. We set

$$\overset{\circ}{\overline{E}}_I := \overline{E}_I \setminus \cup_{j \notin I} \overline{E}_j.$$

Letting a be a closed point of $\overline{Y}$, we put $\mathcal{T}_a := \{i \in \mathcal{T} \mid a \in \overline{E}_i\}$. In the local ring of $\overline{Y}$ at a, we can write

$$\overline{f} \circ \overline{h} = \overline{u} \prod_{i \in \mathcal{T}_a} \overline{g}_i^{N_i},$$

where $\overline{u}$ is a unit, $(\overline{g}_i)_{i \in \mathcal{T}_a}$ is a part of a regular system of parameters and N_i is the corresponding multiplicity defined as above.

3.2.3 Igusa Local Zeta Functions and the Monodromy Conjecture

Recall that L is a p-adic field. Let $f \in L[x_1, \ldots, x_n]$ be a non-constant polynomial in n variables with coefficients in L. Let χ be a multiplicative character of $\mathcal{O}_L^*$ and Φ be a Schwartz-Bruhat function on L^n. Following Weil we associate to the data (L, f, χ, Φ) an Igusa local zeta function

$$Z_{L, \Phi, f}(s, \chi) := \int_{L^n} \Phi(x)\, \chi(\mathrm{ac}(f(x)))\, |f(x)|^s\, |dx|,$$

for $s \in \mathbb{C}$ with $\Re(s) > 0$. One can see that $Z_{L, \Phi, f}(s, \chi)$ is holomorphic in this region and extends to a meromorphic function on $\mathbb{C}$. The following theorem gives basic properties of these zeta functions $Z_{L, \Phi, f}(s, \chi)$.

Theorem 3.2.2 (Igusa [Igu74] and [Igu78]) *We keep the previous notation. Then we have*

(i) $Z_{L, \Phi, f}(s, \chi)$ *is a rational function of* q_L^{-s}.

(ii) *If (Y, h) is an embedded resolution of $f^{-1}(0)$ in $\mathbb{A}_L^n$ with the numerical data $\{(N_i, v_i)_{i \in \mathcal{T}}\}$, then the poles of $Z_{L, \Phi, f}(s, \chi)$ are among the values*

$$s = -\frac{v_i}{N_i} + \frac{2\pi i k}{\log_e q_L}$$

with $k \in \mathbb{Z}$ and $i \in \mathcal{T}$ such that $\chi^{N_i} = 1$.

(iii) *If $\operatorname{Supp}(\Phi) \cap C_f \subset f^{-1}(0)$, then $Z_{L, \Phi, f}(s, \chi) = 0$ for all but finitely many χ. Here recall that $C_f \subset X$ denotes the singular locus of $f : X \to \mathbb{A}_L^1$.*

In the case where we have an embedded resolution having good reduction modulo $\mathcal{M}_L$, the above results could be improved as follows.

Theorem 3.2.3 (Denef [Den91] and [Den87]) *Suppose that there exists an embedded resolution (Y, h) of $f^{-1}(0)$ having good reduction modulo $\mathcal{M}_L$ and $f \neq 0 \mod \mathcal{M}_L$. We suppose further that Φ is a residual Schwartz-Bruhat function on L^n. Then we have*

(i) *If the conductor $c(\chi)$ of χ is at least 2 and that the numerical data $\{(N_i, v_i)_{i \in \mathcal{T}}\}$ of (Y, h) satisfying $N_i \notin \mathcal{M}_L$ for all $i \in \mathcal{T}$, then $Z_{L, \Phi, f}(s, \chi)$ is constant as a function of s. Moreover, if $C_{\overline{f}} \cap \operatorname{Supp}(\overline{\Phi}) \subset \overline{f}^{-1}(0)$, then $Z_{L, \Phi, f}(s, \chi) = 0$.*

(ii) *If $c(\chi) = 1$ and χ is of order d, let $\mathcal{T}_d = \{I \subset \mathcal{T} \mid \forall i \in I : d \mid N_i\}$. Then*

$$Z_{L, \Phi, f}(s, \chi) = q_L^{-n} \sum_{I \in \mathcal{T}_d} c_{I, \Phi, \chi} \prod_{i \in I} \frac{(q_L - 1)q_L^{-N_i s - v_i}}{1 - q_L^{-N_i s - v_i}},$$

where

$$c_{I, \Phi, \chi} = \sum_{a \in \overset{\circ}{\overline{E}}_I(k_L)} \overline{\Phi}(\overline{h}(a)) \Omega_\chi(a),$$

and

$$\Omega_\chi(a) = \chi(\overline{u}(a))$$

for any choice of $\overline{u}$ in the local ring of $\overline{Y}$ at $\overline{a}$ as in Sect. 3.2.1.

$\square$

In many known examples, many of the possible poles are false poles of the zeta function (even if we take the intersection of the sets of possible poles over all embedded resolutions). The monodromy conjecture suggests an explanation for this phenomenon.

Now let us recall some notions about monodromy and Bernstein-Sato polynomials.

Let $f \in \mathbb{C}[x_1, \ldots, x_n]$ be a non-constant polynomial with coefficients in $\mathbb{C}$ and P be a point in $\mathbb{C}^n$ such that $f(P) = a$. Let B be a sufficiently small ball with center P. In [Mil68] Milnor proved that $f|_B$ is a locally trivial C^∞ fibration over a small enough punctured disc $A \subset \mathbb{C} \setminus \{a\}$. Thus the diffeomorphism type of $F_P = f^{-1}(t) \cap B$ of f around P does not depend on $t \in A$. The counter clockwise generator of the fundamental group of A induces an automorphism T of $H^*(F_P, \mathbb{C})$. We call F_P and T the *Milnor fiber* and the *local monodromy* of f at P, respectively.

Let K be a field of characteristic 0 and $f \in K[x_1, \ldots, x_n]$ be a polynomial. Bernstein [Ber72] proved that there exist $P \in K[x, \frac{\partial}{\partial x}, s]$ and a polynomial $b(s) \in K[s] \setminus \{0\}$ such that $Pf^{s+1} = b(s)f^s$. The monic polynomial of smallest degree satisfying this functional equation is called the *Bernstein-Sato polynomial* of f, denoted by b_f. One can show that $(s + 1) \mid b_f(s)$ if f is non-constant. Furthermore, Kashiwara claimed in [Kas76] that all roots of b_f are negative rational numbers. Moreover, Malgrange [Mal83] proved that if α is a root of b_f, then $\exp(2\pi i\alpha)$ is an eigenvalue of the local monodromy of f at some point of $f^{-1}(0)$ and all eigenvalues are obtained in this way.

Igusa suggested that the poles of the Igusa local zeta function associated to f should be described by the roots of the associated Bernstern-Sato polynomial or the eigenvalues of the local monodromy of f.

Conjecture 3.2.4 (Igusa, Monodromy Conjecture) *Let K be a number field and f be a non-constant polynomial in $K[x_1, \ldots, x_n]$. For all but finitely many primes p, if s is a pole of $Z_{L,\Phi,f}(s, \chi)$ where L is a p-adic field containing K, then $\exp(2\pi i\Re(s))$ is an eigenvalue of the local monodromy of f at some complex point of $f^{-1}(0)$.*

Conjecture 3.2.5 (Strong Monodromy Conjecture) *Let K be a number field and f be a non-constant polynomial in $K[x_1, \ldots, x_n]$. For all but finitely many primes p, if s is a pole of $Z_{L,\Phi,f}(s, \chi)$ where L is a p-adic field containing K, then $\Re(s)$ is a root of b_f.*

By the above discussion, if α is a root of b_f, then $\exp(2\pi i\alpha)$ is an eigenvalue of the local monodromy of f at some point. Thus Conjecture 3.2.5 implies Conjecture 3.2.4. Note that Conjecture 3.2.4 only implies that if s is pole of $Z_{L,\Phi,f}(s, \chi)$, then $\Re(s) + a$ is a root of b_f for some integer a.

Both conjectures might be true for all p-adic fields. But it seems very hard for primes for which we cannot find an embedded resolution with good reduction. Although both conjectures have been checked in many cases (see for example [Loe88] for polynomials in two variables), to our knowledge, they are widely open in general.

3.2.4 Exponential Sums and Fiber Integration

In this section we introduce a general form of exponential sums modulo p^m and its relation with Igusa local zeta functions.

Recall that L is a p-adic field. Let f be a non-constant polynomial in $L[x_1, \ldots, x_n]$, Φ be a Schwartz-Bruhat function on L^n and z be an element of L. To this data we associate the exponential sum $E_{L,\Phi,z}(f)$ by

$$E_{L,\Phi,z}(f) := \int_{L^n} \Phi(x)\, \psi_1(zf(x))\, |dx|.$$

It is clear that if $L = \mathbb{Q}_p$, $z = p^{-m}$, $\Phi = \mathbf{1}_{\mathbb{Z}_p}$, then $E_{L,\Phi,z}(f)$ is equal to $E_{p^m}(f)$ introduced in Sect. 3.1.

To describe the relation between exponential sums modulo p^m and Igusa local zeta functions, we need to recall the notion of fiber integration. For each $y \in L$, we set $U_y := f^{-1}(y) \setminus C_f$. Since $f(x) = y$ on U_y, we get

$$\frac{\partial f}{\partial x_1} dx_1 + \cdots + \frac{\partial f}{\partial x_n} dx_n = 0 \tag{3.6}$$

on U_y. Let $a \in U_y$. Since $a \notin C_f$, there exists $1 \le \ell \le n$ such that $\dfrac{\partial f}{\partial x_\ell}(a) \ne 0$. If $j \ne \ell$ and $1 \le j \le n$ such that $\dfrac{\partial f}{\partial x_j}(a) \ne 0$, taking the exterior product on both sides of (3.6) with $\bigwedge_{i \ne j, i \ne \ell} dx_i$ yields

$$(-1)^{j-1} \frac{\partial f}{\partial x_j}(a) \bigwedge_{i \ne \ell} dx_i = (-1)^{\ell-1} \frac{\partial f}{\partial x_\ell}(a) \bigwedge_{i \ne j} dx_i.$$

Thus $d_{f,y} := (-1)^{\ell-1} (\frac{\partial f}{\partial x_\ell})^{-1} \bigwedge_{i \ne \ell} dx_i |_{U_y}$ is a well-defined non-vanishing regular $(n-1)$-form around $a \in U_y$. For each Schwartz-Bruhat function Φ on L^n, we set

$$F_{f,y}(\Phi) := \int_{f^{-1}(y)} \Phi\, |d_{f,y}|.$$

We can show that

$$E_{L,\Phi,z}(f) = \int_L F_{f,y}(\Phi)\, \psi_1(zy)\, |dy|$$

is the Fourier transform of $F_{f,y}(\Phi)$ and

$$Z_{L,\Phi,f}(s, \chi) = \int_L F_{f,y}(\Phi)\, \omega_{\chi,s}(y)\, |dy|$$

is the Mellin transform of $(1 - q_L^{-1}) q_L^{-\operatorname{ord}(y)} F_{f,y}(\Phi)$ where the quasi-character $\omega_{\chi,s}$ is given by $\omega_{\chi,s}(y) = \chi(\operatorname{ac}(y)) q_L^{-\operatorname{ord}(y)s}$.

On the other hand, using Fourier transform we can compute $E_{L,\Phi,z}(f)$ by Igusa local zeta functions.

Proposition 3.2.6 ([Den91], Proposition 1.4.4) *Let $u \in \mathcal{O}_L^\times$, ϖ be a uniformiser of L and $m \in \mathbb{Z}$. Then $E_{L,\Phi,u\varpi^{-m}}(f)$ is equal to*

$$Z_{L,\Phi,f}(0, \chi_{\mathrm{triv}}) + \operatorname{Coeff}_{t^{m-1}}\left(\frac{(t - q_L) Z_{L,\Phi,f}(s, \chi_{\mathrm{triv}})}{(q_L - 1)(1 - t)} \right)$$

$$+ \sum_{\chi \neq \chi_{\mathrm{triv}}} g_{\chi^{-1}} \chi(u) \operatorname{Coeff}_{t^{m-c(\chi)}}\left(Z_{L,\Phi,f}(s, \chi) \right),$$

where g_χ is the Gauss sum given by

$$g_\chi = \frac{q_L^{1-c(\chi)}}{q_L - 1} \sum_{\bar{v} \in (\mathcal{O}_L / \mathcal{M}_L^{c(\chi)})^*} \chi(v) \psi_1(v / \varpi^{c(\chi)}).$$

As a consequence, we obtain the following asymptotic expansion of exponential sums.

Corollary 3.2.7 *Suppose that $C_f \cap \operatorname{Supp}(\Phi) \subset f^{-1}(0)$. Then $E_{L,\Phi,z}(f)$ is a finite $\mathbb{C}$-linear combination of functions of the form*

$$\chi(\operatorname{ac}(z)) |z|^\lambda (\log_{q_L} |z|)^\beta$$

with coefficients independent of z, and with $\lambda \in \mathbb{C}$ a pole of

$$H(L, \chi, s) Z_{L,\Phi,f}(s, \chi)$$

where

$$H(L, \chi, s) = \begin{cases} q_L^{s+1} - 1 & \text{if } \chi = \chi_{\mathrm{triv}}, \\ 1 & \text{otherwise.} \end{cases}$$

and with $\beta \in \mathbb{N}$, $\beta \leq$ (multiplicity of pole λ) $- 1$, provided that $|z|$ is large enough. Moreover, all poles λ appear effectively in this linear combination.

A pole λ appearing in Corollary 3.2.7 will be called a non-trivial pole of the Igusa local zeta function associated to f and Φ. We will denote the set of such poles by $\operatorname{Pol}(f, \Phi)$. For $\lambda \in \operatorname{Pol}(f, \Phi)$ we set

$$m_{f,\Phi}(\lambda) := \max\{m_{f,\Phi,\chi}(\lambda) \mid \lambda \text{ is a pole of } H(L, \chi, s) Z_{L,\Phi,f}(s, \chi)\}$$

where $m_{f,\Phi,\chi}(\lambda)$ is the multiplicity of the pole λ of $H(L,\chi,s)Z_{L,\Phi,f}(s,\chi)$. Moreover, we set

$$\sigma_{f,\Phi} := \min\{-\Re(\lambda) \mid \lambda \in \mathrm{Pol}(f,\Phi)\}$$

and

$$\beta_{f,\Phi} := \max\{m_{f,\Phi}(\lambda) \mid \lambda \in \mathrm{Pol}(f,\Phi),\ \Re(\lambda) = -\sigma_{f,\Phi}\}.$$

It is very useful that the previous asymptotic expansion of exponential sums gives us all the important information about the poles of the Igusa local zeta function associated to f and Φ. If the strong monodromy conjecture (Conjecture 3.2.5) holds, then we would obtain a very deep and mysterious connection between the arithmetic side, the geometric side and the topological side of f.

3.3 Igusa's Conjecture on Exponential Sums Modulo p^m

This section aims to state a general conjecture on exponential sums modulo p^m in spirit of Igusa as we mentioned in Sect. 3.1. To do so we review the notion of a certain Poisson formula (see [Igu78, Igu76] for more details).

3.3.1 Adèles

In what follows, K denotes a number field. Let $\mathcal{O}_K$ be its ring of integers. For each place v of K, we denote by $|.|_v$ the associated absolute value of K and K_v be the completion of K by $|.|_v$. By Ostrowski's theorem, K_v is either $\mathbb{R}$, $\mathbb{C}$ or a p-adic field. We normalize the norms $|.|_v$ where v runs through the set of places of K such that the product formula holds. This formula says that for all $x \in K^*$, $|x|_v = 1$ for all but finitely many places v and we have

$$\prod_v |x|_v = 1$$

where v runs through the set of places of K.

We say that v is an *Archimedean place* of K if $K_v = \mathbb{R}$ or $K_v = \mathbb{C}$. Otherwise, we say that v is a *non-Archimedean place* of K. We denote by S_∞ the set of all Archimedean places of K

$$S_\infty := \{v \mid v \text{ is Archimedean}\}. \tag{3.7}$$

With the notation as in Sect. 3.2, for each non-Archimedean place v, we denote by $\mathcal{O}_v$ the ring of integers of the local field K_v, $\mathcal{M}_v$ the maximal ideal of $\mathcal{O}_v$, k_v the residue field of K_v and p_v the characteristic of K_v. Finally, we fix a uniformizer ϖ_v of K_v and denote by ord_v and ac_v the associated valuation map and the angular component map of K_v, respectively.

Let $X = \mathbb{A}_K^n$ the affine space of dimension n. A subvariety U of X is *locally K-closed* if we can write $U = V \setminus W$ where V and W are closed subvarieties of X defined over K.

Let U be such a subvariety of X. If we write $I(V) = (f_1, \ldots, f_\ell)$ and $I(W) = (g_1, \ldots, g_r)$ with polynomials $f_i, g_j \in K[x_1, \ldots, x_n]$, then $a \in U$ if and only if $f_i(a) = 0$ for all $1 \le i \le \ell$ and $g_j(a) \ne 0$ for some $1 \le j \le r$. For each place v of K, we put

$$U_v := \{x \in K_v^n \mid (\forall i,\ f_i(x) = 0) \wedge (\exists j,\ g_j(x) \ne 0)\}.$$

It is clear that U_v is locally compact. Moreover, if v is non-Archimedean, we set

$$U_v^0 := \{x \in \mathcal{O}_v^n \mid (\forall i,\ f_i(x) = 0) \wedge (\exists g \in I(W) \cap \mathcal{O}_K[x_1, \ldots, x_n],\ g(x) \in \mathcal{O}_v^*)\},$$

then U_v^0 is compact. Let S be a finite set of places of K such that S contains S_∞ defined as in (3.7). Then $\prod_{v \in S} U_v$ is locally compact and $\prod_{v \notin S} U_v^0$ is compact. It implies that

$$U_S = \prod_{v \in S} U_v \times \prod_{v \notin S} U_v^0$$

is also locally compact. It is clear that if $S \subset S'$, then $U_S \subset U_{S'}$. Thus we can take the inductive limit $U_A = \xrightarrow[S]{\lim} U_S$ which is called the adelization of U. The set $U(K)$ of K-points of U can be viewed as a discrete subset of U_A by the diagonal embedding. Note that this construction is functorial.

We suppose further that U is smooth and that there exists an everywhere regular differential form ω of the highest degree on U vanishing nowhere and defined over K. Let Ψ be a non-trivial character of K_A/K, i.e. a homomorphism from K_A to the unit circle which is trivial on K. For each place v of K, there exists a natural embedding $K_v \hookrightarrow K_A$ which sends x to the adèle whose v-th coordinate is x and others coordinates are 0. Via this embedding Ψ induces a character ψ_v on K_v. We can associate a measure $|dx|_v$ on K_v^n which is the n-fold product of the self-dual measure relative to ψ_v on K_v. We observe that for all but finitely many non-Archimedean places v, the character ψ_v is trivial on $\mathcal{O}_v$ but non-trivial on $\mathcal{M}_v^{-1}$, and the measure of $\mathcal{O}_v^n$ is equal to 1. Next, we endow the set U_v with the Borel measure $|\omega|_v$ associated with ω and the measure $|dx|_v$. For each finite set S of places of K such that S contains S_∞ defined as in (3.7), we define the measure $|\omega|_A$ on U_S to be the product of measures

$$|\omega|_A := \bigotimes_{v \in S} |\omega|_v \otimes \bigotimes_{v \notin S} |\omega|_v$$

under the assumption that the product measure $\bigotimes_{v \notin S} |\omega|_v$ exists on $\prod_{v \notin S} U_v^0$. We will call $|\omega|_A$ the Tamagawa measure on U_A. In particular, the Tamagawa measure exists on X_A by taking $U = X$.

Recall that S_∞ is the set of all Archimedean places of K as in (3.7). We set

$$X_\infty := \prod_{v \in S_\infty} X_v,$$

and

$$X_0 := \lim_{\to S} \prod_{v \in S \setminus S_\infty} X_v^0.$$

Viewing X_∞ as a finite product of copies of $\mathbb{R}$, we consider the space $\mathcal{S}(X_\infty)$ of Schwartz-Bruhat functions on X_∞. Since X_0 is a locally compact abelian group with arbitrary large and small compact open subgroups, we can define the space $\mathcal{S}(X_0)$ of Schwartz-Bruhat functions on X_0. The Schwartz-Bruhat functions on X_A is defined to be the tensor product

$$\mathcal{S}(X_A) := \mathcal{S}(X_\infty) \otimes_\mathbb{C} \mathcal{S}(X_0).$$

Each element of $\mathcal{S}(X_A)$ is a $\mathbb{C}$-linear combination of elements of the form $\Phi_\infty \otimes \Phi_0$ with $\Phi_\infty \in \mathcal{S}(X_\infty)$ and $\Phi_0 \in \mathcal{S}(X_0)$. A *tempered distribution T* on X_A is a $\mathbb{C}$-linear form on $\mathcal{S}(X_A)$ such that for all fixed functions $\Phi_0 \in \mathcal{S}(X_0)$, $T(\Phi_\infty \otimes \Phi_0)$ depends continuously on Φ_∞ in $\mathcal{S}(X_\infty)$. We denote by $\mathcal{S}(X_A)'$ the $\mathbb{C}$-vector space of all tempered distributions on X_A.

3.3.2   Poisson Formulas and Formulas of Siegel Type

We continue with the notation of the previous section. Recall that K is a number field and $X = \mathbb{A}_K^n$ is the affine space of dimension n. Let $f \in \mathcal{O}_K[x_1, \ldots, x_n]$ be a non-constant polynomial. We fix a non-trivial character Ψ of K_A/K. For any $z \in K$ we define a tempered distribution $\Psi(zf(x))$ on X_A given by

$$\Psi(zf(x))(\Phi) := \int_{X_A} \Phi(x) \, \Psi(zf(x)) \, |dx|_A.$$

Note that this integral is absolutely convergent.

We say that the Poisson formula holds for f if the following conditions hold:

(i) The infinite sum

$$\sum_{z \in K} \Psi(zf(x))$$

belongs to $\mathcal{S}(X_A)'$. It is equivalent to the fact that the Eisenstein-Siegel series

$$\sum_{z \in K} \int_{X_A} \Phi(x) \, \Psi(zf(x)) \, |dx|_A$$

converges absolutely for every $\Phi \in \mathcal{S}(X_A)$.

(ii) For all $y \in K$, the measure $|d_{f,y}|_A$ exists on $U_{y,A}$.

(iii) If $j : U_{y,A} \to X_A$ is the induced map by $U_y \to X$, then the global singular series $j_*(|d_{f,y}|_A)$ (or simply $|d_{f,y}|_A$) exists in $\mathcal{S}(X_A)'$ or equivalently, the integral

$$\int_{U_{y,A}} \Phi \, |d_{f,y}|_A$$

is absolutely convergent for every $\Phi \in \mathcal{S}(X_A)$.

(iv) The infinite sum

$$\sum_{y \in K} |d_{f,y}|_A$$

belongs to $\mathcal{S}(X_A)'$.

(v) We have the following equality

$$\sum_{z \in K} \Psi(zf(x)) = \sum_{y \in K} |d_{f,y}|_A$$

in $\mathcal{S}(X_A)'$.

Igusa gave a criterion for the existence of Poisson formulas based on his conjecture on exponential sums modulo p^m.

Proposition 3.3.1 (See [Igu78]) *Let f be a form of degree d in $\mathcal{O}_K[x_1, \ldots, x_n]$ (i.e $f \in \mathcal{O}_K[x_1, \ldots, x_n]$ is a homogeneous polynomial of degree d).*

Then the Poisson formula holds for f if the following conditions hold:

(i) $\mathrm{codim}(C_f) \geq 3$, *i.e. the affine hypersurface defined by f is irreducible and normal.*

(ii) *There exist a constant $\sigma > 2$ and a positive constant c such that for all but finitely many non-Archimedean places v and all $z \in K_v \setminus \mathcal{O}_v$, we have*

$$|E_{K_v, \mathbf{1}_{\mathcal{O}_v^n}, z}(f)| \le c \, |z|_v^{-\sigma}. \tag{3.8}$$

There is no reason to restrict (3.8) to homogeneous polynomials and to the condition $\sigma > 2$. Thus we could relax these restrictions to obtain a more general statement. For the constant σ, by Corollary 3.2.7, we should choose

$$\sigma < \liminf_{p_v \to +\infty} \sigma_{f, \mathbf{1}_{\mathcal{O}_v^n}}.$$

We should mention that it may be interesting to investigate (3.8) for families of Schwartz-Bruhat functions $(\Phi_v)_{v \notin S_\infty}$ in the case where there exists a closed subset W defined over $\mathcal{O}_K$ of the affine space $\mathbb{A}_K^n$ such that $\Phi_v = \Phi_{W,v}$ is the characteristic function of the set $\{x \in \mathcal{O}_v^n \mid x \mod \mathcal{M}_v \in W(k_v)\}$ for each place $v \notin S_\infty$.

We are ready to state a general form of Igusa's conjecture on exponential sums.

Conjecture 3.3.2 *Let K be a number field and f be a non-constant polynomial in $\mathcal{O}_K[x_1, \ldots, x_n]$. Let W be a closed subset defined over $\mathcal{O}_K$ of the affine space $\mathbb{A}^n$ such that $f(W(\mathbb{C}))$ contains at most one critical value of f. Let $\Phi_{W,v}$ be the characteristic function of the set $\{x \in \mathcal{O}_v^n \mid x \mod \mathcal{M}_v \in W(k_v)\}$ for each place $v \notin S_\infty$. We set*

$$\sigma := \liminf_{p_v \to +\infty} \sigma_{f, \Phi_{W,v}}$$

and

$$\beta := \limsup_{p_v \to +\infty} \beta_{f, \Phi_{W,v}}$$

as in Sect. 3.2.

Then there exists a positive constant c such that for all but finitely many places v, all $z \in K_v$ with $\mathrm{ord}_v(z) \le -2$, we have

$$|E_{K_v, \Phi_{W,v}, z}(f)| \le c \, |\mathrm{ord}_v(z)|^{\beta-1} |z|_v^{-\sigma}. \tag{3.9}$$

Remark 3.3.3 If $f(W(\mathbb{C}))$ contains no critical values of f, then

$$E_{K_v, \Phi_{W,v}, z}(f) = 0$$

provided that k_v has large enough characteristic and $\mathrm{ord}_v(z) \le -2$ (see [Den91, Remark 4.5.3]). Hence Conjecture 3.3.2 holds in this case.

Remark 3.3.4 We note that in the original statement of Igusa in [Igu78], he only considered the case where f is homogeneous, $W = \mathbb{A}_K^n$ (i.e. $\Phi_{W,v} = \mathbf{1}_{\mathcal{O}_v^n}$ for all finite places v). Further, there are some extra conditions. The first one is that f has an embedded resolution such that $v_i > N_i$ for all exceptional divisors E_i. In this case Igusa chose $\beta = 1$ and an arbitrary real number σ such that

$$\sigma < \min\{\frac{v_i}{N_i} \mid E_i \text{ is an exceptional divisor}\}.$$

The second one is that $\mathrm{ord}_v(z) \leq -1$. When $\mathrm{ord}_v(z) = -1$, the corresponding exponential sums become exponential sums over finite fields and we can apply the method of Deligne and Katz (see for example [Del77, Del74, Del80, Kat89]).

By Remark 3.1.6, the condition $\mathrm{ord}_v(z) \leq -2$ in Conjecture 3.3.2 is necessary.

3.3.3   Some Expected Results

Let $f \in \mathcal{O}_K[x_1, \ldots, x_n]$ be a homogeneous polynomial of degree $d \geq 2$. Suppose that Conjecture 3.3.2 holds for f and $W = \mathbb{A}^n_K$. Further, we suppose that

$$\liminf_{p_v \to +\infty} \sigma_{f, \mathbf{1}_{\mathcal{O}^n_v}} > 1.$$

It follows that f has only rational singularities (see [Clu19, Proposition 3.10]). As a consequence, if we denote by $-\alpha_f$ the biggest root of $(s+1)^{-1}b_f(s)$, then $\alpha_f > 1$.

If the strong monodromy conjecture (Conjecture 3.2.5) also holds for f, then we obtain an upper bound for α_f

$$\alpha_f \leq \liminf_{p_v \to +\infty} (\inf_{\Phi_v \in \mathcal{S}(K^n_v)} \sigma_{f, \Phi_v}). \tag{3.10}$$

A lower bound for this quantity was due to Mustaţă and Popa. In fact, their result holds for any field K of characteristic 0.

Proposition 3.3.5 (Mustaţă and Popa [Mus20]) *With the above notation, we have*

$$\alpha_f \geq \frac{\mathrm{codim}(C_f)}{d}.$$

On the other hand, we have to deal with the case where $\mathrm{ord}(z) = -1$. One of the key ingredients is to have good estimates of exponential sums over finite fields. In this case we have the following result due to Cluckers.

Proposition 3.3.6 (Cluckers [Clu08a]) *Recall that K is a number field and f is a homogeneous polynomial in $\mathcal{O}_K[x_1, \ldots, x_n]$ of degree $d \geq 2$. Then there exists a* constant $c > 0$ such that for all places v of K and all $z \in K_v$ with $\mathrm{ord}_v(z) = -1$, we have

$$|E_{K_v, \mathbf{1}_{\mathcal{O}_v^n}, z}(f)| \leq c |z|^{-\sigma_f}$$

where

$$\sigma_f = \liminf_{p_v \to +\infty} \left(\inf_{\Phi_v \in \mathcal{S}(K_v^n)} \sigma_{f, \Phi_v} \right).$$

As a consequence, we deduce the Poisson formula for f under some conditions.

Proposition 3.3.7 *Let K be a number field and f be a homogeneous polynomial in $\mathcal{O}_K[x_1, \ldots, x_n]$ of degree $d \geq 2$. Suppose that Conjectures 3.2.5 and 3.3.2 hold for f. If $\mathrm{codim}(C_f) \geq 3$ and $\alpha_f > 2$, then the Poisson formula holds for f.*
   In particular, if $\mathrm{codim}(C_f) \geq 2d + 1$, then the Poisson formula holds for f.

Proof The proof follows immediately from (3.10) and Propositions 3.3.1, 3.3.5, 3.3.6.   □

Remark 3.3.8 It is quite tempting to study Conjecture 3.3.2 for $\sigma = \alpha_f$.

3.4   Progress on Igusa's Conjecture

In what follows, let K be a number field with the ring of integers $\mathcal{O}_K$ and let $f \in \mathcal{O}_K[x_1, \ldots, x_n]$ be a non-constant polynomial in n variables. Recall that f is said to be a *form* of degree d if f is a homogeneous polynomial of degree d.

3.4.1   The Non-degenerate Case

Igusa proved his conjecture for strong non-degenerate forms, i.e. homogeneous polynomials with a unique critical point $\{0\}$.

Theorem 3.4.1 (Igusa [Igu78]) *Suppose that f is a form of degree d with $C_f = \{0\}$. Then there exists a positive constant c such that for all non-Archimedean places v of K and all $z \in K_v \setminus \mathcal{O}_v$,*

$$|E_{K_v, \mathbf{1}_{\mathcal{O}_v^n}, z}| \leq c |z|^{-\frac{n}{d}}.$$

In particular, if $n \geq 2d + 1$, then the Poisson formula holds for f.

Denef and Sperber investigated Conjecture 3.3.2 for non-degenerate polynomials (not necessarily homogeneous). We recall first the notion of non-degenerate polynomials.

Let k be a field and $\bar{k}$ be an algebraic closure of k. Let

$$f = f(0) + \sum_{i \in \mathbb{Z}_{\geq 0}^n} c_i x^i \in k[x_1, \ldots, x_n]$$

where we set $x := (x_1, \ldots, x_n)$ and $x^i := x_1^{i_1} \cdots x_n^{i_n}$ with $i = (i_1, \ldots, i_n)$. The Newton polyhedron of f at the origin is defined by

$$\Delta_0(f) = \text{Conv Supp } f + \mathbb{R}_{\geq 0}^n,$$

where $\text{Supp } f = \left\{ i \in \mathbb{Z}_{\geq 0}^n \,\middle|\, c_i \neq 0 \right\}$ denotes the support of f. For all non-empty faces $\tau \subseteq \Delta_0(f)$ of any dimension, ranging from vertices to $\Delta_0(f)$ itself, we write

$$f_\tau = \sum_{i \in \tau \cap \mathbb{Z}_{\geq 0}^n} c_i x^i.$$

We say that f is *non-degenerate* with respect to τ if the system of equations

$$\frac{\partial f_\tau}{\partial x_1} = \ldots = \frac{\partial f_\tau}{\partial x_n} = 0$$

has no solutions in $\bar{k}^{*n}$. It is equivalent to require that the map $\bar{k}^{*n} \to \bar{k}$ given by $\alpha \mapsto f_\tau(\alpha)$ has no critical values. We say that f is non-degenerate with respect to the faces of $\Delta_0(f)$ if it is non-degenerate with respect to all possible choices of τ.

Let $\sigma_{0,f}$ be the biggest real number t such that $(\frac{1}{t}, \ldots, \frac{1}{t}) \in \Delta_0(f)$ and $\beta_{0,f}$ be the codimension of the smallest face $\tau_0(f)$ of $\Delta_0(f)$ containing $(\frac{1}{\sigma_{0,f}}, \ldots, \frac{1}{\sigma_{0,f}})$. Denef and Sperber suggested that certain estimates of exponential sums modulo p^m of a non-degenerate polynomial can follow from those of exponential sums over finite fields. More precisely, they used the work of Adolphson-Sperber (see [Ado89]) on exponential sums over finite fields to obtain the first remarkable result after Igusa's work.

Theorem 3.4.2 (See [Den01]) *Suppose that f is non-degenerate with respect to the faces of its Newton polyhedron $\Delta_0(f)$ at the origin and that $\{0, 1\}^n \cap \tau_0(f) = \emptyset$. Then there exists a positive constant c which depends only on Δ_0 such that for all but finitely many non-Archimedean places v of K and all $z \in K_v \setminus \mathcal{O}_v$, we have*

$$|E_{K_v, 1_{\mathcal{M}_v^n}, z}| \leq c \, |\operatorname{ord}_v(z)|^{\beta_{0,f}-1} \, |z|^{-\sigma_{0,f}}. \tag{3.11}$$

Moreover, if f is homogeneous, then

$$|E_{K_v, 1_{\mathcal{O}_v^n}, z}| \leq c \, |\operatorname{ord}_v(z)|^{\beta_{0,f}-1} \, |z|^{-\sigma_{0,f}}. \tag{3.12}$$

Using the approach of Denef-Sperber, Cluckers replaced the work of Adonphson-Sperber by that of Katz (see [Kat99]) to obtain the same bound as in (3.12) (resp. (3.11)) but for non-degenerate quasi-homogeneous polynomials (resp. all non-degenerate polynomials) without the technical condition $\{0, 1\}^n \cap \tau_0(f) = \emptyset$ (see [Clu08b] and [Clu10]). Recently, Castryck and the author extended Cluckers' results to all non-degenerate polynomials under the condition $\mathrm{ord}_v(z) \leq -2$ (see [Cas19]).

3.4.2 Beyond the Non-degenerate Case

Conjecture 3.3.2 becomes more difficult if we remove non-degenerate conditions for f. Let us mention some results in this direction. On the one hand, Wright proved some results for quasi-homogeneous polynomials in two variables (see [Wri20]). Lichtin rediscovered the results of Wright by another method and extended them to homogeneous polynomials in three variables (see [Lic13], [Lic16]). On the other hand, Cluckers proved some results in the case where $\mathrm{ord}_v(z) = -1$ or $\mathrm{ord}_v(z) = -2$ (see [Clu08a]).

In [Clu16] Cluckers and Veys stated Conjecture 3.3.2 for polynomials f and the function $\mathbf{1}_{\mathcal{O}_v^n}$ (resp. $\mathbf{1}_{\mathcal{M}_v^n}$), $\tilde{\sigma}_f$ (resp. $\sigma = \mathrm{lct}_0(f)$) and $\beta = n$. Here $\mathrm{lct}_0(f)$ denotes the log-canonical threshold of f at 0 and $\tilde{\sigma}_f = \min\{\mathrm{lct}_b(f - f(b)) \mid b \in \mathbb{C}^n\}$. Recall that the log-canonical threshold $\mathrm{lct}_0(f)$ of f at 0 is defined to be the minimum over all the values $\dfrac{v_i}{N_i}$ as in Sect. 3.2.1 with $0 \in h(E_i)$. We refer the reader to [Mus12] for an introduction to log canonical thresholds. Theorem 3.2.2 and the definition of $\tilde{\sigma}_f$ imply

$$\tilde{\sigma}_f \leq \liminf_{p_v \to +\infty} \sigma_{f-a, \Phi_{W,v}} \tag{3.13}$$

for all $a \in \mathbb{C}$ and all choices of W as in the statement of Conjecture 3.3.2 with $f(W(\mathbb{C})) = a$. The above inequality (3.13) becomes an equality for a certain set W if f has non-rational singularities (see [Clu19, Proposition 3.10]). Hence the conjecture of Cluckers and Veys is sharp in case of non-rational singularities.

We mention some results toward the conjecture of Cluckers and Veys. Cluckers and Veys proved their conjecture for some small values of $|\mathrm{ord}_v(z)|$. In [Cha20] Chambille and the author proved this conjecture in the case where $\mathrm{lct}(f)$ (resp. $\mathrm{lct}_0(f)$) is at most $1/2$. Their proof suggested that Conjecture 3.3.2 may hold if we can prove it for each given value of $\mathrm{ord}_v(z)$.

Recently, Cluckers, Mustaţă and the author [Clu19] used a geometric method and proved that the conjecture of Cluckers and Veys holds for all non-constant polynomials f. Moreover, Conjecture 3.3.2 holds fully in the non-rational singularities case. Here are some ideas of the proof. They first gave a so-called power condition for resolutions of singularities to characterize the possible obstruction for Cluckers-Veys' conjecture. If the power condition holds, then they deduce an inequality associated to the numerical data of this resolution which allows to remove the above obstruction. One key ingredient is the existence of some models in the Minimal Model Program. We strongly believe that further developments of the Minimal Model Program could lead to the full resolution of Conjecture 3.3.2 in case of rational singularities. Finally, we mention that Veys obtained a proof of Conjecture 3.3.2 in case of polynomials in two variables in the same line with that of [Clu19]. But he did not use the technique from the Minimal Model Program (see [Vey20]).

To end this section, we state the main result of Cluckers-Mustaţă-Nguyen [Clu19].

Theorem 3.4.3 *Let K be a number field and $f \in \mathcal{O}_K[x_1, \ldots, x_n]$ be a non-constant polynomial, and W be any closed subscheme of $\mathbb{A}^n_{\mathcal{O}_K}$, then there exist $c > 0$ and $M > 0$ such that*

$$|E_{K_v, \Phi_{W,v}, z}| < c \, | \operatorname{ord}_v(z)|^{n-1} \, |z|^{-\tilde{\sigma}_{W,f}} \qquad (3.14)$$

for all finite places v of K with $p_v > M$ and all z with $\operatorname{ord}_v(z) \leq -2$, where $\tilde{\sigma}_{W,f} = \min\{\operatorname{lct}_b(f - f(b)) \mid b \in W\}$. Moreover, c can be chosen to be independent of the number field K containing the coefficients of f.

3.5   A Long History of the Local-Global Principle

3.5.1   The Local-Global Principle

One of the most important techniques in arithmetic geometry is the local-global principle (also known as the Hasse principle). This principle asserts that a certain property is true globally if and only if it is true everywhere locally. This principle reduces certain difficult problems in global fields to those in local fields in which we have more tools. The most famous example of the local-global principle is the Hasse-Minkowski theorem. Minkowski proved that a quadratic form over $\mathbb{Q}$ represents 0 if and only if it represents 0 in any local field containing $\mathbb{Q}$. Hasse generalized Minkowski's theorem to number fields. In fact, the local-global principle for quadratic forms holds for all global fields.

For forms of higher degree (i.e homogeneous polynomials of degree at least 3), the local-global principle does not hold in general and many counterexamples were already constructed (see for example [Mor37, Sel51]). So the question for forms of higher degree is:

How can one characterize forms for which the local-global principle holds?

3.5.2 Progress on the Local-Global Principle

In the case where f is a cubic form in n variables over $\mathbb{Q}$, it is conjectured that f has a non-trivial rational zero as soon as $n \geq 10$. Using the Hardy-Littlewood circle method, Davenport showed in [Dav63] that a cubic form over $\mathbb{Q}$ in at least 16 variables represents 0, so the local-global principle holds trivially in this case. Heath-Brown improved the result of Davenport to cubic forms in at least 14 variables (see [Hea07]) and non-singular cubic forms in at least 10 variables (see [Hea84]). Davenport also proved that cubic forms in at least 10 variables over $\mathbb{Q}$ represent 0 in all p-adic fields (see [Dav05]). Moreover, it is clear that a cubic form over $\mathbb{Q}$ has a non-trivial solution in $\mathbb{R}$. Hence we may ask whether it is possible to remove the non-singular condition in Heath-Brown's work.

There are also results for cubic forms in fewer variables. Hooley proved in [Hoo88] that the local-global principle holds for non-singular cubic forms in at least 9 variables. Recently, Hooley showed that under the validity of the Riemann hypothesis for certain Hasse-Weil L-functions, the local-global principle holds for all non-singular forms in 8 variables (see [Hoo14]). In another approach, Manin suggested that the obstruction of the local-global principle for cubic forms may be explained by the theory of Brauer groups (the so-called Brauer-Manin obstruction) but Skorobogatov showed that the Brauer-Manin obstruction cannot fully explain the failure of the local-global principle in the general case (see [Sko99]). Further, such an obstruction is known to be empty for non-singular cubic forms in at least 5 variables.

For forms of arbitrary degree, by generalizing the method of Davenport, Birch showed in [Bir61] that a form f of degree $d > 2$ in n variables over $\mathbb{Q}$ represents 0 if $f^{-1}(0)$ has a non-singular point over all local fields containing $\mathbb{Q}$ and $n - \dim(C_f) \geq (d-1)2^d$. Recently, Browning and Prendiville improved the second condition of Birch to $n - \dim(C_f) \geq (d - \frac{1}{2}\sqrt{d})2^d$ (see [Bro17b]). In the case where $C_f = \{0\}$, Browning and Heath-Brown conjectured that the local-global principle holds for a form f of degree d in n variables if $n \geq 2d + 1$ (see [Bro17a]). We will see below that this conjecture agrees with the prediction of Igusa. On the other hand, a remarkable result of Birch in [Bir57] stated that for each odd integer $d \geq 1$, there exists a positive integer $N(d)$ such that all forms of degree d in n variables with $n > N(d)$ represent 0. It follows that the local-global principle holds trivially if $n > N(d)$. However, to our knowledge, we do not know any quantitative results in this direction.

We now review basic ideas of the Hardy-Littlewood circle method. Let f be a homogeneous polynomial of degree $d > 1$ in $\mathbb{Z}[x_1, \ldots, x_n]$. Let $\omega : \mathbb{R}^n \to [0, +\infty)$ be a suitable weight function. Our goal is to obtain an asymptotic formula of the function

$$N_\omega(f, B) = \sum_{x \in \mathbb{Z}^n, f(x)=0} \omega(x/B)$$

when $B \to +\infty$. Let us use the identity

$$N_\omega(f, B) = \int_{\mathbb{T}} S(\alpha, B)d\alpha \qquad (3.15)$$

where $\mathbb{T} = \mathbb{R}/\mathbb{Z}$ and

$$S(\alpha, B) = \sum_{x \in \mathbb{Z}^n} \omega(x/B)e^{2\pi i \alpha f(x)}$$

if ω has certain good analytic properties. The Hardy-Littlewood circle method consists of dividing the torus $\mathbb{T}$ into major arcs $\mathfrak{M}$ and minor arcs $\mathfrak{m}$ where for each $\delta > 0$, we set

$$\mathfrak{M}(\delta) := \cup_{q \le B^\delta} \cup_{0 \le a \le q, (a,q)=1} \{\alpha \in \mathbb{T} \mid |\alpha - \frac{a}{q}| \le B^{\delta-d}\}$$

and

$$\mathfrak{m}(\delta) := \mathbb{T} \setminus \mathfrak{M}(\delta).$$

Note that if $3\delta < d$, then $\mathfrak{M}(\delta)$ is in fact a disjoint union of the above arcs provided B is sufficiently large.

To investigate the local-global principle for f, we would like to obtain the following asymptotic formulas

$$\int_{\mathfrak{M}} S(\alpha, B)d\alpha \sim c_f B^{n-d} \qquad (3.16)$$

and

$$\int_{\mathfrak{m}} S(\alpha, B)d\alpha = o(B^{n-d}) \qquad (3.17)$$

where the constant c_f is positive under some good conditions on f and such that f has a smooth solution over every completion of $\mathbb{Q}$ (i.e. f admits a non-singular point of $f^{-1}(0)$ over every completion of $\mathbb{Q}$).

A common way to work with Eq. (3.17) is to use Weyl's bound for $S(\alpha, B)$ and Dirichlet's approximation theorem to control minor arcs (see [Bir61, Bro17b] for more details). Equation (3.17) is in fact very hard to achieve. But the conjecture on exponential sums modulo p^m could improve Eq. (3.16). More precisely, Eq. (3.16) is related to the convergence of certain singular series given by (see [Bir61, Bro17b])

$$\mathfrak{S} = \sum_{1 \le N} N^{-n} \sum_{a \in (\mathbb{Z}/N\mathbb{Z})^*} S_N(a)$$

where

$$S_N(a) = \sum_{y \in (\mathbb{Z}/N\mathbb{Z})^n} e^{\frac{2\pi i a f(y)}{N}}.$$

With the assumption of Remark 3.1.7 we would have

$$|S_N(a)| \leq CN^{-\sigma}$$

for a positive constant C and all $N \geq 1$. A direct calculation implies that $\mathfrak{S}$ converges absolutely for $\sigma > 2$.

3.5.3 Igusa's Approach

Now we sketch another approach given by Igusa to attack the above problem (see [Igu78] and [Har80]). We first recall the idea of Weil on quadratic forms. From the work of Siegel on quadratic forms, Weil gave a general formula called Siegel's formula (see [Wei65]) which relates a theta series to an Eisenstein series. As a consequence, the Hasse-Minkowski theorem follows from Siegel's formula. For forms of higher degree, Igusa expected that we could derive a similar formula and use it to prove the local-global principle for these forms. Inspired by the work of Weil such a formula of Siegel type would follow from a Poisson formula. In fact, Igusa succeeded in proving the following assertion:

For forms of higher degree, if we have a good uniform bound in p and m of exponential sums modulo p^m, then we have a Poisson formula.

Hence the strategy of Igusa breaks into two parts. The first part is to find a good uniform bound in p and m of exponential sums modulo p^m so that we could deduce a Poisson formula. This is exactly the material presented in Sects. 3.2, 3.3 and 3.4. The second part is to use the Poisson formula to derive formulas of Siegel type and then the desired local-global principle.

Let us explain a little bit more about formulas of Siegel type. A formula of Siegel type is an equality between Eisenstein-Siegel series and the integral of a theta series in the space of tempered distributions. In the case of quadratic forms, Weil introduced the notion of metaplectic groups and used their action on the space of Schwartz-Bruhat functions $\mathcal{S}(X_A)$ to construct a theta series and compared its integral with Eisenstein-Siegel series. For forms of higher degree, Igusa pointed out that a good theory of metaplectic groups associated with these forms would be very useful although such a theory is not yet known. But he also remarked that we could use a certain smaller group to obtain similar results. More precisely, let K be a number field and let f be a non-singular form of degree d in n variables with coefficients in $\mathcal{O}_K$. Igusa introduced the group $P = \mathbb{G}_a \times \mathbb{G}_m$ equipped with the law

$$(u, t)(u', t') = (u + t^d u', tt')$$

The action of P_A on $\mathcal{S}(X_A)$ is given by

$$((u, t)(\Phi))(x) = |t|_A^{\frac{n}{2}} \Psi(u f(x)) \Phi(tx)$$

where $|t|_A = \prod_v |t_v|_v$ is the usual norm of t. We consider the tempered distributions E and I_0 given by

$$E(\Phi) = \Phi(0) + \sum_{z \in K} \Psi(z f(x))(\Phi)$$

and

$$I_0(\Phi) = \sum_{\xi \in X_K} \Phi(\xi).$$

If $n \geq 2d + 1$, then Igusa showed that the Poisson formula holds for f (see Theorem 3.4.1). In particular, if $|t|_A > 1$, he proved in [Igu76] that

$$(I_0 - E)((u, t)(\Phi)) = O(|t|_A^{1 - \frac{n}{2d}}) \tag{3.18}$$

as $|t|_A \to +\infty$ and furthermore, if $|t|_A < 1$ but $(u + z)t^{-d}$ remains in a compact subset of K_A for some $z \in K$, then

$$(I_0 - E)((u, t)(\Phi)) = O(|t|_A^{\frac{n}{2d} - 1}) \tag{3.19}$$

as $|t|_A \to 0$. Igusa conjectured that (3.19) is still true without the compactness assumption (or at least we could find some conditions of n and d such that (3.19) holds without compactness). In particular, this conjecture would imply the local-global principle for f (see [Igu76, Har80]). To summarize, under the validity of Igusa's approach, we could prove that the local-global principle holds for any non-singular form of degree d in at least $2d + 1$ variables. This agrees with the conjecture of Browning and Heath-Brown that we mentioned earlier.

Unfortunately, we are in a similar situation as that of (3.17). To our knowledge, (3.19) is out of reach. Even it is not clear that there is a connection between them by looking at the adelic circle method (see [Lac82, Mar73]). Both of them would require a lot of efforts and many new ideas but we can always hope that Igusa's ideas could be realized in the future.

We end this survey with the case of singular forms. If f is a singular form of degree $d \geq 3$, from an observation of Igusa on the work of Birch, the Poisson formula also holds for f if $\text{codim}(C_f) \geq (d-1)2^d$. Moreover, Birch showed that this condition is sufficient to prove the local-global principle for f as we mentioned above. In Proposition 3.3.7, we predicted that the Poisson formula holds for f if $\text{codim}(C_f) \geq 2d+1$. Hence it is tempting to ask whether we could replace the sufficient condition $\text{codim}(C_f) \geq (d-1)2^d$ in the result of Birch by $\text{codim}(C_f) \geq 2d+1$.

Acknowledgments The author is partially supported by the Vietnam Institute for Advanced Study in Mathematics (VIASM) and the Fund for Scientific Research—Flanders (Belgium) (FWO) 12X3519N. The author would like to thank Raf Cluckers, Victoria Cantoral Farfán, Lukas Prader, Le Quy Thuong, Ngo Dac Tuan and the anonymous referees for many useful comments.

References

[Ado89] A. Adolphson, S. Sperber, Exponential sums and Newton polyhedra: cohomology and estimates. Ann. Math. (2) **130**, 367–406 (1989)

[Ber72] I.N. Bernšteĭn, Analytic continuation of generalized functions with respect to a parameter. Funct. Anal. Appl. **6**, 273–285 (1972)

[Bir57] B.J. Birch, Homogeneous forms of odd degree in a large number of variables. Mathematika **4**, 102–105 (1957)

[Bir61] B.J. Birch, Forms in many variables. Proc. Roy. Soc. Ser. A. **265**, 245–263 (1961/1962)

[Bro17a] T.D. Browning, D.R. Heath-Brown, Forms in many variables and differing degrees. J. Eur. Math. Soc. **19**, 357–394 (2017)

[Bro17b] T.D. Browning, S.M. Prendiville, Improvements in Birch's theorem on forms in many variables. J. Reine Angew. Math. **731**, 203–234 (2017)

[Cas19] W. Castryck, K.H. Nguyen, New bounds for exponential sums with a non-degenerate phase polynomial. J. Math. Pures Appl. (9) **130**, 93–111 (2019)

[Cha20] S. Chambille, K.H. Nguyen, Proof of the Cluckers-Veys conjecture on exponential sums for polynomials with log-canonical threshold at most a half. Int. Math. Res. Not. (to appear) https://doi-org.eres.qnl.qa/10.1093/imrn/rnz036

[Clu08a] R. Cluckers, The modulo p and p^2 cases of the Igusa Conjecture on exponential sums and the motivic oscillation index. Int. Math. Res. Not. **2008**, 20 pp. (2008)

[Clu08b] R. Cluckers, Igusa and Denef-Sperber conjectures on nondegenerate p-adic exponential sums. Duke Math. J. **141**(1), 205–216 (2008)

[Clu10] R. Cluckers, Exponential sums: questions by Denef, Sperber, and Igusa. Trans. Am. Math. Soc. **362**(7), 3745–3756 (2010)

[Clu19] R. Cluckers, M. Mustaţă, K.H. Nguyen, Igusa's conjecture for exponential sums: optimal estimates for non-rational singularities. Forum Math. Pi **7**, e3, 28 pp. (2019)

[Clu16] R. Cluckers, W. Veys, Bounds for p-adic exponential sums and log-canonical thresholds. Am. J. Math. **138**(1), 61–80 (2016)

[Coc99] T. Cochrane, Z. Zheng, Pure and mixed exponential sums. Acta Arith. **91**(3), 249–278 (1999)

[Dav63] H. Davenport, Cubic forms in sixteen variables. Proc. R. Soc. Ser. A **272**, 285–303 (1963)

[Dav05] H. Davenport, *Analytic methods for Diophantine equations and Diophantine inequalities* (Cambridge University Press, Cambridge, 2005)

[Del77] P. Deligne, *Cohomologie étale*. Lecture Notes in Mathematics, vol. 569. Séminaire de géométrie algébrique du Bois-Marie SGA $4\frac{1}{2}$ (1977)

[Del74] P. Deligne, La conjecture de Weil I. Inst. Hautes études Sci. Publ. Math. **43**, 273–307 (1974)

[Del80] P. Deligne, La conjecture de Weil II. Inst. Hautes études Sci. Publ. Math. **52**, 137–252 (1980)

[Den87] J. Denef, On the degree of Igusa's local zeta function. Am. J. Math. **109**(6), 991–1008 (1987)

[Den91] J. Denef, Local zeta functions and Euler characteristics. Duke Math. J. **63**(3), 713–721 (1991)

[Den91] J. Denef, Report on Igusa's local zeta function. Séminaire Bourbaki 1990/91 Exp. no. 741 (1991), pp. 359–386

[Den01] J. Denef, S. Sperber, Exponential sums mod p^n and Newton polyhedra. Bull. Belg. Math. Soc.–Simon Stevin (suppl.), 55–63 (2001)

[Har80] S.J. Haris, Number theoretical developments arising from the Siegel formula. Bull. Am. Math. Soc. **2**(3), 417–433 (1980)

[Hea84] D.R. Heath-Brown, Cubic forms in 10 variables. Lect. Notes Math. **1068**, 104–108 (1984)

[Hea07] D.R. Heath-Brown, Cubic forms in 14 variables. Invent. Math. **170**, 199–230 (2007)

[Hir64] H. Hironaka, Resolution of singularities of an algebraic variety over a field of characteristic zero. I. Ann. Math. (2) **79**, 109–203 (1964)

[Hoo88] C. Hooley, On nonary cubic forms. J. Reine Angew. Math. **386**, 32–98 (1988)

[Hoo14] C. Hooley, On octonary cubic forms. Proc. Lond. Math. Soc. (3) **109**, 241–281 (2014)

[Igu74] J.I. Igusa, Complex powers and asymptotic expansions I. J. Reine Angew. Math. **268/269**, 110–130 (1974); Ibid. *II*, **278/279**, 307–321 (1975)

[Igu76] J.I. Igusa, A Poisson formula and exponential sums. J. Fac. Sci. Univ. Tokyo Sect. IA Math. **23**, 223–244 (1976)

[Igu78] J.I. Igusa, *Lectures on Forms of Higher Degree (Notes by S. Raghavan)*. Tata Institute of Fundamental Research, Lectures on Mathematics and Physics, vol. 59 (Springer, Heidelberg/New York/Berlin, 1978)

[Ire90] K. Ireland, M. Rosen, *A Classical Introduction to Modern Number Theory*. Graduate Texts in Mathematics, 2nd edn., vol. 84 (Springer, New York, 1990), xiv+389 pp.

[Kas76] M. Kashiwara, B-functions and holonomic systems. Rationality of roots of B-functions. Invent. Math. (2) **38**, 33–53 (1976)

[Kat85] N.M. Katz, G. Laumon, Transformation de Fourier et majoration de sommes exponentielles. Inst. Hautes études Sci. Publ. Math. **62**, 361–418 (1985)

[Kat89] N.M. Katz, Perversity and exponential sums. Algebraic number theory. Adv. Stud. Pure Math. **17**, 209–259 (1989)

[Kat99] N.M. Katz, Estimates for "singular" exponential sums. Int. Math. Res. Not. **1999**(16), 875–899 (1999)

[Lac82] G. Lachaud, Une présentation adélique de la série singulière et du problème de Waring. Enseign. Math. (2) **28**, 139–169 (1982)

[Lic13] B. Lichtin, On a conjecture of Igusa. Mathematika **59**, 399–425 (2013)

[Lic16] B. Lichtin, On a conjecture of Igusa II. Am. J. Math. **138**, 201–249 (2016)

[Loe88] F. Loeser, Fonctions d'Igusa p-adiques et polynômes de Bernstein. Am. J. Math. **110**(1), 1–21 (1988)

[Mal83] B. Malgrange, Polynômes de Bernstein-Sato et cohomologie évanescente. Astérisque **101**, 243–267 (1983)

[Mar73] J.G.M. Mars, Sur l'approximation du nombre de solutions de certaines équations diophantiennes. Ann. Sci. École Norm. Sup. (4) **6**, 357–387 (1973)

[Mil68] J. Milnor, *Singular Points of Complex Hypersurfaces*. Annals of Mathematics Studies, vol. 61 (Princeton University Press, Princeton, NJ; University of Tokyo Press, Tokyo, 1968)

[Mor37] L. J. Mordell, A remark on indeterminate equations in several variables. J. Lond. Math. Soc. **12**, 127–129 (1937)

[Mus12] M. Mustaţă, IMPANGA lecture notes on log canonical thresholds. Notes by Tomasz Szemberg. EMS Ser. Congr. Rep., Contributions to algebraic geometry. Eur. Math. Soc., Zürich (2012) pp. 407–442

[Mus20] M. Mustaţă, M. Popa, Hodge ideals for Q-divisors, V-filtration, and minimal exponent. Forum Math. Sigma **8**, e19, 41 pp. (2020)

[Poi20a] J. Poineau, D. Turchetti, Berkovich curves and Schottky uniformization I: The Berkovich affine line, in *Arithmetic and Geometry Over Local Fields*, ed. by B. Anglès, T. Ngo Dac (Springer International Publishing, Cham, 2020)

[Poi20b] J. Poineau, D. Turchetti, Berkovich curves and Schottky uniformization II. Analytic uniformization of Mumford curves, in *Arithmetic and Geometry Over Local Fields*, ed. by B. Anglès, T. Ngo Dac (Springer International Publishing, Cham, 2020)

[Sel51] E.S. Selmer, The Diophantine equation $ax^3 + by^3 + cz^3 = 0$. Acta Math. **85**, 203–362 (1951)

[Shi55] G. Shimura, Reduction of algebraic varieties with respect to a discrete valuation of the basis field. Am. J. Math. **77**(1), 134–176 (1955)

[Sko99] A.N. Skorobogatov, Beyond the Manin obstruction. Invent. Math. **135**(2), 399–424 (1999)

[Vey20] W. Veys, On the log canonical threshold and numerical data of a resolution in dimension 2. Manuscr. Math. **163**, 1–11 (2020)

[Wei48] A. Weil, On some exponential sums. Proc. N. A. S. **34**(5), 204–207 (1948)

[Wei65] A. Weil, Sur la formule de Siegel dans la théorie des groupes classiques. Acta Math. **113**, 1–87 (1965)

[Wri20] J. Wright, On the Igusa conjecture in two dimensions. Am. J. Math. **142**(4), 1193–1238 (2020)

Chapter 4
From the Carlitz Exponential to Drinfeld Modular Forms

Federico Pellarin

Abstract This paper contains the written notes of a course the author gave at the VIASM of Hanoi in the Summer 2018. It provides an elementary introduction to the analytic naive theory of Drinfeld modular forms for the simplest 'Drinfeld modular group' $GL_2(\mathbb{F}_q[\theta])$ also providing some perspectives of development, notably in the direction of the theory of vector modular forms with values in certain ultrametric Banach algebras.

4.1 Introduction

The present paper contains the written notes of a course the author gave at the VIASM of Hanoi in the Summer 2018. It provides an elementary introduction to the analytic naive theory of Drinfeld modular forms essentially for the simplest 'Drinfeld modular group' $GL_2(\mathbb{F}_q[\theta])$ also providing some perspectives of development, notably in the direction of the theory of vector modular forms with values in certain ultrametric Banach algebras initiated in [Pel12].

The course was also the occasion to introduce the very first basic elements of the arithmetic theory of Drinfeld modules in a way suitable to sensitize the attendance also to more familiar processes of the classical theory of modular forms and elliptic curves. Most parts of this work are not new and are therefore essentially covered by many other texts and treatises such as the seminal works of Goss [Gos80a, Gos80b, Gos80c] and Gekeler [Gek88]. The present text also has interaction and potential developments along with the contributions to this volume by Poineau-Turchetti and Tavares Ribeiro [Poi20a, Poi20b, Tav20]. It also contains suggestions for further developments, see Problems 4.4.10, 4.4.15, 4.6.5, 4.8.4 and 4.8.9.

This paper will not cover several advanced recent works such that the higher rank theory, including the delicate compactification questions in the path of Basson,

F. Pellarin (✉)
Institut Camille Jordan, UMR 5208, Site de Saint-Etienne, Saint-Etienne, France
e-mail: federico.pellarin@univ-st-etienne.fr

© The Author(s), under exclusive license to Springer Nature Switzerland AG 2021   93
B. Anglès, T. Ngo Dac (eds.), *Arithmetic and Geometry over Local Fields*,
Lecture Notes in Mathematics 2275, https://doi.org/10.1007/978-3-030-66249-3_4

Breuer, Pink [Bas18a, Bas18b, Bas18c], Gekeler [Gek17, Gek19a, Gek18, Gek19b] and it does not even go in the direction of the important arithmetic explorations notably involving the cohomological theory of crystals by Böckle [Boc02, Boc15] or toward several other crucial recent works by several other authors we do not mention here, at once inviting the reader to realise a personal bibliographical research to determine the most recent active areas.

Perhaps, one of the original points of our contribution is instead to consider *exponential functions* from various viewpoints, all along the text, stressing how they interlace with modular forms. The paper describes, for example, a product expansion of the exponential function associated to the lattice $A := \mathbb{F}_q[\theta]$ in the Ore algebra of non-commutative formal series in the Frobenius automorphism which is implicit in Carlitz's work [Car35]. It will be used to give a rather precise description of the analytic structure of the cusp of $\Gamma = GL_2(A)$ acting on the Drinfeld upper-half plane by homographies. We will also use it in connection with local class field theory for the local field $K_\infty = \mathbb{F}_q((\frac{1}{\theta}))$. Another new feature is that, in the last two sections, we explore structures which at the moment have no known analogue in the classical complex setting. Namely, Drinfeld modular forms with values in modules over Tate algebras, following the ideas of [Pel12].

Here is, more specifically, the plan of the paper. In the very elementary Sect. 4.2 the reader familiarises with the rings and the fields which carry the values of the special functions we are going to study in this paper. Instead of the field of complex numbers $\mathbb{C}$, our 'target' field is a complete, algebraically closed field of characteristic $p > 0$. There is an interesting parallel with the classical complex theory where we have the quadratic extension $\mathbb{C}/\mathbb{R}$ and the quotient group $\mathbb{R}/\mathbb{Z}$ is compact, but there are also interesting differences to take into account as the analogue $\mathbb{C}_\infty/K_\infty$ of the extension $\mathbb{C}/\mathbb{R}$ is infinite dimensional, $\mathbb{C}_\infty$ is not locally compact, although the analogue $A := \mathbb{F}_q[\theta]$ of $\mathbb{Z}$ is discrete and co-compact in the analogue $K_\infty = \mathbb{F}_q((\frac{1}{\theta}))$ of $\mathbb{R}$.

We dedicate the whole Sect. 4.3 to exponential functions. More precisely, we give a proof of the correspondence by Drinfeld between *A-lattices* of $\mathbb{C}_\infty$ and *Drinfeld A-modules*. To show that to any Drinfeld module we can naturally associate a lattice we pass by the more general Anderson modules. We introduce Anderson's modules in an intuitive way, privileging one of the most important and useful properties, namely that they are equipped with an exponential function at a very general level. Just like abelian varieties, Anderson modules can be of any dimension. When the dimension is one, one speaks about Drinfeld modules.

In Sect. 4.4 we focus on a particular case of Drinfeld module: the Carlitz module. This is the analogue of the multiplicative group in this theory. We give a detailed account of the main properties of its exponential function denoted by $\exp_C$. We point out that its (multiplicative, rescaled) inverse u is used as uniformiser at infinity to define the analogue of the classical complex 'q-expansions' for our modular forms. In this section we prove, for example, that any generator of the lattice of periods of $\exp_C$ can be expressed by means of a certain convergent product expansion (known to Anderson). To do this, we use the so-called *omega function* of Anderson and Thakur.

In Sect. 4.5 we first study the Drinfeld 'half-plane' $\Omega = \mathbb{C}_\infty \setminus K_\infty$ topologically. We use, to do this, a fundamental notion of distance from the analogue of the real line K_∞. The group $GL_2(A)$ acts on Ω by homographies and we construct a fundamental domain for this action. After a short invitation to the basic notions of rigid analytic geometry, we describe the Bruhat-Tits tree of Ω, the natural action of $GL_2(K_\infty)$ on it, and, after a glimpse on Schottky groups (see [Poi20a, Poi20b] for a more in-depth development), we construct a reasonable analogue of a fundamental domain for the homographic action of $GL_2(\mathbb{F}_q[\theta])$ on Ω.

In Sect. 4.6 we discuss the following question: *find an analogue for the Carlitz module of the following statement: Every holomorphic function which is invariant for the translation by one has a Fourier series.* The answer is: every $\mathbb{F}_q[\theta]$-translation invariant function has a 'u-expansion'. We show why in this section. To do this we introduce the problem of rigid analytic structures on quotient spaces. We mainly focus on the example of the quotient of the rigid affine line $\mathbb{A}_{\mathbb{C}_\infty}^{1,an}$ by the group of translations by the elements of A. The reader will notice how hard things can become without the use of the tool of the analytification functor, also discussed in this section.

In Sect. 4.7 we give a quick account of (scalar) Drinfeld modular forms for the group $GL_2(A)$ (characterised by the u-expansion in $\mathbb{C}_\infty[[u]]$). This appears already in many other references: the main feature is that $\mathbb{C}_\infty$-vector spaces of Drinfeld modular forms are finitely dimensional spaces. Also, non-zero Eisenstein series can be constructed; this was first observed by D. Goss in [Gos80b]. The coefficients of the u-expansions of Eisenstein series are, after normalisation, in $A = \mathbb{F}_q[\theta]$.

The paper also has advanced, non-foundational parts. In Sect. 4.4.3 we apply the developed knowledge of the Carlitz exponential function to give an explicit description of local class field theory for the field K_∞; this subsection is also independent from the rest of the paper. In Sects. 4.8 and 4.9 we revisit Drinfeld modular forms. We introduce vector Drinfeld modular forms with values in other fields and algebras, following [Pel12]; the case we are interested in is that of functions which take values in finite dimensional $\mathbb{K}$-vector spaces where $\mathbb{K}$ is the completion for the Gauss norm of the field of rational functions in a finite set of variables with coefficients in $\mathbb{C}_\infty$. With the use of certain Jacobi-like functions, we deduce an identity relating a matrix-valued Eisenstein series of weight one with certain weak modular forms of weight -1 from which one easily deduces [Pel12, Theorem 8] in a different, more straightforward way.

4.2 Rings and Fields

Before entering the essence of the topic, we first propose the reader to familiarise with certain rings and fields, notably local fields and non-archimedean valued fields. For more about these topics read, for example, the books [Cas86, Ser80b]. The reader must notice that the basic notations of the three other chapters of this volume [Poi20a, Poi20b, Tav20] differ from ours.

Let R be a ring.

Definition 4.2.1 A *real valuation* $|\cdot|$ (or simply a *valuation*) over R is a map $R \xrightarrow{|\cdot|} \mathbb{R}_{\geq 0}$ with the following properties.

(1) For $x \in R$, $|x| = 0$ if and only if $x = 0$.
(2) For $x, y \in R$, $|xy| = |x||y|$.
(3) For $x, y \in R$ we have $|x + y| \leq \max\{|x|, |y|\}$ and if $|x| \neq |y|$, then $|x + y| = \max\{|x|, |y|\}$.

The inequality $|x + y| \leq \max\{|x|, |y|\}$ is usually called the *ultrametric inequality* (the term 'ultrametric' indicates a reinforced triangular inequality). A ring with valuation is called a *valued ring*. A valuation is *non-trivial* if its image is infinite. If the image of a valuation is finite, then it is equal to the set of two elements $\{0, 1\} \subset \mathbb{R}_{\geq 0}$ and all the non-zero elements of R are sent to 1 while 0 is sent to 0. This is the *trivial valuation* of R. A map as above satisfying (2), (3) but not (1) is called a *semi-valuation*.

A valuation over a ring R induces a metric in an obvious way and one easily sees that R, together with this metric, is totally disconnected (the only connected subsets are $\emptyset$ and the points). To any valued ring $(R, |\cdot|)$ we can associate the subset $\mathcal{O}_R = \{x \in R : |x| \leq 1\}$ which is a subring of R, called the *valuation ring* of $|\cdot|$. This ring has the prime ideal $\mathcal{M}_R = \{x \in R : |x| < 1\}$. The quotient ring $k_R := \mathcal{O}_R/\mathcal{M}_R$ is called the *residue ring*. The ring homomorphism $f \in \mathcal{O}_R \mapsto \overline{f} + \mathcal{M}_R \in \mathcal{O}_R/\mathcal{M}_R$ is called the *reduction map*. With R a ring, we denote by $R^\times$ the multiplicative group of invertible elements. The image $|R^\times| = \{|x| : x \in R^\times\}$ is a subgroup of $\mathbb{R}^\times$ called the *valuation group*.

If R is a field, $\mathcal{M}_R$ is a maximal ideal. Two valuations $|\cdot|$ and $|\cdot|'$ over a ring R are *equivalent* if for all $x \in R$, $c_1|x| \leq |x|' \leq c_2|x|$ for some $c_1, c_2 > 0$. Two equivalent valuations induce the same topology. If $(R, |\cdot|)$ is a valued ring, we denote by $\widehat{R}$ (or $\widehat{R}_{|\cdot|}$) the topological space completion of R for $|\cdot|$. It is a ring and if additionally R is a field, $\widehat{R}$ is also a field.

While working over complete valued fields, many properties which are usually quite delicate to check for real numbers, have simple analogues in this context. For instance, the reader can check that in a valued field $(L, |\cdot|)$, a sequence $(x_n)_{n \geq 0}$ is Cauchy if and only if $(x_{n+1} - x_n)_{n \geq 0}$ tends to zero. A series $\sum_{n \geq 0} x_n$ converges if and only if $x_n \to 0$ and an infinite product $\prod_{n \geq 0}(1 + x_n)$ converges if and only if $x_n \to 0$. Another immediate property is that if $(x_n)_{n \geq 0}$ is convergent, then $(|x_n|)_{n \geq 0}$ is ultimately constant.

4.2.1 Local Compactness, Local Fields

Let $(L, |\cdot|)$ be a valued field. Choose $r \in |L^\times|$ and $x \in L$. We set

$$D_L(x, r) = \{y \in L : |x - y| \leq r\}.$$

This is the *disk of center x and radius r*. Some authors like to call r the *diameter* to stress the fact that the metric induced by the valuation makes every point of $D_L(x, r)$ into a center so that it does not really distinguishes between 'radius' and 'diameter'.

Observe that $\mathcal{O}_L = D_L(0, 1)$. Also,

$$\mathcal{M}_L = \bigcup_{\substack{r \in |L^\times| \\ r < 1}} D_L(0, r) =: D_L^\circ(0, 1).$$

More generally we write $D_L^\circ(0, r) = \{x \in L : |x| < r\}$. We use the simpler notation $D(x, r)$ or $D^\circ(0, r)$ when L is understood from the context. Note that $D(x, r) = x + D(0, r)$ and $D(0, r)$ is an additive group. If $|x| \le r$ (that is, $x \in D(0, r)$), then $D(x, r) = D(0, r)$. If $|x| > r$ (that is, $x \notin D(0, r)$), then $D(x, r) \cap D(0, r) = \emptyset$. In other words, if two disks with same radii have a common point, then they are equal. If the radii are not equal, non-empty intersection implies that one is contained in the other.

Now pick $r \in |L^\times|$ and $x_0 \in L^\times$ with $|x_0| = r$. Then, $D(0, r) = x_0 D(0, 1) = x_0 \mathcal{O}_L$. This means that all disks are homeomorphic to $\mathcal{O}_L = D(0, 1)$. This is due to the fact that we are choosing $r \in |L^\times|$.

A complete valued field L is *locally compact* if every disk is compact. We have the following:

Lemma 4.2.2 *A valued field which is complete is locally compact if and only if the valuation group is discrete and the residue field is finite.*

Proof Let L be a field with valuation $|\cdot|$, complete. We first show that $\mathcal{O}_L = D(0, 1)$ is compact if the valuation group is discrete (in this case there exists $r \in]0, 1[\cap |L^\times|$ such that $\mathcal{M}_L = D(0, r)$) and the residue field is finite. Let B be any infinite subset of $\mathcal{O}_L$. We choose a complete set of representatives $\mathcal{R}$ of $\mathcal{O}_L$ modulo $\mathcal{M}_L$. Note the disjoint union

$$\mathcal{O}_L = \bigsqcup_{v \in \mathcal{R}} (v + \mathcal{M}_L).$$

Multiplying all elements of B by an element of $L^\times$ (rescaling), we can suppose that there exists $b_1 \in B$ with $|b_1| = 1$. Then, the above decomposition induces a partition of B and by the fact that k_L is finite and the box principle there is an infinite subset $B_1 \subset B \cap (b_1 + \mathcal{M}_L^{n_1})$ for some integer $n_1 > 0$. We continue in this way and we are led to a sequence $b_1, b_2, \ldots$ in B with $b_{i+1} \in \mathcal{M}_L^{n_i} \setminus \mathcal{M}_L^{n_i - 1}$ with the sequence of the integers n_i which is strictly increasing (set $n_0 = 0$). Hence, $b_{m+1} - b_m \in \mathcal{M}_L^{n_m}$ is a Cauchy sequence, thus converging in L because it is complete.

Let us suppose that k_L is infinite. Then any set of representatives $\mathcal{R}$ of $\mathcal{O}_L$ modulo $\mathcal{M}_L$ is infinite. For all $b, b' \in \mathcal{R}$ distinct, we have $|b - b'| = 1$ and $\mathcal{R}$ has no converging infinite sub-sequence. Let us suppose that the valuation group $G = |L^\times|$ is dense in $\mathbb{R}_{>0}$. There is a strictly decreasing sequence $(r_i)_i \subset G$ with $r_i \to 1$. This means that for all i, there exists $a_i \in \mathcal{O}_L$ such that $|a_i| = r_i$ and for all $i \ne j$ we have that $|a_i - a_j| = \max\{r_i, r_j\}$ so that we cannot extract from $(a_i)_i$ a convergent sequence and $\mathcal{O}_L$ is not compact.   □

Definition 4.2.3 A valued field which is locally compact is called a *local field*.

Note that $\mathbb{R}$ and $\mathbb{C}$, with their euclidean topology, are locally compact, but not valued. Some authors define local fields as locally compact topological field for a non-discrete topology. Then, they distinguish between the non-Archimedean (or ultrametric) local fields, which are the valued ones, and the Archimedean local fields: $\mathbb{R}$ and $\mathbb{C}$.

An important property is the following. Any valued local field L of characteristic 0 is isomorphic to a finite extension of the field of p-adic numbers $\mathbb{Q}_p$ for some p, while any local field L of characteristic $p > 0$ is isomorphic to a local field $\mathbb{F}_q((\pi))$, and with $q = p^e$ for some integer $e > 0$. We say that π is a *uniformiser*. Note that $|L^\times| = |\pi|^{\mathbb{Z}}$ and $|\pi| < 1$. The proof of this result is a not too difficult deduction from the following well known fact: a locally compact topological vector space over a non-trivial locally compact field has finite dimension.

4.2.2 Valued Rings and Fields for Modular Forms

Let $\mathcal{C}$ be a smooth, projective, geometrically irreducible curve over $\mathbb{F}_q$, together with a closed point $\infty \in \mathcal{C}$. We set

$$R = A := H^0(\mathcal{C} \setminus \{\infty\}, \mathcal{O}_{\mathcal{C}}).$$

This is the $\mathbb{F}_q$-algebra of the rational functions over $\mathcal{C}$ which are regular everywhere except, perhaps, at ∞. The choice of ∞ determines an equivalence class of valuations $|\cdot|_\infty$ on A in the following way. Let d_∞ be the degree of ∞, that is, the degree of the extension $\mathbb{F}$ of $\mathbb{F}_q$ generated by ∞ (which is also equal to the least integer $d > 0$ such that $\tau^d(\infty) = \infty$, where τ is a power of the geometric Frobenius endomorphism). Then, for any $a \in A$, the degree

$$\deg(a) := \dim_{\mathbb{F}_q}(A/aA)$$

is a multiple $-v_\infty(a)d_\infty$ of d_∞ and we set $|a|_\infty = c^{-v_\infty(a)}$ for $c > 1$, which is easily seen to be a valuation. It is well known that A is an arithmetic Dedekind domain with $A^\times = \mathbb{F}^\times$. In addition $v_\infty(a) \leq 0$ for all $a \in A \setminus \{0\}$ and $v_\infty(a) = 0$ if and only if $a \in \mathbb{F}^\times$ (as a consequence of the proof of the subsequent Lemma 4.2.4). A good choice to normalise $|\cdot|_\infty$ is $c = q$. We can thus consider the field $K_\infty := \widehat{K}_{|\cdot|_\infty}$ completion of K for $|\cdot|_\infty$ which can be written as the Laurent series field $\mathbb{F}((\pi))$ where π is a *uniformiser element* of K_∞ (such that $v_\infty(\pi) = 1$). K_∞ is a local field with valuation ring $\mathcal{O}_{K_\infty} = \mathbb{F}[[\pi]]$, maximal ideal $\mathcal{M}_{K_\infty} = \pi\mathbb{F}[[\pi]]$, residue field

$\mathbb{F}$ and valuation group $|\pi|_{\infty}^{\mathbb{Z}}$. Note that we have the direct sum of $\mathbb{F}_q$-vector spaces:

$$K_{\infty} = \mathbb{F}[\pi^{-1}] \oplus \mathcal{M}_{K_{\infty}}.$$

The case of $\mathcal{C} = \mathbb{P}^1_{\mathbb{F}_q}$ with its point at infinity ∞ (defined over $\mathbb{F}_q$) is the simplest one. Let θ be any rational function having a simple pole at infinity, regular away from it. Then, $A = \mathbb{F}_q[\theta]$, $K = \mathbb{F}_q(\theta)$ and we can take $\pi = \theta^{-1}$ so that $K_{\infty} = \mathbb{F}_q((\frac{1}{\theta}))$ the completion of K for the valuation $|\cdot|_{\infty} = q^{\deg_{\theta}(\cdot)}$. Note that for all $\pi = \lambda\theta^{-1} + \sum_{i>1} \lambda_i \theta^{-i} \in K_{\infty}$ with $\lambda \in \mathbb{F}_q^{\times}$ and $\lambda_i \in \mathbb{F}_q$, we have $K_{\infty} = \mathbb{F}_q((\pi))$. The field K_{∞} has an advantage over the field $\mathbb{R}$: it has uniformisers. But there also is a disadvantage: there is no canonical choice in the uncountable subset of uniformisers.

We come back to the case of A general. Let U be a subset of K_{∞}. We say that U is *strongly discrete* if any disk

$$D_{K_{\infty}}(x, r) = \{y \in K_{\infty} : |x - y|_{\infty} \leq r\} \subset K_{\infty}$$

only contains finitely many $y \in U$ for every $r \geq 0$. Note that, transposing the definition to the case of $\mathbb{R}$, the ring $\mathbb{Z}$ is discrete and co-compact in $\mathbb{R}$ (this is well known).

Analogously:

Lemma 4.2.4 *The $\mathbb{F}_q$-algebra A is strongly discrete and co-compact in K_{∞}.*

Proof of the First Part of Lemma 4.2.4 That A is strongly discrete in K_{∞} can be seen by using the *Liouville inequality*, asserting that for any $x \in A \setminus \{0\}$, $|x|_{\infty} \geq 1$. The fraction field K of A is an extension of $\mathbb{F}$ of transcendence degree one, and $\mathbb{F}$ is algebraically closed in K. The closed points P of $\mathcal{C}$ correspond to the classes of equivalence of multiplicative valuations over K which have discrete image in $\mathbb{R}_{>0}$ (discrete valuations), and which are trivial over $\mathbb{F}$. There is a set of valuations $|\cdot|_P$ (associated to the closed points of $\mathcal{C}$ different from ∞) such that for all $a \in A \setminus \{0\}$, $|a|_P \leq 1$ and $|a|_P = 1$ for all but finitely many P, and such that

$$|a|_{\infty} \prod_P |a|_P = 1,$$

see the axiomatic theory of Artin and Whaples and [Art45, Theorem 2]. This is the *product formula* for A. Let us consider $x \in A \setminus \{0\}$. We cannot have $|x|_{\infty} < 1$ because this would violate the product formula. Therefore, $|x|_{\infty} \geq 1$ and this suffices to show strong discreteness. $\qquad\square$

We deduce that $A \cap \mathcal{M}_{K_{\infty}} = \{0\}$. The next Lemma tells us that, as a 'valued vector space over $\mathbb{F}$', A is not too different from $\mathbb{F}[\pi^{-1}]$. This can be used to show co-compactness.

Lemma 4.2.5 *There exists a finite dimensional vector space $V \subset \mathbb{F}[\pi^{-1}]$ over $\mathbb{F}$ such that, isometrically, $K_\infty \cong \mathbb{F}A \oplus V \oplus \mathcal{M}_{K_\infty}$.*

Proof We can invoke Weierstrass' gap Theorem. It can be seen as one of the consequences of the Theorem of Riemann-Roch and it is nicely presented in Stichtenoth's book in the case of $d_\infty = 1$, [Sti08, Theorem 1.6.8]. If $d_\infty > 1$ we can use [Mat05, Proposition 2.1]. Let $H(\infty)$ be the subset of $\mathbb{N}$ whose elements are the nonnegative integers k such that there exists an element f in A with polar divisor $k[\infty]$. The Weierstrass gap Theorem asserts that $\mathbb{N} \setminus H(\infty) = \{n_1, \ldots, n_h\} \subset [1, \frac{2g-1}{d_\infty}]$ (so that if the genus g of $\mathcal{C}$ is zero, this set is empty). Moreover, if $d_\infty = 1$ this set contains exactly g elements, where g is the genus of $\mathcal{C}$. We set $V := \oplus_{i=1}^{h} \mathbb{F}\pi^{-n_i}$ and if $g = 0$ we set $V = \{0\}$. Note that $A \cap V = V \cap \mathcal{M}_{K_\infty} = A \cap \mathcal{M}_{K_\infty} = \{0\}$. Then, every element f of K_∞ can be decomposed in a unique way as $f = a \oplus v \oplus m$ with $a \in \mathbb{F}A$, $v \in V$ and $m \in \mathcal{M}_{K_\infty}$. □

Proof of the Second Part of Lemma 4.2.4 Co-compactness is equivalent to the property that, for the metric induced on the quotient K_∞/A, every sequence contains a convergent sequence. We have an isometric isomorphism

$$\frac{K_\infty}{\mathbb{F}A} \cong V \oplus \mathcal{M}_{K_\infty}$$

where V is a vector space as in Lemma 4.2.5 and we deduce that K_∞/A, with the induced metric, is compact. □

Up to a certain extent, the tower of rings $A \subset K \subset K_\infty$ associated to the datum $(\mathcal{C}, \infty)$ can be viewed in analogy with the tower of rings $\mathbb{Z} \subset \mathbb{Q} \subset \mathbb{R}$.

Here is a fact which encourages to 'think ultrametrically'. We cannot cover a disk of radius q (e.g. $D_L(0, q)$) of a non locally compact field L, with finitely many disks of radius 1. Of course, this is possible, by local compactness, for the disk $D_{K_\infty}(0, q)$ in K_∞. Explicitly, in the case $\mathcal{C} = \mathbb{P}^1_{\mathbb{F}_q}$:

$$D_{K_\infty}(0, q) = D_{K_\infty}(0, 1) \oplus \mathbb{F}_q\theta = \sqcup_{\lambda \in \mathbb{F}_q^\times} D_{K_\infty}(\lambda\theta, 1) \sqcup D_{K_\infty}(0, 1).$$

4.2.3 Algebraic Extensions

We start with an example in the local field $L = \mathbb{F}_q((\pi))$ (with $|\pi| < 1$). Let M be an element of L such that $|M| < 1$. We want to solve the equation

$$X^q - X = M. \tag{4.1}$$

Assuming that there exists a solution $x \in L$ we have $x = x^q - M$ so that inductively for all n:

$$x = x^{q^{n+1}} - \sum_{i=0}^{n} M^{q^i}.$$

The series $\sum_{i=0}^{n} M^{q^i}$ converges to H in $\mathcal{M}_L$ by the hypothesis on M and $|H| = |M|$. But $H^q - H = M$ and $x = H$ is a solution of (4.1) and the polynomial $X^q - X - M$ totally splits in $L[X]$ as all the roots are in $\{H + \lambda : \lambda \in \mathbb{F}_q\}$. If $|M| = 1$ we could think of writing $M = M_0 + M'$ with $M_0 \in \mathbb{F}_q^\times$ and $|M'| < 1$ but Eq. (4.1) with $M = M_0$ has no roots in $\mathbb{F}_q$. One easily sees that the Eq. (4.1) has no roots in L if $|M| \geq 1$. What makes the above algorithm of approximating a solution in the case $|M| < 1$ is that the equation $X^q - X = 0$ has solutions in $\mathbb{F}_q$. These arguments can be generalised and formalised in what is called *Hensel's lemma*. It can be used to show the following property, which is basic and will be used everywhere. Let L be a valued field with valuation $|\cdot| = c^{-v(\cdot)}$ (with a map $v : L \rightarrow \mathbb{R} \cup \{\infty\}$), complete, and let us consider F/L a finite extension (necessarily complete). Then, setting

$$N_{F/L}(x) = \left(\prod_{\sigma \in S} \sigma(x) \right)^{[F:L]_i}, \quad x \in F,$$

where S is the set of embeddings of F in an algebraic closure of L and $[F : L]_i$ is the inseparable degree of the extension F/L, the map $w : F \rightarrow \mathbb{R} \cup \{\infty\}$ determined by $w(0) = \infty$ and

$$w(x) = \frac{v(N_{F/L}(x))}{[F : L]}, \quad x \in F^\times$$

defines a valuation $|\cdot|_w := c^{-w(\cdot)}$ extending $|\cdot|$ over F in the only possible way. Coming back to the local field $L = \mathbb{F}_q((\pi))$, denoting by L^{ac} an algebraic closure of L, there is a unique valuation over L^{ac} extending the one of L; we will denote it by $|\cdot|$ by abuse of notation. The valuation group is $|\pi|^{\mathbb{Q}} = \{|\pi|^\rho : \rho \in \mathbb{Q}\}$ therefore dense in $\mathbb{R}_{>0}$ and the residue field is the algebraic closure $\mathbb{F}_q^{\mathrm{ac}}$ of $\mathbb{F}_q$. It is easy to see that L^{ac} is not complete, although each intermediate finite extension is so.

Lemma 4.2.6 *The completion $\widehat{L^{\mathrm{ac}}}$ of L^{ac} is algebraically closed.*

Proof We follow [Gos96, Proposition 2.1]. Let $F/\widehat{L^{\mathrm{ac}}}$ be a finite extension. Then, as seen previously, F carries a unique extension of the valuation $|\cdot|$ of $\widehat{L^{\mathrm{ac}}}$. Let x be an element of F. We want to show that $x \in \widehat{L^{\mathrm{ac}}}$. For a polynomial $P = \sum_i P_i X^i \in \widehat{L^{\mathrm{ac}}}[X]$ we set $\|P\| := \sup\{|P_i|\}$. It is easy to see that $\|\cdot\|$ is a valuation over $\widehat{L^{\mathrm{ac}}}[X]$, called the *Gauss valuation* (to see the multiplicativity it suffices to study the image of a polynomial in $\mathcal{O}_{\widehat{L^{ac}}}[X]$ by the residue map

$$\mathcal{O}_{\widehat{L^{ac}}}[X] \to k_{\widehat{L^{ac}}}[X]$$

which is a ring homomorphism). Let $P \in \widehat{L^{ac}}[X]$ be the minimal polynomial of x over $\widehat{L^{ac}}$. For $\| \cdot \|$, P is a limit of polynomials of the same degree, which split completely. It is easy to show that for all $\epsilon > 0$, there exists $N \geq 0$ with the property that for all $i \geq N$, a root $x_i \in K_\infty^{ac}$ of P_i satisfies $|x - x_i|_\infty < \epsilon$. This shows that x is a limit of a sequence of $\widehat{L^{ac}}$ and therefore, $x \in \widehat{L^{ac}}$. $\square$

4.2.4   Analytic Functions on Disks

To introduce the next discussions we recall here some basic facts about ultrametric analytic functions in disks, following [Gos96, Chapter 3]. In this subsection, L denotes a valued field which is algebraic closed and complete for a valuation $| \cdot |$. We consider a map $v : L^\times \to \mathbb{R}$ such that $| \cdot | = c^{-v(\cdot)}$ for some $c > 1$. We consider a formal power series

$$f = \sum_{i \geq 0} f_i X^i \in L[[X]]. \tag{4.2}$$

The *Newton polygon* $\mathcal{N}$ of f is the lower convex hull in $\mathbb{R}^2$ of the set $\mathcal{S} = \{(i, v(f_i)) : i \geq 0\}$. It is equal to $\bigcap_{\mathcal{H}} \mathcal{H}$ where $\mathcal{H}$ runs over all the closed half-planes of $\mathbb{R}^2$ which contain at once $\mathcal{S}$ and a half-line $\{(x, y) : y \gg 0\}$ for some $x \in \mathbb{R}$, where $y \gg 0$ ('large enough') means that $y \geq y_0$ for some $y_0 \in \mathbb{R}$.

Here is a practical method of constructing the Newton polygon $\mathcal{N}$ of a formal series $f \in L[[X]]$, if you have on-hand a wooden board, a pencil, nails, a hammer, string and a compass. Draw the axes coordinates i and v on the board, with the positive direction of the latter pointed toward the north, as indicated by the compass. Mark the coordinate points $(i, v(i))$ with the pencil, then hammer nails into the points. Place yourself in front of the wooden board pointing north. Take the string and pull it tautly between your hands, then begin winding it from south to north (being careful to not choose $f = 0$, meaning you must have hammered in at least one nail!). A polygon figure will appear, which, transferred on the board, represents the Newton polygon of f.

Note that if $f \neq 0$, there is always a vertical side on the left of $\mathcal{N}$. If f is a non-zero polynomial, there is also a vertical side on the right. If $x \in L$ and $|f_i x^i| \to 0$ then the series $\sum_i f_i x^i$ converges in L to an element that we denote by $f(x)$. There exists $r \in |L|$ such that $f(x)$ is defined for all $x \in D(0, r) := D_L(0, r)$ and we have thus defined a function

$$D(0, r) \xrightarrow{f} L$$

that we call *analytic function* on the disk $D(0, r)$ (note the abuse of language).

Proposition 4.2.7 *The following properties hold.*

(1) *The sequence of slopes of $\mathcal{N}$ is strictly increasing and its limit is $-\rho(f) = \limsup_{i \to \infty} v(f_i)$. The real number $\rho(f)$ is unique with the property that the series $f(x)$ converges for $x \in L$ such that $v(x) > \rho(f)$, and $f(x)$ diverges if $v(x) < \rho(f)$.*

(2) *If there is a side of the Newton polygon of f which has slope $-m$ and such that it does not contain any point of the Newton polygon in its interior, then f has exactly $r(m)$ zeroes x counted with multiplicity, with $v(x) = m$, where $r(m)$ is the length of the projection of this side of slope $-m$ onto the horizontal line. There are no other zeroes of f with this property.*

(3) *If $\rho(f) = -\infty$, assuming that f is not identically zero, we can expand, in a unique way (Weierstrass product expansion):*

$$f(X) = cX^n \prod_i \left(1 - \frac{X}{\alpha_i}\right)^{\beta_i}$$

with $c \in L^\times$, where $\alpha_i \to \infty$ is the sequence of zeroes such that $v(\alpha_i) > v(\alpha_{i+1})$ (with multiplicities $\beta_i \in \mathbb{N}^*$).

By (2) of the proposition, if we set $r = c^{-\rho(f)} \in \mathbb{R}_{\geq 0}$, f is analytic on $D(0, r')$ for all $r' \in |L|$ such that $r' < r$ and r is maximal with this property. If $\rho(f) = -\infty$ then we say that f is *entire*. We can show easily that if f is entire and non-constant, then it is surjective, and furthermore, an entire function without zeroes is constant. Also, if f as above is non-entire and non-constant, in general it is not surjective, but we have a reasonable description of the image of disks by it, given by the next corollary, the proof of which is left to the reader.

Corollary 4.2.8 *Let f be as in (4.2) with $f_0 = 0$ and let us suppose that it converges on $D_L(0, r)$ with $r \in |L^\times|$. Then, $f(D_L(0, r)) = D_L(0, s)$ for some $s \in |L|$.*

To be brief: an analytic function sends disks to disks.

4.2.5   Further Properties of the Field $\mathbb{C}_\infty$

We consider as in Sect. 4.2.2 the local field K_∞. Then, $K_\infty = \mathbb{F}((\pi))$ for some uniformiser π and by Lemma 4.2.6, the field

$$\mathbb{C}_\infty := \widehat{K_\infty^{\mathrm{ac}}}$$

is algebraically closed and complete. It will be used in the sequel as an alternative to $\mathbb{C}$ 'for silicon-based mathematicians',[1] but there are many important differences. For instance, note that $\mathbb{C}/\mathbb{R}$ has degree 2, while $\mathbb{C}_\infty/K_\infty$ is infinite dimensional, as the reader can easily see by observing that $\mathbb{F}$-linear elements of $\mathbb{F}_q^{ac}$ are also K_∞-linearly independent (in fact, this K_∞-vector space is uncountably-dimensional and the group of automorphisms is an infinite, profinite group).

Complex analysis makes heavy use of local compactness so that we can cover a compact analytic space with finitely many disks. For example, we can cover an annulus with finitely many disks so that the union does not contain the center, which is very useful in path integration of analytic functions over $\mathbb{C} \setminus \{0\}$. The ultrametric counterparts of this and other familiar and intuitive statements are false in $\mathbb{C}_\infty$ as well as in other non-locally compact fields. We cannot use 'partially overlapping disks' to 'move' in $\mathbb{C}_\infty$, or, more generally, in a non-Archimedean space. The intuitive idea of 'moving' itself is different even thought it is not too different, as two annuli, or a disk and an annulus, may overlap somewhere without being one included in the other.

On another hand, the field $\mathbb{C}_\infty$ also has 'nice' properties. Let us review some of them; we denote by L^{sep} the separable closure of a field L.

Lemma 4.2.9 *We have* $\mathbb{C}_\infty = \widehat{K_\infty^{sep}}$.

Proof This is consequence of simple metric properties of Artin-Schreier extensions. We follow [Ax70]. First look at the equation

$$X^{q'} - X = M$$

with $M \in K_\infty$ and where $q' = p^{e'}$ for some $e' > 0$. Then, if $|M|_\infty > 1$, all the solutions $\gamma \in \mathbb{C}_\infty$ of the equation are such that $|\gamma|_\infty^{q'} = |M|_\infty$ and $|\gamma^{q'} - M|_\infty < |M|_\infty$. This also is a very simple consequence of Proposition 4.2.7: the reader can study the Newton polygon of $f(X) = X^{q'} - X - M$ inspecting the three different cases $|M|_\infty < 1, |M|_\infty = 1$ and $|M|_\infty > 1$. Here, with $|M_\infty| > 1$, the extension $K_\infty(\gamma)/K_\infty$ is clearly separable but ramified as by Proposition 4.2.7, the polynomial $X^{q'} - X - M$ has q' distinct roots x in K_∞^{ac} with valuation $|x|_\infty = |M|_\infty^{\frac{1}{q'}}$. It is in fact a *wildly ramified* extension: this means that the characteristic p of $\mathbb{F}_q$ divides the index of ramification.

We now consider $\alpha \in K_\infty^{ac}$. We want to show that α is a limit of K_∞^{sep}. There exists $q' = p^{e'}$ with $a := \alpha^{q'} \in K_\infty^{sep}$. For instance, we can take $q' = [K_\infty(\alpha) : K_\infty]_i$ (inseparable degree). Consider $b \in K_\infty^{\times}$ and a root $\beta \in K_\infty^{ac}$ of the polynomial equation $X^{q'} - bX - a = 0$. Clearly, $\beta \in K_\infty^{sep}$. Let $\lambda \in K_\infty^{sep}$ be such that $\lambda^{q'-1} = b$.

---

[1]Opposed to 'carbon-based mathematicians', following David Goss.

Then, setting $\gamma = \frac{\beta}{\lambda}$, we have $\gamma^{q'} = \frac{\beta^{q'}}{\lambda^{q'}} = \frac{\beta^{q'}}{b\lambda}$ so that

$$\gamma^{q'} - \gamma = \frac{a}{b\lambda} =: M.$$

We can choose $b \in K_\infty^\times$ such that $|b|_\infty$ is small enough so that $|M|_\infty > 1$. If this is the case, then $|\gamma|_\infty^{q'} = |\frac{a}{b\lambda}|_\infty$ so that

$$|\beta|_\infty^{q'} = |a|_\infty.$$

Since $(\beta - \alpha)^{q'} = \beta^{q'} - a = b\beta$,

$$v_\infty(\beta - \alpha) = \frac{1}{q'} v_\infty(\beta^{q'} - a) = \frac{1}{q'} (v_\infty(b) + v_\infty(\beta)) = \frac{1}{q'} \left(v_\infty(b) + \frac{1}{q'} v_\infty(a) \right).$$

We choose a sequence $(b_i)_i \subset K_\infty^\times$ with $b_i \to 0$. For all i, let $\beta_i \in K_\infty^{\text{sep}}$ be such that $\beta_i^{q'} = \beta_i b_i + a$ and $\beta_i \to \alpha$. Then, $v_\infty(\beta_i - \alpha) \to \infty$ as $v_\infty(b_i) \to \infty$ so that $\beta_i \to \alpha$. $\square$

We deduce that $|\mathbb{C}_\infty^\times|_\infty = |\pi|_\infty^\mathbb{Q}$ with π a uniformiser of K_∞, and the residue field of $\mathbb{C}_\infty$ is $\mathbb{F}_q^{ac}$ the algebraic closure of $\mathbb{F}_q$ in $\mathbb{C}_\infty$.

The next results are not used in the rest of the text but mentioning them is helpful in understanding important subtleties lying in the bases of the theory of Drinfeld modular forms.

Lemma 4.2.10 *The group $\mathbb{C}_\infty^\times$ contains a subgroup $\pi^\mathbb{Q} \cong (\mathbb{Q}, +)$ which is totally ordered for $|\cdot|_\infty$. There are group epimorphisms*

$$\mathbb{C}_\infty^\times \xrightarrow{\varpi} \pi^\mathbb{Q}, \quad \mathbb{C}_\infty^\times \xrightarrow{\text{sgn}} (\mathbb{F}_q^{ac})^\times$$

such that ϖ induces the identity on $\pi^\mathbb{Z}$, sgn induces the identity on $(\mathbb{F}_q^{ac})^\times$, and for all $x \in \mathbb{C}_\infty^\times$,

$$|x - \varpi(x) \, \text{sgn}(x)|_\infty < |x|_\infty.$$

One can see that a choice of $\pi^\mathbb{Q}$, ϖ etc. corresponds to an embedding of $\mathbb{C}_\infty$ in a *maximal immediate extension* of it (that is to say, a field extension which is maximal with same valuation group and same residue field) or, equivalently, in a certain type of field of *Hahn generalised series*, *spherically complete*. Read Poineau and Turchetti's contribution [Poi20a, Definition I.2.17, Theorem I.2.18, Example I.2.20]. Read also Kedlaya's [Ked01].

The group $G := \mathrm{Gal}(K_\infty^{\mathrm{sep}}/K_\infty)$ acts on $\mathbb{C}_\infty$ by continuous K_∞-linear auto-morphisms. Then the following important result holds, where the completion on the right is that of the perfect closure of K_∞ in $\mathbb{C}_\infty$ (see for example [Ax70]):

Theorem 4.2.11 (Ax-Sen-Tate) $\mathbb{C}_\infty^G := \{x \in \mathbb{C}_\infty : g(x) = x, \forall g \in G\} = \widehat{K_\infty^{\mathrm{perf}}}.$

4.3 Drinfeld Modules and Uniformisation

Drinfeld modules are also at the hearth of Tavares Ribeiro contribution to this volume, read [Tav20, §1.4]. Let R be an $\mathbb{F}_q$-algebra and $\tau : R \to R$ be an $\mathbb{F}_q$-linear endomorphism. We denote by $R[\tau]$ the left R-module of the finite sums $\sum_i f_i \tau^i$ ($f_i \in R$) equipped with the R-algebra structure given by $\tau b = \tau(b)\tau$ for $b \in R$.[2]

Let $f = \sum_{i=0}^n f_i \tau^i$ be in $R[\tau]$. For any $b \in R$ we can evaluate f in b by setting

$$f(b) = \sum_{i=0}^n f_i \tau^i(b) \in R.$$

This gives rise to an $\mathbb{F}_q$-linear map $R \to R$. Note that the element $f = \sum_i f_i \tau^i$ and the associated evaluation map $f : R \to R$ are two completely different objects. However, in this text, we will denote them with the same symbols.

We choose R by returning to the notations of Sect. 4.2.2. In particular considering the $\mathbb{F}_q$-algebra $A = H^0(\mathcal{C} \setminus \{\infty\}, \mathcal{O}_\mathcal{C})$ we construct the tower of rings

$$A \subset K \subset K_\infty \subset \mathbb{C}_\infty$$

arising from Sect. 4.2.3 which is analogous of $\mathbb{Z} \subset \mathbb{Q} \subset \mathbb{R} \subset \mathbb{C}$.

4.3.1 Drinfeld A-Modules and A-Lattices

We show here the crucial correspondence between Drinfeld A-modules and A-lattices, due to Drinfeld [Dri74]. The definition of Drinfeld module that we give here is not the most general one but it will nevertheless be enough for our purposes. Remember that, in the construction of the tower of rings $A \subset K \subset K_\infty \subset K_\infty^{\mathrm{ac}} \subset \mathbb{C}_\infty$ we have in fact chosen an embedding $A \subset \mathbb{C}_\infty$.

---

[2]It would be more appropriate, to define this R-algebra, to choose an indeterminate X and consider as the underlying R-module the polynomial ring $R[X]$ setting the product to be $Xb = \tau(b)X$. This is an Ore algebra and the standard notation for it is $R[X; \tau]$. For the purposes we have in mind, the abuse of notation $R[\tau]$ is harmless.

Definition 4.3.1 An injective $\mathbb{F}_q$-algebra morphism $\phi : A \to \operatorname{End}_{\mathbb{F}_q}(\mathbb{G}_a(\mathbb{C}_\infty)) \cong \mathbb{C}_\infty[\tau]$ is a *Drinfeld A-module of rank* $r > 0$ if for all $a \in A$

$$\phi_a := \phi(a) = a + (a)_1 \tau + \cdots + (a)_{rd_\infty \deg(a)} \tau^{r \deg(a)} \in \mathbb{C}_\infty[\tau],$$

where the coefficients $(a)_i$ are in $\mathbb{C}_\infty$ and depend on a, and where $\deg(a) = \dim_{\mathbb{F}_q}(A/(a))$. If R is an $\mathbb{F}_q$-subalgebra of $\mathbb{C}_\infty$ containing A and the coefficients $(a)_i$ with $1 \leq i \leq r \deg(a)$ and $a \in A$, we say that the Drinfeld A-module ϕ is *defined over* R *and we write* ϕ/R.

Note that geometrically, a Drinfeld module defined over $\mathbb{C}_\infty$ is just $\mathbb{G}_a$ over $\mathbb{C}_\infty$. What makes the theory interesting is the fact that there are many embeddings of A in $\operatorname{End}_{\mathbb{F}_q}(\mathbb{G}_a(\mathbb{C}_\infty))$. The case of the *Carlitz module*, which can be viewed as the 'simplest' Drinfeld module of rank one, is analysed in Sect. 4.4.

The set of Drinfeld A-modules of rank r is equipped with a natural structure of small category. If φ and ψ are two Drinfeld A-modules, we say that they are *isogenous* if there exists $\nu \in \mathbb{C}_\infty[\tau]$ such that $\varphi_a \nu = \nu \psi_a$ for all $a \in A$. If ν, seen as a non-commutative polynomial in τ, is constant, then we say that φ and ψ are *isomorphic*. Being isogenous induces an equivalence relation on Drinfeld A-modules and isogenies are the morphisms connecting Drinfeld A-modules of same rank in our category.

We prove that the category of Drinfeld A-modules of rank r is equivalent to another category, that of A-*lattices*.

Definition 4.3.2 An A-*lattice* in $\mathbb{C}_\infty$ is a finitely generated strongly discrete A-submodule $\Lambda \subset \mathbb{C}_\infty$ and two A-lattices Λ and Λ' are *isogenous* if there exists $c \in \mathbb{C}_\infty^\times$ such that $c\Lambda \subset \Lambda'$ with $c\Lambda$ of finite index in Λ'.

Isogenies are the morphisms connecting lattices. Clearly, this also defines an equivalence relation. If two A-lattices Λ and Λ' are such that there exists $c \in \mathbb{C}_\infty$ with $c\Lambda = \Lambda'$, then we say that Λ and Λ' are *isomorphic*.

Since A is a Dedekind ring, any A-lattice Λ is projective and has a rank $r = \operatorname{rank}_A(\Lambda)$. We have the following lemma, the proof of which is left to the reader.

Lemma 4.3.3 *Let* Λ *be a projective A-module of rank* r. *Then* Λ *is an A-lattice if and only if the* K_∞-*vector space generated by* Λ *has dimension* r.

Observe that, in contrast with the complex case, for all $r > 1$ there exist infinitely many non-isomorphic A-lattices (this can be deduced from the fact that $\mathbb{C}_\infty$ is not locally compact). We choose an A-lattice Λ of rank r as above.

By Proposition 4.2.7 the following product (where the dash $(\cdot)'$ indicates that the factor corresponding to $\lambda = 0$ is omitted)

$$\exp_\Lambda(Z) := Z \prod_{\lambda \in \Lambda}{}' \left(1 - \frac{Z}{\lambda}\right)$$

converges to an entire function $\mathbb{C}_\infty \to \mathbb{C}_\infty$ (hence surjective) called the *exponential function* associated to Λ. Note that this is an $\mathbb{F}_q$-linear entire function with kernel Λ, and we can write

$$\exp_\Lambda(Z) = \sum_{i \geq 0} \alpha_i \tau^i(Z), \quad \alpha_i \in \mathbb{C}_\infty, \quad \alpha_0 = 1, \quad \forall Z \in \mathbb{C}_\infty.$$

In particular, $\frac{d}{dZ} \exp_\Lambda(Z) = 1$, and the 'logarithmic derivative' (defined in the formal way) of $\exp_\Lambda$ coincides with its multiplicative inverse and is equal to the series

$$\sum_{\lambda \in \Lambda} \frac{1}{Z - \lambda}, \quad Z \in \mathbb{C}_\infty \setminus \Lambda.$$

We refer to [Gek88, §2] for an account on the properties of this fundamental class of analytic functions.

It is not always an easy task to construct explicitly Drinfeld A-modules for a given $A = H^0(\mathcal{C} \setminus \{\infty\}, \mathcal{O}_\mathcal{C})$, if $\mathcal{C} \neq \mathbb{P}^1_{\mathbb{F}_q}$. The following result is due to Drinfeld [Dri74] and shows the depth of the problem.

Theorem 4.3.4 *There is an equivalence of small categories*

$$\{A - lattices\ of\ rank\ r\} \to \{Drinfeld\ A\text{-modules of rank }r\text{ defined over }\mathbb{C}_\infty\}.$$

Proof The proof that we propose is essentially self-contained except for the use of Theorem 4.3.7 which is the crucial tool, showing how to associate to any Drinfeld A-module an exponential function. We postpone this result and its proof to Sect. 4.3.2.

Let Λ be a lattice of rank r (so that it is a projective A-module). The $\mathbb{F}_q$-linear entire map $\exp_\Lambda$ gives rise to the exact sequence of $\mathbb{F}_q$-vector spaces

$$0 \to \Lambda \to \mathbb{C}_\infty \xrightarrow{\exp_\Lambda} \mathbb{C}_\infty \to 0.$$

For any $a \in A$ there is a unique $\mathbb{F}_q$-linear map $\mathbb{C}_\infty \xrightarrow{\phi_a} \mathbb{C}_\infty$ such that

$$\exp_\Lambda(aZ) = \phi_a(\exp_\Lambda(Z))$$

for all $Z \in \mathbb{C}_\infty$ and we want to show that the family $(\phi_a)_{a \in A}$ gives rise to a Drinfeld A-module of rank r. By the snake lemma we get $\ker(\phi_a) \cong \Lambda/a\Lambda \cong (A/(a))^r$. Note also that $\ker(\phi_a) = \exp_\Lambda(a^{-1}\Lambda)$. We set

$$P_a(Z) := aZ \prod_{\alpha \in \ker(\phi_a)}' \left(1 - \frac{Z}{\alpha}\right) = aZ + (a)_1 Z^q + \cdots + (a)_{r\deg(a)} Z^{q^{r\deg(a)}}.$$

Note that the functions $P_a(\exp_\Lambda(Z))$ and $\exp_\Lambda(aZ)$ are both entire with divisor $a^{-1}\Lambda$ and the coefficient of Z in their entire series expansions are equal. Hence these functions are equal and we can write

$$\phi_a(Z) = aZ + (a)_1 Z^q + \cdots + (a)_{r\deg(a)} Z^{q^{r\deg(a)}}, \quad \forall a \in A, \quad Z \in \mathbb{C}_\infty.$$

This defines a Drinfeld A-module ϕ of rank r such that $\exp_\Lambda(aZ) = \phi_a(\exp_\Lambda(Z))$ for all $a \in A$ so we have defined a map associating to Λ an A-lattice of rank r a Drinfeld module ϕ_Λ of rank r.

The next step is to show that the map $\Lambda \mapsto \phi_\Lambda$ that we have just constructed, from the set of A-lattices of rank r to the set of Drinfeld A-modules of rank r, is surjective. From the proof it will be possible to derive that it is also injective. Let ϕ be a Drinfeld A-module of rank r. We want to construct Λ an A-lattice of rank r such that $\phi = \phi_\Lambda$. By the subsequent Theorem 4.3.7, there exists a unique entire $\mathbb{F}_q$-linear function $\exp_\phi : \mathbb{C}_\infty \to \mathbb{C}_\infty$ such that for all $a \in A$, $\exp_\phi(aZ) = \phi_a(\exp_\phi(Z))$, and this, for all $Z \in \mathbb{C}_\infty$. We set $\Lambda = \mathrm{Ker}(\exp_\phi)$. Then Λ is a strongly discrete A-module in $\mathbb{C}_\infty$. The snake lemma implies that $\Lambda/a\Lambda \cong \mathrm{Ker}(\phi_a)$, which is an $\mathbb{F}_q$-vector space of dimension $r\deg(a)$. Let $\epsilon > 0$ be a real number and let V_ϵ be the K_∞-subvector space of $\mathbb{C}_\infty$ generated by $\Lambda \cap D(0, \epsilon)$. We also set $\Lambda_\epsilon := V_\epsilon \cap \Lambda$. Observe that Λ_ϵ is an A-lattice (it is a finitely generated A-module because of the finiteness of the dimension of V_ϵ) which is saturated by construction. Hence $\Lambda_\epsilon/a\Lambda_\epsilon$ injects in $\Lambda/a\Lambda$ and this for all $\epsilon > 0$ which means $\mathrm{rank}_A(\Lambda_\epsilon) = \dim_{K_\infty}(V_\epsilon) \leq r$ for all $\epsilon > 0$. Setting $V = \cup_\epsilon V_\epsilon$ we see that $\dim_{K_\infty}(V) \leq r$. From this we easily deduce that Λ is finitely generated and since $\Lambda/a\Lambda \cong (A/(a))^r$ we derive that Λ is an A-lattice of rank r.

Hence the map $\Lambda \mapsto \phi_\Lambda$ is surjective and one sees easily that it is also injective by looking at $\exp_\Lambda$. Finally, the map is in fact an equivalence of small categories with the natural notions of morphisms between A-lattices and Drinfeld A-modules that we have introduced. We leave the details of these verifications to the reader. $\quad\square$

4.3.2   From Drinfeld Modules to Exponential Functions

In order to complete the proof of Theorem 4.3.4 it remains to show how to associate to a Drinfeld A-module an exponential function. This is the object of the present subsection and we will take the opportunity to present things in a rather more general setting, by introducing Anderson's A-modules. We recall here the definition of Hartl and Juschka in [Har20].

Definition 4.3.5 An *Anderson A-module* of dimension d (over $\mathbb{C}_\infty$) is a pair $\underline{E} = (E, \varphi)$ where E is an $\mathbb{F}_q$-module scheme isomorphic to $\mathbb{G}_a(\mathbb{C}_\infty)^d$, together with a ring homomorphism $\varphi : A \to \mathrm{End}_{\mathbb{F}_q}(E)$, such that for all $a \in A$, $(\mathrm{Lie}(\varphi(a)) - a)^d = 0$.

If R is a ring, we denote by $R^{m \times n}$ the set of matrices with m rows and n columns with entries in R. Note that there is an $\mathbb{F}_q$-isomorphism $\mathrm{End}_{\mathbb{F}_q}(E) \cong \mathbb{C}_\infty[\tau]^{d \times d}$. If $d = 1$ we are brought to Definition 4.3.1 of Drinfeld A-modules.

Anderson modules fit in a category which can be compared to that of commutative algebraic groups; this category is of great importance for the study of global function field arithmetic. A remarkable feature which allows to track similarities with commutative algebraic groups is the fact that we can associate, to every such module, an exponential function. In [Boc07, Proposition 8.7] (see also Anderson in [And86, Theorem 3]) Böckle and Hartl proved that every Anderson's A-module $\underline{E}$ possesses a unique exponential function

$$\exp_{\underline{E}} : \mathrm{Lie}(\underline{E}) \to E(\mathbb{C}_\infty)$$

in the following way (compare also with [Tav20, Proposition 1.11]). Identifying $\mathrm{Lie}(\underline{E})$ (defined factorially) with $\mathbb{C}_\infty^{d \times 1}$, $\exp_{\underline{E}}$ is an entire function of d variables $z = {}^t(z_1, \ldots, z_d) \in \mathbb{C}_\infty^{d \times 1}$ (${}^t(\cdots)$ denotes the transposition)

$$z \mapsto \exp_{\underline{E}}(z) = \sum_{i \geq 0} E_i z^{q^i}$$

with $E_0 = I_d$ and $E_i \in \mathbb{C}_\infty^{d \times d}$ such that, for all $a \in A$ and $z \in \mathbb{C}_\infty^d$,

$$\exp_{\underline{E}}(\mathrm{Lie}(\varphi_a)z) = \varphi_a(\exp_{\underline{E}}(z)).$$

We show how to construct $\exp_{\underline{E}}$ in a slightly more general setting. Let B be any commutative integral countably dimensional $\mathbb{F}_q$-algebra. We follow [Gaz19] and we define $\| \cdot \|_\infty$ on $A \otimes_{\mathbb{F}_q} B$ by setting, for $x \in A \otimes_{\mathbb{F}_q} B$, $\|x\|_\infty$ to be the infimum of the values $\max_i |a_i|_\infty$, running over any finite sum decomposition

$$x = \sum_i a_i \otimes b_i$$

with $a_i \in A$ and $b_i \in B \setminus \{0\}$. Then, $\| \cdot \|_\infty$ is a valuation of $A \otimes_{\mathbb{F}_q} B$ extending the valuation of A via $a \mapsto a \otimes 1$. The $\mathbb{F}_q$-algebra $A \otimes_{\mathbb{F}_q} B$ is equipped with the B-linear endomorphism τ defined by $a \otimes b \mapsto a^q \otimes b$ (thus extending the q-th power map $a \mapsto a^q$ which is an $\mathbb{F}_q$-linear endomorphism of A). Similarly, we can consider the $\mathbb{C}_\infty$-algebra

$$\mathbb{T} = \mathbb{C}_\infty \widehat{\otimes}_{\mathbb{F}_q} B,$$

the completion of $\mathbb{C}_\infty \otimes_{\mathbb{F}_q} B$ for $\| \cdot \|_\infty$ defined accordingly, and we also have a B-linear extension of τ. Let $d > 0$ be an integer. We allow τ to act on $d \times d$ matrices of $\mathbb{T}^{d \times d}$ with entries in $\mathbb{T}$ on each coefficient. Then, $\mathbb{T}[\tau]$ acts on $\mathbb{T}$ by evaluation and $\mathbb{T}[\tau]^{d \times d} \subset \mathrm{End}_B(\mathbb{T}^{d \times 1})$. If $f \in \mathbb{T}[\tau]^{d \times d}$ we can write $f = \sum_{i=0}^n f_i \tau^i$ with

$f_i \in \mathbb{T}^{d \times d}$ and we set $\mathrm{Lie}(f) := f_0$ which provides a $\mathbb{T}$-algebra morphism

$$\mathrm{Lie}(f) : \mathbb{T}[\tau]^{d \times d} \to \mathbb{T}^{d \times d}.$$

Definition 4.3.6 An *Anderson* $A \otimes_{\mathbb{F}_q} B$-module φ of dimension d is an injective B-algebra homomorphism

$$A \otimes_{\mathbb{F}_q} B \xrightarrow{\varphi} \mathbb{T}[\tau]^{d \times d}$$

such that for all $a \in A$, $(\mathrm{Lie}(\varphi(a)) - a)^d = 0$.

We prefer to write φ_a in place of $\varphi(a)$.

We now revisit the proof of Proposition 8.7 of [Boc07] and the method is flexible enough to adapt to the setting of Definition 4.3.6. Note also that later in this text, we will be interested in the case $B = \mathbb{F}_q$ only, case in which we essentially recover [And86, Theorem 3]. In the following, the non-commutative ring $\mathbb{T}[[\tau]]$ is defined in the obvious way with $\mathbb{T}[\tau]$ as a subring. In the following, we denote by $\|M\|_\infty$ the supremum of $\|x\|_\infty$ where x varies in the entries of a matrix $M \in \mathbb{T}^{m \times n}$. We show:

Theorem 4.3.7 *Given an Anderson* $A \otimes_{\mathbb{F}_q} B$-module φ, *there exists a unique series*

$$\exp_\varphi = \sum_{i \geq 0} E_i \tau^i \in \mathbb{T}[[\tau]]^{d \times d}$$

with the coefficients $E_i \in \mathbb{T}^{d \times d}$ *and with* $E_0 = I_d$, *such that the evaluation series* $\exp_\varphi(z)$ *is convergent for all* $z \in \mathbb{T}^{d \times 1}$, *and such that*

$$\varphi_a(\exp_\varphi(z)) = \exp_\varphi(\mathrm{Lie}(\varphi_a)z),$$

for all $z \in \mathbb{T}^{d \times 1}$ *and* $a \in A \otimes_{\mathbb{F}_q} B$. *For all* $a \in A \setminus \mathbb{F}$ *we have that* $\exp_\varphi$ *is the limit for* $n \to \infty$ *of the sequence of entire functions* $\varphi_{a^n} a^{-n} \in \mathbb{C}_\infty[[\tau]]^{d \times d}$, *uniformly convergent on every subset of* $\mathbb{T}[[\tau]]^{d \times 1}$, *bounded for the norm* $\|\cdot\|_\infty$.

Before proving this result, we need two lemmas.

Lemma 4.3.8 *Let us consider* $\mathcal{L}, \mathcal{M} \in \mathbb{T}[\tau]^{d \times d}$ *with* $\mathcal{L} = \alpha + \mathcal{N}$, *with* $\alpha \in \mathrm{GL}_d(\mathbb{T})$ *such that* $\|\alpha\|_\infty > 1 > \|\alpha^{-1}\|_\infty$ *and* $\mathcal{M}, \mathcal{N} \in (\mathbb{T}[\tau]\tau)^{d \times d}$. *Then, for all* $R \in \|\mathbb{T}^\times\|_\infty$, *the sequence of functions given by the evaluation of* $(\mathcal{L}^N \mathcal{M} \alpha^{-N})_{N \geq 0}$ *converges uniformly on* $D_\mathbb{T}(0, R)^{d \times 1}$ *to the zero function.*

Proof The multiplication defining $\mathcal{L}^N \mathcal{M} \alpha^{-N}$ is that of $\mathbb{T}[[\tau]]^{d \times d}$. Locally near the origin, $\alpha^{-1}\mathcal{L}$ is an isometric isomorphism and there exists $R_0 \in \|\mathbb{T}^\times\|_\infty$ with $0 < R_0 < 1$ such that for all $x \in D_\mathbb{T}(0, R_0)^{d \times 1}$, $\|\mathcal{L}(x)\|_\infty = \|\alpha x\|_\infty \leq \|\alpha\|_\infty \|x\|_\infty$. Hence, for $N \geq 0$, if $\|x\|_\infty \leq \|\alpha\|_\infty^{-N} R_0$ ($< R_0$ because of the hypothesis on α), we have $\|\mathcal{L}^N(x)\|_\infty \leq \|\alpha\|_\infty^N \|x\|_\infty$.

We can choose R_0 small enough so that $\|\mathcal{M}(x)\|_\infty \leq \beta \|x\|_\infty^{q^l}$ for some $\beta \in \|\mathbb{T}^\times\|_\infty$ and $l > 0$. Let R be in $\|\mathbb{T}^\times\|_\infty$ fixed, and let us suppose that N is large enough so that $\|\alpha\|_\infty^{-N} R \leq R_0$. Then, for all $x \in D_\mathbb{T}(0, R)^d$, $\|\mathcal{M}(\alpha^{-N}x)\|_\infty \leq \beta(\|\alpha\|_\infty^{-N}R)^{q^l}$. If N is large enough, we can also suppose that

$$\beta(\|\alpha\|_\infty^{-N}R)^{q^l} < \|\alpha\|_\infty^{-N}R_0$$

(because $l > 0$). Therefore, $\|(\mathcal{L}^N\mathcal{M})(\alpha^{-N}x)\|_\infty \leq \|\alpha\|_\infty^N \beta(\|\alpha\|_\infty^{-N}R)^{q^l} \to 0$ as $N \to \infty$, for all $x \in D_\mathbb{T}(0, R)^{d\times 1}$. $\qquad\square$

We consider an Anderson $A \otimes B$-module φ and we recall that $\mathrm{Lie}(\varphi_a)$ is the coefficient in $\mathbb{T}^{d\times d}$ of $\tau^0 I_d$ in the expansion of $\varphi_a \in \mathbb{T}[\tau]^{d\times d}$ along powers of $I_d\tau$. If $a \in A \otimes B \setminus \mathbb{F}_q \times B$, $\mathrm{Lie}(\varphi_a) = aI_d + N_a$ with N_a nilpotent. Then, $\alpha = \mathrm{Lie}(\varphi_a) \in \mathrm{GL}_d(\mathbb{T})$ is such that $\|\alpha\|_\infty > 1$. Indeed otherwise $N_a - \alpha - aI_d$ would be invertible.

Let us consider $a, b \in A \otimes B$, $\|a\|_\infty > 1$. We construct the sequence of B-linear functions $\mathbb{T}^{d\times 1} \xrightarrow{\mathcal{F}_N^a} \mathbb{T}^{d\times 1}$ defined by

$$\mathcal{F}_N^a = \varphi_{a^N b} \, \mathrm{Lie}(\varphi_{a^N b})^{-1}, \quad N \geq 0.$$

Lemma 4.3.9 *The sequence $(\mathcal{F}_N^a)$ converges uniformly on every polydisk $D_\mathbb{T}(0, R)^{d\times 1}$ and the limit function $\mathbb{T}^{d\times 1} \to \mathbb{T}^{d\times 1}$ is independent of the choice of b.*

Proof We set $\mathcal{G}_N^a = \mathcal{F}_{N+1}^a - \mathcal{F}_N^a$. Then,

$$\mathcal{G}_N^a = \underbrace{\varphi_{a^N}}_{=:\mathcal{L}^N} \underbrace{\varphi_b(\varphi_a \, \mathrm{Lie}(\varphi_a)^{-1} - I_d) \, \mathrm{Lie}(\varphi_b)^{-1}}_{=:\mathcal{M}} \, \underbrace{\mathrm{Lie}(\varphi_a)^{-N}}_{=:\alpha^{-N}}$$

and by Lemma 4.3.8, the sequence converges uniformly to the zero function on every polydisk $D_\mathbb{T}(0, R)^{d\times 1}$ which ensures the uniform convergence of the sequence $\mathcal{F}_N^a$). Observe now that, writing momentarily $\mathcal{F}_N^{a,b}$ to designate the above function associated to the choice of a, b,

$$\mathcal{F}_{a^N}^{a,b} - \mathcal{F}_{a^N}^{a,1} = \underbrace{\varphi_{a^N}}_{=:\mathcal{L}^N} \underbrace{(\varphi_b \, \mathrm{Lie}(\varphi_b)^{-1} - I_d)}_{=:\mathcal{M}} \underbrace{\mathrm{Lie}(\varphi_{a^N})^{-1}}_{=:\alpha^{-N}},$$

so that, again by Lemma 4.3.8 this sequence tends to zero uniformly on every polydisk, and the limit $\mathcal{F}^a$ of the sequence $\mathcal{F}_N^a$ is uniquely determined, independent of b. $\qquad\square$

Proof of Theorem 4.3.7 Let us denote by $\mathcal{F}^a$ the continuous B-linear map which, by Lemma 4.3.9 is the common limit of all the sequences $(\mathcal{F}_N^{a,b})_N$ (that can be identified with a formal series $x \mapsto \sum_{i\geq 0} E_i \tau^i(x) \in \mathbb{T}^{d\times d}[[\tau]]$). First of all, note that $E_0 = I_d$ so that this map is not identically zero. Moreover, observe that, for all $b \in A \otimes B$:

$$\varphi_b \mathcal{F}^a = \varphi_b \lim_{N \to \infty} \mathcal{F}_N^{a,1}$$

$$= \varphi_b \lim_{N \to \infty} \varphi_{a^N} \operatorname{Lie}(\varphi_{a^N})^{-1}$$

$$= \lim_{N \to \infty} \varphi_{ba^N} \operatorname{Lie}(\varphi_{ba^N})^{-1} \operatorname{Lie}(\varphi_b)$$

$$= \lim_{N \to \infty} \mathcal{F}_N^{a,b} \operatorname{Lie}(\varphi_b)$$

$$= \mathcal{F}^a \operatorname{Lie}(\varphi_b).$$

Hence we see that for all a, $\mathcal{F}^a$ satisfies the property of the theorem. Now, let $\mathcal{F}_1$ and $\mathcal{F}_2$ be two elements of $\mathbb{T}^{d \times d}[[\tau]]$ such that $\varphi_b(\mathcal{F}_i(z)) = \mathcal{F}_i(bz)$ for all $b \in A \otimes B$ and $i = 1, 2$, and with the property that $\mathcal{F}_3 = \mathcal{F}_1 - \mathcal{F}_2 \in \mathbb{T}^{d \times d}[[\tau]]\tau$. Suppose by contradiction that $\mathcal{F}_3$ is non-zero. Then we can write $\mathcal{F}_3 = \sum_{i \geq i_0} F_i \tau^i$ with $F_i \in \mathbb{T}^{d \times d}$ and F_{i_0} non-zero. Since $\mathcal{F}_3$ also satisfies the same functional identities of both $\mathcal{F}_1, \mathcal{F}_2$ (for $b \in A \otimes B$), we get $\operatorname{Lie}(\varphi_b) F_{i_0} = F_{i_0} \tau^{i_0}(\operatorname{Lie}(\varphi_b))$ for all b. Let w be an eigenvector of F_{i_0} with non-zero eigenvalue, defined over some algebraic closure of the fraction field of $\mathbb{T}$. We consider $b \in A \otimes B$ with $\|b\|_\infty > 1$. Writing $\operatorname{Lie}(\varphi_b) = b + N_b$ with N_b nilpotent, we see that $\operatorname{Lie}(\varphi_b) w = \tau^{i_0}(\operatorname{Lie}(\varphi_b)) w$ which implies $(b - \tau^{i_0}(b)) w = (\tau^{i_0}(N_b) - N_b) w = Mw$ and M is nilpotent. Hence, there is a power c of $b - \tau^{i_0}(b)$ such that $cw = 0$ which means that $b = \tau^{i_0}(b)$; a contradiction because the valuations do not agree. This means that $\mathcal{F}_1 = \mathcal{F}_2$. In particular, $\mathcal{F} = \mathcal{F}^a$ does not depend on the choice of a and the theorem is proved.

$\square$

4.4   The Carlitz Module and Its Exponential

In this section we set

$$A = H^0(\mathbb{P}^1_{\mathbb{F}_q} \setminus \{\infty\}, \mathcal{O}_{\mathbb{P}^1_{\mathbb{F}_q}}) = \mathbb{F}_q[\theta],$$

θ being a rational function over $\mathbb{P}^1_{\mathbb{F}_q}$ having a simple pole at ∞ and no other singularity. The simplest example of Anderson's A-module is the *Carlitz module* which is discussed here; it has rank one and it is perhaps the only one with which we can make very simple computations so it is legitimate to spend some time on it. In order to simplify our notations, we write

$$|\cdot| = |\cdot|_\infty = q^{-v_\infty(\cdot)}, \qquad \|\cdot\| = \|\cdot\|_\infty$$

from now on; this will not lead to confusion.

Definition 4.4.1 (Cf. Example 1.9 of [Tav20]) The *Carlitz A-module* is the Drinfeld A-module $A \xrightarrow{C} \mathbb{C}_\infty[\tau]$ uniquely defined by $C_\theta = C(\theta) = \theta + \tau$.

Let a be in A. Then, $C_a \in A[\tau]$ has degree $\deg_\theta(a)$ in τ and the rank is 1. Note also that C is defined over the $\mathbb{F}_q$-algebra A.

We give an example of computation where we can see how this A-module structure over an A-algebra R works. We suppose $q = 2$. Let 1 be the identity of $R^\times$. We have $C_\theta(1) = \theta + 1$. Hence,

$$C_{\theta^2 + \theta}(1) = C_{\theta + 1}(C_\theta(1)) = (\theta + 1)^2 + \theta^2 + 1 = 0.$$

This means that 1 is a $(\theta^2 + \theta)$-torsion point for this A-module structure given by the Carlitz module.

By Theorem 4.3.7, the limit series

$$\exp_C := \lim_{N \to \infty} C_{\theta^N} \theta^{-N} \in \mathbb{C}_\infty[[\tau]],$$

not identically zero and which can be identified with an entire $\mathbb{F}_q$-linear endomorphism of $\mathbb{C}_\infty$, satisfies

$$\exp_C a = C_a \exp_C \tag{4.3}$$

for all $a \in A$ and has constant term (with respect to the expansion in powers of τ) equal to one. By Theorem 4.3.4, the Carlitz module C corresponds to a rank one lattice $\nu A \subset \mathbb{C}_\infty$, with generator $\nu \in \mathbb{C}_\infty$, and we have

$$\exp_C(Z) = \exp_{\nu A}(Z) = Z \prod_{\lambda \in \nu A}{}' \left(1 - \frac{Z}{\lambda}\right), \quad Z \in \mathbb{C}_\infty.$$

Our next purpose is to compute ν explicitly. To do this, we are going to use properties of the Newton polygon of $\exp_C$. Indeed, staring at (4.3) it is a simple exercise to show that there is a unique solution $Y \in \mathbb{C}_\infty[[\tau]]$ of $C_\theta Y = Y \theta$ with the coefficient of τ^0 equal to one, and by uniqueness, we find

$$\exp_C = \sum_{i \geq 0} d_i^{-1} \tau^i,$$

where

$$d_i = (\theta^{q^i} - \theta^{q^{i-1}}) \cdots (\theta^{q^i} - \theta^q)(\theta^{q^i} - \theta) = (\theta^{q^i} - \theta) d_{i-1}^q$$

(if $i > 0$ and with $d_0 = 1$). From $v_\infty(d_i) = -iq^i$ we observe again that $\exp_C$ defines an $\mathbb{F}_q$-linear entire function which is therefore also surjective over $\mathbb{C}_\infty$ (use Proposition 4.2.7). We have the normalisation of $|\cdot|$ by $|\theta| = q$.

Proposition 4.4.2 *There exists an element $v \in \mathbb{C}_\infty$ with $v_\infty(v) = -\frac{q}{q-1}$, such that the kernel of $\exp_C$ is equal to the $\mathbb{F}_q$-vector space vA. The element v is defined up to multiplication by an element of $\mathbb{F}_q^\times$.*

Proof We know already from Theorem 4.3.4 that the kernel of $\exp_C$ has rank one over A. The novelty here is that we can compute the valuation of its generators, a property which is not available from the theorem. The Newton polygon of $\exp_C$ is the lower convex hull in $\mathbb{R}^2$ of the set whose elements are the points (q^i, iq^i). Since

$$(q^{i+1}, (i+1)q^{i+1}) - (q^i, iq^i) = (q^i(q-1), iq^i(q-1) + q^{i+1})$$

for $i \geq 0$, the sequence (m_i) of the slopes of the Newton polygon is

$$\frac{iq^i(q-1) + q^{i+1}}{q^i(q-1)} = i + \frac{q}{q-1}.$$

Projecting this polygon on the horizontal axis we deduce that for all $i \geq 0$, $\exp_C$ has exactly $q^i(q-1)$ zeroes x such that $v_\infty(x) = -i - \frac{q}{q-1}$ (counted with multiplicity) and no other zeroes. In particular, we have $q-1$ distinct zeroes such that $v_\infty(x) = -\frac{q}{q-1}$. The multiplicity of any such zero is one (note that $\frac{d}{dX}\exp_C(X) = 1$) so they are all distinct. Now, since $\exp_C$ is $\mathbb{F}_q$-linear, we have that all the roots x such that $v_\infty(x) = -1 - \frac{1}{q-1}$ are multiple, with a factor in $\mathbb{F}_q^\times$, of a single element v (there are $q-1$ choices). We denote by $A[d]$ the set of polynomials of A of exact degree d. For all $a \in A[d]$, $0 = C_a(\exp_C(v)) = \exp_C(av)$ and $v_\infty(av) = -d - \frac{q}{q-1}$. This defines an injective map from $A[d]$ to the set of zeroes of $\exp_C$ of valuation $-d - \frac{q}{q-1}$. But this set has cardinality $q^d(q-1)$ which also is the cardinality of $A[d]$. This means that $\exp_C(x) = 0$ if and only if $x \in vA$. $\qquad\square$

Corollary 4.4.3 *We have $\exp_C(X) = X \prod_{a \in A \setminus \{0\}} \left(1 - \frac{X}{av}\right)$ and $\exp_C$ induces an exact sequence of A-modules*

$$0 \to vA \to \mathbb{C}_\infty \xrightarrow{\exp_C} C(\mathbb{C}_\infty) \to 0.$$

4.4.1 A Formula for v

We have seen that if $\Lambda \subset \mathbb{C}_\infty$ is the kernel of $\exp_C$, then Λ is a free A-module of rank one generated by $v \in \mathbb{C}_\infty$ with $v_\infty(v) = -\frac{q}{q-1}$, defined up to multiplication by an element of $\mathbb{F}_q^\times$. Let us choose a $(q-1)$-th root $(-\theta)^{\frac{1}{q-1}}$ of $-\theta$; this is also defined up to multiplication by an element of $\mathbb{F}_q^\times$, and the valuation is $-\frac{1}{q-1}$. We want to prove the following formula:

$$v = \theta(-\theta)^{\frac{1}{q-1}} \prod_{i > 0} \left(1 - \frac{\theta}{\theta^{q^i}}\right)^{-1}.$$

To do this, we will use Theorem 4.3.7. We recall that this result implies that the sequence

$$f_n(z) = \exp_C(z) - C_{\theta^n}(z\theta^{-n})$$

converges uniformly on every bounded disk of $\mathbb{C}_\infty$ to the zero function. To continue further, we need to introduce the *function ω of Anderson and Thakur*. This function is defined by the following product expansion:

$$\omega(t) = (-\theta)^{\frac{1}{q-1}} \prod_{i \geq 0} \left(1 - \frac{t}{\theta^{q^i}}\right)^{-1}.$$

The convergence of this product is easily seen to hold for any $t \in \mathbb{C}_\infty \setminus \{\theta, \theta^q, \theta^{q^2}, \ldots\}$. Also, for all $n \neq 1$, the function

$$(t - \theta)(t - \theta^q) \cdots (t - \theta^{q^{n-1}})\omega(t)$$

extends to an analytic function over $D_{\mathbb{C}_\infty}(0, q^{n-1})$ (we can also say that ω defines a meromorphic function over $\mathbb{C}_\infty$ having simple poles at the singularities defined above). To study the arithmetic properties of ω, it is useful to work in *Tate algebras*. However, at this level of generality, this is not necessary, strictly speaking. For the purposes we have in mind now, it will suffice to work with formal Newton-Puiseux series. Let y, t be two variables, choose a $(q - 1)$-th root of y and define:

$$F(y, t) = (-y)^{\frac{1}{q-1}} \prod_{i \geq 0} \left(1 - \frac{t}{y^{q^i}}\right)^{-1} \in \mathbb{F}_q((y^{-\frac{1}{q-1}}))((t)).$$

Then,

$$F(y^q, t) = (t - y)F(y, t).$$

Writing the series expansion

$$\omega(t) = \sum_{i \geq 0} \lambda_{i+1} t^i \in \mathbb{C}_\infty[[t]],$$

we deduce, from the uniqueness of the series expansion of an analytic function in $D_{\mathbb{C}_\infty}(0, 1)$, that the sequence $(\lambda_i)_{i \geq 0}$ can be defined by setting $\lambda_0 = 0$ and the algebraic relations

$$C_\theta(\lambda_{i+1}) = \lambda_{i+1}^q + \theta\lambda_{i+1} = \lambda_i$$

which include $\lambda_1 = (-\theta)^{\frac{1}{q-1}}$. Now set $\mu_i = \theta^i \lambda_i$, $i \geq 0$.

Lemma 4.4.4 *For all $i \geq 1$, $|\mu_i| = q^{\frac{q}{q-1}}$ and $(\mu_i)_{i \geq 0}$ is a Cauchy sequence.*

Proof Developing the product defining ω we see that $|\lambda_i| = q^{\frac{q}{q-1}-i}$. To see that (μ_i) is a Cauchy sequence, it suffices to show that $\mu_{i+1} - \mu_i \to 0$. But

$$\mu_{i+1} - \mu_i = \theta^{i+1}\lambda_{i+1} - \theta^i \lambda_i = \theta^i(\lambda_i - \lambda_{i+1}^q) - \theta^i \lambda_i = -\theta^i \lambda_{i+1}^q \to 0.$$

$\square$

Let $\mu \in \mathbb{C}_\infty$ be the limit of (μ_i).

Lemma 4.4.5 *We have $\mu = -\lim_{t \to \theta}(t-\theta)\omega(t) = \theta(-\theta)^{\frac{1}{q-1}} \prod_{i>0}(1-\theta^{1-q^i})^{-1}$.*

Proof From the functional equation of $F(y,t)$ we see that $\lim_{t \to \theta}(t - \theta)\omega(t) = (-\theta)^{\frac{q}{q-1}} \prod_{i>0}(1 - \theta^{1-q^i})^{-1} = \sum_{i \geq 0} \theta^i \lambda_{i+1}^q$, the latter series being convergent. Using that $C_\theta(\lambda_{i+1}) = \lambda_i$ we see that the last sum is:

$$\sum_{i \geq 0} \theta^i(\lambda_i - \theta\lambda_{i+1}) = \sum_{i=0}^{N-1} \theta^i(\lambda_i - \theta\lambda_{i+1}) + \sum_{i \geq N} \theta^i \lambda_{i+1}^q, \quad \forall N.$$

The first sum telescopes to $-\theta^N \lambda_N$ while the second being a tail series of a convergent series, it converges and the sum depending on N tends to 0 as $N \to \infty$.

$\square$

Hence μ is the residue of $-\omega$ at $t = \theta$. We can write

$$\mu = -\operatorname{Res}_{t=\theta}(\omega).$$

This is the analogue of a well known lemma sometimes called Appell's Lemma: if (a_n) is a converging sequence of complex numbers, then $\lim_n a_n = \lim_{x \to 1^-}(1 - x)\sum_n a_n x^n$.

We are now ready to prove the following well known and classical result:

Theorem 4.4.6 *The kernel Λ of $\exp_C$ is generated, as an A-module, by*

$$\mu = v = \theta(-\theta)^{\frac{1}{q-1}} \prod_{i>0}\left(1 - \theta^{1-q^i}\right)^{-1}.$$

Proof Since $\Lambda = vA$ for some $v \in \mathbb{C}_\infty$ such that $|v| = q^{\frac{q}{q-1}}$ and since $|\mu| = q^{\frac{q}{q-1}}$, it suffices to show that $\exp_C(\mu) = 0$. Now, we can write $\mu = \mu_n + \epsilon_n$ where $\epsilon_n \to 0$ and $|\epsilon_n| < q^{\frac{q}{q-1}}$. Also, we have $\exp_C(z) = f_n(z) + C_{\theta^n}(\theta^{-n}z)$ and we have that the sequence of entire functions (f_n) converges uniformly to the zero function on any bounded subset of $\mathbb{C}_\infty$. We have:

$$\exp_C(\mu) = (C_{\theta^n}\theta^{-n} + f_n)(\mu_n + \epsilon_n)$$
$$= \underbrace{C_{\theta^n}(\lambda_n)}_{=0} + \underbrace{f_n(\mu_n)}_{\to 0} + \underbrace{\exp_C(\epsilon_n)}_{\to 0}.$$

Hence, $\mu = \nu$. $\square$

Remark 4.4.7 The formula of Theorem 4.4.6 can be easily derived from the following result of Carlitz in [Car35] that also appears in [Gos96, Theorem 3.2.8]. Let η be a $(q-1)$-th root of $\theta - \theta^q$ in the algebraic closure K^{ac} of K in $\mathbb{C}_\infty$. We set:

$$\xi = \eta \prod_{j\geq 1}\left(1 - \frac{\theta^{q^j} - \theta}{\theta^{q^{j+1}} - \theta}\right) \in K_\infty^{\text{ac}}.$$

Then $\mu \in \mathbb{F}_q^\times \xi$. To see this, observe the identity:

$$\prod_{j=1}^{d-1}\left(1 - \frac{\theta^{q^j} - \theta}{\theta^{q^{j+1}} - \theta}\right) = \prod_{j=0}^{d-1}(1 - \theta^{q^j(1-q)})\prod_{i=1}^{d}(1 - \theta^{1-q^i})^{-1}, \quad d > 1.$$

Both products on d converge in K_∞ for $d \to \infty$. If we set $H = \eta\prod_{j\geq 0}(1 - \theta^{q^j(1-q)}) \in K_\infty^{\text{ac}}$ we see that H is algebraic over K by the relations $H^q = (\theta - \theta^q)\eta(1 - \theta^{1-q})^{-1}\prod_{j\geq 0}(1 - \theta^{q^j(1-q)}) = -\theta^q H$. Since $-\theta^q = \theta^{q-1}(-\theta)$, we deduce that $H \in \mathbb{F}_q^\times\theta(-\theta)^{\frac{1}{q-1}}$. The formulation that we adopt in our text is that of Anderson, Brownawell and Papanikolas in [And04, §5.1]. In fact, the proof of Theorem 4.4.6 that we gave above is inspired by that of these authors.

One of the most used notations for our μ is $\widetilde{\pi}$. This is suggestive due to the resemblance between the exact sequence of Corollary 4.4.3 and $0 \to 2\pi i\mathbb{Z} \to \mathbb{C} \xrightarrow{\exp} \mathbb{C} \to 1$; there is an analogy between $\widetilde{\pi} \in \mathbb{C}_\infty$ and $2\pi i \in \mathbb{C}$. It can be proved, by the product expansion we just found, that $\widetilde{\pi}$ in transcendental over $K = \mathbb{F}_q(\theta)$. The first transcendence proof of it is that of Wade in [Wad41] but there are several others, very different from each other. See for example [And04, §3.1.2]. There are proofs which make use of computations of dimensions of 'motivic Galois groups' which connect to the topics of Di Vizio's contribution to this volume [DiV20] and which are the roots of a vast program in transcendence and algebraic independence inaugurated by Anderson, Brownawell and Papanikolas in [And04], and later by Papanikolas in [Pap08].

4.4.2 A Factorization Property for the Carlitz Exponential

In Corollary 4.4.3, we described the Weierstrass product expansion of the entire function $\exp_C : \mathbb{C}_\infty \to \mathbb{C}_\infty$. We now look again at $\exp_C$ as a formal series and we provide it with another product expansion, this time in $\mathbb{C}_\infty[[\tau]]$; see Proposition 4.4.9. This result is implicit in Carlitz's [Car35, (1.03), (1.04) and (5.01)]. The function we factorise is not $\exp_C$ but a related one:

$$\exp_A(z) = z \prod_{a \in A \setminus \{0\}} \left(1 - \frac{z}{a}\right) = \tilde{\pi}^{-1} \exp_C(\tilde{\pi} z),$$

so that

$$\exp_A = \sum_{i \geq 0} d_i^{-1} \tilde{\pi}^{q^i - 1} \tau^i \in K_\infty[[\tau]].$$

Before going on we must discuss the *Carlitz logarithm*. It is easy to see that in $\mathbb{C}_\infty[[\tau]]$, there exists a unique formal series $\log_C$ with the following properties: (1) $\log_C = 1 + \cdots$ (the constant term in the power series in τ is the identity $1 = \tau^0$) and (2) for all $a \in A$, $a \log_C = \log_C C_a$, a condition which is equivalent to $\theta \log_C = \log_C C_\theta$ by the fact that $A = \mathbb{F}_q[\theta]$. Writing $\log_C = \sum_{i \geq 0} l_i^{-1} \tau^i$ and using this remark one easily shows that

$$l_i = (\theta - \theta^q)(\theta - \theta^{q^2}) \cdots (\theta - \theta^{q^i}),$$

$i \geq 0$. We note that $v_\infty(l_i) = -q \frac{q^i - 1}{q - 1}$. This means that the series $\log_C$ does not converge to an entire function but for all $R \in |\mathbb{C}_\infty^\times|$ such that $R < |\tilde{\pi}|$, $\log_C$ defines an $\mathbb{F}_q$-linear function on $D_{\mathbb{C}_\infty}(0, R)$. We also note, reasoning with the Newton polygons of $\exp_C$ and $\log_C$, that

$$|\exp_C(z)| = |z| = |\log_C(z)|, \quad \forall z \in D^\circ_{\mathbb{C}_\infty}(0, |\tilde{\pi}|), \tag{4.4}$$

which implies that the Carlitz's exponential induces an isometric automorphism of $D^\circ_{\mathbb{C}_\infty}(0, |\tilde{\pi}|)$. More generally, the exponential function of a Drinfeld module induces, locally, an isometric automorphism, see [Tav20, Corollary 1.12]. We observe that the series $U = \exp_C \log_C$ and $V = \log_C \exp_C$ in $K_\infty[[\tau]]$ satisfy $Ua = aU$ and $Va = aV$ for all $a \in A$. Since they further satisfy $U = 1 + \cdots$ and $V = 1 + \cdots$, we deduce that $\log_C$ is the inverse of $\exp_C$ in $K_\infty[[\tau]]$. In particular,

$$C_a = \exp_C a \log_C \in K_\infty[\tau], \quad \forall a \in A.$$

We define:

$$C_z = \exp_C z \log_C \in \mathbb{C}_\infty[[\tau]], \quad z \in \mathbb{C}_\infty.$$

Then,

$$C_z = \sum_{i\geq 0} d_i^{-1}\tau^i z \sum_{j\geq 0} l_j^{-1}\tau^j$$

$$= \sum_{i\geq 0} d_i^{-1} z^{q^i}\tau^i \sum_{j\geq 0} l_j^{-1}\tau^j$$

$$= \sum_{k\geq 0} \underbrace{\left(\sum_{i=0}^{k} d_i^{-1} l_{k-i}^{-q^i} z^{q^i} \right)}_{=:E_k(z).} \tau^k$$

We can thus expand, for all $z \in \mathbb{C}_\infty$:

$$C_z = \sum_{k\geq 0} E_k(z)\tau^k \in \mathbb{C}_\infty[[\tau]]$$

with the coefficients

$$E_k(z) = \sum_{i=0}^{k} d_i^{-1} l_{k-i}^{-q^i} z^{q^i} = \frac{z}{l_k} + \cdots + \frac{z^{q^k}}{d_k} \in K[z]$$

which are $\mathbb{F}_q$-linear polynomials of degree q^k in z for $k \geq 0$. They are called the *Carlitz' polynomials*. In the next proposition we collect some useful properties of these polynomials.

Proposition 4.4.8 *The following properties hold:*

(1) For all $k \geq 0$ we have

$$E_k(z) = d_k^{-1} \prod_{\substack{a\in A \\ |a|<q^k}} (z - a).$$

(2) For all $k \geq 0$ and $z \in \mathbb{C}_\infty$ we have

$$E_k(z)^q = E_k(z) + (\theta^{q^{k+1}} - \theta)E_{k+1}(z).$$

(3) We have $l_k E_k(z) \to \exp_A(z)$ uniformly on every bounded subset of $\mathbb{C}_\infty$.

Proof

(1) Since $C_a \in A[\tau]$ has degree in τ which is equal to $\deg_\theta(a)$, E_k vanishes on $A(< k)$ the $\mathbb{F}_q$-vector space of the polynomials of A which have degree $< k$. Since the cardinality of this set is equal to the degree of E_k, this vector space exhausts the zeroes of E_k, and the leading coefficient is clearly d_k^{-1}.

(2) This is a simple consequence of the relation $C_a C_z = C_z C_a$.

(3) We note that

$$\frac{l_k}{d_k} \prod_{|a|<q^k} (z-a) = \frac{l_k}{d_k} z \prod_{\substack{a \neq 0 \\ |a|<q^k}} (-a) \left(1 - \frac{z}{a}\right).$$

Now, it is easy to see that

$$\prod_{0 \neq |a|<q^k} (-a) = \prod_{0 \neq |a|<q^k} a = \frac{d_k}{l_k}. \tag{4.5}$$

(see [Gos96, §3.2]). The uniform convergence is clear.

□

We come back to the series $\exp_A = \sum_{i \geq 0} d_i^{-1} \tilde{\pi}^{q^i-1} \tau^i \in K_\infty[[\tau]]$. We now show that

$$\exp_A = \cdots \left(1 - \frac{\tau}{l_n^{q-1}}\right) \left(1 - \frac{\tau}{l_{n-1}^{q-1}}\right) \cdots \left(1 - \frac{\tau}{l_1^{q-1}}\right) (1 - \tau) =$$

$$= \cdots l_n (1-\tau) \frac{1}{\theta^{q^n} - \theta} (1-\tau) \cdots \frac{1}{\theta^{q^2} - \theta} (1-\tau) \frac{1}{\theta^q - \theta} (1-\tau). \tag{4.6}$$

in $K_\infty[[\tau]]$ with its (τ)-topology. We have in fact more:

Proposition 4.4.9 *On every bounded subset of* $\mathbb{C}_\infty$, *the entire function* $\exp_A(z)$ *is the uniform limit of the sequence of* $\mathbb{F}_q$-*linear polynomials*

$$\left(z - \frac{z^q}{l_n^{q-1}}\right) \circ \left(z - \frac{z^q}{l_{n-1}^{q-1}}\right) \circ \cdots \circ \left(z - \frac{z^q}{l_1^{q-1}}\right) \circ (z - z^q),$$

where $\circ$ *is the composition.*

Proof We write:

$$\widetilde{\mathcal{E}}_n = \left(1 - \frac{\tau}{l_{n-1}^{q-1}}\right) \cdots \left(1 - \frac{\tau}{l_1^{q-1}}\right) (1 - \tau) \in K[\tau].$$

We also denote by $\mathcal{E}_n \in K[\tau]$ the unique element such that for all $z \in \mathbb{C}_\infty$, $\mathcal{E}_n(z) = E_n(z)$ (evaluation). Part (3) of Proposition 4.4.8 implies that $l_k E_k$ converges uniformly to $\exp_A(z)$ on every bounded subset of $\mathbb{C}_\infty$. Hence, we are done if we show that the evaluations agree: $\widetilde{\mathcal{E}}_n = l_n \mathcal{E}_n$ for all $n \geq 0$. This is certainly true if $n = 0$. We continue by induction. From part (2) of Proposition 4.4.8 we see that $\tau \mathcal{E}_n = \mathcal{E}_n + (\theta^{q^{n+1}} - \theta)\mathcal{E}_{n+1}$ for all $n \geq 0$. Therefore:

$$
\begin{aligned}
\widetilde{\mathcal{E}}_{n+1} &= \left(1 - \frac{\tau}{l_n^{q-1}}\right) \widetilde{\mathcal{E}}_n \\
&= \left(1 - \frac{\tau}{l_n^{q-1}}\right) l_n \mathcal{E}_n \\
&= l_n \mathcal{E}_n - l_n^q l_n^{-q+1} \tau \mathcal{E}_n \\
&= l_n \mathcal{E}_n - l_n (\mathcal{E}_n + (\theta^{q^{n+1}} - \theta)\mathcal{E}_{n+1}) \\
&= \underbrace{l_n (\theta - \theta^{q^{n+1}})}_{=l_{n+1}} \mathcal{E}_{n+1},
\end{aligned}
$$

and we are done.   □

Proposition 4.4.9 was essentially known by Carlitz; it can be derived easily with elementary manipulations on the left-hand side of [Car35, (5.01)]. It is interesting to note the two rationality properties for $\exp_C = \exp_{\widetilde{\pi}A}$ and $\exp_A$ which follow from the above result: the terms of the series defining $\exp_C$ are defined over K (the coefficients d_i^{-1}) and the factors of the infinite product of $\exp_A$ we just considered are also defined over K (the coefficients are l_i^{1-q}).

Problem 4.4.10 Generalise Lemma 4.4.14 and Proposition 4.4.9 to the framework of *Drinfeld-Hayes A-modules of rank one* considered in [Hay74] for a general $\mathbb{F}_q$-algebra of regular functions A and highlight a connection to the *shtuka functions* in the sense of [Gos96, §7.11] in this context, see also [Tav20, §4.2].

Remark 4.4.11 This can be viewed as a digression. There is a simple connection with Thakur's multiple zeta values, defined by:

$$
\zeta_A(n_1, n_2, \ldots, n_r) := \sum_{\substack{a_1, \ldots, a_n \in A^+ \\ |a_1| > \cdots > |a_r|}} a_1^{-n_1} \cdots a_r^{-n_r} \in K_\infty, \quad n_1, \ldots, n_r \in \mathbb{N}^*, \quad r \geq 1,
$$

where A^+ denotes the subset of monic polynomials of A. Indeed, one sees directly that the coefficient of τ^r in (4.6) is equal to

$$
(-1)^r \sum_{i_1 > \cdots > i_r \geq 0} l_{i_1}^{1-q} l_{i_2}^{q-q^2} \cdots l_{i_r}^{q^{r-1}-q^r}.
$$

One proves easily $\sum_{\substack{a \in A^+ \\ |a|=q^i}} a^{-l} = l_i^{-l}$ for $1 \leq l \leq q$ and we deduce that

$$
\exp_A = \sum_{r \geq 0} (-1)^r \zeta_A(q-1, q(q-1), \ldots, q^{r-1}(q-1))\tau^r.
$$

Therefore, equating the corresponding coefficients of the powers of τ we reach the formula:

$$\zeta_A(q-1, q(q-1), \ldots, q^{r-1}(q-1)) = (-1)^r \frac{\tilde{\pi}^{q^r-1}}{d_r}, \quad r \geq 0,$$

with the convention $\zeta_A(\emptyset) = 1$. Note that the identity derived by the specialisation $t = \theta$ in [Pel16a, (22)] rather involves the 'reversed' multiple zeta values $\zeta_A^*(q^{r-1}(q-1), \ldots, q(q-1), q-1)$, the $*$ denoting the variant of multiple zeta value involving sums with non-strict inequalities $|a_1| \geq \cdots \geq |a_r|$.

4.4.3   The Function $\exp_A$ and Local Class Field Theory

This subsection is not logically related to the other topics of the text. Just as the Euler exponential function, the Carlitz exponential function has an important role in explicit class field theory for the field K (see Hayes [Hay74] for the rational function field $K = \mathbb{F}_q(\theta)$, [Hay79] and the more recent work of Zywina [Zyw13], for the general case). Note that even more recently, a direct link between the explicit class field theory of $K = \mathbb{F}_q(\theta)$ and the function ω of Anderson and Thakur has been found in [Ang15]. It does not belong to our purposes to describe these results here. In this subsection we are going to achieve a more modest objective which is to apply, in the case $A = \mathbb{F}_q[\theta]$, the properties of the function $\exp_A$ we have reviewed so far, in relation with the local class field theory for $K_\infty = \mathbb{F}_q((\frac{1}{\theta}))$. Interestingly, these properties do not seem to have simple analogues in the theory of Euler's exponential function.

Let $L \subset \mathbb{C}_\infty$ be an algebraic extension of K_∞. Then, $\exp_A$ defines an $\mathbb{F}_q$-linear map $L \to L$. Indeed, for all $x \in L$, $K_\infty(x)/K_\infty$ is a finite extension, hence complete, and $\exp_A(K_\infty(x)) \subset K_\infty(x)$.

Definition 4.4.12 We say that L is *uniformised by* $\exp_A$ if the map $\exp_A : L \to L$ is surjective.

For example, $L = \mathbb{C}_\infty$ is uniformised by $\exp_A$, thanks to Proposition 4.2.7. Observe that if $L, L' \subset \mathbb{C}_\infty$ are two algebraic extensions of K_∞ which are uniformised by $\exp_A$, then also $L \cap L'$ is uniformised by $\exp_A$. Indeed, let x be an element of $L \cap L'$ and let $y \in L$, $y' \in L'$ be such that $\exp_A(y) = \exp_A(y') = x$. Then $y - y' \in A = \text{Ker}(\exp_A) \subset K_\infty$ so that $y, y' \in L \cap L'$. Hence, there is a *minimal algebraic extension* L/K_∞ in $\mathbb{C}_\infty$ that is uniformised by $\exp_A$; this is what we want to study here.

We denote by K_∞^{ab} the *maximal abelian extension* of K_∞ in $K_\infty^{\text{sep}} \subset \mathbb{C}_\infty$, that is, the maximal extension of K_∞ which is Galois, with abelian Galois group. We also choose λ_θ a $(q-1)$-th root of $-\theta \in K_\infty^{\text{sep}}$ and we note that if L/K_∞ is an algebraic extension, then $L[\lambda_\theta]$ is an algebraic extension of K_∞. The aim of this subsection is to prove:

Theorem 4.4.13 *Let L be the minimal algebraic extension of K_∞ in $\mathbb{C}_\infty$ which is uniformised by $\exp_A$. Then, $L[\lambda_\theta] = K_\infty^{ab}$.*

In the complex setting, and for the Eulerian exponential, we would have the analogue but deceiving result: the minimal algebraic extension of $\mathbb{R}$ which is uniformised by $z \mapsto e^z$ is $\mathbb{C}$. Theorem 4.4.13 confirms that in some sense, function field arithmetic is more transparent and allows to see more structure in the watermark. We need the next:

Lemma 4.4.14 *Let n be a non-negative integer. For every $r \in |\mathbb{C}_\infty^\times|$ with $r < |l_n|$ the product*

$$\mathcal{F}_n := \cdots \left(1 - \frac{\tau}{l_{n+1}^{q-1}}\right)\left(1 - \frac{\tau}{l_n^{q-1}}\right) \in K[[\tau]]$$

defines an entire function $\mathbb{C}_\infty \to \mathbb{C}_\infty$ and induces an isometric bi-analytic isomorphism of the disk $D_{\mathbb{C}_\infty}(0, r)$.

Proof This is easy to verify by using Proposition 4.2.7 and Corollary 4.2.8. Indeed, if we set

$$\psi_m := 1 - \frac{\tau}{l_m^{q-1}}, \quad m \geq 0$$

we see that for all $z \in \mathbb{C}_\infty$ such that $|z| < |l_n|$, $\psi_m(z) = z + z'$ with $z' \in \mathbb{C}_\infty$ depending on m and $|z'| < |z|$, for all $m \geq n$. $\square$

Proof of Theorem 4.4.13 We have a well defined $\mathbb{F}_q$-linear map $\exp_A : K_\infty^{ab} \to K_\infty^{ab}$. We first show that this map is surjective so that if L is the minimal algebraic extension of K_∞ which is uniformised by $\exp_A$, then $L \subset K_\infty^{ab}$. To do this, we note that we have, for all $n \geq 0$, a well defined $\mathbb{F}_q$-linear algebraic map $E_n : \mathbb{A}_{K_\infty^{ab}}^1 \to \mathbb{A}_{K_\infty^{ab}}^1$ given by the Carlitz polynomials ($\mathbb{A}_L^n$ denotes the affine space of dimension n over a field L). By the proof of Proposition 4.4.9, E_n is surjective. Indeed, for all $y' \in K_\infty^{ab}$, the splitting field of the polynomial $E_n(X) - y' \in K_\infty(y')[X]$ is an abelian extension of $K_\infty(y')$ which can be constructed by iterating Artin-Schreier extensions. Let x be an element of K_∞^{ab}. There exists $n \geq 0$ such that $|x| < |l_n|$. By Lemma 4.4.14, $\mathcal{F}_n^{-1}(x) \in K_\infty^{ab}$ is well defined. Let $x' \in K_\infty^{ab}$ be such that

$$l_n E_n(x') = \mathcal{F}_n^{-1}(x).$$

Then we have, by Proposition 4.4.9, $\exp_A(x') = \mathcal{F}_n(l_n E_n(x')) = \mathcal{F}_n(\mathcal{F}_n^{-1}(x)) = x$ and we have proved that K_∞^{ab} is uniformised by $\exp_A$. Now let $L \subset \mathbb{C}_\infty$ be an algebraic extension of K_∞ that is uniformised by $\exp_A$. To show that $L[\lambda_\theta]$ contains K_∞^{ab} we proceed in two steps.

In the first step, we show that K_∞^{un}, the maximal abelian extension of K_∞ which is unramified at the ∞-place, is contained in L. To do this it suffices to show that

the algebraic closure $\mathbb{F}_q^{\mathrm{ac}}$ of $\mathbb{F}_q$ in $\mathbb{C}_\infty$ is contained in L. Indeed, it is easy to see that
$K_\infty^{\mathrm{un}} = \mathbb{F}_q^{\mathrm{ac}}((\frac{1}{\theta}))$.

By using Proposition 4.2.7 we see that for every $y \in \mathbb{C}_\infty$ such that $|y| = 1$ there
exists a unique $x \in \mathbb{C}_\infty$ with $|x| = 1$, such that $\exp_A(x) = y$, and of course if
$y \in L$, then $x \in L$ because we have supposed that L is uniformised by $\exp_A$. Since
$\mathbb{F}_q \subset K_\infty \subset L$, if $y \in \mathbb{F}_q^\times$, there exists $x \in L$, $|x| = 1$, such that $\exp_A(x) = y$.
Now observe with Proposition 4.4.9 that $\exp_A(x) = (\mathcal{F}_1 \circ \mathcal{E}_1)(x) = y$ and applying
Lemma 4.4.14

$$x - x^q = \mathcal{E}_1(x) = \mathcal{F}^{-1}(y) = y + y'$$

where $y' \in \frac{1}{\theta}\mathbb{F}_q[[\frac{1}{\theta}]]$. Setting $x' = \sum_{i \geq 0}(y')^{q^i} \in \frac{1}{\theta}\mathbb{F}_q[[\frac{1}{\theta}]]$ we deduce that $x - x' \in$
$\mathbb{F}_{q^2} \setminus \mathbb{F}_q \subset \mathbb{C}_\infty$ is an element of L, and $\mathbb{F}_{q^2} \subset L$. This shows that $\mathbb{F}_{q^2}((\frac{1}{\theta})) \subset L$
because $\mathbb{F}_{q^2}((\frac{1}{\theta})) = \mathbb{F}_q((\frac{1}{\theta}))[\mathbb{F}_{q^2}]$. We can of course repeat this argument with $y \in$
$\mathbb{F}_{q^2} \subset L$ etc. to show that, inductively, $\mathbb{F}_{q^d} \subset L$ for all $d \geq 1$ so that $\mathbb{F}_{q^d}((\frac{1}{\theta})) =$
$K_\infty[\mathbb{F}_{q^d}] \subset L$ for all $d \geq 1$ and with a little additional work we conclude that
$K_\infty^{\mathrm{un}} \subset L$.

Before passing to the second step we need a little bit of terminology. We say that
a sequence $(x_i)_{i \geq 0}$ in K_∞^{ab} is a *Lubin-Tate sequence* if $\frac{1}{\theta}x_0 + x_0^q = 0$ and

$$\frac{1}{\theta}x_i + x_i^q = x_{i-1}, \quad i > 0.$$

We note that $x_0\lambda_\theta \in \mathbb{F}_q^\times$. Similarly, we say that a sequence $(y_i)_{i \geq 0}$ of K_∞^{ab} is an
Artin-Schreier sequence if $y_0 = 1$ and

$$\mathcal{E}_1(y_i) = y_i - y_i^q = \theta y_{i-1}, \quad i > 0.$$

By a simple application of Proposition 4.2.7 we see that $|y_i| = |\theta|^{\frac{1}{q}+\cdots+\frac{1}{q^i}}$ for all
$i > 0$. Moreover,

$$\frac{1}{\theta}x_0 y_i + (x_0 y_i)^q = x_0 y_{i-1}, \quad i > 0$$

so that if $(y_i)_{i \geq 0}$ is an Artin-Schreier sequence and x_0 satisfies the previous equation,
then $(x_0 y_i)_{i \geq 0}$ is a Lubin-Tate sequence and if $(x_i)_{i \geq 0}$ is a Lubin-Tate sequence with
$x_0\lambda_\theta = 1$, then $(\frac{x_i}{x_0})_{i \geq 0}$ is an Artin-Schreier sequence.

The second step of the proof of our theorem is to show that L contains an Artin-
Schreier sequence. First of all, we note that for any Artin-Schreier sequence $(y_i)_{i \geq 0}$,
$\theta y_i \in D_{K_\infty^{\mathrm{ab}}}(0, r)$ for all $r \in |\mathbb{C}_\infty^\times|$ such that $r < |\theta|^{\frac{q}{q-1}}$ so that $|\theta y_i| < |l_1|$ for all
$i \geq 0$. We fix $i \geq 0$. Let $a_{i+1} \in K_\infty^{\mathrm{ab}}$ be such that

$$a_{i+1} - a_{i+1}^q = \mathcal{F}_1^{-1}(\theta y_i).$$

We have that

$$\exp_A(a_{i+1}) = \mathcal{F}_1(\mathcal{F}^{-1}(\theta y_i)) = \theta y_i.$$

Since by hypothesis, L is uniformised by $\exp_A$, we have that $a_{i+1} \in L$ if $y_i \in L$. It is easy to see that $\mathcal{F}_1^{-1}(\theta y_i) = \theta y_i + y_i'$ where $|y_i'| < 1$. In particular, $a_{i+1}' = \sum_{j \geq 0}(y_i')^{q^j}$ converges to an element of L such that $a_{i+1}' - (a_{i+1}')^q = y_i'$. If we set $b_{i+1} = a_{i+1} - a_{i+1}'$ we can conclude, under the hypothesis that $y_i \in L$, that $b_{i+1} \in L$ is such that

$$b_{i+1} - b_{i+1}^q = \theta y_i.$$

By induction over $i \geq 0$ we obtain that L contains an Artin-Schreier sequence $(y_i)_{i \geq 0}$.

We can now conclude the proof of the theorem. By what written earlier, $L[\lambda_\theta]$ contains a Lubin-Tate sequence $(x_i)_{i \geq 0}$. We set $\widetilde{K} := K_\infty[x_i : i \geq 0]$. By Lubin-Tate theory (see [Lub65]) K_∞^{ab} is the compositum in $\mathbb{C}_\infty$ of $\widetilde{K}$ and K_∞^{un} and therefore, $L[\lambda_\theta]$ contains K_∞^{ab}. $\square$

We are not going to deepen the facts outlined below, but the main theorem of local class field theory asserts, in the special case of our local field K_∞ (it holds for any local field with appropriate modifications) the existence of an isomorphism of profinite groups

$$\widehat{\theta}_{K_\infty} : \widehat{K}_\infty^\times \to \mathrm{Gal}(K_\infty^{\mathrm{ab}}/K_\infty),$$

the *local Artin homomorphism*, where $\widehat{K}_\infty^\times$ is the profinite group completion of $K_\infty^\times \cong \mathbb{F}_q[[\frac{1}{\theta}]]^\times \times \mathbb{Z}$, non-canonically isomorphic to the profinite group $\mathbb{F}_q[[\frac{1}{\theta}]]^\times \times \widehat{\mathbb{Z}}$. The non-canonical isomorphism depends on the choice of a uniformiser π of K_∞. If we set K_π to be the subfield of K_∞^{ab} which is fixed by $\widehat{\theta}_{K_\infty}(\pi) \in \mathrm{Gal}(K_\infty^{\mathrm{ab}}/K_\infty)$, then K_∞^{ab} is the compositum $K_\pi K_\infty^{\mathrm{un}}$, and we have isomorphisms $\mathrm{Gal}(K_\infty^{\mathrm{un}}/K_\infty) \cong \widehat{\mathbb{Z}}$ and $\mathrm{Gal}(K_\pi/K_\infty) \cong \mathbb{F}_q[[\frac{1}{\theta}]]^\times$. Choosing a Lubin-Tate sequence in $K_\infty^{\mathrm{ab}}/K_\infty$ is therefore equivalent to the choice of a uniformiser π of K_∞. One can see, along these remarks (but we will not give full details), that the minimal algebraic extension $L \subset K_\infty^{\mathrm{ab}}$ of K_∞ that is uniformised by $\exp_A$ is determined by $\mathrm{Gal}(K_\infty^{\mathrm{ab}}/L) \cong \mathbb{F}_q^\times$.

Problem 4.4.15 The notion of minimal field extension of K_∞ which is uniformised by the exponential $\exp_A$ can be generalised to e.g. Drinfeld A-modules via Theorem 4.3.4 in a natural way, but it is unclear how this field can be characterised in the light of local class field theory so that the role of a statement like Theorem 4.4.13 must be clarified in this more general setting.

4.5   Topology of the Drinfeld Upper-Half Plane

We go back to the settings and notations of Sect. 4.2.2, considering the $\mathbb{F}_q$-algebra $A = H^0(C \setminus \{\infty\}, \mathcal{O}_C)$ with C a smooth projective curve over $\mathbb{F}_q$ and ∞ a closed point. We therefore have the tower of inclusions of $\mathbb{F}_q$-algebras $A \subset K \subset K_\infty \subset \mathbb{C}_\infty$. In this section we give an explicit topological description of what is called the *Drinfeld upper-half plane* Ω. It goes back to Drinfeld, in [Dri74]. D. Goss called it the *'algebraist's upper-half plane'* in [Gos80a]. It can be viewed as an analogue of the complex upper-half plane that can be constructed by cutting $\mathbb{C}$ in two along the real line and taking one piece only. As a set, Ω is very simple:

$$\Omega = \mathbb{C}_\infty \setminus K_\infty,$$

but subtracting K_∞ results in a different operation than cutting; this is what we are going to show here. We begin by presenting some elementary properties following [Ger80]. We recall that $\mathbb{C}_\infty = \widehat{K_\infty^{\mathrm{ac}}}$, where $K_\infty = \mathbb{F}((\pi))$ for some uniformiser π. First of all, there is an action of $\mathrm{GL}_2(K_\infty)$ on Ω by homographies. If $\gamma = \left(\begin{smallmatrix} a & b \\ c & d \end{smallmatrix}\right) \in \mathrm{GL}_2(K_\infty)$, then we have the automorphism of $\mathbb{P}^1_{\mathbb{F}_q}(\mathbb{C}_\infty)$ uniquely defined by

$$z \mapsto \gamma(z) := \frac{az + b}{cz + d}$$

if $z \notin \{\infty, -\frac{d}{c}\}$. Observe that if F/L is a field extension, then $\mathrm{GL}_2(L)$ acts by homographies on the set $F \setminus L$. For instance, $\mathrm{GL}_2(\mathbb{R})$ acts on $\mathbb{C} \setminus \mathbb{R} = \mathcal{H}^+ \sqcup \mathcal{H}^-$ (disjoint union of the complex upper- and lower-half planes).

It is well known that the imaginary part $\Im(z)$ of a complex number z, the *distance* of z from the real axis, is submitted to the following transformation rule under the action by homographies. If $\gamma = \left(\begin{smallmatrix} a & b \\ c & d \end{smallmatrix}\right) \in \mathrm{GL}_2(\mathbb{R})$:

$$\Im(\gamma(z)) = \frac{\Im(z)\det(\gamma)}{|cz + d|^2}, \quad z \in \mathbb{C} \setminus \mathbb{R}. \tag{4.7}$$

There is an analogous notion of *distance from K_∞* in $\mathbb{C}_\infty$. We set:

$$|z|_\Im := \inf\{|z - x| : x \in K_\infty\}, \quad z \in \mathbb{C}_\infty.$$

We have the following result.

Proposition 4.5.1

(1) For all $z \in \mathbb{C}_\infty$, $|z|_\Im$ is a minimum, and $|z|_\Im = 0$ if and only if $z \in K_\infty$.

(2) Let z be an element of Ω. Then, there exist $z_0 = \pi^m(\alpha_0 + \cdots + \alpha_n \pi^{-n}) \in \mathbb{F}_q[\pi, \pi^{-1}]$ and $z_1 \in \Omega$ with $|z_1| = |z_1|_\Im < |\pi|^m$, uniquely determined, with $n \in \mathbb{N} \cup \{-\infty\}$ and $\alpha_0 \neq 0$ if $z_0 \neq 0$, such that $z = z_0 + z_1$.

Proof

(1) If $z \in K_\infty$, there is nothing to prove. Assume thus that $z \in \Omega \subset \mathbb{C}_\infty$ is fixed. Define the map $K_\infty \xrightarrow{f} |\mathbb{C}_\infty^\times|$, $f(x) = |z - x|$. Then, f is locally constant, hence continuous. But K_∞ is locally compact so there is $x_0 \in D_{K_\infty}(0, |z|)$ (not uniquely determined) such that $f(x_0)$ is a minimum and $|z|_\Im = |z - x_0|$.

(2) For all $x \in K_\infty$, $|x| > |z|$, we have $|z - x| = |x|$. Then, we have two cases.

    (a) For all $x \in D_{K_\infty}(0, |z|)$, $|z - x| = |z|$. In this case, $|z|_\Im = |z|$ and $|z|_\Im$ is a minimum. We thus get $n = -\infty$, $z_0 = 0$ and $z = z_1$.

    (b) There exists $x \in D_{K_\infty}(0, |z|) \setminus \{0\}$ such that $|z| = |x|$ and $|z - x| < |z|$. This implies that the image of z/x in the residue field of $\mathbb{C}_\infty$ is 1. We can therefore write $z = \lambda_1 \pi^{-n_1} + \eta_1$ with $\lambda_1 \in \mathbb{F}$ and $\eta_1 \in \Omega$, $|\eta_1| < |z| = |\theta|^{n_1}$.

    We can iterate by studying now η_1 at the place of z. Either the procedure stops and we get a decomposition $z = \lambda_1 \pi^{-n_1} + \cdots + \lambda_k \pi^{-n_k} + \eta_k$ with $n_1 > \cdots > n_k$, $|z|_\Im = |\eta_k| = |\eta_k|_\Im$ and there exists $z_0 \in K_\infty$ such that $|z - z_0| = |z|_\Im > 0$ as claimed in the statement, or the procedure does not stop but in this case we have $z \in K_\infty$ which is excluded.

$\square$

In particular, either $|z_1| \notin |K_\infty^\times|$, or $|z_1| = |\pi^m|$ but the image of $z_1 \pi^{-m}$ in the residue field of $\mathbb{C}_\infty$ is not one of the elements of $\mathbb{F}^\times$. Part (2) of Proposition 4.5.1 implies that for all $x = z_0 + y$ with $y \in D_{K_\infty}(0, |z_1|)$, $|z - x| = |z|_\Im = |z_1| = |z_1|_\Im$.

We also have the following elementary consequences of the above proposition. First of all, if $c \in K_\infty$, then $|cz|_\Im = |c||z|_\Im$ for all $z \in \Omega$. Moreover, if $v_\infty(z) \notin \mathbb{Z}$, then $|z|_\Im = |z|$. Also, if $|z| = 1$, we have $|z|_\Im = 1$ if and only if the image of z in the residue field of $\mathbb{C}_\infty$ is not in $\mathbb{F}$.

The next property is the analogous of (4.7):

Lemma 4.5.2 *For all $z \in \Omega$ and $\gamma = \left(\begin{smallmatrix} * & * \\ c & d \end{smallmatrix} \right) \in GL_2(K_\infty)$,*

$$|\gamma(z)|_\Im = \frac{|\det(\gamma)||z|_\Im}{|cz + d|^2}.$$

Proof First of all, suppose that we have proved that

$$|\gamma(z)|_\Im \leq \frac{|\det(\gamma)||z|_\Im}{|cz + d|^2}, \quad \forall \gamma = \left(\begin{smallmatrix} * & * \\ c & d \end{smallmatrix} \right) \in GL_2(K_\infty), \quad \forall z \in \Omega. \tag{4.8}$$

In particular, for all $\widetilde{z} \in \Omega$, and with γ replaced by $\gamma^{-1} = \delta^{-1} \left(\begin{smallmatrix} * & * \\ -c & a \end{smallmatrix} \right)$ (where $\delta = \det(\gamma)$), we get

$$|\gamma^{-1}(\widetilde{z})|_\Im \leq \frac{|\delta||\widetilde{z}|_\Im}{|-c\widetilde{z} + a|^2}.$$

We set $\widetilde{z} = \gamma(z)$. Then, $-c\widetilde{z} + a = \frac{\delta}{cz+d}$ and therefore,

$$|z|_\Im = |\gamma^{-1}(\widetilde{z})|_\Im \leq |\widetilde{z}|_\Im |\delta| \left|\frac{cz+d}{\delta}\right|^2 = |\delta|^{-1}|cz+d|^2|\widetilde{z}|_\Im = |\delta|^{-1}|cz+d|^2|\gamma(z)|_\Im,$$

so that

$$\frac{|\delta||z|_\Im}{|cz+d|^2} \leq |\gamma(z)|_\Im,$$

and we get the identity we are looking for. All we need is therefore to show that (4.8) holds.

Now, let $x \in \mathbb{C}_\infty$ be such that x is not a pole of γ. An easy calculation shows that

$$\gamma(z) - \gamma(x) = \frac{\det(\gamma)(z-x)}{(cz+d)(cx+d)}.$$

Hence, if $x \in K_\infty$ is not a pole of γ, we have

$$|\gamma(z) - \gamma(x)| = \frac{|\det(\gamma)||z-x|}{|cz+d|^2} \frac{|cz+d|}{|cx+d|}. \tag{4.9}$$

We can find $x \in K_\infty$ such that $|z - x| = |z|_\Im$ and with the property that x is not a pole of γ (we have noticed that there are infinitely many such elements). We claim that $|cx+d| \leq |cz+d|$. If $c = 0$ this is clear. Otherwise, if this were false we would have $|cx+d| > |cz+d|$ and

$$|c||z|_\Im = |c||z - x| = |cz + d - (cx+d)| = |cx+d| > |cz+d|_\Im = |c||z|_\Im$$

which would be impossible. Hence, with the claim in mind, we deduce from (4.9):

$$|\gamma(z)|_\Im \leq |\gamma(z) - \gamma(x)| \leq \frac{|\det(\gamma)||z-x|}{|cz+d|^2} = \frac{|\det(\gamma)||z|_\Im}{|cz+d|^2}$$

by our choice of x and we are done. $\square$

4.5.1   Rigid Analytic Spaces

The notion of rigid analytic space originates in ideas of Tate in the years 1960'. We do not want to go in very precise details because there is already a plethora of important references, among which [Bos84, Fre04]. A more recent introduction to rigid analytic spaces is the chapter 'Several approaches to non-archimedean geometry' by Conrad, see [Bak08, Chapter 2] (the whole volume is close, in many aspects, to the topics of the present text). Important is also Berkovich's viewpoint which is outlined in this volume, [Poi20a, Poi20b]. We discuss, in a rather informal way, the nature of these structures before making use of some very particular special cases. Let L be a field with valuation $|\cdot|$, complete, algebraically closed.

We are going to describe a rigid analytic space over L (or analytic space over L) as a triple

$$(X, G, \mathcal{O}_X)$$

where X is a non-empty set, G a Grothendieck topology on X, $\mathcal{O}_X$ a sheaf, satisfying several natural conditions. A *Grothendieck topology* G on X can be outlined as a set $\mathcal{S}$ of subsets U of X and, for all $U \in G$, a 'covering' $\mathrm{Cov}(U)$ of U again by elements of G. If $\mathcal{C}$ is the family of all such coverings,[3] then G is the datum $(\mathcal{S}, \mathcal{C})$ and the quality of being a Grothendieck topology results in a collection of properties we shall not give here, refining the simpler notion of topology (see [Fre04] for the precise collection of conditions). If a Grothendieck topology $G = (\mathcal{S}, \mathcal{C})$ on X is given, then the elements of $\mathcal{S}$ are called the *admissible subsets* of X and the elements of $\mathcal{C}$ are called the *admissible coverings*. This refines the notion of topology because if we forget the coverings, the conditions we are left on $\mathcal{S}$ are precisely those of a topology on X so that right at the beginning we could have said that X is a topological space, and the admissible sets are just the open sets for this topology. We have of course a corresponding notion of morphism of Grothendieck's topological spaces which strengthens that of continuous maps of topological spaces: pre-images of admissible sets (resp. coverings) are again admissible.

What is a *sheaf* on a Grothendieck topological space? If we choose a ring R, a sheaf $\mathcal{F}$ of R-algebras (R-modules...) is a contravariant functor from $\mathcal{S}$ (with inclusion) to the category of R-algebras (or R-modules... this is called a *pre-sheaf*) which satisfy certain compatibility conditions. For instance, if $f, g \in \mathcal{F}(U)$, $U \in \mathcal{S}$ and $f|_V = g|_V$ for all $V \in \mathrm{Cov}(U) \in \mathcal{C}$, then $f = g$. Furthermore, if we choose $\mathrm{Cov}(U) = (U_i)_{i \in I} \in \mathcal{C}$ and for all i, $f_i \in \mathcal{F}(U_i)$ are such that $f_i|_{U_i \cap U_j} = f_j|_{U_i \cap U_j}$, then there exists a 'continuation' $f \in \mathcal{F}(U)$ with $f|_{U_i} = f_i$ for all i (this is an abstract formalisation of 'analytic continuation'). Every pre-sheaf can be embedded in a sheaf canonically, but checking that a given pre-sheaf is itself a sheaf might result in subtle problems. The datum of $(X, G, \mathcal{F})$ with G a Grothendieck topology and $\mathcal{F}$ a sheaf of R-algebras on (X, G) is called a *Grothendieck ringed space of R-algebras* and there is a natural notion of morphism of such structures which mimics the more familiar notion of morphism of ringed spaces of algebraic geometry. Say

---

[3]Do not mix up with the curve $\mathcal{C}$ of the previous sections.

for commodity that X, Y are two Grothendieck topological spaces with respective
sheaves $\mathcal{F}$ and $\mathcal{G}$, then a morphism of Grothendieck ringed spaces of R-algebras

$$(X, \mathcal{F}) \xrightarrow{(f, f^{\sharp})} (Y, \mathcal{G})$$

is the datum of a morphism of Grothendieck topological spaces f and for all
$U \subset Y$ admissible, an R-algebra morphism $f^{\sharp} : \mathcal{G}(U) \to \mathcal{F}(f^{-1}(U))$. So far,
we discussed Grothendieck topological spaces, sheaves etc. But now, what is a
rigid analytic variety? A rigid analytic variety over L, our valued field, complete,
algebraically closed (say, $L = \mathbb{C}_{\infty}$, the most relevant in our notes), is a particular
kind of Grothendieck ringed space; let us see how. We still need a few more tools.
We have the unit disk

$$D_L(0, 1) = \{z \in L : |z| \le 1\}$$

playing the role of a basic brick for constructing rigid analytic spaces, just as the
affine line does for algebraic varieties. For this reason, we focus on affinoid algebras.
An *affinoid algebra over* L is any quotient of a *Tate algebra*

$$\mathbb{T}_n(L) = \widehat{L[\underline{t}]}_{\|\cdot\|}$$

by an ideal, where the Tate algebra $\mathbb{T}_n(L)$ of dimension n is the completion $\widehat{}$
of the polynomial ring $L[\underline{t}]$ in n indeterminates $\underline{t} = (t_i)_{1 \le i \le n}$ for the Gauss
valuation $\| \cdot \|$ that we recall it is defined, for elements $a_{i_1, \ldots, i_n} \in L$, by
$\| \sum_{i_1, \ldots, i_n} a_{i_1, \ldots, i_n} t_1^{i_1} \cdots t_n^{i_n} \| = \sup |a_{i_1, \ldots, i_n}|$. It is known that it is noetherian, with
unique factorization, of Krull dimension the number of variables n. The resulting
quotient $\mathcal{A}$ of $\mathbb{T}_n(L)$ (by an ideal) is endowed with a structure of L-Banach algebra.
In other words, the Gauss norm of $\widehat{L[\underline{t}]}$ induces a (sub-multiplicative) norm on $\mathcal{A}$,
and it is complete. In fact, any L-Banach algebra $\mathcal{A}$ together with a continuous
epimorphism $\mathbb{T}_n(L) \to \mathcal{A}$ for some n, making $\mathcal{A}$ into a finitely generated $\mathbb{T}_n(L)$-
algebra, is an affinoid algebra. Affinoid algebras over L are the basic bricks to
construct a rigid analytic variety.

The maximal spectrum $\mathrm{Spm}(R)$ of an affinoid L-algebra R can be made into
a Grothendieck ringed space $(X, G, \mathcal{F})$; this is called an *affinoid variety* over L.
If $X = \mathrm{Spm}(R)$ and $Y = \mathrm{Spm}(R')$, an L-algebra morphism $R \to R'$ defines
a morphism of ringed spaces $Y \to X$ which is called a *morphism of affinoid
algebras*. This serves to describe the other pieces of $(X, G, \mathcal{F})$. The admissible sets
in $\mathcal{S}$ (recall that $G = (\mathcal{S}, \mathcal{C})$) are exactly the images in X of *open immersions* of
affinoid varieties and similarly, we define the coverings in $\mathcal{C}$. This gives rise to a
Grothendieck topology G on $X = \mathrm{Spm}(\mathcal{A})$. Furthermore, we have the pre-sheaf
$\mathcal{O}_X$ defined by associating to $U \subset X$ an admissible set the L-algebra $\mathcal{O}_X(U) = R'$
where $U = \mathrm{Spm}(R')$. Thanks to *Tate's acyclicity theorem* one shows that this
is in fact a sheaf (see [Tat71], see also [Fre04, Theorem 4.2.2]). This result was

generalised by Grauert and Gerritzen [Bos84, 7.3.5 and 8.2]). *Dulcis in fundo*, we have:

Definition 4.5.3 A Grothendieck ringed space $X = (X, G, \mathcal{F})$ is a *rigid analytic variety over L* if X has an admissible covering of admissible subsets U which have the property that $(U, \mathcal{F}|_U)$ is an affinoid variety over L for all U.

4.5.1.1   Analytification

An important process to construct rigid analytic spaces is the analytification of an algebraic variety. Let X/L be a scheme of finite type. The *analytification X^{an}* of X is a rigid analytic space over L that can be defined by an affinoid covering starting from the geometric data as follows. We consider affine Zariski open subsets $U = \mathrm{Spec}(\mathcal{A}) \hookrightarrow X$ and embeddings $U \hookrightarrow \mathbb{A}_L^N$ which correspond, on the algebraic side, to surjective L-algebra maps $L[\underline{t}] \to \mathcal{A}$ (where $\underline{t}$ denotes the set of independent variables $t_1, \ldots, t_N$) endowing $\mathcal{A}$ with a structure of $L[\underline{t}]$-algebra, for some N. Taking the completion for the Gauss valuation yields a surjective morphism:

$$\widehat{L[\underline{t}]} \to \mathcal{A} \otimes_{L[\underline{t}]} \widehat{L[\underline{t}]}$$

which gives rise to a map $V := \mathrm{Spm}(\mathcal{A} \otimes_{L[\underline{t}]} \widehat{L[\underline{t}]}) \hookrightarrow D_L(0,1)^N = \mathrm{Spm}(\widehat{L[\underline{t}]})$. We can proceed similarly for polydisks of different radii in $|L^\times|$ and this is used to construct a rigid analytic space U^{an} such that $V = U^{an} \cap D_L(0,1)^N$. Glueing, we construct the rigid analytic space X^{an}. For example, the *rigid affine line over* L, $\mathbb{A}_L^{1,an}$ is obtained by glueing together the rigid analytic spaces $D_L(0,r)$ along the inclusions with $r \in |L^\times|$. Similarly, the *rigid projective line over* L, $\mathbb{P}_L^{1,an}$, can be constructed by glueing two copies of $D_L(0,1)$ along the set $\{z \in L : |z| = 1\}$, or also glueing two copies of $\mathbb{A}_L^{1,an}$, see also Berkovich's construction in [Poi20b, Definition II.1.5]. The *Berkovich's affine line* is described in detail in ibid. See [Poi20a, Definition I.1.1].

Rigid analytification defines a functor, called the 'GAGA functor' from the category of L-schemes of finite type to the category of rigid analytic spaces over L. Note that we can also consider analytifications of morphisms, coherent sheaves etc. Finally, there is an alternative way to define the analytification functor over an affine variety X over L, introduced by Berkovich, which makes the underlying topological space particularly easy to compute as it is defined over the set of multiplicative seminorms over the coordinate ring of X satisfying certain compatibility conditions with the valuation of L. See [Poi20b, Definition II.1.1] for the construction of the Berkovich spectrum of an algebra of finite type over L. See also Temkin's [Tem15, Chapter 1] for a nice survey in the area.

4.5.1.2   The Rigid Analytic Variety Ω

We now focus on $L = \mathbb{C}_\infty$ with $A = H^0(C \setminus \{\infty\}, \mathcal{O}_C)$ in our usual notation. We discuss a structure of rigid analytic space over $\mathbb{C}_\infty$ on $\Omega = \mathbb{C}_\infty \setminus K_\infty$. Note that

$$\Omega = \bigcup_{M>1} U_M,$$

where $U_M = \{z \in \Omega : M^{-1} \leq |z|_\Im \leq |z| \leq M\}$, the filtered union being over the elements $M \in |\mathbb{C}_\infty| \setminus |K_\infty|$ with $M > 1$. Observe now:

Lemma 4.5.4 *With $M \in |\mathbb{C}_\infty| \setminus |K_\infty|$ we have*

$$U_M = D_{\mathbb{C}_\infty}(0, M) \setminus \bigsqcup_{\substack{\lambda \in \mathbb{F}[\pi, \pi^{-1}] \\ \lambda = \lambda_{-\beta}\pi^\beta + \cdots + \lambda_\beta \pi^{-\beta} \\ 1 \leq |\pi|^{-\beta} \leq M}} D_{\mathbb{C}_\infty}^\circ(\lambda, M^{-1}).$$

Proof This easily follows from the fact that K_∞ is locally compact in combination with the ultrametric inequality. $\qquad\square$

Hence, U_M is admissible and carries a structure of affinoid variety $U_M = \mathrm{Spm}(\mathcal{A}_M)$ where $\mathcal{A}_M$ is an integral affinoid algebra. We say that U_M is a *connected affinoid* of $\mathbb{P}_{\mathbb{C}_\infty}^{1,an}$ (as in the language introduced in [Fre04], motivated by the integrality of $\mathcal{A}_M$). In particular Ω can be covered (in fact filled) with connected affinoids and the analytic structure of Ω arises from viewing it as the complementary in $\mathbb{C}_\infty$ of smaller and smaller disks located over certain elements of K_∞ which is close to the familiar view that we have also for the set $\mathbb{C} \setminus \mathbb{R}$. This gives the Grothendieck topology on Ω, and the sheaf $\mathcal{O}_\Omega$ is that of *rigid analytic functions* over Ω. Practically, a rigid analytic function $f : \Omega \to \mathbb{C}_\infty$ is a function such that the restriction on every set U_M is the uniform limit of a sequence of rational functions on U_M without poles in U_M.

4.5.2   Fundamental Domains for $\Gamma \backslash \Omega$

This subsection is motivated by an essential construction in the theory of Schottky groups, that of fundamental domains. Schottky groups have been first introduced by Schottky in 1877 in the complex setting; they are useful to analytically uniformise compact Riemann surfaces. In the years 1970, after the work of Tate on p-adic uniformisation of elliptic curves with split multiplicative reduction, Mumford discovered how to p-adically uniformise smooth projective curves of genus $g \geq 2$ with 'split degenerate stable reduction' by using p-adic Schottky groups Γ acting on non-archimedean variants Ω_Γ of the classical complex upper-half plane. The reader is encouraged to read the modern contribution of Poineau-Turchetti to this volume

[Poi20a, Poi20b]. An older reference is [Ger80]; it also contains determinant tools to explore this profound theory. Consequently, we will not give all the details, this would bring us too far away from our path.

Let us recall that, given a local field L with valuation $|\cdot|$, the group $\mathrm{PGL}_2(L)^4$ acts on the rigid analytic projective line $\mathbb{P}^{1,\mathrm{an}}_F$ where F is the completion of an algebraic closure of L (see [Ger80, Fre04]). A *Schottky group* over L is a finitely generated subgroup Γ of $\mathrm{PGL}_2(L)$ which is discrete and such that no element but the identity has finite order. Schottky groups are free (see [Ger80, Theorem (3.1)]) this being an important consequence of the fact that they act freely on certain rigid analytic spaces. Every Schottky group Γ over L has a compact *limit set* $\mathcal{L}_\Gamma \subset \mathbb{P}^{1,\mathrm{an}}_F$ so that Γ acts freely over $\Omega_\Gamma := \mathbb{P}^{1,\mathrm{an}}_F \setminus \mathcal{L}_\Gamma$. The quotient space $\Gamma \backslash \Omega_\Gamma$ naturally carries a structure of rigid analytic space over L which is associated with a smooth, geometrically connected, projective curve X_Γ over L, of genus g the rank of Γ. We learn from [Ger80, Theorem (4.3)] that every Schottky group Γ in $\mathrm{PGL}_2(L)$ admits a *good fundamental domain* $\mathfrak{F}_\Gamma$. Without entering the details, for every element $z \in \Omega_\Gamma$ the set of $\gamma \in \Gamma$ such that $\gamma(z) \in \mathfrak{F}_\Gamma$ is non-empty and finite. In fact, if $\gamma \in \Gamma$, then $\mathfrak{F}_\Gamma \cap \gamma(\mathfrak{F}_\Gamma) \neq \emptyset$ if and only if $\gamma \in \{1, \gamma_1^{\pm 1}, \ldots, \gamma_g^{\pm 1}\}$, where $\gamma_1, \ldots, \gamma_g$ freely generate Γ. Moreover, $\mathfrak{F}_\Gamma$ can be written as

$$\mathfrak{F}_\Gamma := \mathbb{P}^{1,\mathrm{an}}_F \setminus \bigsqcup_{i=1}^{2g} D_i$$

where the D_i's are the rigid analytic spaces associated to disks $D^\circ_F(a_i, r_i) = \{z \in F : |z - a_i| < r_i\}$ with $r_i \in |L^\times|$ for all i, such that the disks $D_F(a_i, r_i) = \{z \in F : |z - a_i| \leq r_i\}$ are pairwise disjoint. One can therefore see easily that $\mathfrak{F}_\Gamma$ carries a structure of rigid analytic variety over F (read also [Poi20b, §II.3.1] along with its more general settings and the theory of uniformisation of Mumford curves).

The interesting point in this discussion is that if we set $L = K_\infty = \mathbb{F}((\pi))$, $A = H^0(C \setminus \{\infty\}, \mathcal{O}_C) \subset K_\infty$, $F = \mathbb{C}_\infty$ etc. the group $\mathrm{PGL}_2(A)$ acts on $\Omega = \mathbb{P}^{1,an}_F \setminus \mathbb{P}^{1,an}_{K_\infty}$ but the action is in general not free; there usually are elliptic points (this happens, for instance, when $[\mathbb{F} : \mathbb{F}_q]$ is odd, see [Mas15]). Even more seriously, the group itself is not finitely generated (see Serre's book [Ser80a] for more details), so that $\mathrm{PGL}_2(A)$ *is not* a Schottky group.

4.5.2.1   Some Structural Properties of $\Gamma = \mathrm{GL}_2(A)$

For the purposes of the present paper, we will be content to study the case in which C has genus 0, so that in Lemma 4.2.5 we have $V = \{0\}$ and therefore, $K_\infty \cong A \oplus \mathcal{M}_{K_\infty}$. It is easy to see that there exists a uniformiser π of K_∞ such that

---

[4]Projective linear group over L, defined as the quotient of $\mathrm{GL}_2(L)$ by its center.

$\mathbb{F}A = \mathbb{F}[\pi^{-1}]$. We can indeed choose $\pi = \theta^{-1}$ where θ is any element of A with a simple pole at ∞. In particular, $\mathbb{F}A = \mathbb{F}[\theta]$.

It is not difficult to show that the group $GL_2(\mathbb{F}[\theta])$ is generated by its subgroups $GL_2(\mathbb{F})$ (finite) and the Borel subgroup $B(*) = \{\left(\begin{smallmatrix} * & * \\ 0 & * \end{smallmatrix}\right)\}$. In fact, a Theorem of Nagao in [Nag59] asserts that, given a field k and an indeterminate t,

$$GL_2(k[t]) = GL_2(k) *_{B(k)} B(k[t]), \qquad (4.10)$$

where $*_{B(k)}$ denotes the *amalgamated product* along $B(k)$, which is by definition the quotient of the free product $GL_2(k) * B(k[t])$ by the normal subgroup generated by those elements arising from the natural identifications existing between the elements of $B(k) * 1$ and $1 * B(k)$ coming from the maps

$$GL_2(k) \to GL_2(k) * B(k[t]) \leftarrow B(k[t])$$

(a gluing along compatibility conditions). Note that $B(k[t])$ is not finitely generated, so that $GL_2(k[t])$ is not finitely generated (this is trivial if k is infinite) in contrast with a theorem of Livingston, asserting that $GL_n(k[t])$ is finitely generated if $n \geq 3$, and also with the more familiar result that $SL_2(\mathbb{Z}) \cong \mathbb{Z}/2\mathbb{Z} * \mathbb{Z}/3\mathbb{Z}$ so that it is, in particular, finitely presented.

Corollary 4.5.5 $PGL_2(\mathbb{F}[\theta])$ *is not a Schottky group.*

4.5.2.2   Bruhat-Tits Trees and 'Good Fundamental Domains'

We recall that $K_\infty = \mathbb{F}((\pi))$ for a uniformiser π, with $\mathbb{F}$ a finite extension of $\mathbb{F}_q$. Our first task is to describe a combinatorial structure which allows to 'move inside' Ω, the Bruhat-Tits tree; in practice, we can 'move along annuli'. The second task, in the case $A = \mathbb{F}[\theta]$, is to construct a subset of Ω that we can qualify as a 'good fundamental domain' for the homography action of $GL_2(A)$ over Ω, being understood that $GL_2(A)$ is not a Schottky group.

We recall that if $x \in \mathbb{C}_\infty$, $D^\circ_{\mathbb{C}_\infty}(x, r) = \{z \in \mathbb{C}_\infty : |z - x| < r\}$. Let S be a subset of $\mathbb{C}_\infty^\times$ such that if $x, x' \in S$ are distinct, $|x - x'| = \max\{|x|, |x'|\}$. Then, with $x \in S$, the sets

$$D_x := D^\circ_{\mathbb{C}_\infty}(x, |x|) = x + D^\circ_{\mathbb{C}_\infty}(0, |x|)$$

are pairwise distinct subsets of $\mathbb{C}_\infty^\times$. Indeed, clearly, they do not contain 0. Moreover, if $x \neq x'$ we have $y \in D^\circ_{\mathbb{C}_\infty}(x, |x|) \cap D^\circ_{\mathbb{C}_\infty}(x', |x'|)$ if and only if we can find $z \in D^\circ_{\mathbb{C}_\infty}(x, |x|), z' \in D^\circ_{\mathbb{C}_\infty}(x', |x'|)$, such that $y = x + z = x' + z'$ with $|z| < |x|$ and $|z'| < |x'|$, so that $|z - z'| < \max\{|x|, |x'|\}$. This means that $\max\{|x|, |x'|\} > |z - z'| = |x' - x| = \max\{|x|, |x'|\}$ which is impossible.

We choose, for any element $r \in \mathbb{Z}_{>1}$, an element, denoted by $\pi^{\frac{1}{r}} \in \mathbb{C}_\infty$, with the property that $(\pi^{\frac{1}{r}})^r = \pi$, which exists because $\mathbb{C}_\infty$ is algebraically closed. The set $\Sigma := \{(\pi^{\frac{1}{r}})^s\}$ inherits the total order of $\mathbb{R}$ by $\frac{s}{r} \in \mathbb{Q} \subset \mathbb{R}$.[5] We have observed (after Lemma 4.2.9) that the valuation group of $|\cdot|$ is $|\pi|^{\mathbb{Q}}$. Hence if $z \in \mathbb{C}_\infty^\times$, we can find r, s relatively prime, unique, such that $|z(\pi^{\frac{1}{r}})^s| = 1$. Since the residue field of $\mathbb{C}_\infty$ is $\mathbb{F}^{ac}$, we obtain that there exists a unique $\zeta \in (\mathbb{F}^{ac})^\times$ such that

$$|z - \zeta(\pi^{\frac{1}{r}})^s| < |z|.$$

If we set $S = \{\zeta(\pi^{\frac{1}{r}})^s : \zeta \in (\mathbb{F}^{ac})^\times, r > 1, s \in \mathbb{Z}$ such that r, s are relatively prime$\}$, then for all $x, x' \in S$ distinct we have $|x - x'| = \max\{|x|, |x'|\}$ and we obtain a partition of $\mathbb{C}_\infty^\times$:

$$\mathbb{C}_\infty^\times = \bigsqcup_{x \in S} D_x. \tag{4.11}$$

Let us now consider the subset

$$\widetilde{S} := \{x \in S : |x| \notin |\pi|^{\mathbb{Z}}\} \sqcup \{\zeta \pi^n : \zeta \in \mathbb{F}^{ac} \setminus \mathbb{F}, n \in \mathbb{Z}\} \subset S.$$

With it, we can still somehow reconstruct $\mathbb{C}_\infty$. Indeed, the reader can easily see that if $x \in \widetilde{S}$, $D_x \cap K_\infty = \emptyset$ and

$$\mathbb{C}_\infty = \left(K_\infty + \bigsqcup_{x \in \widetilde{S}} D_x\right) \sqcup K_\infty.$$

As a consequence we have

$$\Omega = K_\infty + \bigsqcup_{x \in \widetilde{S}} D_x$$

and

$$\bigsqcup_{x \in \widetilde{S}} D_x = \{z \in \Omega : |z| = |z|_\Im\}.$$

---

[5]Thanks to Lemma 4.2.10 we can even additionally suppose that the elements $\pi^{\frac{1}{r}}$ are chosen in such a way that $\Sigma = \pi^{\mathbb{Q}}$ is a subgroup of $\mathbb{C}_\infty^\times$.

We observe that if $\lambda \in \mathbb{Q} \setminus \mathbb{Z}$, then

$$\bigsqcup_{\substack{x \in S \\ |x|=|\pi|^\lambda}} D_x = \bigsqcup_{\substack{x \in \widetilde{S} \\ |x|=|\pi|^\lambda}} D_x = \{z \in \mathbb{C}_\infty : |z| = |\pi|^\lambda\} =: C_\lambda.$$

We also set, for $\lambda \in \mathbb{Z}$,

$$C_\lambda := \bigsqcup_{\zeta \in \mathbb{F}^{ac} \setminus \mathbb{F}} D^\circ_{\mathbb{C}_\infty}(\zeta \pi^\lambda, |\pi|^\lambda).$$

Note that $C_\lambda = \{z \in \Omega : |z| = |z|_\Im = |\pi|^\lambda\}$ for all $\lambda \in \mathbb{Q}$. For all λ, the set C_λ is invariant by translation of elements in $D_{K_\infty}(0, |\pi|^{\lceil \lambda \rceil})$, where $\lceil \cdot \rceil$ denotes the smallest of the integers which are larger than $(\cdot)$. If $\alpha \in K_\infty \setminus D_{K_\infty}(0, |\pi|^{\lceil \lambda \rceil}) = \oplus_{i \le \lfloor \lambda \rfloor} \mathbb{F}\pi^i$ (with $\lfloor \cdot \rfloor$ the largest integer which is smaller than $(\cdot)$) then $C_\lambda \cap (\alpha + C_\lambda) = \emptyset$. We have obtained the next result.

Lemma 4.5.6 *The following partition of Ω holds:*

$$\Omega = \bigsqcup_{\substack{\lambda \in \mathbb{Q} \\ \alpha \in K_\infty \setminus D_{K_\infty}(0, |\pi|^{\lceil \lambda \rceil})}} \alpha + C_\lambda.$$

Note that this can be very easily used to construct admissible coverings of Ω. The above is the crucial statement which allows to construct the Bruhat-Tits tree associated to Ω. It relies on the existence of a natural partial ordering on the set $\mathcal{T} := \{\alpha + C_\lambda : \alpha \in K_\infty \setminus D_{K_\infty}(0, |\pi|^{\lceil \lambda \rceil}), \lambda \in \mathbb{Q}\}$. We declare that $\alpha + C_\lambda \succ \alpha' + C_{\lambda'}$ if $C_\lambda \succ \alpha' - \alpha + C_{\lambda'}$ and $C_\lambda \succ \alpha' + C_{\lambda'}$ if $\lambda < \lambda'$ and $\alpha' + C_\lambda = C_\lambda$. For example, for $\lambda \notin \mathbb{Z}$, $\alpha + C_\lambda \succ C_\lambda$ if and only if $\alpha + C_\lambda = C_\lambda$ if and only if $|\alpha| \le |\pi|^\lambda$. Then $\mathcal{T}$ can be enriched with the structure of a *tree*, the *Bruhat-Tits tree*. We recall that a tree $\mathcal{T}$ is a metric space such that, on one side, for any distinct points P, P' of $\mathcal{T}$ there exists one and only one topological arc in $\mathcal{T}$ of extremities P, P' and, on the other side, this arc is isometric to an interval of $\mathbb{R}$ (this definition is due to Tits). A tree has edges and vertices. The *vertices* of our Bruhat-Tits tree $\mathcal{T}$ are represented by the subsets $\alpha + C_\lambda$ of $\mathbb{C}_\infty$ with $\lambda \in \mathbb{Z}$ and the *edges* are represented by real intervals $]n - 1, n[$ with $n \in \mathbb{Z}$, with the extremities given by a couple of vertices $(\alpha + C_{n-1}, \alpha' + C_n)$ such that $\alpha' + C_{n-1} = C_{n-1}$. The intervals are oriented and our tree itself acquires an orientation. The upper direction is that of the negative λ's or, alternatively, of the larger $|z|_\Im$'s. The edges are therefore organised so that at every lower (for the ordering) extremity the vertex is a $q^{d_\infty} + 1$ branching point with q^{d_∞} edges below and one above (with respect to the orientation).

The next picture represents a small piece of $\mathcal{T}$ for $q^{d_\infty} = 2$.

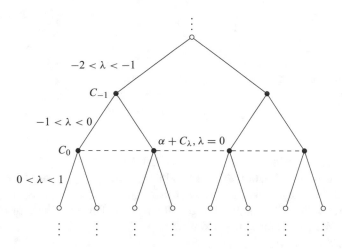

Also note that the euclidean closure of the image in $\mathcal{T}$ of any set $\alpha + \sqcup_{\lambda \in \mathbb{Q}} C_\lambda$ for $\alpha \in K_\infty$ fixed is isometric to $\mathbb{R}$ and any two such sets, if distinct, have a upper half-line in common. Any element of Ω is K_∞-translation equivalent to finitely many elements in $\sqcup_{\lambda \in \mathbb{Q}} C_\lambda$ and finally, the homography action of $GL_2(K_\infty)$ over Ω is compatible with a continuous action over $\mathcal{T}$ in a way that can be made completely explicit.

The structure of the spaces $\mathbb{C}_\infty$ and Ω may look topologically very complicate but the Bruhat-Tits tree is some kind of 'central nervous system' which allows to obtain a combinatorial picture of these spaces (or rather, their admissible coverings) and to move in their interior, by means of the *reduction map*, which is $GL_2(K_\infty)$-equivariant

$$\mathrm{red} : \Omega \to \mathcal{T},$$

defined by $z \mapsto \alpha + C_\lambda \in \mathcal{T}$ where $\alpha + C_\lambda$ is the unique element of the partition of Lemma 4.5.6 such that $z \in \alpha + C_\lambda$. This presentation may look different, it is in fact essentially equivalent to that of Teitelbaum in [Tei91, Preliminaries] (see also Teitelbaum's chapter in [Bak08]). To help the reader to connect with the formalism of Teitelbaum, which also is that of [Ger80], note that the set $U(1)$ of [Tei91, p. 492] plays the role of our disjoint union $\sqcup_{\lambda \in]-1,1[} C_\lambda$ and that the set V introduced one page later is equal to our $\sqcup_{\lambda \in]-1,0[} C_\lambda$. The sets $\gamma(U(1))$ for $\gamma \in GL_2(K_\infty)$ define an admissible covering of Ω and $\mathcal{T}$ can be alternatively constructed defining edges and vertices by a criterion of overlapping for the various $\gamma(U(1))$'s and an identification between the set $\{\gamma(U(1)) : \gamma \in GL_2(K_\infty)\}$ and the quotient $GL_2(\mathcal{O}_{K_\infty}) \backslash GL_2(K_\infty)$, corresponding to the vertices.

We set $\mathfrak{F} := \{z \in \Omega : |z| = |z|_{\Im} \geq 1\}$. By Lemma 4.5.6, we have $\mathfrak{F} = \sqcup_{\lambda \leq 0} C_\lambda$ and $\mathrm{red}(\mathfrak{F})$ is an upper half-line in $\mathcal{T}$. We deduce that

$$\mathfrak{F} = \mathbb{C}_\infty \setminus \left(\bigsqcup_{\zeta \in \mathbb{F}} D_{\mathbb{C}_\infty}^\circ(\zeta, 1) \sqcup \bigsqcup_{n \geq 1} \bigsqcup_{\zeta \in \mathbb{F}^\times} D_{\mathbb{C}_\infty}^\circ(\zeta \pi^{-n}, |\pi|^{-n}) \right).$$

We now focus on the case $A = \mathbb{F}[\theta]$. For $z \in \Omega$ we denote by $\mathfrak{F}_z$ the set $\{z' \in \mathfrak{F} :$ there exists $\gamma \in \mathrm{GL}_2(A)$ such that $\gamma(z) = z'\} \subset \mathfrak{F}$. We show:

Proposition 4.5.7 *For all z, the set $\mathfrak{F}_z$ is non-empty and finite.*

Proof If $z \in \mathfrak{F}$ then there exists $x \in \widetilde{S}$ such that $|x| \geq 1$ and $z \in D_x$ and we see that the set of $a \in A$ such that $z - a \in \mathfrak{F}$ is finite. Note that $1/z \in D_{1/x}$ so that $1/z \notin \mathfrak{F}$ (in fact, if $\gamma = \left(\begin{smallmatrix} 0 & 1 \\ 1 & 0 \end{smallmatrix}\right)$, $\gamma(D_x) = D_{\gamma(x)}$). From Nagao's Theorem we deduce that the set $\{\gamma \in \mathrm{GL}_2(A) : \gamma(z) \in \mathfrak{F}\}$ is finite so that, for all $z \in \Omega$, $\mathfrak{F}_z$ is finite (but note that the cardinality is not uniformly bounded in terms of z). Let z be in Ω. If $|z|_{\Im} \geq 1$ there exists $a \in A$ such that $|z - a| = |z|_{\Im}$ and $\mathfrak{F}_z$ is non-empty. All we need to show is that if $z \in \Omega$ is such that $|z|_{\Im} < 1$, then there exists $\gamma \in \mathrm{GL}_2(A)$ such that $\gamma(z) \in \mathfrak{F}$. To see this, there is no loss of generality in supposing that $|z| < 1$. Indeed, we can replace z with $z - a$ for $a \in A$. We can therefore write:

$$z = w + x + y$$

where $w \in \oplus_{i=1}^{\lfloor \lambda \rfloor} \mathbb{F}\pi^i \in \mathcal{M}_{K_\infty}$, $x \in \widetilde{S}$ with $|x| = |\pi|^\lambda$ and $y \in D_x$. Applying $\gamma = \left(\begin{smallmatrix} 0 & 1 \\ 1 & 0 \end{smallmatrix}\right)$ we see that z is $\mathrm{GL}_2(A)$-equivalent to an element $z' \in \alpha' + C_{\lambda'}$ with λ' such that $\lambda - \lambda' \in \mathbb{Z}_{>1}$ and $\alpha' \in \oplus_{i \leq \lfloor \lambda' \rfloor} \mathbb{F}\pi^i$, so that, in particular, $|z'|_{\Im} > |z|_{\Im}$. We can iterate this process with z' playing the role of z. The fact that $\lambda - \lambda' \in \mathbb{Z}_{\geq 1}$ implies that z is $\mathrm{GL}_2(A)$-equivalent to an element of $\mathfrak{F}$ and $\mathfrak{F}_z$ is non-empty. $\qquad \square$

This seems enough to allow us calling $\mathfrak{F}$ a 'good fundamental domain' for $\Gamma \backslash \Omega$ with $A = \mathbb{F}[\theta]$, even though it is undoubtedly not as well behaved as the good fundamental domains in the framework of Schottky groups. Note that $\Gamma \backslash \mathcal{T}$ contains an 'end': this metric space is not compact, but can be made compact with the addition of one point represented by one of the upper half-lines contained by $\mathcal{T}$ which, at the level of $\Gamma \backslash \Omega$, corresponds to a 'cusp'.

Similar constructions are possible for $\Gamma = \mathrm{GL}_2(A)$ with a more general projective curve $\mathcal{C}$ but we do not describe them here. In this broader case it is possible to show that $\Gamma \backslash \mathcal{T}$ has the structure of a finite graph with finitely many ends attached to it. More general 'fundamental domains' can be constructed from the Bruhat-Tits tree of Ω and constructed by Serre (see [Ser80a, Theorem 10]) thanks to a more refined interpretation of the elements of $\Gamma \backslash \Omega$ as classes of rank two vector bundles over $\mathcal{C}$. We refer to ibid. for the details.

4.5.3 An Elementary Result on Translation-Invariant Functions Over Ω

We recall that $\mathcal{H}$ denotes the complex upper-half plane. Let $f : \mathcal{H} \to \mathbb{C}$ be a holomorphic function such that for all $n \in \mathbb{Z}$ and for all $z \in \mathcal{H}$, $f(z+n) = f(z)$. Then, we can expand

$$f(z) = \sum_{n \in \mathbb{Z}} f_n e^{2\pi i n z}, \quad f_n \in \mathbb{C},$$

a series which is convergent for $q(z) = e^{2\pi i z}$ in

$$\dot{D}^{\circ}_{\mathbb{C}}(0, 1) = \{z \in \mathbb{C} : 0 < |z| < 1\}$$

the punctured open unit disk centered at 0 of $\mathbb{C}$ or equivalently, for z in every horizontal strip of finite height in $\mathcal{H}$ (note that they are invariant by horizontal translation).

4.5.3.1 A Digression

The proof of the above statement for f is simple and we can afford a short digression. The function $z \mapsto q(z)$ does not allow a global holomorphic section $\mathcal{H} \leftarrow \dot{D}^{\circ}_{\mathbb{C}}(0, 1)$. But we can cover $\mathbb{C}^{\times}$ with say, three open half-planes U_1, U_2, U_3, and there are sections s_1, s_2, s_3 defined and holomorphic over U_1, U_2, U_3 such that $s_i - s_j \in \mathbb{Z}$ over $U_i \cap U_j$ for all i, j. Let f be holomorphic on $\mathcal{H}$ such that $f(z + 1) = f(z)$ for all $z \in \mathcal{H}$. Define $g_i(q) = f(s_i(q))$ for all $i = 1, 2, 3$. Then, the compatibility conditions and the fact that the pre-sheaf of holomorphic functions over any open set is a sheaf (the well known principle of analytic continuation) ensure that this defines a holomorphic function $g(q)$ over $\dot{D}^{\circ}_{\mathbb{C}}(0, 1)$. But the ring of holomorphic functions over $\dot{D}^{\circ}_{\mathbb{C}}(0, 1)$ is precisely that of the convergent double series $\sum_{n \in \mathbb{Z}} f_n q^n$, as one can easily see, and our claim follows. One also deduces that there is an isomorphism of Riemann's surfaces

$$\mathcal{H}/\mathbb{Z} \cong \dot{D}^{\circ}_{\mathbb{C}}(0, 1)$$

induced by $e^{2\pi i z}$, concluding the digression.

We now come back to our characteristic $p > 0$ setting and we suppose, from now on, that

$$A = H^0(\mathbb{P}^1_{\mathbb{F}_q} \setminus \{\infty\}, \mathcal{O}_{\mathbb{P}^1_{\mathbb{F}_q}}).$$

We note that Ω is invariant by translations of $a \in A$ and the function

$$\exp_A(z) = z \prod_{a \in A \setminus \{0\}} \left(1 - \frac{z}{a}\right) = \tilde{\pi}^{-1} \exp_C(\tilde{\pi} z)$$

is an entire function $\mathbb{C}_\infty \to \mathbb{C}_\infty$, $\mathbb{F}_q$-linear, surjective, of kernel $A = \mathbb{F}_q[\theta]$, hence also invariant by translations by elements of A. It is thus natural to ask for an analogue statement of the above, complex one. Consider $R \in |\mathbb{C}_\infty^\times|$. Now, note that A acts on $\Omega_R = \{z \in \Omega : |z|_\Im \geq R\}$ by translations. Giving $A \setminus \Omega_R$ the quotient topology we have:

Lemma 4.5.8 *There is $S \in |\mathbb{C}_\infty^\times|$ such that the function $\exp_A$ induces a homeomorphism of topological spaces*

$$A \setminus \Omega_R \to \{z \in \mathbb{C}_\infty : |z| \geq S\}.$$

Proof From the Weierstrass product expansion we see that, setting

$$S := \max_{z \in D_{\mathbb{C}_\infty}(0,R)} |\exp_A(z)| =: \| \exp_A \|_R = \|z\|_R \prod_{\substack{a \in A \\ a \neq 0}} \left\| 1 - \frac{z}{a} \right\|_R = R \prod_{\substack{a \in A \\ a \neq 0 \\ |a| < R}} \frac{R}{|a|},$$

$\exp_A(D(0, R)) = D(0, S)$ by Corollary 4.2.8. Hence, $D^\circ(0, S) = D^\circ_{\mathbb{C}_\infty}(0, S) = \exp_A(D^\circ(0, R))$ from which we deduce that

$$\{z \in \mathbb{C}_\infty : |\exp_A(z)| < S\} = A + D^\circ(0, R).$$

Recall that $K_\infty = A \oplus \mathcal{M}_{K_\infty}$. If $R \geq 1$, we have $D^\circ(0, R) \supset \mathcal{M}_{K_\infty}$. Now observe that

$$\{z \in \mathbb{C}_\infty : |z|_\Im < R\} = \cup_{a \in K_\infty} D^\circ(a, R) = \cup_{a \in A} D^\circ(a, R).$$

Therefore we have the chain of identities

$$A + D^\circ(0, R) = K_\infty + D^\circ(0, R) = \cup_{a \in K_\infty} D^\circ(a, R) = \{z \in \mathbb{C}_\infty : |z|_\Im < R\} = \Omega \setminus \Omega_R,$$

and taking complementaries, we see that

$$\Omega_R = \{z \in \mathbb{C}_\infty : |\exp_A(z)| \geq S\}, \quad R \geq 1.$$

$\square$

4.6 Some Quotient Spaces

Our topologies are totally disconnected and Lemma 4.5.8 is weaker if compared with analogous statements in the complex setting. Fortunately there is a structure of *quotient analytic space* over Ω_R/A, and it is isomorphic to the analytic structure of the complementary of the disk $D^\circ(0, S)$.

4.6.1 A-Periodic Functions Over Ω

We suppose that $A = \mathbb{F}_q[\theta]$ all along this subsection. The analogue for $\Omega = \mathbb{C}_\infty \setminus K_\infty$ of the simple claim over $\mathbb{C}$ of the beginning of Sect. 4.5.3 and the proof in Sect. 4.5.3.1 is not as easy to prove but it is true, and not too difficult. In fact, the following result holds:

Proposition 4.6.1 *Let* $f : \Omega \to \mathbb{C}_\infty$ *be an analytic function such that* $f(z + a) = f(z)$ *for all* $a \in A$. *Then, there exists* $S \in |\mathbb{C}_\infty^\times|$, $S < 1$, *such that*

$$f(z) = \sum_{n \in \mathbb{Z}} f_n \exp_A(z)^n, \quad f_n \in \mathbb{C}_\infty,$$

the series being uniformly convergent for $\exp_A(z)^{-1}$ *in every annulus of* $\dot{D}^\circ_{\mathbb{C}_\infty}(0, S) = \{x \in \mathbb{C}_\infty : 0 < |x| < S\}$, $S \in |\mathbb{C}_\infty^\times|$, *small enough.*

To prove this result and to motivate the proof we are giving, we need some preparation.

4.6.1.1 Analytification and quotients

Let $\mathcal{X}$ be a rigid analytic variety over a valued field L, complete and algebraically closed. Let us consider a group Γ acting on $\mathcal{X}$ with 'admissible action'. 'Admissible action' means that $\mathcal{X}$ can be covered by Γ-stable admissible subsets and that Γ acts through an embedding ι of Γ in $\mathrm{Aut}(\mathcal{X})$, topological group, and the image is discrete. We are interested in such triples

$$(\mathcal{X}, \Gamma, \iota).$$

For example, we can take $\Gamma = A$ acting on Ω or $\mathbb{A}^1_{\mathbb{C}_\infty}$ by translations (the theme of Proposition 4.6.1) or $\Gamma = \mathrm{GL}_2(A)$ acting on Ω by homographies (the theme of the text).

The quotient map

$$\mathcal{X} \to \Gamma \backslash \mathcal{X}$$

can be used to define a structure of Grothendieck ringed space on the quotient $\Gamma\backslash\mathcal{X}$. A subset of $\Gamma\backslash\mathcal{X}$ is admissible if its pre-image is admissible, and the sections are Γ-invariant $\mathbb{C}_\infty$-valued functions over pre-images of Γ-invariant subsets. One needs conditions under which the quotient acquires a structure of rigid analytic space. For example, a finite group Γ acting on $\mathcal{X} = \mathrm{Spm}(\mathcal{A})$ affinoid variety which allows a covering by invariant admissible subsets gives rise to an isomorphism of affinoid varieties $\Gamma\backslash\mathrm{Spm}(\mathcal{A}) \to \mathrm{Spm}(\mathcal{A}^\Gamma)$, where $\mathcal{A}^\Gamma$ is the sub-algebra of Γ-invariant elements of $\mathcal{A}$; see [Bos84, §6.3.3]. See also Hansen's more general [Han20, Theorem 1.3].

We invoke the analytification functor in Sect. 4.5.1.1 by choosing $\mathcal{X} = X^{an}$. If X is a scheme of finite type over L with an 'admissible action' of a finite group Γ 'admissible', now in the algebraic sense that there is a covering with Γ-invariant affine sub-schemes, it can be proved that there exists a unique scheme structure (of finite type over L) on the ringed quotient space

$$p : X \to \Gamma\backslash X.$$

The following proposition is due to Amaury Thuillier: we warmly thank him for having brought our attention to it.

Proposition 4.6.2 *The canonical map* $\Gamma\backslash X^{an} \to (\Gamma\backslash X)^{an}$ *is an isomorphism of rigid analytic varieties.*

Proof We can suppose, without loss of generality, $X = \mathrm{Spec}(\mathcal{A})$ affine, so that $\Gamma\backslash X = \mathrm{Spec}(\mathcal{A}^\Gamma)$. In terms of algebras, we have (horizontal arrows are surjective and vertical arrows injective, and $\widehat{L[\underline{t}]}$ is the standard Tate L-algebra in the variables $\underline{t} = (t_1, \ldots, t_N)$ for some N):

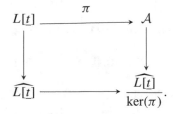

Then we have:

$$\mathcal{A}_V := \frac{\widehat{L[\underline{t}]}}{\ker(\pi)} = \mathcal{A} \otimes_{L[\underline{t}]} \widehat{L[\underline{t}]} = H^0(V, \mathcal{O}_{X^{an}})$$

where $V := \mathrm{Spm}(\mathcal{A} \otimes_{L[\underline{t}]} \widehat{L[\underline{t}]}) \subset (\Gamma\backslash X)^{an}$.

The L-algebra $\mathcal{B} = \mathcal{A} \otimes_{\mathcal{A}^\Gamma} \mathcal{A}_V$ is finite over $\mathcal{A}_V$, hence it inherits a structure of affinoid L-algebra. We deduce, with $p^{an} : X^{an} \to (\Gamma\backslash X)^{an}$ the analytification of p, that $W = (p^{an})^{-1}(V)$ is a Γ-invariant affinoid domain of X^{an} and $\mathcal{A}_W = H^0(W, \mathcal{O}_{X^{an}}) = \mathcal{B}$. The quotient space $\Gamma\backslash W$ is also affinoid, of algebra $\mathcal{B}^\Gamma$ (see

[Bos84, §6.3.3]). Therefore, all we need to show is that the canonical morphism

$$\mathcal{A}_V \to \mathcal{B} = \mathcal{A} \otimes_{A^\Gamma} \mathcal{A}_V$$

induces an isomorphism $\mathcal{A}_V \to \mathcal{B}^\Gamma = (\mathcal{A} \otimes_{A^\Gamma} \mathcal{A}_V)^\Gamma$.

The morphism $\mathcal{A} \to \mathcal{A}_V$ is flat [Ber90, Theorem 3.4.1, (ii)]. Therefore the exact sequence

$$0 \to \mathcal{A}^\Gamma \to \mathcal{A} \xrightarrow{\oplus(g - \mathrm{Id}_\mathcal{A})} \bigoplus_{g \in \Gamma} \mathcal{A}$$

yields an exact sequence

$$0 \to \mathcal{A}_V = \mathcal{A}^\Gamma \otimes_{A^\Gamma} \mathcal{A}_V \to \mathcal{A} \otimes_{A^\Gamma} \mathcal{A}_V \xrightarrow{\oplus(g - \mathrm{Id}_\mathcal{A})} \bigoplus_{g \in \Gamma} \mathcal{A} \otimes_{A^\Gamma} \mathcal{A}_V.$$

We have thus that $\mathcal{A}_V$ is equal to the kernel of the last arrow, which is just $\mathcal{B}^\Gamma$.   □

We consider $L = \mathbb{C}_\infty$ and we denote by $A(n)$ the $\mathbb{F}_q$-vector space $\{a \in A : |a| < |\theta|^n\}$ (dimension n and cardinality q^n). If $X = \mathbb{A}^1_{\mathbb{C}_\infty}$ and we look at $\Gamma = A(n)$ acting on X by translations, we have the quotient scheme $\Gamma \backslash X = \mathrm{Spec}(\mathbb{C}_\infty[x]^\Gamma)$. Note that $\mathbb{C}_\infty[x]^\Gamma = \mathbb{C}_\infty[E_n(x)]$ with E_n characterised by Proposition 4.4.8, by Euclidean division. Proposition 4.6.2 applies.

We introduce the sets for $n \geq 1$

$$\mathcal{B}_n = D^\circ_{\mathbb{C}_\infty}(0, |\theta|^n) \setminus \bigcup_{a \in A(n)} D^\circ_{\mathbb{C}_\infty}(a, 1).$$

We define, in parallel, with $l_n = (\theta - \theta^q) \cdots (\theta - \theta^{q^n})$:

$$\mathcal{C}_n = D^\circ_{\mathbb{C}_\infty}(0, |l_n|) \setminus D^\circ_{\mathbb{C}_\infty}(0, 1).$$

Each of these sets has an admissible covering by affinoid subsets so that it is a rigid analytic sub-variety of $\mathbb{A}^{1,an}_{\mathbb{C}_\infty}$. A function $f : \mathcal{B}_n \to \mathbb{C}_\infty$ is analytic if its restriction to every affinoid subset is analytic. Note that $\mathcal{B}_n \subset \mathcal{B}_{n+1}$ and $\mathcal{C}_n \subset \mathcal{C}_{n+1}$ for all $n \geq 1$. We set

$$\psi_m := 1 - \frac{\tau}{l_m^{q-1}}, \quad m \geq 0$$

(recall that $\tau(x) = x^q$ for $x \in \mathbb{C}_\infty$). It is easy to see that ψ_n induces an isometric biholomorphic isomorphism of $\mathcal{C}_m$ for all $n \geq m$. In particular the non-commutative infinite product

$$\mathcal{F}_n := \cdots \left(1 - \frac{\tau}{l_{n+1}^{q-1}}\right)\left(1 - \frac{\tau}{l_n^{q-1}}\right) \in K[[\tau]]$$

induces an isometric biholomorphic isomorphism of $\mathcal{C}_n$ (for every n).

In a similar vein, Proposition 4.6.2 implies:

Corollary 4.6.3 *The function $\mathcal{E}_n = l_n E_n$ is a degree q^n étale covering $\mathcal{B}_n \to \mathcal{C}_n$ which induces an isomorphism of rigid analytic spaces*

$$A(n)\backslash\mathcal{B}_n \to \mathcal{C}_n,$$

where the analytic structure on the pre-image is induced by the analytification of $\mathrm{Spec}(\mathbb{C}_\infty[x]^{A(n)})$.

4.6.1.2 Proof of Proposition 4.6.1

A global section g_n of $\mathcal{O}_{\mathcal{C}_n}$ can be identified, in a unique way, with a convergent series

$$\sum_{k\in\mathbb{Z}} g_k^{(n)} x^k, \quad g_k^{(n)} \in \mathbb{C}_\infty.$$

Let $f : \Omega \to \mathbb{C}_\infty$ be a rigid analytic function with the property that for all $a \in A$, $f(z+a) = f(z)$. We fix $m > 0$, let n be such that $n \geq m$. Then, $f : \mathcal{B}_n \to \mathbb{C}_\infty$ is holomorphic such that $f(z+a) = f(z)$ for all $a \in A(n)$ and therefore there exists a unique $g_n \in \mathcal{O}_{\mathcal{C}_n}$ such that $f(z) = g_n(\mathcal{E}_n(z))$ over $\mathcal{C}_n$ and we can write:

$$f(z) = \sum_{k\in\mathbb{Z}} g_k^{(n)} (\mathcal{E}_n(z))^k.$$

We observe that $\mathcal{B}_m \subset \mathcal{B}_n$. Thus, we have the following commutative diagram for $n > m$, where the left vertical arrows are the identity, and the bottom right vertical arrow is ψ_m, while the top one is $\psi_{m+1,n}$, where $\psi_{m,n}$ is the composition $\psi_{m,n} := \psi_{n-1} \circ \cdots \circ \psi_m$:

$$\begin{array}{ccc} \mathcal{B}_m & \xrightarrow{\mathcal{E}_n} & \mathcal{C}_m \\ \uparrow & & \uparrow \\ \mathcal{B}_m & \xrightarrow{\mathcal{E}_{m+1}} & \mathcal{C}_m \\ \uparrow & & \uparrow \\ \mathcal{B}_m & \xrightarrow{\mathcal{E}_m} & \mathcal{C}_m, \end{array}$$

and there also exists a unique $g_m \in \mathcal{O}_{\mathcal{C}_m}$ such that $f(z) = g_m(\mathcal{E}_m(z))$, this time over $\mathcal{C}_m \subset \mathcal{C}_n$ so that, noticing that $\psi_{m,n}$ induces an isometric biholomorphic isomorphism of $\mathcal{C}_m$, we must have:

$$g_n(\psi_{m,n}(x)) = g_m(x), \quad x \in \mathcal{C}_m.$$

In particular, we have the equality

$$g_{n+1}(\psi_n(x)) = g_n(x), \quad x \in \mathcal{C}_m.$$

Since $\psi_n(x) = x(1 - (\frac{x}{l_n})^{q-1})$ and $\psi_n(x)^k = x^k(1 + \sigma_{n,k}(x))$ with $|\sigma_{n,k}(x)| \le |\frac{x}{l_n}|^{q-1} < 1$ for all $n \ge m$, $k \in \mathbb{Z}$, we deduce that the function $g_{n+1} - g_n$ tends to zero uniformly on every admissible subset of $\mathcal{C}_m$, for $n \ge m$. This means that the sequence of functions $(g_n)_{n \ge m}$ converges to an element $g \in \mathcal{O}_{\mathcal{C}_m}$ uniformly on every admissible subset of $\mathcal{C}_m$.

With this new function g the existence of which is given by Cauchy convergence criterion, we can write:

$$g_m(x) = g(\mathcal{F}_m(x)), \quad x \in \mathcal{C}_m.$$

We use the results of Sect. 4.4.2 and more precisely Proposition 4.4.9, or with a more manageable notation, (4.6). We thus recall the identity of entire functions:

$$\exp_A = \mathcal{F}_n \underbrace{\left(1 - \frac{\tau}{l_{n-1}^{q-1}}\right) \cdots \left(1 - \frac{\tau}{l_1^{q-1}}\right) (1 - \tau)}_{\mathcal{E}_n}.$$

In particular, by uniqueness:

$$f(z) = g(\exp_A(z)), \quad z \in \mathcal{B}_m, \quad \forall m.$$

Since the sets $\mathcal{B}_n$ cover the set $\Omega_1 := \{z \in \mathbb{C}_\infty : |z|_\Im \ge 1\}$, the result follows.

Restated in more geometric, but essentially equivalent language, the arguments of the proof of Proposition 4.6.1 lead to:

Proposition 4.6.4 *For all $M \in [1, \infty[\cap |\mathbb{C}_\infty^\times|$, the function $z \mapsto \frac{1}{\exp_A}$ yields an isomorphism of rigid analytic spaces $A \backslash \Omega_M \cong \dot{D}_{\mathbb{C}_\infty}(0, S) = D_{\mathbb{C}_\infty}(0, S) \backslash \{0\}$ for some $S \ge 1$ depending on M.*

Problem 4.6.5 The above proof, although simple, is longer than the one we gave in the digression 4.5.3.1, in the complex case. This leads to the following question: is it possible to construct explicitly an admissible covering $(U_i)_i$ of an annulus $D_{\mathbb{C}_\infty}(0, R) \backslash D_{\mathbb{C}_\infty}^\circ(0, r)$ and local inverses $g_i \in \mathcal{O}_{U_i}$ of the function $\exp_A$ or even better, the function $\frac{1}{\exp_A}$, delivering a simpler proof of Proposition 4.6.1 and making no use of the process of analytification?

Also, note that the fact that the Grothendieck ringed space $A\backslash\mathbb{A}^{1,an}$ carries a structure of rigid analytic variety and much more general results in this vein can be also deduced from Simon Häberli's thesis [Hab18, Proposition 2.34].

4.6.1.3 The Bruhat-Tits Tree and $\exp_A$

As a complement for the previous discussions, in this subsection we describe how the Bruhat-Tits tree of Sect. 4.5.2.2 can be used to study the function $\exp_A$. We are going to see that somewhat, $\exp_A$ defines a covering $\mathbb{A}_{\mathbb{C}_\infty}^{1,an} \to \mathbb{A}_{\mathbb{C}_\infty}^{1,an}$ 'ramified of degree q^∞'; the reader is invited to compare with the results of Sect. 4.4.3. To give more strength to this, we use again Proposition 4.4.9. We are therefore led to analyse the image of $\mathcal{E}_n = l_n E_n$ on $D_{\mathbb{C}_\infty}^\circ(0, |\theta|^n)$ and then, take the limit for $n \to \infty$. We note that

$$D_{\mathbb{C}_\infty}^\circ(0, |\theta|^n) \setminus K_\infty = \bigsqcup_{\substack{\lambda \in \mathbb{Q} \cap]-n,\infty[\\ \alpha \in \oplus_{-\lambda \leq i < n} \mathbb{F}_q \theta^i}} \alpha + C_\lambda.$$

Since

$$\mathcal{E}_n(z) = \frac{l_n}{d_n} \prod_{a \in A(n)} (z - a)$$

is $\mathbb{F}_q$-linear of kernel $A(n)$, it suffices to study how $\mathcal{E}_n$ behaves over

$$\mathcal{T}_n = \bigsqcup_{\substack{\lambda \in \mathbb{Q} \cap]-n,\infty[\\ \alpha \in \oplus_{-\lambda \leq i < 0} \mathbb{F}_q \theta^i}} \alpha + C_\lambda.$$

Note that if $\lambda \leq 0$, the direct sum over i is empty. This means that in the Bruhat-Tits tree $\mathcal{T}$, $\mathcal{T}_n$ entails a very simple subtree which can be obtained by glueing in 0 a segment $]-n, 0]$ (the subtree $\mathcal{T}_n^-$) with the union of q disjoint copies of a complete q-ary tree equating $\mathcal{T}_0^+ = \text{red}(D_{\mathbb{C}_\infty}^\circ(0, 1) \setminus K_\infty)$ (independent of n), so that $\mathcal{T}_n = \mathcal{T}_n^- \sqcup \mathcal{T}_0^+$ and $\mathcal{T}_0 = \mathcal{T}_0^+$. Since $\mathcal{E}_n$ induces an isometric isomorphism of $D_{\mathbb{C}_\infty}^\circ(0, 1)$ such that for all $z \in D_{\mathbb{C}_\infty}^\circ(0, 1)$, $\mathcal{E}_n(z) = z + z'$ with $|z'| < |z|$, it induces the identity on $\mathcal{T}_0^+$, and this, for all $n \geq 0$. The action of the maps $\mathcal{E}_n$ are all equal to the action of $\mathcal{E}_0(z) = z$ on $\mathcal{T}_0^+$. We now choose $n > 0$ and we look at the behaviour of $\mathcal{E}_n$ on $\mathcal{T}_n^-$, which is the most interesting part of the story.

Consider x such that $\mathrm{red}(x) \in \mathcal{T}_n^-$. Then, there exists $i > 0$ maximal with the property that $x \in \mathcal{T}_i^- \setminus \mathcal{T}_{i-1}^-$ ($\mathcal{T}_0^-$ is empty by definition) and there exists a unique $\lambda \in \mathbb{Q}$ with $-\lambda \in [i-1, i[$ such that $x \in C_\lambda$. We recall that $\prod_{0 \neq a \in A(n)} a = \frac{d_n}{l_n}$, see (4.5). We have:

$$\mathcal{E}_n(x) = \frac{l_n}{d_n} \prod_{a \in A(i)} (x - a) \prod_{a \in A(n) \setminus A(i)} (x - a)$$

$$= \mathcal{E}_i(x) \frac{l_n}{d_n} \frac{d_i}{l_i} \prod_{a \in A(n) \setminus A(i)} (-a) \prod_{a \in A(n)} \left(1 - \frac{x}{a}\right)$$

$$= (1 + \xi)\mathcal{E}_i(x),$$

where $\xi \in D^\circ_{\mathbb{C}_\infty}(0, 1)$ (because $\frac{|x|}{|a|} < 1$ for all $a \in A(n) \setminus A(i)$). If $y \in D^\circ_{\mathbb{C}_\infty}(0, |x|)$ we get

$$\mathcal{E}_n(x + y) = \mathcal{E}_i(x) + \underbrace{\xi \mathcal{E}_i(x) + (1 + \xi)\mathcal{E}_i(y)}_{\text{element of } C^\circ_{\mathbb{C}_\infty}(0, |\mathcal{E}_i(x)|)}.$$

We deduce that the map

$$D^\circ_{\mathbb{C}_\infty}(x, |x|) \xrightarrow{\mathcal{E}_n} D^\circ_{\mathbb{C}_\infty}(\mathcal{E}_i(x), |\mathcal{E}_i(x)|)$$

is an étale covering of degree q^i. Hence, the image by $\mathcal{E}_n$ of $\mathrm{res}^{-1}(\mathcal{T}_i^- \setminus \mathcal{T}_{i-1}^-)$ (annulus) is an étale covering of degree q^i of the annulus

$$\frac{l_i}{d_i} \left[D^\circ_{\mathbb{C}_\infty}(0, |\theta|^{iq^i}) \setminus D^\circ_{\mathbb{C}_\infty}(0, |\theta|^{(i-1)q^i}) \right] = D^\circ_{\mathbb{C}_\infty}(0, |l_i|) \setminus D^\circ_{\mathbb{C}_\infty}(0, |l_{i-1}|).$$

From this it is not difficult to deduce that $\mathcal{E}_n$ defines a covering $D^\circ_{\mathbb{C}_\infty}(0, |\theta|^n) \to D^\circ_{\mathbb{C}_\infty}(0, |l_n|)$ ramified of degree q^n at the points of $A(n)$ and étale on the complementary of these points but we get even more. Namely, that for any $z \in \Omega$, $\mathrm{res}(\exp_A(z))$ can be very easily computed. If $|z|_\Im < 1$ then $|\exp_A(z)| < 1$ and if $z \notin K_\infty$, $\mathrm{res}(\exp_A(z))$ is equal to $\mathrm{res}(z - a)$ where $a \in A$ is the unique element such that $\mathrm{res}(z - a) \in \mathcal{T}_0^+$. If $|z|_\Im \geq 1$ then $\mathrm{res}(\exp_A(z)) = \mathrm{res}(\mathcal{E}_n(z))$ for all but finitely many n (depending on how large is $|z|_\Im$).

We consider $\mathcal{T}_\infty^- = \cup_{n \geq 1} \mathcal{T}_n^-$ (homeomorphic to $\mathbb{R}_{\leq 0}$) and $\mathcal{T}_\infty = \mathcal{T}_\infty^- \sqcup \mathcal{T}_0^+$. Note that $\mathrm{res}(\mathfrak{F}) = \mathcal{T}_\infty^-$ and $\mathrm{res}(\mathfrak{F} \sqcup \{z \in \Omega : |z|, |z|_\Im < 1\}) = \mathcal{T}_\infty$. In the terminology of §4.5.2.2, $\mathfrak{F} \sqcup \{z \in \Omega : |z|, |z|_\Im < 1\}$ can be viewed as a 'good fundamental domain' for the action of A over Ω by translations. We ultimately get, with a few more details to develop which are left to the reader:

Proposition 4.6.6 *The map* $\exp_A$ *induces a surjective, A-periodic map* $\mathfrak{F} \rightarrow$ $\mathbb{A}^{1,an}_{\mathbb{C}_\infty} \setminus D^\circ_{\mathbb{C}_\infty}(0,1)$ *and rigid analytic isomorphisms* $A \backslash \mathfrak{F} \rightarrow \mathbb{A}^{1,an}_{\mathbb{C}_\infty} \setminus D^\circ_{\mathbb{C}_\infty}(0,1)$ *and* $A \backslash \mathbb{A}^{1,an}_{\mathbb{C}_\infty} \rightarrow \mathbb{A}^{1,an}_{\mathbb{C}_\infty}$.

Note that $\mathfrak{F}$ is not, properly speaking, invariant by A-translations, but A-translations define an equivalence relation on $\mathfrak{F}$. The above statement needs to be interpret in the light of the richer combinatorial structure described earlier. In the classical setting we have, of course, the classical well known properties that the Eulerian exponential $z \mapsto e^z$ induces analytic isomorphisms $\mathbb{Z} \backslash \mathcal{H} \rightarrow D^\circ_{\mathbb{C}}(0,1)$ and $\mathbb{Z} \backslash \mathbb{C} \rightarrow \mathbb{C}^\times$. Interestingly too, we note that, just as $\mathbb{C} = \mathcal{H} \sqcup \mathbb{R} \sqcup \mathcal{H}^-$ (the latter is the lower complex half-plane), here we have an analogous decomposition

$$\mathbb{C}_\infty = \Omega \sqcup K_\infty = \Omega_1 \sqcup \Omega^- \sqcup K_\infty$$

with $\Omega_1 = \{z \in \Omega : |z|_\Im \geq 1\}$, $\Omega^- = \{z \in \Omega : |z|_\Im < 1\}$.

We hope that, with this description, we have convinced the reader that the functions $\exp_A$ and the Carlitz's exponential carry an extraordinary structural richness. We now complete our discussion with the quick exposition of some properties of the quotient $\mathrm{GL}_2(A) \backslash \Omega$ and then we move our attention to Drinfeld modular forms.

4.6.2 The Quotient $\mathrm{GL}_2(A) \backslash \Omega$

In the previous subsection we gave, in the most explicit way, but also in compatibility with the purposes of this text, a description of the analytic structure of the quotient space ($A = \mathbb{F}_q[\theta]$ acting by translations) $A \backslash \Omega_1$. Following [Ger80, Chapter 10]), we now describe the action of $\mathrm{GL}_2(\mathbb{F}_q)$ on certain admissible subsets of Ω. We consider $M \in |\mathbb{C}^\times_\infty|$ and we set

$$\Omega_M := \{z \in \Omega : |z|_\Im \geq M\}.$$

Note that this set, which is called *horocycle neighbourhood of* ∞, is non-empty and is invariant by translations by elements of K_∞. The multiplication by elements of $\mathbb{F}^\times_q$ induce bijections of Ω_M. Here is a lemma that will be useful later.

Lemma 4.6.7 *If* $M > 1$ *and if* $\gamma \in \mathrm{GL}_2(A)$ *is such that* $\gamma(\Omega_M) \cap \Omega_M \neq \emptyset$, *then* γ *belongs to the Borel subgroup* $\left(\begin{smallmatrix} * & * \\ 0 & * \end{smallmatrix}\right)$ *of* $\mathrm{GL}_2(A)$.

Proof Let $\gamma = \left(\begin{smallmatrix} a & b \\ c & d \end{smallmatrix}\right) \in \mathrm{GL}_2(A)$. By Lemma 4.5.2, $|\gamma(z)|_\Im = \frac{|z|_\Im}{|cz+d|^2}$. Let us suppose that $z, \gamma(z) \in \Omega_M$, and that $c \neq 0$. Then, since $|c| \geq 1$ if $c \in A \setminus \{0\}$,

$$|cz + d| \geq |cz + d|_\Im = |c||z|_\Im \geq |z|_\Im.$$

Then, $\gamma(z) \in \Omega_M$ implies that $|z|_\Im \geq M|cz+d|^2 \geq M|z|_\Im^2$ so that $M^{-1} \geq |z|_\Im$. Now, if $M > 1$, from $|z|_\Im \geq M$ we get a contradiction. $\qquad\square$

We set, with $M \in |\mathbb{C}_\infty^\times| \cap]1, \infty[$:

$$\mathcal{D}_M := D_{\mathbb{C}_\infty}(0, M) \setminus (\mathbb{F}_q + D_{\mathbb{C}_\infty}^\circ(0, M^{-1})) \subset \Omega.$$

This is the complementary in $\mathbb{P}_{\mathbb{F}_q}^{1,an}(\mathbb{C}_\infty)$ of the union of $q+1$ disjoint disks and is an affinoid subset of Ω. In the following, we can choose $M = |\theta|^{\frac{1}{2}}$. It is easy to see that the group $GL_2(\mathbb{F}_q)$ acts by homographies on $\mathcal{D}_M$ (note that more generally, the subsets $\{z \in \mathbb{C}_\infty : |z| \leq q^n, |z|_\Im \geq q^{-n}\}$, which also are affinoid subsets, are invariant under the action by homographies of the subgroups of $GL_2(A)$ finitely generated by $GL_2(\mathbb{F}_q)$ and $\{(\begin{smallmatrix} \lambda & \theta^i \\ 0 & \mu \end{smallmatrix}) : \lambda, \mu \in \mathbb{F}_q^\times, i \leq n\}$, the union of which is $GL_2(A)$). Further, if $\gamma \in GL_2(A)$, one easily sees that if $\gamma(\mathcal{D}_M) \cap \mathcal{D}_M \neq \emptyset$, then $\gamma \in GL_2(\mathbb{F}_q)$. It is also easily seen that

$$\Omega = \bigcup_{\gamma \in GL_2(A)} \gamma(\mathcal{D}_M).$$

We can apply Proposition 4.6.2 to the isomorphism of affine varieties

$$GL_2(\mathbb{F}_q) \backslash \mathbb{A}_{\mathbb{C}_\infty}^1 \xrightarrow{j_0} \mathbb{A}_{\mathbb{C}_\infty}^1,$$

where

$$j_0(z) = -\frac{(1+z^{q-1})^{q+1}}{z^{q-1}}$$

(this is the *finite j-invariant* of Gekeler in [Gek01]) to obtain an isomorphism of analytic spaces

$$GL_2(\mathbb{F}_q) \backslash \mathcal{D}_M \cong D_{\mathbb{C}_\infty}(0, 1).$$

In parallel, we have the Borel subgroup $B = B(A) = \{(\begin{smallmatrix} * & * \\ 0 & * \end{smallmatrix})\}$ which acts on Ω_M and the isomorphism of analytic spaces $B \backslash \Omega_M \cong \dot{D}_{\mathbb{C}_\infty}(0, S)$ induced by the map $\exp_A(z)^{-1}$ (Proposition 4.6.4). We recall from Lemma 4.6.7 that $\gamma \in GL_2(\mathbb{F}_q)$ is such that $\gamma(\Omega_M) \cap \Omega_M \neq \emptyset$ if and only if γ is in B.

There is a procedure of gluing two quotient rigid analytic spaces with such compatibility boundary conditions, into a new rigid analytic space, along with (4.10) for $k = \mathbb{F}_q$ and $t = \theta$. Note that $\mathcal{D}_M \cap \Omega_M = \{z \in \mathbb{C}_\infty : |z|_\Im = |z| = M\}$ and the two actions of B over Ω_M and of $GL_2(\mathbb{F}_q)$ on $\mathcal{D}_M$ agree with the action of $B \cap GL_2(\mathbb{F}_q)$ on $\mathcal{D}_M \cap \Omega_M$ and the gluing of these two quotient spaces is a well defined analytic space whose underlying topological space is homeomorphic to the quotient topological space $GL_2(A) \backslash \Omega$ which also carries a natural structure of analytic space. Additionally, this quotient space is isomorphic to the gluing of $D_{\mathbb{C}_\infty}(0, 1)$ and $\mathbb{C}_\infty \setminus D^\circ_{\mathbb{C}_\infty}(0, 1)$ along $\{z \in \mathbb{C}_\infty : |z| = 1\}$, which is in turn isomorphic to $\mathbb{C}_\infty$. This construction finally yields:

Theorem 4.6.8 *There is an isomorphism between the quotient rigid analytic space* $\mathrm{GL}_2(A)\backslash\Omega$ *and the rigid analytic affine line* $\mathbb{A}^{1,an}_{\mathbb{C}_\infty}$.

4.7   Drinfeld Modular Forms

We give a short synthesis on Drinfeld modular forms for the group $\Gamma = \mathrm{GL}_2(A)$ in the simplest case where $A = \mathbb{F}_q[\theta]$, so that we can prepare the next part of this paper, where we construct modular forms for Γ with (vector) values in certain $\mathbb{C}_\infty$-Banach algebras.

The map

$$\mathrm{GL}_2(K_\infty) \times \Omega \to \mathbb{C}^\times_\infty$$

defined by $(\gamma, z) \mapsto J_\gamma(z) = cz + d$ if $\gamma = \left(\begin{smallmatrix} * & * \\ c & d \end{smallmatrix}\right)$ behaves like the classical factor of automorphy for $\mathrm{GL}_2(\mathbb{R})$. Indeed we have the cocycle condition:

$$J_{\gamma\delta}(z) = J_\gamma(\delta(z)) J_\delta(z), \quad \gamma, \delta \in \mathrm{GL}_2(K_\infty).$$

Note that the image is indeed in $\mathbb{C}^\times_\infty$, as $z, 1$ are K_∞-linearly independent if $z \in \Omega$.

Definition 4.7.1 Let $f : \Omega \to \mathbb{C}_\infty$ be an analytic function. We say that f is *modular-like* of weight $w \in \mathbb{Z}$ if for all $z \in \Omega$,

$$f(\gamma(z)) = J_\gamma(z)^w f(z), \quad \forall \gamma \in \mathrm{GL}_2(A).$$

It is a simple exercise to verify that w is uniquely determined.

We say that a modular-like function of weight w is:

(1) *weakly modular* (of weight w) if there exists $N \in \mathbb{Z}$ such that the map $z \mapsto |\exp_A(z)^N f(z)|$ is bounded over Ω_M for some $M > 1$,
(2) a *modular form* if the map $z \mapsto |f(z)|$ is bounded over Ω_M for some $M > 1$.
(3) a *cusp form* if it is a modular form and $\max_{z \in \Omega_M} |f(z)| \to 0$ as $M \to \infty$.

Let f be modular like (of weight $w \in \mathbb{Z}$). Taking $\gamma = \left(\begin{smallmatrix} 1 & a \\ 0 & 1 \end{smallmatrix}\right)$ we see that $f(z + a) = f(z)$ for all $a \in A$. Therefore, by Proposition 4.6.1, there is a convergent series expansion of the type

$$f(z) = \sum_{i \in \mathbb{Z}} f_i \exp_A(z)^i, \quad f_i \in \mathbb{C}_\infty.$$

There is a rigid analytic analogue of Riemann's principle of removable singularities due to Bartenwerfer (see [Bar76]) in virtue of which we see that the $\mathbb{C}_\infty$-vector space $M_w^!$ of weak modular forms of weight w embeds in the field of Laurent series $\mathbb{C}_\infty((u))$ with the discrete valuation given by the order in u, where $u = u(z)$ is the *uniformiser at infinity*

$$u(z) = \frac{1}{\tilde{\pi} \exp_A(z)} = \frac{1}{\tilde{\pi}} \sum_{a \in A} \frac{1}{z - a},$$

which is an analytic function $\Omega \to \mathbb{C}_\infty$. Since $M_w^! \cap M_{w'}^! = \{0\}$ if $w \neq w'$ we have a $\mathbb{C}_\infty$-algebra $M^! = \oplus_w M_w^!$ which also embeds in the field of Laurent series $\mathbb{C}_\infty((u))$. Denoting by M_w the $\mathbb{C}_\infty$-vector space of modular forms of weight w and by $M = \oplus_w M_w$ the $\mathbb{C}_\infty$-algebra of modular forms, we also have an embedding $M \to \mathbb{C}_\infty[[u]]$ and cusp forms generate an ideal whose image in $\mathbb{C}_\infty[[u]]$ is contained in the ideal generated by u.

It is easy to deduce, from the modularity property, that $M_w^! \neq \{0\}$ implies $q - 1 \mid w$. Furthermore, for all w such that $M_w \neq \{0\}$, M_w can be embedded via u-expansions in $\mathbb{C}_\infty[[u^{q-1}]]$ and therefore the $\mathbb{C}_\infty$-vector space of cusp forms S_w can be embedded in $u^{q-1}\mathbb{C}_\infty[[u^{q-1}]]$.

4.7.1   u-Expansions

We have seen that we can associate in a unique way to any Drinfeld modular form f a formal series $\sum_{i \geq 0} f_i u^i \in \mathbb{C}_\infty[[u]]$ which is analytic in some disk $D(0, R)$, $R \in |\mathbb{C}_\infty^\times| \cap]0, 1[$. This is the analogue of the 'Fourier series' of a complex-valued modular form for $SL_2(\mathbb{Z})$; for such a function $f : \mathcal{H} \to \mathbb{C}$ we deduce, from $f(z + 1) = f(z)$, a Fourier series expansion

$$f = \sum_{i \geq 0} f_i q^i, \quad f_i \in \mathbb{C},$$

converging for $q = q(z) = e^{2\pi i z} \in D_{\mathbb{C}}^\circ(0, 1)$. We want to introduce some useful tools for the study of u-expansions of Drinfeld modular forms.

For $n \geq 0$ we introduce the $\mathbb{C}_\infty$-linear map $\mathbb{C}_\infty[z] \xrightarrow{\mathcal{D}_n} \mathbb{C}_\infty[z]$ uniquely determined by

$$\mathcal{D}_n(z^m) = \binom{m}{n} z^{m-n}.$$

Note that we have Leibniz's formula $\mathcal{D}_n(fg) = \sum_{i+j=n} \mathcal{D}_i(f)\mathcal{D}_j(g)$. The linear operators $\mathcal{D}_n$ extend in a unique way to $\mathbb{C}_\infty(z)$ and further, on the $\mathbb{C}_\infty$-algebra of analytic functions over any rational subset of Ω therefore inducing linear endomorphisms of the $\mathbb{C}_\infty$-algebra of analytic functions $\Omega \to \mathbb{C}_\infty$. Additionally, if $f : \Omega \to \mathbb{C}_\infty$ is analytic and A-periodic, $\mathcal{D}_n(f)$ has this same property, and for all n, $\mathcal{D}_n$ induces $\mathbb{C}_\infty$-linear endomorphisms of $\mathbb{C}_\infty[[u]]$ (this last property follows from the fact that $\mathcal{D}_n(u)$ is bounded on Ω_M as one case easily see distributing $\mathcal{D}_n$ on $u = \frac{1}{\widetilde{\pi}}\sum_{a\in A}\frac{1}{z-a}$, which gives $(-1)^n \frac{1}{\widetilde{\pi}}\sum_{a\in A}\frac{1}{(z-a)^{n+1}}$). We normalise $\mathcal{D}_n$ by setting:

$$D_n = (-\widetilde{\pi})^{-n}\mathcal{D}_n.$$

Lemma 4.7.2 For all $n \geq 0$, $D_n(K[u]) \subset u^2 K[u]$.

Proof It suffices to show that for all $n \geq 0$, $D_n(u) \in u^2 K[u]$. We proceed by induction on $n \geq 0$; there is nothing to prove for $n = 0$. Recall that $u(z) = \frac{1}{\exp_C(\widetilde{\pi}z)}$. Then, by Leibniz's formula:

$$0 = D_n(1) = D_n(u\exp_C(\widetilde{\pi}z))$$

$$= D_n(u)\exp_C(\widetilde{\pi}z) + \sum_{\substack{i+q^k=n\\k\geq 0}} D_i(u)D_{q^k}(\exp_C(\widetilde{\pi}z)),$$

because $\exp_C$ is $\mathbb{F}_q$-linear. In fact, $D_{q^k}(\exp_C(\widetilde{\pi}z))$ is constant and equals the coefficient of z^{q^k} in the z-expansion of $\exp_C$, which is $\frac{1}{d_k}$. We can therefore use induction to conclude that

$$D_n(u) = -u\left(-\sum_{\substack{i+q^k=n\\k\geq 0}} D_i(u)d_k^{-1}\right) \in u^2 K[u].$$

$\square$

The polynomials $G_{n+1}(u) := D_n(u) \in K[u]$ $(n \geq 1)$ are called the *Goss polynomials* (see [Gek88, §3]). It is easy to deduce from the above proof that $D_j(u) = u^{j+1}$ as $j = 1, \ldots, q-1$. There is no general formula currently available to compute $D_j(u)$ for higher values of j.

4.7.1.1 Constructing Drinfeld Modular Forms

The first non-trivial examples of Drinfeld modular forms have been described by Goss in his Ph. D. Thesis. To begin this subsection, we follow Goss [Gos80b] and we show how to construct non-zero Eisenstein series by using that $Az+A$ is strongly discrete in $\mathbb{C}_\infty$ if $z \in \Omega$. We set:

$$E_w(z) = \sum_{a,b\in A}' \frac{1}{(az+b)^w}.$$

There are many sources where the reader can find a proof of the following lemma (see for instance [Gek88, (6.3)]), but we prefer to give full details.

Lemma 4.7.3 *The series E_w defines a non-zero element of M_w if and only if $w > 0$ and $q - 1 \mid w$.*

Proof The above series converges uniformly on every set Ω_M and this already gives that E_w is analytic over Ω. The first property, that E_w is modular-like of weight w, follows from a simple rearrangement of the sum defining $E_w(\gamma(z))$ for $\gamma \in \Gamma$ and its (unconditional) convergence, which leaves it invariant by permutation of its terms. Additionally, it is very easy to see that all terms involved in the sum are bounded on Ω_M for every M which, by the ultrametric inequality, implies that E_w itself is bounded on Ω_M for every M. It remains to describe when the series are zero identically, or non-zero.

For the non-vanishing property, we give an explicit evidence why E_w has a u-expansion in $\mathbb{C}_\infty[[u]]$, and we derive from partial knowledge of its shape the required property (but we are not able to compute in limpid way the coefficients of the u-expansion!). First note that

$$D_n(u) = \frac{1}{\tilde{\pi}^{n+1}} \sum_{b \in A} \frac{1}{(z - b)^{n+1}},$$

so that we can use the Goss' polynomials $G_{n+1}(u) = D_n(u)$ as a 'model' to construct the u-expansion of E_w. Now, observe, for $w > 0$:

$$E_w(z) = \sum_{b \in A} \frac{1}{b^w} + {\sum_{a \in A}}' \sum_{b \in A} \frac{1}{(az + b)^w}.$$

If $(q - 1) \mid w$, we note that

$$\sum_{b \in A} \frac{1}{b^w} = -\prod_P \left(1 - P^{-w}\right)^{-1} =: -\zeta_A(w),$$

where the product runs over the monic irreducible polynomials $P \in A$ and therefore is non-zero. Then, if $(q - 1) \mid w$ and if A^+ denotes the subset of monic polynomials in A:

$$E_w(z) = -\zeta_A(w) - \sum_{a \in A^+} \sum_{b \in A} \frac{1}{(az + b)^w}$$

$$= -\zeta_A(w) - \tilde{\pi}^w \sum_{a \in A^+} G_w(u(az)),$$

a series which converges uniformly on every affinoid subset of Ω. Note that for $a \in A \setminus \{0\}$, the function $u(az)$ can be expanded as a formal series u_a of $u^{|a|} K[[u]]$

(normalise $|\cdot|$ by $|\theta| = q$) locally converging at $u = 0$ (in a disk of positive radius r independent of a). This yields the explicit series expansion (convergent for the u-valuation, or for the sup-norm over the disk $D(0, r)$ in the variable u):

$$E_w(z) = -\zeta_A(w) - \widetilde{\pi}^w \sum_{a \in A^+} G_w(u_a). \tag{4.12}$$

This also shows that E_w is, in this case, not identically zero. Indeed $\zeta_A(w)$ is non-zero, while the part depending on u in the above expression tends to zero as $|z|_\Im$ tends to ∞. On the other hand, if $(q - 1) \nmid w$, the factor of automorphy J_γ^w does not induce a factor of automorphy for the group $\mathrm{PGL}_2(A)$ defined as the quotient of $\mathrm{GL}_2(A)$ by scalar matrices and this implies that any modular form of such weight w vanishes identically, and so it happens that E_w vanishes in this case. $\qquad\square$

Remark 4.7.4 It is instructive at this point to compare our observations with the settings of the original, complex-valued Eisenstein series. Indeed, it is well known, classically, that if $w > 2$, $2 \mid w$ and $q = e^{2\pi i z}$:

$$E_w(z) = \sideset{}{'}\sum_{a,b \in \mathbb{Z}} \frac{1}{(az + b)^w} = 2\zeta(w) + 2 \frac{(2\pi i)^{\frac{w}{2}}}{(\frac{w}{2} - 1)!} \sum_{n \geq 1} \frac{n^{\frac{w}{2}-1} q^n}{1 - q^n}, \quad \Im(z) > 0.$$

The analogy is therefore between the series

$$\sum_{a \in A^+} G_w(u_a)$$

and

$$\sum_{n \geq 1} \frac{n^{\frac{w}{2}-1} q^n}{1 - q^n}.$$

However, it is well known that the latter series can be further expanded as follows, with $\sigma_k(n) = \sum_{d \mid n} d^k$:

$$\sum_{n \geq 1} \sigma_{\frac{w}{2}-1}(n) q^n.$$

For the series $\sum_{a \in A^+} G_w(u_a)$, this aspect is missing, and there is no available intelligible recipe to compute the coefficients of the u-expansion of E_w directly, at the moment.

4.7.2   Construction of Non-trivial Cusp Forms

We have constructed non-trivial modular forms, but they are not cusp forms. We construct non-zero cusp forms in this section. Let z be an element of Ω. Then, $\Lambda = \Lambda_z = Az + A$ is an A-lattice of rank 2 of $\mathbb{C}_\infty$. By Theorem 4.3.4, we have the Drinfeld A-module $\phi := \phi_\Lambda$ which is of rank 2. Hence, we can write

$$\phi_\theta(Z) = \theta Z + \widetilde{g}(z)Z^q + \widetilde{\Delta}(z)Z^{q^2}, \quad \forall (z, Z) \in \Omega \times \mathbb{C}_\infty$$

for functions $\widetilde{g}, \widetilde{\Delta} : \Omega \to \mathbb{C}_\infty$.

We consider the function $\Omega \times \mathbb{C}_\infty \xrightarrow{(z,Z) \mapsto \mathbb{E}(z,Z)} \mathbb{C}_\infty$ which associates to (z, Z) the value

$$\mathbb{E}(z, Z) := \exp_\Lambda(Z) = \sum_{i \geq 0} \alpha_i(z) Z^{q^i} = Z \prod_{\lambda \in \Lambda}' \left(1 - \frac{Z}{\lambda}\right) \qquad (4.13)$$

at Z of the exponential series $\exp_\Lambda$ associated to the A-lattice $\Lambda = \Lambda_z$ of $\mathbb{C}_\infty$. It is an analytic function and we have $\phi_a(\exp_\Lambda(Z)) = \exp_\Lambda(aZ)$ for all $a \in A$.

The following result collects the various functional properties of $\mathbb{E}(z, Z)$; proofs rely on simple computations that we leave to the reader.

Lemma 4.7.5 *For all* $z \in \Omega$, $Z \in \mathbb{C}_\infty$, $\gamma \in \Gamma$ *and* $a \in A$:

(1) $\phi_\Lambda(a)(\mathbb{E}(z, Z)) = \mathbb{E}(z, aZ)$,
(2) $\mathbb{E}(\gamma(z), Z) = J_\gamma(z)^{-1}\mathbb{E}(z, J_\gamma(z)Z)$.
(3) $\mathbb{E}(z, Z + az + b) = \mathbb{E}(z, Z)$, *for all* $a, b \in A$.

Remark 4.7.6 Loosely, we can say that $\mathbb{E}$ is a 'non-commutative modular form of weight $(-1, 1)$'. The second formula can be also rewritten as:

$$\mathbb{E}\left(\gamma(z), \frac{Z}{J_\gamma(z)}\right) = J_\gamma(z)^{-1}\mathbb{E}(z, Z), \quad \gamma \in \mathrm{GL}_2(A),$$

so that $\mathbb{E}$ functionally plays the role of a Jacobi form of level 1, weight -1 and index 0 (this is in close analogy with the Weierstrass $\wp$-functions).

By taking the formal logarithmic derivative in the variable Z of the Weierstrass product expansion of $\exp_\Lambda(Z)$ (for z fixed) we note that

$$\frac{Z}{\mathbb{E}(z, Z)} = 1 - \sum_{\substack{k \geq 0 \\ (q-1)|k}} E_k(z) Z^k$$

so that the coefficients in this expansion in powers of Z are analytic functions on Ω, from which we deduce, by inversion, that the coefficient functions $\alpha_i : \Omega \to \mathbb{C}_\infty$ of $\mathbb{E}$ are analytic. By Lemma 4.7.3 and the homogeneity of the algebraic expressions expressing the functions α_i in terms of the Eisenstein series E_k we see that $\alpha_i \in M_{q^i-1}$ for all $i \geq 0$. As $|z|_\Im \to \infty$ we have $E_k(z) \to -\zeta_A(k)$, after a simple computation we see that

$$\mathbb{E}(z, Z) \to \exp_A(Z)$$

uniformly for $Z \in D$ for every disk $D \subset \mathbb{C}_\infty$. This means that the functions α_i are not cusp forms (the coefficients of $\exp_A \in K_\infty[[\tau]]$ are all non-zero). To construct cusp forms, we now look at the coefficients $\widetilde{g}, \widetilde{\Delta}$ of ϕ_θ which are functions of the variable $z \in \Omega$. By (1) and (2) of Lemma 4.7.5, for $\gamma \in \Gamma$, writing now $\phi_{\Lambda_z}(\theta)$ in place of ϕ_θ:

$$\phi_{\Lambda_{\gamma(z)}}(\theta)(J_\gamma(z)^{-1}\mathbb{E}(z, J_\gamma(z)Z)) = \phi_{\Lambda_{\gamma(z)}}(\theta)(\mathbb{E}(\gamma(z), Z))$$
$$= \mathbb{E}(\gamma(z), \theta Z)$$
$$= J_\gamma(z)^{-1}\mathbb{E}(z, \theta J_\gamma(z)Z).$$

Hence, $\phi_{\Lambda_{\gamma(z)}}(\theta)(J_\gamma(z)^{-1}\mathbb{E}(z, W)) = J_\gamma(z)^{-1}\mathbb{E}(z, \theta W) = J_\gamma(z)^{-1}\phi_{\Lambda_z}(\mathbb{E}(z, W))$ for $W \in \mathbb{C}_\infty$. Since it is obvious that the coefficient functions $\widetilde{g}, \widetilde{\Delta}$ are analytic on Ω, they are in this way respectively modular-like functions of respective weights $q - 1$ and $q^2 - 1$. Furthermore:

Lemma 4.7.7 $\widetilde{g} \in M_{q-1} \setminus S_{q-1}$ and $\widetilde{\Delta} \in S_{q^2-1} \setminus \{0\}$. Additionally, $\widetilde{\Delta}(z) \neq 0$ for all $z \in \Omega$.

Proof The modularity of $\widetilde{g}$ and $\widetilde{\Delta}$ follows from the previously noticed fact that $\exp_{\Lambda_z}(Z) \to \exp_A(Z)$ uniformly with Z in disks as $|z|_\Im \to \infty$. Indeed, this implies that $\phi_\theta(Z) \to \theta Z + \widetilde{\pi}^{q-1}Z^q$ (uniformly on every disk) so that $\widetilde{g} \to \widetilde{\pi}^{q-1}$ and $\widetilde{\Delta} \to 0$ as $|z|_\Im \to \infty$ and we see that $\widetilde{g}$ is a modular form of weight $q - 1$ which is not a cusp form, and $\widetilde{\Delta}$ is a cusp form.

We still need to prove that $\widetilde{\Delta}$ is not identically zero; to do this, we prove now the last property of the lemma, which is even stronger. Assume by contradiction that there exists $z \in \Omega$ such that $\widetilde{\Delta}(z) = 0$. Then

$$\phi_{\Lambda_z}(\theta) = \theta + \widetilde{g}(z)\tau$$

which implies that the exponential $\exp_{\Lambda_z}$ induces an isomorphism of A-modules $\exp_{\Lambda_z} : \mathbb{C}_\infty/\Lambda_z \to C(\mathbb{C}_\infty)$ (the Carlitz module). But this disagrees with Theorem 4.3.4 which would deliver an isomorphism $\Lambda_z \cong A$ between lattices of different ranks. This proves that $\widetilde{\Delta}$ does not vanish on Ω. □

Following Gekeler in [Gek88], we define the modular forms g, Δ of respective weights $q - 1$ and $q^2 - 1$ by $\widetilde{g} = \widetilde{\pi}^{q-1}g$ and $\widetilde{\Delta} = \widetilde{\pi}^{q^2-1}\Delta$. The reason for choosing these normalisations is that it can be proved that the u-expansions of g, Δ have coefficients in A. We are not far from a complete proof of the following (see [Gek88, (5.12)] for full details):

Theorem 4.7.8 $M = \oplus_{w \in \mathbb{Z}} M_w = \mathbb{C}_\infty[g, \Delta]$

The proof rests on three crucial properties (1) existence of Eisenstein series (2) existence of the cusp form Δ which additionally is nowhere vanishing on Ω, and (3) modular forms of weight 0 for Γ are constant, which follows from the fact that a modular form of weight 0 can be identified with a holomorphic function over $\mathbb{P}^1_{\mathbb{F}_q}(\mathbb{C}_\infty)$ by Theorem 4.6.8, which is constant. We omit the details.

4.7.2.1 Drinfeld Modular Forms and the Bruhat-Tits Tree

We briefly sketch the interaction between Drinfeld modular forms and the Bruhat-Tits tree, mainly inviting the reader, yet in quite an informal way, to read the important work of Teitelbaum in [Tei91]. A simple computation indicates that if f is a rigid analytic function over the annulus $V = \sqcup_{-1 < \lambda < 0} C_\lambda$ (or on a more general annulus in Ω) so that f is defined by a convergent series $\sum_{i \in \mathbb{Z}} f_i z^i$ with the coefficients f_i in $\mathbb{C}_\infty$, then the residue

$$\mathrm{Res}_V(f(z)dz) := f_{-1}$$

does not depend on the local coordinate chosen to express the differential form $\omega = f(z)dz$. Namely, if t is another local coordinate and $z = z(t) = \sum_{i>0} z_i t^i$ with $z_i \in \mathbb{C}_\infty$ and $z_1 \in \mathbb{C}_\infty^\times$ (with suitable convergence conditions), then the coefficient of $t^{-1}dt$ in $\omega(z(t)) = f(z(t))dz(t) = f(z(t))\frac{dz}{dt}dt$ is also equal to f_{-1}, and in particular, $\mathrm{Res}_V(f dz)$ does not depend on the choice of the 'center' of the annulus.

We consider $\mathcal{T}^e$ the set of the oriented edges of the Bruhat-Tits tree. The elements are in one-to-one correspondence with the disjoint subsets of Ω:

$$V_{n,\alpha} := \alpha + \bigsqcup_{\lambda \in]n-1,n[} C_\lambda, \quad n \in \mathbb{Z}, \quad \alpha \in \oplus_{i \leq n-1} \mathbb{F}\pi^i.$$

Note that $V_{n,\alpha} = \{z \in \mathbb{C}_\infty : |\pi|^n < |z - \alpha| < |\pi|^{n-1}\}$, which is an annulus centered at elements of K_∞ with inner radius $|\pi|^n$ and outer radius $|\pi|^{n-1}$, n varying in $\mathbb{Z}$. Moreover, $V = V_{0,0}$. If $f : \Omega \to \mathbb{C}_\infty$ is a rigid analytic function, then f is rigid analytic on every $V_{n,\alpha}$ and we have a well defined residue map

$$\mathcal{T}^e \xrightarrow{\mathrm{res}(f)} \mathbb{C}_\infty$$

which is a 'harmonic function' in virtue of the *ultrametric residue theorem* (see [Ger80, §3]; we do not give full details and definitions of 'harmonic functions' etc., this would bring us too far away from the objectives of this paper). Of course, we do not expect the map $\mathrm{res}(f)$ to reproduce faithfully the behaviour of f. For example, if f is entire over $\mathbb{C}_\infty$ then all the residues of the differential form $f dz$ are clearly zero and $\mathrm{res}(f)$ vanishes identically, which might not be the case for f.

Where the map res(f) becomes really useful is with rigid analytic functions f which are determined by more elaborate patching of local data than just entire functions. Typically, functions defined by globally non uniform convergent series over Ω. If f is a Drinfeld modular form, Teitelbaum proved, in a much more general setting (Γ arithmetic subgroup of $GL_2(A)$), that a suitable variant of the residue map provides us with an isomorphism of $\mathbb{C}_\infty$-vector spaces

$$S_w(\Gamma) \to C_{\mathrm{har}}(\Gamma, w),$$

where $S_w(\Gamma)$ is the space of Drinfeld cusp forms of weight w for Γ as defined in ibid. and generalising our space S_w for $\Gamma = GL_2(A)$, and where $C_{\mathrm{har}}(\Gamma, w)$ is the space of 'weight w harmonic cocycles' for Γ. This map can be defined also over $M_w(\Gamma)$, the space of Drinfeld modular forms of weight w for Γ. Then, the kernel is spanned by the Eisenstein series of weight w. For this and other deep properties such as a homological interpretation of the residue map and an interesting and yet mysterious analysis of the Fourier series of cusp forms, see the paper [Tei91].

4.8   Eisenstein Series with Values in Banach Algebras

The final purpose of this and the next more advanced sections of the present paper is to show certain identities for a variant-generalisation of Eisenstein series (see Theorem 4.9.9). We recall that $A = \mathbb{F}_q[\theta]$. Let B be a $\mathbb{C}_\infty$-Banach algebra with sub-multiplicative norm $\| \cdot \|^6$ norm $\| \cdot \|$ (extending the norm $| \cdot |$ of $\mathbb{C}_\infty$) with the property that $\|B\| = |\mathbb{C}_\infty|$. Let X be a rigid analytic variety. We set

$$\mathcal{O}_{X/B} = \mathcal{O}_X \widehat{\otimes}_{\mathbb{C}_\infty} B,$$

with $\mathcal{O}_X$ the structural sheaf of X, of $\mathbb{C}_\infty$-algebras. In other words, if $U \subset X$ is an affinoid subset of X, then $\mathcal{O}_X(U)$ carries the supremum norm $\| \cdot \|_U$ and we define $\mathcal{O}_{X/B}(U)$ to be the completion of $\mathcal{O}_X \otimes_{\mathbb{C}_\infty} B$ for the norm induced by $\|f \otimes b\| = |f|_U$, for $f \in \mathcal{O}_X(U)$ and $b \in B$. If B has a countable orthonormal basis $\mathcal{B} = (b_i)_{i \in \mathcal{I}}$, an element $f \in \mathcal{O}_{X/B}(U)$ has a convergent series expansion

$$f = \sum_{i \in \mathcal{I}} f_i b_i,$$

where $f_i \in \mathcal{O}_X(U)$, with $|f_i|_U \to 0$ for the Fréchet filter on $\mathcal{I}$.

---

[6]That is, $\|ab\| \leq \|a\| \|b\|$ for all $a, b \in B$. We adopt the simpler notations $\| \cdot \|$ and $| \cdot |$ at the place of $| \cdot |_\infty$ etc. that we have used in the first few sections of our text.

One sees that that Tate's acyclicity Theorem extends to this setting, namely, if X is an affinoid variety, $\mathcal{O}_{X/B}$ is a sheaf of B-algebras. The global sections are the *analytic functions* $X \to B$.

We will mainly use the cases $X = \Omega$ and $X = \mathbb{A}_{\mathbb{C}_\infty}^{s,an}$. If $X = \mathbb{A}_{\mathbb{C}_\infty}^{s,an}$, an element of $\mathcal{O}_{X/B}$ is a *B-valued entire function of s variables.* We can identify it with a map $\mathbb{C}_\infty^s \to B$ allowing a series expansion in $B[[\underline{t}]]$ with $\underline{t} = (t_1, \ldots, t_s)$ converging on $D(0, R)^s$ for all $R > 0$. A bounded entire function $\mathbb{C}_\infty \to B$ is constant (this is a generalisation of Liouville's theorem which uses the hypothesis that $\|B\| = |\mathbb{C}_\infty|$ is not discrete, see [Pel16b]).

We work with B-valued analytic functions where $B = \mathbb{K}$ is the completion of $\mathbb{C}_\infty(\underline{t})$ for the Gauss norm $\| \cdot \| = \| \cdot \|_\infty$, where $\underline{t} = (t_1, \ldots, t_s)$. We have $\|\mathbb{K}\| = |\mathbb{C}_\infty|$ and the residue field is $\mathbb{F}_q^{ac}(\underline{t})$. In all the following, we consider matrix-valued analytic functions and we extend norms to matrices in the usual way by taking the supremum of norms of the entries of a matrix.

We extend the $\mathbb{F}_q$-automorphism $\tau : \mathbb{C}_\infty \to \mathbb{C}_\infty$, $x \mapsto x^q$, $\mathbb{F}_q(\underline{t})$-linearly and continuously on $\mathbb{K}$. The subfield of the fixed elements $\mathbb{K}^{\tau=1} = \{x \in \mathbb{K} : \tau(x) = x\}$ is easily seen to be equal to $\mathbb{F}_q(\underline{t})$ by a simple variant of Mittag-Leffler theorem. Let $\lambda_1, \ldots, \lambda_r \in \mathbb{C}_\infty$ be K_∞-linearly independent. This is equivalent to saying that the A-module

$$\Lambda = A\lambda_1 + \cdots + A\lambda_r \subset \mathbb{C}_\infty$$

is an A-lattice. In this way, the exponential function $\exp_\Lambda$ induces a continuous open $\mathbb{F}_q(\underline{t})$-linear endomorphism of $\mathbb{K}$, the kernel of which contains $\Lambda \otimes_{\mathbb{F}_q} \mathbb{F}_q(\underline{t})$ (it can be proved that $\exp_\Lambda$ is surjective over $\mathbb{K}$ and the kernel is exactly $\Lambda \otimes_{\mathbb{F}_q} \mathbb{F}_q(\underline{t})$ but we do not need this in the present paper). The Drinfeld A-module $\phi = \phi_\Lambda$ gives rise to a structure of $\mathbb{F}_q(\underline{t})^{nr \times n}[\theta]$-module

$$\phi(\mathbb{K}^{nr \times n})$$

by simply using the $\mathbb{F}_q(\underline{t})$-vector space structure of $\mathbb{K}$ and defining the multiplication ϕ_θ by θ with the above extension of τ.

We consider an injective $\mathbb{F}_q$-algebra morphism

$$A \xrightarrow{\chi} \mathbb{F}_q(\underline{t})^{n \times n}$$

and we set, with $(\lambda_1, \ldots, \lambda_r)$ an A-basis of Λ (the exponential now applied coefficientwise):

$$\omega_\Lambda = \exp_\Lambda \left((\theta I_n - \chi(\theta))^{-1} \begin{pmatrix} \lambda_1 I_n \\ \vdots \\ \lambda_r I_n \end{pmatrix} \right) \in \mathbb{K}^{rn \times n}.$$

Lemma 4.8.1 *For all $a \in \mathbb{F}_q(t)[\theta]$ we have the identity $\phi_a(\omega_\Lambda) = \chi(a)\omega_\Lambda$ in $\mathbb{K}^{rn \times n}$.*

Proof Since the variables t_i are central for τ and $\mathbb{F}_q(t)[\theta]$ is euclidean, it suffices to show that $\phi_\theta(\omega_\Lambda) = \chi(t)\omega_\Lambda$. Now observe, for $a \in A$:

$$\phi_\Lambda(a)(\omega_\Lambda) = \exp_\Lambda((\theta I_n - \chi(\theta))^{-1} \begin{pmatrix} (aI_n - \chi(a) + \chi(a))\lambda_1 \\ \vdots \\ (aI_n - \chi(a) + \chi(a))\lambda_r \end{pmatrix}$$

$$= \chi(a)\omega_\Lambda,$$

because $(\theta I_n - \chi(\theta))^{-1}(aI_n - \chi(a)) \in \mathbb{F}_q(t)[\theta]^{n \times n}$ so that $(\theta I_n - \chi(\theta))^{-1}(aI_n - \chi(a))\lambda_i$ lies in the kernel of $\exp_\Lambda$ (applied coefficientwise). $\square$

Hence, ω_Λ is a particular instance of *special function* as defined and studied in [Ang17, Gaz19]. Note also that the map

$$\Phi_\Lambda : Z \mapsto \exp_\Lambda((\theta I_n - \chi(\theta))^{-1}Z)$$

defines an entire function $\mathbb{C}_\infty \to \mathbb{K}^{n \times n}$. An easy variant of the proof of Lemma 4.8.1 delivers:

Lemma 4.8.2 *We have the functional equation $\tau(\Phi_\Lambda(Z)) = (\chi(\theta) - \theta I_n)\Phi_\Lambda(Z) + \exp_\Lambda(Z)I_n$ in $\mathbb{K}^{n \times n}$.*

We now introduce a 'twist' of the logarithmic derivative of $\exp_\Lambda$. We recall that $A \xrightarrow{\chi} \mathbb{F}_q(t)^{n \times n}$ is an injective $\mathbb{F}_q$-algebra morphism. We introduce the *Perkins' series* (introduced in a slightly narrower setting by Perkins in his Ph. D. thesis [Per13]):

$$\psi_\Lambda(Z) := \sum_{a_1,\ldots,a_r \in A} \frac{1}{Z - a_1\lambda_1 - \cdots - a_r\lambda_r}(\chi(a_1), \ldots, \chi(a_r)), \quad Z \in \mathbb{C}_\infty$$

(depending on the choice of the basis of Λ as well as on the choice of the algebra morphism χ). The series converges for $Z \in \mathbb{C}_\infty \setminus \Lambda$ to a function $\mathbb{C}_\infty \setminus \Lambda \to \mathbb{K}^{n \times rn}$. We have (after elementary rearrangement of the terms):

$$\psi_\Lambda(Z - b_1\lambda_1 - \cdots - b_r\lambda_r) = \psi_\Lambda(Z) - (\chi(b_1), \ldots, \chi(b_r))\exp_\Lambda(Z)^{-1}, \quad b_1, \ldots, b_r \in A.$$

$$(4.14)$$

The next proposition explains why we are interested in the Perkins' series: they can be viewed as generating series of certain $\mathbb{K}$-vector-valued Eisenstein series that we introduce below. Determining identities for the Perkins' series results in determining identities for such Eisenstein series.

Proposition 4.8.3 *There exists $r \in |\mathbb{C}_\infty^\times|$ such that the following series expansion, convergent for Z in $D(0, r)$, holds:*

$$\psi_\Lambda(Z) = - \sum_{\substack{j \geq 1 \\ j \equiv 1(q-1)}} Z^{j-1} \mathcal{E}_\Lambda(j; \chi),$$

where for $j \geq 1$,

$$\mathcal{E}_\Lambda(j; \chi) := \sideset{}{'}\sum_{a_1,\dots,a_r \in A} \frac{1}{(a_1\lambda_1 + \cdots + a_r\lambda_r)^j} (\chi(a_1), \dots, \chi(a_r)) \in \mathbb{K}^{n \times rn}.$$

The series $\mathcal{E}_\Lambda(j; \chi)$ is the *Eisenstein series of weight j* associated to Λ and χ. Note that this is in deep correspondence with the *canonical deformations* of the Carlitz module in Tavares Ribeiro's contribution to this volume, [Tav20, §4.2]. The reader can make these connections deeper with an accurate analysis on which we skip here.

Problem 4.8.4 Develop the appropriate generalisation of the theory of harmonic cocycles of Teitelbaum [Tei91] and construct the residue map along the notion of $\mathbb{K}$-vector-valued modular form which naturally includes the above Eisenstein series as in [Pel18].

Proof of Proposition 4.8.3 Since Λ is strongly discrete, $D(0, r) \cap (\Lambda \setminus \{0\}) = \emptyset$ for some $r \neq 0$. Then, we can expand, for the coefficients a_i not all zero,

$$\frac{1}{Z - a_1\lambda_1 - \cdots - a_r\lambda_r} = \frac{-1}{a_1\lambda_1 + \cdots + a_r\lambda_r} \sum_{i \geq 0} \left(\frac{Z}{a_1\lambda_1 + \cdots + a_r\lambda_r} \right)^i.$$

The result follows from the fact that $\mathcal{E}_\Lambda(j; \chi)$, which is always convergent for $j > 0$, vanishes identically for $j \not\equiv 1 \pmod{q-1}$ which is easy to check observing that $\Lambda = \lambda\Lambda$ for all $\lambda \in \mathbb{F}_q^\times$, and reindexing the sum defining $\mathcal{E}_\Lambda(j; \chi)$. $\square$

Lemma 4.8.5 *The function $F^\sharp(Z) := \exp_\Lambda(Z)\psi_\Lambda(Z)$ defines an entire function $\mathbb{C}_\infty \to \mathbb{K}^{n \times rn}$ such that, for all $\lambda = a_1\lambda_1 + \cdots + a_r\lambda_r \in \Lambda$, $F^\sharp(\lambda) = (\chi(a_1), \dots, \chi(a_r)) \in \mathbb{F}_q(t)^{n \times nr}$.*

Proof This easily follows from the fact that ψ_Λ converges at $Z = 0$, and (4.14). $\square$

The function ψ_Λ is intimately related to the exponential $\exp_\Lambda$ by means of the following result, where $\exp_\Lambda$ on the right is the unique continuous map $\mathbb{K}^{n \times n} \to \mathbb{K}^{n \times n}$ which induces a $\mathbb{F}_q(t)^{n \times n}[\theta]$-module morphism $\mathbb{K}^{n \times n} \to \phi_\Lambda(\mathbb{K}^{n \times n})$.

Lemma 4.8.6 *We have the identity of entire functions $\mathbb{C}_\infty \to \mathbb{K}^{n \times n}$ of the variable Z:*

$$\exp_\Lambda(Z)\psi_\Lambda(Z)\omega_\Lambda = \exp_\Lambda((\theta I_n - \chi(\theta))^{-1}Z).$$

Proof By Lemma 4.8.5, the function

$$F(Z) := F^\sharp(Z) \cdot \omega_\Lambda : \mathbb{C}_\infty \to \mathbb{K}^{n \times n}$$

is an entire function such that

$$F(\lambda) = (\chi(a_1), \ldots, \chi(a_r))\omega_\Lambda \in \mathbb{K}^{n \times n}, \quad \forall \lambda = a_1 \lambda_1 + \cdots + a_r \lambda_r \in \Lambda.$$

We set

$$G(Z) = \exp_\Lambda((\theta I_n - \chi(\theta))^{-1} Z).$$

Let $\lambda = a_1 \lambda_1 + \cdots + a_r \lambda_r \in \Lambda$. We have, by Lemma 4.8.1,

$$G(\lambda) = \exp_\Lambda((\theta I_n - \chi(\theta))^{-1}((a_1 I_n - \chi(a_1) + \chi(a_1))\lambda_1 + \cdots + (a_r I_n - \chi(a_r) + \chi(a_r))\lambda_r)$$

$$= (\chi(a_1), \ldots, \chi(a_r))\omega_\Lambda.$$

Hence, the entire functions F, G agree on Λ. The function $F - G$ is an entire function $\mathbb{C}_\infty \to \mathbb{K}^{n \times n}$ which vanishes over Λ. Hence,

$$H(Z) = \frac{F(Z) - G(Z)}{\exp_\Lambda(Z)}$$

defines an entire function over $\mathbb{C}_\infty$. Now, it is easy to see that

$$\lim_{|Z| \to \infty} \|H(Z)\| = 0.$$

Since the valuation group of $\mathbb{K}$ is dense in $\mathbb{R}^\times$, the appropriate generalisation of Liouville's theorem [Pel16b, Proposition 8] for entire functions holds in our settings and $H = 0$ identically. $\square$

Remark 4.8.7 More generally, we can study A-module maps

$$\Lambda \xrightarrow{\chi} \mathbb{K}^{n \times n}$$

with bounded image (the A-module structure on $\mathbb{K}^{n \times n}$ being induced by an injective algebra homomorphism $A \hookrightarrow \mathbb{F}_q(t) \hookrightarrow \mathbb{K}^{n \times n}$) and Perkins' series

$$\psi_\Lambda(n; \chi) := \sum_{\lambda \in \Lambda} \frac{\chi(\lambda)}{(Z - \lambda)^n}.$$

Lemma 4.8.6 delivers an identity for ψ_Λ in terms of certain analytic functions of the variable Z which are explicitly computable in terms of $\exp_\Lambda$. To see this, observe that the $\mathbb{K}$-algebra of analytic functions $D(0, r) \to \mathbb{K}$ is stable by the

$\mathbb{K}$-linear divided higher derivatives $\mathcal{D}_{Z,n}$ defined by $\mathcal{D}_{Z,n}(Z^m) = \binom{m}{n}Z^{m-n}$. In particular, $\mathcal{D}_{Z,n}(\psi_\Lambda)$ is well defined for any $n > 0$. We write $f^{(k)}$ for $\tau^k(f)$, $f \in \mathbb{K}$ or for f more generally a $\mathbb{K}^{r \times s}$-valued map for arbitrary integers r, s. If $f = \sum_{i \geq 0} f_i Z^i$ is an analytic function over a disk $D(0, r)$ in the variable Z, then $f^{(k)} = \sum_{i \geq 0} \tau(f_i)Z^{q^k i}$ is again analytic if $k \geq 0$. Observe that in particular,

$$\psi_\Lambda(Z)^{(k)} = \mathcal{D}_{q^k-1}(\psi_\Lambda(Z)), \quad k \geq 0.$$

Lemma 4.8.6 implies

$$\psi_\Lambda(Z)\omega_\Lambda = \mathcal{H}(Z) := \exp_\Lambda(Z)^{-1}\exp_\Lambda((\theta I_n - \chi(\theta))^{-1}Z),$$

and we note that on the right we have an analytic function $D(0, r) \to \mathbb{K}^{n \times n}$ for some $r \in |\mathbb{C}_\infty^\times|$. Applying $\mathcal{D}_{q^k-1}$ on both sides of this identity and observing that ω_Λ does not depend on Z, we deduce:

$$\psi_\Lambda(Z)^{(k)}\omega_\Lambda = \mathcal{D}_{q^k-1}(\mathcal{H})(Z), \quad k \geq 0.$$

Now, since the function $\psi_\Lambda(Z)^{(k)}$ is in fact an analytic function of the variable Z^{q^k}, this is also true for the function $\mathcal{D}_{q^k-1}(\mathcal{H})(Z)$ so that

$$\mathcal{H}_k(Z) = (\mathcal{D}_{q^k-1}(\mathcal{H})(Z))^{(-k)}, \quad k \geq 0$$

are all analytic functions $D(0, r) \to \mathbb{K}^{n \times n}$ (note that $\mathcal{H}_0 = \mathcal{H}$). We introduce the matrices

$$\boldsymbol{\Omega}_\Lambda = (\omega_\Lambda, \omega_\Lambda^{(-1)}, \ldots, \omega_\Lambda^{(1-r)}) \in \mathbb{K}^{rn \times rn}, \quad \boldsymbol{\mathcal{H}}_\Lambda(Z) = (\mathcal{H}_0, \ldots, \mathcal{H}_{r-1}),$$

where the latter is an $n \times rn$-matrix of analytic functions $D(0, r) \to \mathbb{K}$. Then,

$$\psi_\Lambda(Z)\boldsymbol{\Omega}_\Lambda = \boldsymbol{\mathcal{H}}_\Lambda(Z).$$

But a simple variant of the Wronskian lemma (see [Pel08, §4.2.3]) implies that $\boldsymbol{\Omega}_\Lambda$ is invertible. We have reached:

Theorem 4.8.8 *The identity* $\psi_\Lambda(Z) = \boldsymbol{\mathcal{H}}_\Lambda(Z)\boldsymbol{\Omega}_\Lambda^{-1}$ *holds, for functions locally analytic at $Z = 0$.*

The identity of the previous theorem connects the 'twisted logarithmic derivative' $\psi_\Lambda(Z)$ to the inverse Frobenius twists of the divided higher derivatives of the mysterious function $\mathcal{H}$, which are certainly not always easy to compute, unless $r = 1$, where there is no higher derivative to compute at all. If we set, additionally, $\chi = \chi_t$ where $\chi_t(a) = a(t)$ so that $n = 1$, then we reach a known identity, which was first discovered by R. Perkins in [54] (that we copy below adapting it to our notations):

$$\exp_A(Z)\omega(t) \sum_{a\in A} \frac{a(t)}{Z-a} = \exp_A\left(\frac{Z}{\theta-t}\right),$$

with ω Anderson-Thakur's function and $\exp_A(Z) = Z\prod'_{a\in A}(1-\frac{Z}{a})$. This formula is expressed in [Pel16b, Theorem 1] in a slightly different manner by using Papanikolas' deformation of the Carlitz logarithm. Note that these references also contain other types of generalisation. The above formula can be viewed as an analogue of [Kat91, Lemma 1.3.21] (the analogy can be pursued further). We owe this remark to Lance Gurney that we thankfully acknowledge.

Problem 4.8.9 This should be considered as a starting point for an extension of Kato's arguments related to the connection between the zeta-values phenomenology and Iwasawa's theory appearing in [Kat91]. One may ask how far a parallel with Kato's viewpoint can go.

4.9 Modular Forms with Values in Banach Algebras

In this section, more technical than the previous ones, we suppose that B is a Banach $\mathbb{C}_\infty$-algebra with norm $\|\cdot\|$ such that $\|B\| = |\mathbb{C}_\infty|$ and we suppose that it is endowed with a countable orthonormal basis $\mathcal{B} = (b_i)_{i\in\mathcal{I}}$. The example on which we are focusing here is that of $B = \mathbb{K}$, the completion of the field $\widehat{\mathbb{C}_\infty(t)}$ for the Gauss valuation $\|\cdot\|$. Any basis of $\mathbb{F}_q^{ac}(t)$ as a vector space over $\mathbb{F}_q^{ac}$ is easily seen to be an orthonormal basis of $\mathbb{K}$. We recall that we have considered, in Sect. 4.8, a notion of B-valued analytic function. The main purpose of this section is to show, through some examples, that if $N > 1$, there is a generalisation

$$\Omega \to \mathbb{K}^{N\times 1}$$

of Drinfeld modular form which cannot by studied by using just 'scalar' Drinfeld modular forms.

We consider a representation

$$\rho : \Gamma \to \mathrm{GL}_N(\mathbb{F}_q(t)) \subset \mathrm{GL}_N(\mathbb{K}).$$

Definition 4.9.1 Let $f : \Omega \to \mathbb{K}^{N\times 1}$ be an analytic function. We say that f is *modular-like* (for ρ) of weight $w \in \mathbb{Z}$ if for all $\gamma \in \mathrm{GL}_2(A)$,

$$f(\gamma(z)) = J_\gamma(z)^w \rho(\gamma) f(z), \quad \gamma \in \mathrm{GL}_2(A).$$

We say that a modular-like function of weight w is:

(1) *weakly modular* (of weight w) if there exists $L \in \mathbb{Z}$ such that the map $z \mapsto \| \exp_A(z)^L f(z) \|$ is bounded over Ω_M for some $M > 1$,
(2) a *modular form* if the map $z \mapsto \| f(z) \|$ is bounded over Ω_M for some $M > 1$.
(3) a *cusp form* if it is a modular form and $\max_{z \in \Omega_M} \| f(z) \| \to 0$ as $M \to \infty$.

We denote by $M_w^!(\rho)$, $M_w(\rho)$, $S_w(\rho)$ the $\mathbb{K}$-vector spaces of weak modular, modular, and cusp forms of weight w for ρ. Note that these notations are loose, in the sense that these vector spaces strongly depend of the choice of $\mathbb{K}$ (in particular, of the variables $\underline{t} = (t_i)$).

We now describe a very classical example with $N = 1$ and $B = \mathbb{C}_\infty$ (no variables $\underline{t}$ at all). If $\rho : \Gamma \to \mathbb{C}_\infty^\times$ is a representation, there exists $m \in \mathbb{Z}/(q-1)\mathbb{Z}$ unique, such that $\rho(\gamma) = \det(\gamma)^{-m}$ for all γ. We write

$$\rho = \det{}^{-m}$$

(note that this is well defined). Gekeler constructed a cusp form $h \in S_{q+1}(\det^{-1}) \setminus \{0\}$; see [Gek88, (5.9)]. The first few terms of its u-expansion in $\mathbb{C}_\infty$ can be computed explicitly by various methods (including the explicit formulas (4.16) and (4.17) below):

$$h(z) = -u(1 + u^{(q-1)^2} + \cdots). \qquad (4.15)$$

We deduce that $h^{q-1}\Delta^{-1}$ is a Drinfeld modular form of weight zero which is constant by Theorem 4.7.8. The factor of proportionality is easily seen to be -1: $\Delta = -h^{q-1}$.

The computation in (4.15) can be pushed to coefficients of higher powers of the uniformiser u by using two formulas that we describe here. The first formula is due to López [Lop10]. We have the convergent series expansion (in both $K[[u]]$ for the u-adic metric and in $D(0, r)$ for some $r \in |\mathbb{C}_\infty| \cap]0, 1[$ for the norm of the uniform convergence)

$$h = - \sum_{\substack{a \in A \\ \text{monic}}} a^q u_a \in A[[u]]. \qquad (4.16)$$

The second formula is due to Gekeler [Gek85] and is an analogue of Jacobi's product formula

$$\Delta = q \prod_{n \geq 0} (1 - q^n)^{24} \in q\mathbb{Z}[[q]]$$

for the classical complex-valued normalised discriminant cusp form Δ (we have an unfortunate and unavoidable conflict of notation here!). Gekeler's formula is the following u-convergent product expansion:

$$h = -u \prod_{\substack{a \in A \\ \text{monic}}} \left(u^{|a|} C_a \left(\frac{1}{u} \right) \right)^{q^2-1} \in A[[u]], \qquad (4.17)$$

with C_a the multiplication by a for the Carlitz module structure. Note that $(u^{|a|} C_a(\frac{1}{u}))^{q^2-1} \in 1 + K[[u]]$ and the u-valuation of

$$\left(u^{|a|} C_a(u^{-1}) \right)^{q^2-1} - 1$$

goes to infinity as a runs in $A \setminus \{0\}$. One deduces, from Gekeler's result [Gek88, Theorem (5.13)], that $M_w(\det^{-m}) = h^m M_{w-m_0(q+1)}$ if $m_0 = m \cap \{0, \ldots, q-2\}$ (m is a class modulo $q-1$).

4.9.1   Weak Modular Forms of Weight -1

We analyse another class of representations, this time in higher dimension and we construct a new kind of modular form associated to it. Let

$$A \xrightarrow{\chi} \mathbb{F}_q(\underline{t})^{n \times n}$$

be an injective $\mathbb{F}_q$-algebra morphism. Then, the map

$$\rho_\chi : \Gamma \to GL_{2n}(\mathbb{F}_q(\underline{t})) \subset GL_{2n}(\mathbb{K})$$

defined by

$$\rho_\chi \begin{pmatrix} a & b \\ c & d \end{pmatrix} = \begin{pmatrix} \chi(a) & \chi(b) \\ \chi(c) & \chi(d) \end{pmatrix}$$

is a representation of Γ. We denote by ρ_χ^* the contragredient representation

$$\rho_\chi^* = {}^t \rho_\chi^{-1}.$$

We shall study the case $\rho = \rho_\chi$ or ρ_χ^*. We also set $N = 2n$.

We construct weak modular forms of weight -1 associated to the representations ρ_χ; the main result is Theorem 4.9.3 where we show that a certain matrix function defined in (4.19) has its columns which are weak modular forms of weight -1. We think that this construction is interesting because there seems to be no analogue of it in the settings of complex-vector-valued modular forms for $SL_2(\mathbb{Z})$.

Before going on, we need the next lemma, where we give a uniform bound for the valuations of the coefficients of the u-expansions $\sum_{m\geq 0} c_{i,m} u^m$ of the modular forms α_i appearing in (4.13).

Lemma 4.9.2 *There exists a constant $C > 0$ such that for all $i, m \geq 0$,*

$$|c_{i,m}| \leq q^{-iq^i} |\widetilde{\pi}|^{q^i-1} C^m.$$

Proof This is [Pel14, Lemma 2.1]. Although the statement presented in this reference is correct, there is a typographical problem in (2.17) so that, to avoid confusion, we give full details here. We set without loss of generality $|\theta| = q$. We recall ([Pel14, (2.14)]) that

$$\alpha_i = \frac{1}{\theta^{q^i} - \theta} (\widetilde{g} \alpha_{i-1}^q + \widetilde{\Delta} \alpha_{i-2}^{q^2}), \quad i > 0,$$

with the initial values $\alpha_0 = 1$ and $\alpha_{-1} = 0$. Now, writing additionally the u-expansions:

$$\widetilde{g} = \sum_{i\geq 0} \widetilde{\gamma}_i u^i, \quad \widetilde{\Delta} = \sum_{i\geq 0} \widetilde{\delta}_i u^i,$$

we find (as in ibid.)

$$c_{i,m} = \frac{1}{\theta^{q^i} - \theta} \left(\sum_{j+qk=m} \widetilde{\gamma}_j c_{i-1,k}^q + \sum_{j'+q^2k'=m} \widetilde{\delta}_{j'} c_{i-2,k'}^{q^2} \right), \quad i > 0, \quad m \geq 0$$

with the initial values $c_{i,0} = \frac{\widetilde{\pi}^{q^i-1}}{d_i}$ and $c_{-1,m} = 0$. Clearly, we can choose $C > 0$ such that $|\widetilde{\delta}_j| \leq C^j$ and $|\widetilde{\gamma}_j| \leq C^j |\widetilde{\pi}^{q-1}|$ for all $j \geq 0$, and additionally, we can suppose that the inequality of the Lemma is true for $|c_{i,m}|$ with $i = 0, 1$. We now prove the inequality by induction over i. Indeed, note that if $j + qk = m$, then, by induction hypothesis, $|\widetilde{\gamma}_j c_{i-1,k}^q| \leq C^j q^{-(i-1)q^{i-1}q} C^{kq} |\widetilde{\pi}|^{q^i-q} |\widetilde{\pi}|^{q-1} \leq C^m q^{-(i-1)q^i} |\widetilde{\pi}|^{q^i-1}$ and similarly, if $j + q^2 k = m$, then we have $|\widetilde{\delta}_j c_{i-2,k}^{q^2}| \leq C^m q^{-(i-2)q^i} |\widetilde{\pi}|^{q^i-2}-1$, and the inequality follows. $\qquad\square$

We write $\vartheta = \chi(\theta)$. If we set

$$W = (\theta I_n - \vartheta)^{-1} \in GL_n(\mathbb{K}),$$

we have that for all $a \in A$:

$$(\chi(a) - aI_n)W \in \mathbb{F}_q(\underline{t}_\Sigma)[\theta]^{n \times n}. \tag{4.18}$$

Now, we consider, for χ and W as in (4.18), the matrix function $Q(z) = \binom{zW}{W}$, which is a holomorphic function $\Omega \to \mathbb{K}^{N \times n}$. We observe that if $\gamma = \binom{a\ b}{c\ d} \in \Gamma$, then

$$Q(\gamma(z)) = J_\gamma(z)^{-1} \binom{(az+b)W}{(cz+d)W} \equiv J_\gamma(z)^{-1} \rho_\chi(\gamma)Q(z) \pmod{\Lambda_z^{N \times n}}.$$

Hence, if we set

$$\mathbb{F}(z) := \mathbb{E}(z, Q(z)), \tag{4.19}$$

then, by the fact that $\Lambda_z \otimes \mathbb{F}_q(\underline{t})$ is contained in the kernel of $\exp_{\Lambda_z}$,

$$\mathbb{F}(\gamma(z)) = J_\gamma(z)^{-1} \mathbb{E}(z, J_\gamma(z)J_\gamma(z)^{-1} \rho_\chi(\gamma)Q(z)) = J_\gamma(z)^{-1} \rho_\chi(\gamma)\mathbb{F}(z), \quad \forall \gamma \in \Gamma.$$

This means that the function $\mathbb{F} : \Omega \to \mathbb{K}^{N \times n}$ is modular-like of weight -1 for ρ_χ. We are going to describe this function $\mathbb{F}$ in more detail.

Theorem 4.9.3 *We have* $\mathbb{F} \in M_{-1}^!(\rho_\chi)^{1 \times n}$.

Proof We set $e_C(z) = \exp_C(\widetilde{\pi}z)$ so that $u(z) = \frac{1}{e_C(z)}$. Lemma 4.8.2 implies:

$$\tau(e_C(W)) = (\vartheta - \theta I_n)e_C(W), \quad \tau(e_C(zW)) = (\vartheta - \theta I_n)e_C(zW) + e_C(z).$$

The subset $\mathcal{W} \subset \mathbb{R}_{>0}$ of the $r \in |\mathbb{C}_\infty|$ such that the elements $|d_i^{-1}r^{q^i}|$ are all distinct for $i \geq 0$ is dense in $\mathbb{R}_{>0}$. Let $z \in \mathbb{C}_\infty$ be such that $r = |\widetilde{\pi}z| \in \mathcal{W}$. Then:

$$|e_C(z)| = \max_i \{q^{-iq^i}|\widetilde{\pi}|^{q^i}|z|^{q^i}\}.$$

We write $\mathbb{F} = \binom{\mathcal{F}_1}{\mathcal{F}_2}$ with $\mathcal{F}_i : \Omega \to \mathbb{K}^{n \times n}$. We first look at the matrix function

$$\mathcal{F}_1 = \exp_\Lambda(zW) = \sum_{i \geq 0} \alpha_i(z)z^{q^i}\tau^i(W).$$

We suppose that $|u(z)| < \frac{1}{B}$ with B as in Lemma 4.9.2. Then

$$\mathcal{F}_1 = \sum_{i \geq 0} z^{q^i}\tau^i(W) \sum_{j \geq 0} c_{i,j}u^j$$

so that if $\|zW\widetilde{\pi}\| = r \in \mathcal{W}$ with $|u| < \frac{1}{B}$, then

$$\|\mathcal{F}_1\| = \max_{i,j}\{|z|^{q^i} q^{-iq^i} |\widetilde{\pi}|^{q^i-1} \underbrace{(C|u|)^j}_{<1}\}$$

$$= \|\exp_C(\widetilde{\pi}zW)\|$$

$$= \|e_C(z/\theta)\|,$$

and $\frac{\mathcal{F}_1}{e_C(z/\theta)} - \widetilde{\pi}^{-1}I_n$ is bounded as $|z|_{\Im}$ is bounded from below.

We now look at the matrix function $\mathcal{F}_2 = e_\Lambda(W)$. Since $\mathcal{F}_2 = \sum_{i\geq 0}\alpha_i(z)\tau^i(W)$, for $|u| < \frac{1}{B}$ we get in a similar way that $\mathcal{F}_2 - \widetilde{\pi}^{-1}e_C(W)$ goes to zero as $|z|_{\Im} \to \infty$. Hence, the n columns of the matrix function $\mathbb{F}$, which are modular-like of weight -1 are weak modular forms of $M^!_{-1}(\rho_\chi)$.   □

We set

$$\mathfrak{F} = (\mathbb{F}, \tau(\mathbb{F})) = \begin{pmatrix} \mathcal{F}_1 & \tau(\mathcal{F}_1) \\ \mathcal{F}_2 & \tau(\mathcal{F}_2) \end{pmatrix}.$$

Then, $\mathfrak{F}$ is an analytic function $\Omega \to \mathbb{K}^{N\times N}$ and the first n columns are weak modular forms of weight -1, while the last n columns are weak modular forms of weight $-q$ (for the representation ρ_χ).

Lemma 4.9.4 *We have the difference equation $\tau(\mathfrak{F}) = \mathfrak{F}\Phi$ where*

$$\Phi = \begin{pmatrix} 0 & \widetilde{\Delta}^{-1}(\chi(\theta) - \theta I_n) \\ 1 & -\widetilde{\Delta}^{-1}\widetilde{g}I_n \end{pmatrix}.$$

Proof For any choice of $n, m > 0$, we extend the function $\mathbb{E}(z, Z)$ of Lemma 4.7.5 to

$$\Omega \times \mathbb{K}^{n\times m} \overset{\mathbb{E}}{\to} \mathbb{K}^{n\times m}$$

by setting $\mathbb{E}(z, Z) = \sum_{i\geq 0}\alpha_i(z)\tau^i(Z)$ (so τ acts diagonally). Lemma 4.7.5 holds in this generalised setting, where the Drinfeld modules ϕ_Λ now acts on $\mathbb{K}^{n\times m}$ (case of $\Lambda = \Lambda_z$). The present statement follows from (1) of Lemma 4.7.5 with $a = \theta$ in a manner which is sensibly similar to that of [Pel14, Theorem 1.3]. Indeed, note that, with $\phi_\Lambda(\theta) = \theta + \widetilde{g}\tau + \widetilde{\Delta}\tau^2$, we have $\phi_\Lambda(\theta)(\mathbb{F}) - \chi(\theta)\mathbb{F} = 0$.   □

Lemma 4.9.5 *We have that $\sup_{z\in\Omega_M} \|\mathfrak{F} - \mathcal{X}\mathcal{Y}\mathcal{Z}\| \to 0$ as $M \to \infty$, where*

$$\mathcal{X} = \begin{pmatrix} I_n & 0 \\ 0 & e_C(W) \end{pmatrix}, \quad \mathcal{Y} = \begin{pmatrix} e_C(zW) & \tau(e_C(zW)) \\ I_n & \vartheta - \theta I_n \end{pmatrix}, \quad \mathcal{Z} = \begin{pmatrix} \widetilde{\pi}^{-1}I_n & 0 \\ 0 & \widetilde{\pi}^{-q}I_n \end{pmatrix}.$$

Proof We observe (recall that $\vartheta = \chi(\theta)$):

$$\mathcal{X}\mathcal{Y}\mathcal{Z} = \begin{pmatrix} \widetilde{\pi}^{-1}e_C(zW) & \widetilde{\pi}^{-q}((\vartheta - \theta I_n)e_C(zW) + e_0 I_n) \\ \widetilde{\pi}^{-1}e_C(W) & \widetilde{\pi}^{-q}(\vartheta - \theta I_n)e_C(W) \end{pmatrix}.$$

Since the second block column of $\mathfrak{F}$ is the image by τ of the first block column, all we need to show is that $\sup_{z \in \Omega_M} \|\mathbb{F} - \binom{\widetilde{\pi}^{-1}e_C(zW)}{\widetilde{\pi}^{-1}e_C(W)})\| \to 0$ as $M \to \infty$. We note that

$$\mathcal{F}_1 = e_\Lambda(zW) = \widetilde{\pi}^{-1}e_C(zW) + \underbrace{\sum_{i \geq 0} z^{q^i}\tau^i(W)\sum_{j>0}c_{i,j}u^j}_{=:\Upsilon}.$$

We show that $\|\Upsilon\|$ tends to zero when $|z|_\Im \to \infty$. We suppose that $|z|_\Im$ is large so that $|u|C < 1$. then, the double series defining Υ is convergent and we can write

$$\Upsilon = \sum_{j>0}\sum_{i\geq 0}u^j c_{i,j}z^{q^i}\tau^i(W).$$

The general term of this series, $\Upsilon_{i,j} := u^j c_{i,j}z^{q^i}\tau^i(W)$, has absolute value which satisfies:

$$\|\Upsilon_{i,j}\| \leq q^{-iq^i}|\widetilde{\pi}|^{q^i-1}(|u|C)^j|z|^{q^i}\|W\|^{q^i}$$

$$\leq |u|C \max_i\{|z|^{q^i}\|W\|^{q^i}|\widetilde{\pi}|^{q^i-1}\}$$

$$\leq |\widetilde{\pi}|^{-1}C\left|\frac{e_C(z/\theta)}{e_C(z)}\right|$$

and tends to zero as $|z|_\Im \to \infty$. In a similar way, one proves that $\|\mathcal{F}_2 - \widetilde{\pi}^{-1}e_C(W)\|$ tends to zero in the same way, we leave the details to the reader. $\square$

Lemma 4.9.6 We have $\|\det(\mathfrak{F}) - (-1)^n e_C(z)^n \widetilde{\pi}^{-n(q+1)}\det(e_C(W))\| \to 0$ as $|z|_\Im \to \infty$, and $\det(e_C(W))$ is non-zero.

Proof The formula follows directly from the expression for $\mathcal{X}\mathcal{Y}\mathcal{Z}$. The non-vanishing of $\det(e_C(W))$ is easy to show. $\square$

This result implies that the columns of $\mathfrak{F}$ are linearly independent. Moreover, it is plain that $\sup_{z \in \Omega_M} \|\det(\mathfrak{F}^{-1}) - (-1)^n u^n \widetilde{\pi}^{(q+1)n}\det(e_C(W))^{-1}\| \to 0$ as $M \to \infty$. Since at once the scalar function $F = \det(\mathfrak{F}^{-1})$ satisfies $F(\gamma(z)) = J_\gamma(z)^{n(q+1)}\det(\gamma)^{-n}F(z)$ for all $z \in \Omega$ and $\gamma \in \Gamma$, we get $F \in M_{n(q+1)}(\det^{-n}) \otimes_{\mathbb{C}_\infty} \mathbb{K}$. Now, Fh^{-n} is a modular form of weight 0, therefore equal to an element of $\mathbb{K}^\times$. We obtain:

Corollary 4.9.7 *We have* $\det(\mathfrak{F}^{-1}) = (-1)^n \widetilde{\pi}^{-(q+1)n} h^n \det(e_C(W))^{-1}$ *and, writing* $\mathfrak{H} := {}^t\mathfrak{F}^{-1} = (\mathfrak{H}_1, \mathfrak{H}_2)$ *with* $\mathfrak{H}_i : \Omega \to \mathbb{K}^{n \times n}$, *we have that the n columns of* $\mathfrak{H}_1$ *are linearly independent modular forms of weight 1 and the n columns of* $\mathfrak{H}_2$ *are linearly independent modular forms of weight q for the representation* ρ_χ^*.

What can be further proved is, by setting

$$M(\rho_\chi^*) = \bigoplus_w M_w(\rho_\chi^*)$$

the weight-graded $(M \otimes_{\mathbb{C}_\infty} \mathbb{K})$-module of modular forms for ρ_χ^*, where $M = \bigoplus_w M_w(\mathbf{1})$ is the $\mathbb{C}_\infty$-algebra of scalar modular forms ($\mathbf{1}$ is the trivial representation):

Theorem 4.9.8 $M(\rho_\chi^*) = (M \otimes_{\mathbb{C}_\infty} \mathbb{K})^{1 \times N} \mathfrak{H}$.

We will not give the details of the deduction of the proof of this theorem from Corollary 4.9.7, since it rests on an easy generalisation and modification of [Pel18, Theorem 3.9]. Instead of this, we insist on the result of Gekeler [Gek88, Theorem (5.13)], which implies that

$$M_w(\det^{-m}) = M_{w-m(q+1)} h^m, \quad m \le q - 1$$

with h the Poincaré series of weight $q + 1$ and 'type 1' defined in ibid. (5.11) (with u-expansion (4.15)) so that, with $M(\det^{-m}) = \oplus_w M_w(\det^{-m})$,

$$M(\det^{-m}) = M h^m.$$

In view of this, we can think about $\mathfrak{H}$ (up to normalisation) as to a matrix-valued generalisation of the Poincaré series h.

4.9.2 Jacobi-Like Forms

We consider the series

$$\Psi(z, Z) := \psi_{\Lambda_z}(Z) = \sum_{a,b \in A} \frac{1}{Z - az - b} (\chi(a), \chi(b)),$$

converging for $Z \in \mathbb{C}_\infty \setminus \Lambda$ where $\Lambda = \Lambda_z = Az + A$, $z \in \Omega$. We have the following functional identities

$$\Psi\left(\gamma(z), \frac{Z}{J_\gamma(z)}\right) = J_\gamma(z) \Psi(z, Z) \rho(\gamma)^{-1}, \quad \gamma \in \Gamma,$$

together with the identities arising from (4.14). Proposition 4.8.3 implies that, for $Z \in D(0, r)$ for some $r \in |\mathbb{C}_\infty| \cap]0, 1[$,

$$^t\Psi(z, Z) = - \sum_{\substack{j>0 \\ j \equiv 1(q-1)}} Z^{j-1}\mathcal{E}(j; \chi)$$

where $\mathcal{E}(j; \chi)$ is the Eisenstein series (non-vanishing if $j \equiv 1 \pmod{q-1}$)

$$\mathcal{E}(j; \chi) := \sideset{}{'}\sum_{a,b\in A} \frac{1}{(az+b)^j} \begin{pmatrix} \chi(a) \\ \chi(b) \end{pmatrix},$$

which satisfies

$$\mathcal{E}(j; \chi)(\gamma(z)) = J_\gamma(z)^j \rho_\chi^*(\gamma)\mathcal{E}(j; \chi), \quad \gamma \in \Gamma, \quad z \in \Omega.$$

Since it is also apparent that $\|\mathcal{E}(j; \chi)(z)\|$ is bounded on Ω_M for $M > 1$ and $j > 0$, we deduce that the n columns of $\mathcal{E}(j; \chi)$ are modular forms of weight j for ρ_χ^* in the sense of Definition 4.9.1 (see [Pel18, §3.2.1] for a special case). By Theorem 4.8.8 we obtain

$$\Psi(z, Z) = [\mathcal{H}(Z), \mathcal{D}_{q-1}(\mathcal{H})(Z)^{(-1)}]\mathbf{\Omega}_\Lambda(z)^{-1} \tag{4.20}$$

which allows to explicitly compute the Eisenstein series $\mathcal{E}(j; \chi)$ in terms of the function $\mathcal{H}(Z)$. To make this interesting relation a little bit more transparent, we give below an explicit expression of the matrix $\mathbf{\Omega}_\Lambda(z)^{-1}$. We have:

$$\mathbf{\Omega}_\Lambda(z)^{-1} = \begin{pmatrix} 0 & 1 \\ 1 & 0 \end{pmatrix} \tau^{-1}(\Phi)\mathfrak{F}^{-1} = \begin{pmatrix} 1 & -\left(\frac{\tilde{g}}{\tilde{\Lambda}}\right)^{\frac{1}{q}} \\ 0 & (\chi(\theta) - \theta^{\frac{1}{q}})\tilde{\Delta}^{-\frac{1}{q}} \end{pmatrix} \mathfrak{F}^{-1}, \tag{4.21}$$

with Φ the matrix defined in Lemma 4.9.4. To see this, observe that in the notation of Theorem 4.8.8,

$$\mathbf{\Omega}_\Lambda(z) = (\mathbb{F}, \tau^{-1}(\mathbb{F})) = \tau^{-1}(\mathfrak{F}) \begin{pmatrix} 0 & 1 \\ 1 & 0 \end{pmatrix},$$

with $\Lambda = \Lambda_z$ as above. By Lemma 4.9.4, $\tau(\mathfrak{F}) = \mathfrak{F}\Phi$, so that $\tau^{-1}(\mathfrak{F}) = \mathfrak{F}(\tau^{-1}(\Phi))^{-1}$ which yields

$$\mathbf{\Omega}_\Lambda = \tau^{-1}(\mathfrak{F}) \begin{pmatrix} 0 & 1 \\ 1 & 0 \end{pmatrix} = \mathfrak{F} \begin{pmatrix} 0 & 1 \\ 1 & 0 \end{pmatrix} \begin{pmatrix} 0 & 1 \\ 1 & 0 \end{pmatrix} (\tau^{-1}(\Phi))^{-1} \begin{pmatrix} 0 & 1 \\ 1 & 0 \end{pmatrix},$$

which implies (4.21) by the (licit) inversion of the two sides.

Substituting in (4.20) and transposing, we get:

$$
- \sum_{\substack{j \geq 1 \\ j \equiv 1(q-1)}} \mathcal{E}(j; \chi) Z^{j-1} = \mathfrak{H} \begin{pmatrix} 1 & 0 \\ -\left(\frac{\widetilde{g}}{\Delta}\right)^{\frac{1}{q}} & \widetilde{\Delta}^{-\frac{1}{q}}({}^t\chi(\theta) - \theta^{\frac{1}{q}}) \end{pmatrix} \begin{pmatrix} {}^t\mathcal{H}(Z) \\ \mathcal{D}_{q-1}({}^t\mathcal{H})(Z)^{(-1)} \end{pmatrix}.
$$

For example, the Eisenstein series of weight one $\mathcal{E}(1; \chi)$ arises as the coefficient of Z^0 in the left-hand side and the above yields an explicit formula for it. Note that the constant term of the Z-expansion of ${}^t[\mathcal{H}(Z), \mathcal{D}_{q-1}(\mathcal{H})(Z)^{(-1)}]$ is

$$
{}^t[(\theta I_n - \chi(\theta))^{-1}, \alpha_1(z)^{\frac{1}{q}}((\theta I_n - \chi(\theta))^{-1} - (\theta^{\frac{1}{q}} I_n - \chi(\theta))^{-1})].
$$

The formula that we get is this one:

$$
-\mathcal{E}_1(1; \chi) = \mathfrak{H} \begin{pmatrix} 1 & 0 \\ -\left(\frac{\widetilde{g}}{\Delta}\right)^{\frac{1}{q}} & \widetilde{\Delta}^{-\frac{1}{q}}({}^t\chi(\theta) - \theta^{\frac{1}{q}}) \end{pmatrix} \begin{pmatrix} {}^t(\theta I_n - \chi(\theta))^{-1} \\ \alpha_1(z)^{\frac{1}{q}}{}^t((\theta I_n - \chi(\theta))^{-1} - (\theta^{\frac{1}{q}} I_n - \chi(\theta))^{-1}) \end{pmatrix},
$$

and what looks as a miracle at first sight is that it greatly simplifies, by using the explicit computation of α_1 which arises from [Pel14, (2.14)], and which is $\alpha_1 = \frac{g}{\theta^q - \theta}$, we reach the following:

Theorem 4.9.9 *The following identity holds*

$$
\mathcal{E}_1(1; \chi) = -\mathfrak{H} \begin{pmatrix} {}^t(\theta I_n - \chi(\theta))^{-1} \\ 0_n \end{pmatrix},
$$

involving $N \times n$ matrices whose columns are modular forms of weight 1.

In fact, this is not a miracle; it is just due to the fact that the left-hand side must be bounded at the infinity; this is only possible if the second matrix entry of the column above is identically zero, because it is anyway a multiple by a constant matrix of the weak modular form $\widetilde{g}/\widetilde{\Delta}$ (this somewhat forces α_1 to be equal to the above multiple of $\widetilde{g}$, giving this artificial impression of miraculous simplification). It is easy from here to deduce [Pel12, Theorem 8] in the special case of $N = 2, n = 1$ and $\chi = \chi_t$.

Acknowledgments The author is thankful to the VIASM of Hanoi for the very nice conditions that surrounded the development of the course and the stimulating environment in which he was continuously immersed all along his visit in June 2018. Part of this text was written during a stay at the MPIM of Bonn in April 2019 and the author wishes to express gratitude for the very good conditions of work there. The author is thankful to A. Thuillier for the proof of Proposition 4.6.2, to L. Gurney for fruitful discussions, and to H. Furusho for several corrections on the final version of the manuscript. He expresses his gratitude to the reviewers that, by means of a careful reading and interesting suggestions, allowed to improve it. This work was supported by the ERC ANT.

This work is dedicated to the memory of Velia Stassano, mother of the author.

References

[And86] G. Anderson, t-motives. Duke Math. J. **53**, 457–502 (1986)

[And04] G.W. Anderson, W.D. Brownawell, M.A. Papanikolas, Determination of the algebraic relations among special Γ-values in positive characteristic. Ann. Math. **160**, 237–313 (2004)

[Ang15] B. Angles, F. Pellarin, Universal Gauss-Thakur sums and L-series. Invent. Math. **200**(2), 653–669 (2015)

[Ang17] B. Anglès, F. Tavares Ribeiro, Arithmetic of function fields units. Math. Annalen **367**, 501–579 (2017)

[Art45] E. Artin, G. Whaples, Axiomatic characterization of fields by the product formula for valuations. Bull. Am. Math. Soc. **51**, 469–492 (1945)

[Ax70] J. Ax, Zeros of polynomials over local fields. J. Algebra **15**, 417–428 (1970)

[Bak08] M. Baker, B. Conrad, S. Dasgupta, K.S. Kedlaya, J. Teitelbaum. p-adic geometry: lectures from the 2007 Arizona Winter School p-adic geometry lectures. AMS University Lecture Series, vol. 45 (2008)

[Bar76] W. Bartenwerfer, Der erste Riemannsche Hebbarkeitssatz im nichtarchimedischen Fall. J. Reine Angew. Math. **286/87**, 144–163 (1976)

[Bas18a] D. Basson, F. Breuer, R. Pink, Drinfeld modular forms of arbitrary rank, Part I: Analytic Theory (2018). Preprint. arXiv:1805.12335

[Bas18b] D. Basson, F. Breuer, R. Pink, Drinfeld modular forms of arbitrary rank, Part II: Comparison with Algebraic Theory (2018). Preprint. arXiv:1805.12337

[Bas18c] D. Basson, F. Breuer, R. Pink, Drinfeld modular forms of arbitrary rank, Part III: Examples (2018). Preprint. arXiv:1805.12339

[Ber90] V. Berkovich, *Spectral Theory and Analytic Geometry Over Non-archimedean Fields*. Mathematical Surveys and Monographs, vol. 33 (American Mathematical Society, Providence, 1990)

[Boc02] G. Böckle, An Eichler-Shimura isomorphism over function fields between Drinfeld modular forms and cohomology classes of crystals. Manuscript, currently available on the webpage of the author

[Boc07] G. Böckle, U. Hartl, Uniformizable families of t-motives. Trans. Am. Math. Soc. **359**, 3933–3972 (2007)

[Boc15] G. Böckle, Hecke characters associated to Drinfeld modular forms. With an appendix by the author and T. Centeleghe. Compositio Math. **151**, 2006–2058 (2015)

[Bos84] S. Bosch, U. Güntzer, R. Remmert, Non-Archimedean analysis. Grundlehren der Mathematischen Wissenschaften, vol. 261 (Springer, Berlin, 1984)

[Car35] L. Carlitz, On certain functions connected with polynomials in a Galois field. Duke Math. J. **1**, 137–168 (1935)

[Cas86] J.W.S. Cassels, *Local Fields* (Cambridge University Press, Cambridge, 1986)

[DiV20] L. Di Vizio, Difference Galois theory for the 'applied' mathematician, in *Arithmetic and Geometry Over Local Fields*, ed. by B. Anglès, T. Ngo Dac (Springer International Publishing, Cham, 2020)

[Dri74] V. Drinfeld, Elliptic modules. Math. Sb. **94**, 594–627 (1974) (in Russian). English translation in Math. USSR Sb. (1974)

[Fre04] J. Fresnel, M. van der Put, *Rigid Analytic Geometry and its Applications* (Birkhäuser, Boston, 2004)

[Gaz19] Q. Gazda, A. Maurischat, Special Functions and Gauss-Thakur Sums in Higher Rank and Dimension (2019). Preprint. arXiv:1903.07302

[Gek85] E.-U. Gekeler, A product expansion for the discriminant function of Drinfeld modules of rank two. J. Number Theory **21**, 135–140 (1985)

[Gek88] E.-U. Gekeler, On the coefficients of Drinfeld modular forms. Invent. Math. **93**, 667–700 (1988)

[Gek01] E.-U. Gekeler. Finite modular forms. Finite Fields Appl. **7**, 553–572 (2001)

[Gek17] E.U. Gekeler, On Drinfeld modular forms of higher rank. J. de Théor. Nombres Bordeaux **29**, 875–902 (2017)

[Gek18] E.U. Gekeler, On Drinfeld modular forms of higher rank III: The analogue of the $k/12$-formula. J. Number Theory **192**, 293–306 (2018)

[Gek19a] E.U. Gekeler, On Drinfeld modular forms of higher rank II. J. Number Theory (2019). doi.org/10.1016/j.jnt.2018.11.01

[Gek19b] E.U. Gekeler, On Drinfeld modular forms of higher rank IV: modular forms with level. J. Number Theory (2019) doi.org/10.1016/j.jnt.2019.04.019

[Ger80] L. Gerritzen, M. van der Put, *Schottky Groups and Mumford Curves*. Springer Lectures Notes in Mathematics, vol. 817 (Springer, Berlin, Heidelberg, 1980)

[Gos80a] D. Goss, The algebraist's upper half-plane. Bull. Am. Math. J. **2**, 391–415 (1980)

[Gos80b] D. Goss. π-adic Eisenstein series for Function Fields. Compos. Math. **41**, 3–38 (1980)

[Gos80c] D. Goss, Modular forms for $\mathbb{F}_r[T]$. J. Reine Angew. Math. **317**, 16–39 (1980)

[Gos96] D. Goss, *Basic Structures of Function Field Arithmetic* (Springer, Berlin, 1996)

[Hab18] S. Häberli, Satake compactification of analytic Drinfeld modular varieties. ETH thesis dissertation 25544, Zürich (2018)

[Han20] D. Hansen, Quotients of adic spaces by finite groups. Math. Res. Lett. (to appear)

[Har20] U. Hartl, A.-K. Juschka, Pink's theory of Hodge structures and the Hodge conjecture over function fields, in *Hodge Structures, Transcendence and Other Motivic Aspects*, ed. by G. Böckle, D. Goss, U. Hartl, M. Papanikolas. Series of Congress Reports (European Mathematical Society, Berlin, 2020)

[Hay74] D.R. Hayes, Explicit class field theory for rational function fields. Trans. AMS **189**, 77–91 (1974)

[Hay79] D.R. Hayes, Explicit class field theory in global function fields, in *Studies in Algebra and Number Theory*. Advances in Mathematics: Supplementary Studies, vol. 6 (Academic, New York, 1979), pp. 173–217

[Kat91] K. Kato, Lectures on the approach to Iwasawa theory for Hasse-Weil L-functions via B_{dR}, in *Arithmetic Algebraic Geometry*. Springer Lectures Notes in Mathematics, vol. 1553 (CIME, Trento, 1991)

[Ked01] K. Kedlaya, The algebraic closure of the power series field in positive characteristic. Proc. Am. Math. Soc. **129**, 3461–3470 (2001)

[Lop10] B. López, A non-standard Fourier expansion for the Drinfeld discriminant function. Arch. Math. **95**, 143–150 (2010)

[Lub65] J. Lubin, J. Tate, Formal complex multiplication in local fields. Ann. Math. **81**, 380–387 (1965)

[Mas15] A.W. Mason, A. Schweizer, Elliptic points of the Drinfeld modular groups. Math. Z. **279**, 1007–1028 (2015)

[Mat05] G.L. Matthews, T. W. Michel, One-point codes using places of higher degree. IEEE Trans. Inf. Theory **51**, 1590–1593 (2005)

[Nag59] H. Nagao, On GL(2, $K[x]$). J. Inst. Polytechn. Osaka City Univ. Ser. A. **10**, 117–121 (1959)

[Pap08] M.A. Papanikolas, Tannakian duality for Anderson-Drinfeld motives and algebraic independence of Carlitz logarithms. Invent. Math. **171**, 123–174 (2008)

[Pel08] F. Pellarin, Aspects de l'indépendance algébrique en caractéristique non nulle. Bourbaki seminar. Volume 2006/2007. Exposés 967–981. Astérisque **317**, 205–242 (2008). Société Mathématique de France, Paris

[Pel12] F. Pellarin, Values of certain L-series in positive characteristic. Ann. Math. **176**, 2055–2093 (2012)

[Pel14] F. Pellarin, Estimating the order of vanishing at infinity of Drinfeld quasi-modular forms. J. Reine Angew. Math. **687**, 1–42 (2014)

[Pel16a] F. Pellarin, A note on multiple zeta values in Tate algebras. Riv. Mat. Univ. Parma **7**, 71–100 (2016)

[Pel16b] F. Pellarin, R.B. Perkins, On certain generating functions in positive characteristic. Monat. Math. **180**, 123–144 (2016)

[Per13] R.B. Perkins, On special values of Pellarin's L-series. Ph.D. Dissertation. The Ohio State University (2013)

[54] R.B. Perkins, Explicit formulae for L-values in positive characteristic. Math. Z. **248**, 279–299 (2014)

[Pel18] F. Pellarin, R.B. Perkins, On vectorial Drinfeld modular forms over Tate algebras. Int. J. Numb. Theory **14**, 1729–1783 (2018)

[Poi20a] J. Poineau, D. Turchetti, Berkovich curves and Schottky uniformisation I: the Berkovich affine line, in *Arithmetic and Geometry Over Local Fields*, ed. by B. Anglès, T. Ngo Dac (Springer International Publishing, Cham, 2020)

[Poi20b] J. Poineau, D. Turchetti, Berkovich curves and Schottky uniformisation II: analytic uniformization of Mumford curves, in *Arithmetic and Geometry Over Local Fields*, ed. by B. Anglès, T. Ngo Dac (Springer International Publishing, Cham, 2020)

[Ser80a] J.-P. Serre. *Trees* (Springer, Berlin, New York, 1980)

[Ser80b] J.-P. Serre, *Local Fields* (Springer, Berlin, New York, 1980)

[Sti08] H. Stichtenoth, *Algebraic Function Fields and Codes*. Graduate Texts in Mathematics, vol. 254 (Springer, Berlin, Heidelberg, 2008)

[Tat71] J. Tate, Rigid analytic spaces. Invent. Math. **12**, 257–289 (1971)

[Tav20] F. Tavares Ribeiro, On the Stark units of Drinfeld modules, in *Arithmetic and Geometry Over Local Fields*, ed. by B. Anglès, T. Ngo Dac (Springer International Publishing, Cham, 2020)

[Tei91] J. Teitelbaum, The Poisson kernel for Drinfeld modular curves. J. AMS **4**, 491–511 (1991)

[Tem15] M. Temkin, Introduction to Berkovich analytic spaces, in *Berkovich Spaces and Applications*, ed. by A. Ducros, Ch. Favre, J. Nicaise. Springer Lecture Notes in Mathematics, vol. 2119 (2015)

[Wad41] L. Wade, Certain quantities transcendental over $GF(p^n, x)$. Duke Math. J. **8**, 707–720 (1941)

[Zyw13] D. Zywina, Explicit class field theory for global function fields. J. Numb. Theory **133**, 1062–1078 (2013)

Chapter 5
Berkovich Curves and Schottky Uniformization I: The Berkovich Affine Line

Jérôme Poineau and Daniele Turchetti

Abstract This is the first part of a survey on the theory of non-Archimedean curves and Schottky uniformization from the point of view of Berkovich geometry. This text is of an introductory nature and aims at giving a general idea of the theory of Berkovich spaces by focusing on the case of the affine line. We define the Berkovich affine line and present its main properties, with many details: classification of points, path-connectedness, metric structure, variation of rational functions, etc. Contrary to many other introductory texts, we do not assume that the base field is algebraically closed.

5.1 Introduction

The purpose of the present notes is to provide an introduction to non-Archimedean analytic geometry from the perspective of uniformization of curves. The main characters of this compelling story are *analytic curves over a non-Archimedean complete valued field* $(k, |\cdot|)$, and *Schottky groups*. The main difficulty in establishing a theory of non-Archimedean analytic spaces over k is that the natural topology induced over k by the absolute value $|\cdot|$ gives rise to totally disconnected spaces, that are therefore not suitable for defining analytic notions, such as that of a function locally expandable in power series.

However, in the late 1950s, J. Tate managed to develop the basics of such a theory (see [Tat71]), and christened the resulting spaces under the name *rigid analytic spaces*. To bypass the difficulty mentioned above, those spaces are not defined as usual topological spaces, but as spaces endowed with a so-called Grothendieck topology: some open subsets and some coverings are declared admissible and are

J. Poineau (✉)
Laboratoire de mathématiques Nicolas Oresme, Université de Caen - Normandie, Caen, France
e-mail: jerome.poineau@unicaen.fr; https://poineau.users.lmno.cnrs.fr/

D. Turchetti
Dalhousie University, Department of Mathematics & Statistics, Halifax, NS, Canada
e-mail: daniele.turchetti@dal.ca; https://www.mathstat.dal.ca/~dturchetti/

© The Author(s), under exclusive license to Springer Nature Switzerland AG 2021   179
B. Anglès, T. Ngo Dac (eds.), *Arithmetic and Geometry over Local Fields*,
Lecture Notes in Mathematics 2275, https://doi.org/10.1007/978-3-030-66249-3_5

the only ones that may be used to define local notions. For instance, one may define an analytic function by prescribing its restrictions to the members of an admissible covering. We refer to [Pel20, Sect. 5.1] in this volume for a short introduction to rigid analytic spaces.

Towards the end of the 1980s, V. Berkovich provided another definition of non-Archimedean analytic spaces. One of the advantages of his approach is that the resulting spaces are true topological spaces, endowed with a topology that makes them especially nice: they are Hausdorff, locally compact, and locally path-connected. This is the theory that we will use.

In the present text, which forms the first part of the survey, we introduce the Berkovich affine line over a non-Archimedean valued field k and study its properties. Contrary to several introductory texts, we do not assume that k is algebraically closed. The text is meant to be completely introductory, includes many details, and could be read by an undergraduate student with a minimal knowledge of abstract algebra and valuation theory. We develop the theory of the Berkovich affine line $\mathbb{A}_k^{1,\mathrm{an}}$ over k starting from scratch: definition (Sect. 5.2), classification of points with several examples (Sect. 5.3), basic topological properties, such as local compactness or a description of bases of neighborhoods of points (Sect. 5.4), and definition of analytic functions (Sect. 5.5). We then move to more subtle aspects of the theory such as extensions of scalars, including a proof that the Berkovich affine line over k is the quotient of the Berkovich affine line over $\widehat{k^a}$ (the completion of an algebraic closure of k) by the absolute Galois group of k (Sect. 5.6) and connectedness properties, culminating with the tree structure of $\mathbb{A}_k^{1,\mathrm{an}}$ (Sect. 5.7). We finally investigate even finer aspects of $\mathbb{A}_k^{1,\mathrm{an}}$ by considering virtual discs and annuli, and their retractions onto points and intervals respectively (Sect. 5.8), defining canonical lengths of intervals inside $\mathbb{A}_k^{1,\mathrm{an}}$ (Sect. 5.9), and ending with results on variations of rational functions, which are the very first steps of potential theory (Sect. 5.10).

The second part [PT20] of the survey, which is more advanced, contains a review of the theory of Berkovich analytic curves and a treatment of uniformization in this setting, as well as references to further developments and applications in the literature.

Notation Let $(\ell, |\cdot|)$ be a non-Archimedean valued field.

We set $\ell^\circ := \{a \in \ell : |a| \leqslant 1\}$. It is a subring of ℓ with maximal ideal $\ell^{\circ\circ} := \{a \in \ell : |a| < 1\}$. We denote the quotient by $\tilde{\ell}$ and call it the *residue field* of ℓ.

We set $|\ell^\times| := \{|a|, a \in \ell^\times\}$. It is a multiplicative subgroup of $\mathbb{R}_{>0}$ that we call the *value group* of ℓ. We denote its divisible closure by $|\ell^\times|^{\mathbb{Q}}$. It is naturally endowed with a structure of $\mathbb{Q}$-vector space.

Once and for all the paper, we fix a non-Archimedean complete valued field $(k, |\cdot|)$, a separable closure k^s of k and the corresponding algebraic closure k^a. The absolute value $|\cdot|$ on k extends uniquely to an absolute value on k^a, thanks to the fact that it extends uniquely to any given finite extension.[1] We denote by $\widehat{k^a}$ the completion of k^a: it is algebraically closed and coincides with the completion $\widehat{k^s}$ of k^s. We still denote by $|\cdot|$ the induced absolute value on $\widehat{k^a}$. We have $|\widehat{k^a}^\times| = |k^\times|^{\mathbb{Q}}$.

The first object we introduce in our exposition of non-Archimedean analytic geometry is the Berkovich affine line. This is already an excellent source of knowledge of properties of Berkovich curves, such as local path-connectedness, local compactness, classification of points, and behaviour under base change. Other properties, such as global contractibility, do not generalize, but will be useful later to study curves that "locally look like the affine line" (see [PT20, Section 2.1]).

Our main reference for this section is V. Berkovich's foundational book [Ber90]. We have also borrowed regularly from A. Ducros's thorough manuscript [Duc].

5.2 The Underlying Set

Definition 5.2.1 The *Berkovich affine line* $\mathbb{A}_k^{1,\mathrm{an}}$ is the set of multiplicative seminorms on $k[T]$ that induce the given absolute value $|\cdot|$ on k.

In more concrete terms, a point of $\mathbb{A}_k^{1,\mathrm{an}}$ is a map $|\cdot|_x \colon k[T] \to \mathbb{R}_+$ satisfying the following properties:

(i) $\forall P, Q \in k[T], \ |P + Q|_x \leqslant \max(|P|_x, |Q|_x)$;
(ii) $\forall P, Q \in k[T], \ |PQ|_x = |P|_x|Q|_x$;
(iii) $\forall \alpha \in k, \ |\alpha|_x = |\alpha|$.

With a slight abuse of notation, we set

$$\ker(|\cdot|_x) := \{P \in k[T] \colon |P|_x = 0\}.$$

It follows from the multiplicativity of $|\cdot|_x$ that $\ker(|\cdot|_x)$ is a prime ideal of $k[T]$.

In the following, we often denote a point of $\mathbb{A}_k^{1,\mathrm{an}}$ by x and by $|\cdot|_x$ the corresponding seminorm. This is purely for psychological and notational comfort since x and $|\cdot|_x$ are really the same thing.

Example 5.2.2 Each element α of k gives rise to a point of $\mathbb{A}_k^{1,\mathrm{an}}$ *via* the seminorm

$$|\cdot|_\alpha \colon P \in k[T] \longmapsto |P(\alpha)| \in \mathbb{R}_+.$$

We denote it by α again. Such a point is called a *k-rational point* of $\mathbb{A}_k^{1,\mathrm{an}}$.

---

[1] The reader can find a proof of this classical result in many textbooks, for example in [Cas86, Chapter 7].

Note that, conversely, the element α may be recovered from $|\cdot|_\alpha$ since $\ker(|\cdot|_\alpha) = (T - \alpha)$. It follows that the construction provides an injection $k \hookrightarrow \mathbb{A}_k^{1,\mathrm{an}}$.

Example 5.2.3 The construction of the previous example still makes sense if we start with a point $\alpha \in k^a$ and consider the seminorm

$$|\cdot|_\alpha : P \in k[T] \longmapsto |P(\alpha)| \in \mathbb{R}_+.$$

Such a point is called a *rigid point* of $\mathbb{A}_k^{1,\mathrm{an}}$.

However, it is no longer possible to recover α from $|\cdot|_\alpha$ in general. Indeed, in this case, we have $\ker(|\cdot|_a) = (\mu_\alpha)$, where μ_α denotes the minimal polynomial of α over k and, if σ is a k-linear automorphism of k^a, then, by uniqueness of the extension of the absolute value, we get $|\cdot|_{\sigma(\alpha)} = |\cdot|_\alpha$. One can check that we obtain an injection $k^a / \mathrm{Aut}(k^a/k) \hookrightarrow \mathbb{A}_k^{1,\mathrm{an}}$.

Readers familiar with scheme theory will notice that the rigid points of $\mathbb{A}_k^{1,\mathrm{an}}$ correspond exactly to the closed points of the schematic affine line $\mathbb{A}_k^1$. However the Berkovich affine line contains many more points, as the following examples show.

Example 5.2.4 Each element α of $\widehat{k^a}$ gives rise to a point of $\mathbb{A}_k^{1,\mathrm{an}}$ *via* the seminorm

$$|\cdot|_\alpha : P \in k[T] \longmapsto |P(\alpha)| \in \mathbb{R}_+.$$

This is similar to the construction of rigid points but, if α is transcendental over k, then we have $\ker(|\cdot|_\alpha) = (0)$ (i.e. $|\cdot|_\alpha$ is an absolute value) and the set of elements α' in $\widehat{k^a}$ such that $|\cdot|_{\alpha'} = |\cdot|_\alpha$ is infinite.

There also are examples of a different nature: points that look like "generic points" of discs.

Lemma 5.2.5 *Let $\alpha \in k$ and $r \in \mathbb{R}_{>0}$. The map*

$$|\cdot|_{\alpha,r} : \quad \begin{array}{ccc} k[T] & \longrightarrow & \mathbb{R}_{\geqslant 0} \\ \sum_{i \geqslant 0} a_i (T - \alpha)^i & \longmapsto & \max_{i \geqslant 0}(|a_i| r^i) \end{array}$$

is an absolute value on $k[T]$.

For $\alpha, \beta \in k$ and $r, s \in \mathbb{R}_{>0}$, we have $|\cdot|_{\alpha,r} = |\cdot|_{\beta,s}$ if, and only if, $|\alpha - \beta| \leqslant r$ and $r = s$.

Proof It is easy to check that $|\cdot|_{\alpha,r}$ is a norm. It remains to prove that it is multiplicative.

Let $P = \sum_{i \geqslant 0} a_i T^i$ and $Q = \sum_{j \geqslant 0} b_j T^j$. We may assume that $PQ \neq 0$. Let i_0 be the minimal index such that $|a_{i_0}| r^{i_0} = |P|_r$ and j_0 be the minimal index such that $|b_{j_0}| r^{j_0} = |Q|_r$.

For $\ell \in \mathbb{N}$, the coefficient of degree ℓ in PQ is

$$c_\ell := \sum_{i+j=\ell} a_i b_j,$$

hence we have

$$|c_\ell| r^\ell \leqslant \max_{i+j=\ell} (|a_i| r^i |a_j| r^j) \leqslant |P|_r |Q|_r.$$

For $\ell = \ell_0 := i_0 + j_0$, we find

$$c_{\ell_0} = a_{i_0} b_{j_0} + \sum_{\substack{i+j=\ell_0 \\ (i,j) \neq (i_0, j_0)}} \cdot \ a_i b_j.$$

For each $(i, j) \neq (i_0, j_0)$ with $i + j = \ell_0$, we must have $i < i_0$ or $j < j_0$, hence $|a_i| r^i < |P|_r$ or $|b_j| r^j < |Q|_r$ and, in any case, $|a_i b_j| r^{\ell_0} < |P|_r |Q|_r$. We now deduce from the equality case in the non-Archimedean triangle inequality that $|c_{\ell_0}| r^{\ell_0} = |P|_r |Q|_r$. The result follows.

Let $\alpha, \beta \in k$ and $r, s \in \mathbb{R}_{>0}$. Assume that we have $|\cdot|_{\alpha,r} = |\cdot|_{\beta,s}$. Applying the equality to $T - \alpha$ and $T - \beta$, we get

$$r = \max(|\alpha - \beta|, s) \text{ and } \max(|\alpha - \beta|, r) = s.$$

We deduce that $r = s$ and $|\alpha - \beta| \leqslant r$, as claimed.

Conversely, assume that we have $r = s$ and $|\alpha - \beta| \leqslant r$. Arguing by symmetry, it it enough to prove that, for each $P \in k[T]$, we have $|P|_{\beta,r} \leqslant |P|_{\alpha,r}$. Let $P = \sum_{i \geqslant 0} a_i (T - \alpha)^i \in k[T]$. We have

$$|T - \alpha|_{\beta,r} = \max(|\alpha - \beta|, r) = r$$

and, since $|\cdot|_{\beta,r}$ is multiplicative, $|(T - \alpha)^i|_{\beta,r} = r^i$ for each $i \geqslant 0$. Applying the non-Archimedean triangle inequality, we now get

$$|P|_{\beta,r} \leqslant \max_{i \geqslant 0}(|a_i| r^i) = |P|_{\alpha,r}.$$

The result follows.   □

Example 5.2.6 Let $\alpha \in k$ and $r \in \mathbb{R}_{>0}$. The map

$$|\cdot|_{\alpha,r} : \sum_{i \geqslant 0} a_i (T - \alpha)^i \mapsto \max_{i \geqslant 0}(|a_i| r^i)$$

Fig. 5.1 The Berkovich
affine line $\mathbb{A}_k^{1,\text{an}}$ when k is an
algebraically closed,
complete, valued field

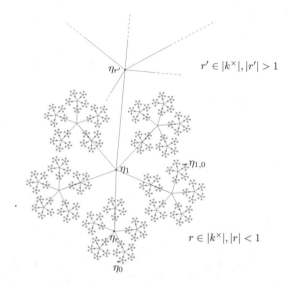

$r' \in |k^\times|, |r'| > 1$

$r \in |k^\times|, |r| < 1$

from Lemma 5.2.5 is an absolute value, hence gives rise to a point of $\mathbb{A}_k^{1,\text{an}}$, which we will denote by $\eta_{\alpha,r}$. To ease notation, we set $|\cdot|_r := |\cdot|_{0,r}$ and $\eta_r := \eta_{0,r}$. For a graphical representation of these points see Fig. 5.1.

Note that the relation characterizing the equality between $\eta_{\alpha,r}$ and $\eta_{\beta,s}$ is the same that characterizes the equality between the disc with center α and radius r and the disc with center β and radius s in a non-Archimedean setting. This is no coincidence and one can actually prove that, if k is not trivially valued, the absolute value $|\cdot|_{\alpha,r}$ is equal to the supremum norm on the closed disc of radius α and center r in the algebraic closure k^a of k.

Remark 5.2.7 The definitions of $|\cdot|_r$ and $|\cdot|_{\alpha,r}$ from Example 5.2.6 still make sense for $r = 0$. In this case, the points η_0 and $\eta_{\alpha,0}$ that we find are the rational points associated to 0 and α respectively. It will sometimes be convenient to use this notation.

Note that we could combine the techniques of Examples 5.2.3 and 5.2.6 to define even more points.

5.3   Classification of Points

In this section, we give a classification of the points of the Berkovich affine line $\mathbb{A}_k^{1,\text{an}}$. Let us first introduce a definition.

Definition 5.3.1 Let $x \in \mathbb{A}_k^{1,\text{an}}$. The *completed residue field* $\mathscr{H}(x)$ of x is the completion of the fraction field of $k[T]/\ker(|\cdot|_x)$ with respect to the absolute value induced by $|\cdot|_x$. It is a complete valued extension of k.

We will simply denote by $|\cdot|$ the absolute value induced by $|\cdot|_x$ on $\mathscr{H}(x)$.

The construction provides a canonical morphism of k-algebras $\chi_x : k[T] \to \mathscr{H}(x)$. We think of it as an evaluation morphism (into a field that varies with the point x). For $P \in k[T]$, we set $P(x) := \chi_x(P)$. It then follows from the definition that we have $|P(x)| = |P|_x$.

Example 5.3.2 Let $x \in \mathbb{A}_k^{1,\mathrm{an}}$. If x is a k-rational point, associated to some element $\alpha \in k$, we have $\mathscr{H}(x) = k$ and the morphism χ_x is nothing but the usual evaluation morphism $P \in k[T] \mapsto P(\alpha) \in k$.

If x is a rigid point, associated to some element $\alpha \in k^a$, we have an isomorphism $\mathscr{H}(x) \simeq k(\alpha)$. Conversely, if $\mathscr{H}(x)$ is a finite extension of k, then we have $\ker(|\cdot|_x) = (P)$ for some irreducible polynomial P, hence the point x is rigid (associated to any root of P).

Example 5.3.3 Let $\alpha \in k$ and $r \in \mathbb{R}_{>0} - |k^\times|^{\mathbb{Q}}$. Then $\mathscr{H}(\eta_{\alpha,r})$ is isomorphic to the field

$$k_r := \left\{ f = \sum_{i \in \mathbb{Z}} a_i (T - \alpha)^i : \lim_{i \to \pm\infty} |a_i| r^i = 0 \right\}$$

endowed with the absolute value $|f| = \max_{i \in \mathbb{Z}}(|a_i| r^i)$.

In the previous example, there exists a unique $i_0 \in \mathbb{Z}$ for which the quantity $|a_i| r^i$ is maximal. The fact that k_r is a field follows. This is no longer true for $r \in |k^\times|^{\mathbb{Q}}$ and the completed residue field $\mathscr{H}(\eta_{\alpha,r})$ is more difficult to describe (see [Chr83, Theorem 2.1.6] for instance).

Definition 5.3.4 A *character* of $k[T]$ is a morphism of k-algebras $\chi : k[T] \to K$, where K is some complete valued extension of k.

Two characters $\chi' : k[T] \to K'$ and $\chi'' : k[T] \to K''$ are said to be equivalent if there exists a character $\chi : k[T] \to K$ and isometric embeddings $i' : K \to K'$ and $i'' : K \to K''$ that make the following diagram commutative:

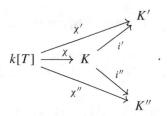

We have already explained how a point $x \in \mathbb{A}_k^{1,\mathrm{an}}$ gives rise to a character $\chi_x : k[T] \to \mathscr{H}(x)$. Conversely, to each character $\chi : k[T] \to K$, where $(K, |\cdot|)$ is a complete valued extension of $(k, |\cdot|)$, we may associate the multiplicative seminorm

$$|\cdot|_\chi : P \in k[T] \to |\chi(P)| \in \mathbb{R}.$$

Any equivalent character would lead to the same seminorm.

Lemma 5.3.5 *The map $x \mapsto \chi_x$ is a bijection from $\mathbb{A}_k^{1,\mathrm{an}}$ to the set of equivalences classes of characters of $k[T]$. Its inverse is the map $\chi \mapsto |\cdot|_\chi$.*   □

We mention the following related standard fact (see [Ked15, Lemma 2.8 and Remark 2.9] for instance).

Lemma 5.3.6 *Any two complete valued extensions of k may be isometrically embedded in a common one.*   □

Even if we do not have a explicit description of the completed residue fields associated to the points of $\mathbb{A}_k^{1,\mathrm{an}}$, we can use them to introduce some invariants.

Notation 5.3.7 For each valued extension $(\ell, |\cdot|)$ of $(k, |\cdot|)$, we set

$$s(\ell) := \mathrm{tr.\,deg.}(\tilde{\ell}/\tilde{k})$$

and

$$t(\ell) := \dim_\mathbb{Q}(|\ell^\times|^\mathbb{Q}/|k^\times|^\mathbb{Q}).$$

For $x \in \mathbb{A}_k^{1,\mathrm{an}}$, we set

$$s(x) := s(\mathscr{H}(x)) \text{ and } t(x) := t(\mathscr{H}(x)).$$

This invariants are related by the Abhyankar inequality (see [Bou06, VI, §10.3, Cor 1]).

Theorem 5.3.8 *Let ℓ be a valued extension of k. Then, we have*

$$s(\ell) + t(\ell) \leqslant \mathrm{tr.\,deg.}(\ell/k).$$

Moreover, if ℓ/k is a finitely generated extension for which equality holds, then $|\ell^\times|/|k^\times|$ is a finitely generated abelian group and $\tilde{\ell}/\tilde{k}$ is a finitely generated field extension.

For each $x \in \mathbb{A}_k^{1,\mathrm{an}}$, the fraction field of $k[T]/\ker(|\cdot|_x)$ has degree of transcendence 0 or 1 over k. Since its invariants s and t coincide with that of $\mathscr{H}(x)$, it follows from Abhyankar's inequality that we have

$$s(x) + t(x) \leqslant 1.$$

We can now state the classification of the points of the Berkovich affine line.

Definition 5.3.9 Let $x \in \mathbb{A}_k^{1,\mathrm{an}}$.

The point x is said to be of *type 1* if it comes from a point in $\widehat{k^a}$ in the sense of Example 5.2.4. In this case, we have $s(x) = t(x) = 0$.

The point x is said to be of *type 2* if we have $s(x) = 1$ and $t(x) = 0$.

The point x is said to be of *type 3* if we have $s(x) = 0$ and $t(x) = 1$.

The point x is said to be of *type 4* otherwise. In this case, we have $s(x) = t(x) = 0$.

Example 5.3.10 Let $\alpha \in k$ and $r \in \mathbb{R}_{>0}$.

Assume that $r \in |k^\times|^\mathbb{Q}$. There exist $n, m \in \mathbb{N}_{\geq 1}$ and $\gamma \in k$ with $r^n = |c|^m$. Consider such an equality with n minimal. Denote by t the image of $(T - \alpha)^n / c^m$ in $\widetilde{\mathcal{H}(x)}$. It is transcendental over $\tilde{k}$ and we have $\tilde{k}(t) = \widetilde{\mathcal{H}(x)}$. We deduce that $\eta_{\alpha,r}$ has type 2.

Assume that $r \notin |k^\times|^\mathbb{Q}$. Then, we have $\widetilde{\mathcal{H}(\eta_{\alpha,r})} = \tilde{k}$, so $\eta_{\alpha,r}$ has type 3.

The classification can be made more explicit when k is algebraically closed. Note that, in this case, we have $|k^\times|^\mathbb{Q} = |k^\times|$.

Lemma 5.3.11 *Assume that k is algebraically closed. Then x has type 2 (resp. 3) if, and only if, there exist $\alpha \in k$ and $r \in |k^\times|^\mathbb{Q}$ (resp. $r \notin |k^\times|^\mathbb{Q}$) such that $x = \eta_{\alpha,r}$.*

Proof Assume that x is of type 2. Since $s(x) = 1$, there exists $P \in k[T]$ such that $|P(x)| = 1$ and $\tilde{P}$ is transcendental over $\tilde{k}$. Since k is algebraically closed, we have $|k^\times| = |k^\times|^\mathbb{Q} = |\mathcal{H}(x)^\times|^\mathbb{Q}$, hence we may write P as a product of linear polynomials, all of which have absolute value 1. One of these linear polynomials has an image in $\widetilde{\mathcal{H}(x)}$ that is transcendental over $\tilde{k}$. Write it as $c(T - \alpha)$, with $c \in k^\times$ and $\alpha \in k$. We then have $x = \eta_{\alpha,|c|^{-1}}$.

Assume that x is of type 3. Since $t(x) = 1$, there exists $P \in k[T]$ such that $r := |P(x)| \notin |k^\times|^\mathbb{Q}$. As before, we may assume that $P = T - \alpha$ with $\alpha \in k$. We then have $x = \eta_{\alpha,r}$.

The converse implications are dealt with in Example 5.3.10. $\qquad\square$

Proposition 5.3.12 *Assume that k is algebraically closed. Let $x \in \mathbb{A}_k^{1,\mathrm{an}}$. There exist a set I, a family $(\alpha_i)_{i \in I}$ of k and a family $(r_i)_{i \in I}$ of $\mathbb{R}_{\geq 0}$ such that, for each $i, j \in I$, we have*

$$\max(|\alpha_i - \alpha_j|, r_i) \leq r_j \text{ or } \max(|\alpha_i - \alpha_j|, r_j) \leq r_i$$

and, for each $P \in k[T]$,

$$|P|_x = \inf_{i \in I}(|P|_{\alpha_i, r_i}).$$

Proof Our set I will be the underlying set of k. For each $a \in k$, we set $\alpha_a := a$ and $r_a := |T - a|_x$.

Let $a, b \in k$. We have

$$|a - b| = |a - b|_x = |a - T + T - b|_x \leqslant \max(|a - T|_x, |T - b|_x),$$

so the first condition of the statement is satisfied. It implies that we have

$$\forall P \in k[T], \ |P|_{a,r_a} \leqslant |P|_{b,r_b} \ \text{or} \ \forall P \in k[T], \ |P|_{b,r_b} \leqslant |P|_{a,r_a}.$$

It follows that the map $v \colon P \in k[T] \mapsto \inf_{a \in k}(|P|_{a,r_a})$ is multiplicative, hence a multiplicative seminorm.

Since k is algebraically closed, every polynomial factors as a product of monomials. As a consequence, to prove that v and $|\cdot|_x$ coincide, it is enough to prove that they coincide on monomials, because of multiplicativity.

Let $\alpha \in k$. We have $|T - \alpha|_{\alpha,r_\alpha} = r_\alpha = |T - \alpha|_x$, hence $v(T - \alpha) \leqslant |T - \alpha|_x$. On the other hand, for each $a \in k$, we have

$$|T - \alpha|_x = |T - a + a - \alpha|_x \leqslant \max(|T - a|_x, |a - \alpha|_x) = \max(r_a, |a - \alpha|) = |T - \alpha|_{a,r_a},$$

hence $|T - \alpha|_x \leqslant v(T - \alpha)$. □

Remark 5.3.13 One should think of the families $(\alpha_i)_{i \in I}$ and $(r_i)_{i \in I}$ in the statement of Proposition 5.3.12 as a single family of discs in k (with center α_i and radius r_i). Then, the condition of the statement translates into the fact that, for each pair of discs of the family, one is contained in the other.

Moreover, it is not difficult to check that, if the intersection of this family of discs contains a point α of k, then we have $|\cdot|_x = |\cdot|_{\alpha,r}$, where $r = \inf_{i \in I}(r_i) \geqslant 0$.

On the other hand, if the family of discs has empty intersection, we find a new point, necessarily of type 4. Note that we must have $\inf_{i \in I}(r_i) > 0$ is this case. Otherwise, the completeness of k would ensure that the intersection of the discs contains an element of k.

Definition 5.3.14 Assume that k is algebraically closed. For each $x \in \mathbb{A}_k^{1,\mathrm{an}}$, we define the *radius of the point x* to be

$$r(x) := \inf_{c \in k}(|T - c|_x).$$

It can be thought of as the distance from the point to k.

Example 5.3.15 Assume that k is algebraically closed. Let $x \in \mathbb{A}_k^{1,\mathrm{an}}$.

If x has type 1, then $r(x) = 0$.

If x has type 2 or 3, then, by Lemma 5.3.11, it is of the form $x = \eta_{\alpha,r}$ and we have $r(x) = r$.

If x has type 4, then, with the notation of Proposition 5.3.12, we have $r(x) = \inf_{i \in I}(r_i)$. Indeed, for each $i \in I$, we have $|T - \alpha_i|_x \leqslant |T - \alpha_i|_{\alpha_i,r_i} = r_i$. It follows that $r(x) \leqslant \inf_{i \in I}(r_i)$. On the other hand, let $c \in k$. For i big enough, c is

Fig. 5.2 The points $\eta_{\alpha,r}$ with $r \in |k^\times|$ are of type 2, the points $\eta_{\alpha,s}$ with $s \notin |k^\times|$ are of type 3, and the points $\eta_{\alpha,0}$ are of type 1. If k is not spherically complete, points of type 4 will occur

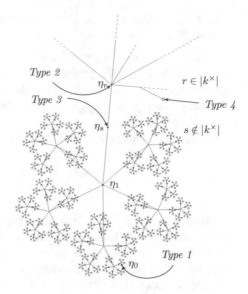

not contained in the disc of center α_i and radius r_i, that is to say $|\alpha_i - c| > r_i$, from which it follows that $|T - c|_{\alpha_i, r_i} = |\alpha_i - c|$.

We deduce that $|T - c|_x = \inf_{i \in I}(|\alpha_i - c|) \geqslant \inf_{i \in I}(r_i)$; see Fig. 5.2 for a representation of these types of points on the affine line.

Remark 5.3.16 The radius of a point of type different from 1 is not intrinsically attached to the point in the sense that it depends on the chosen coordinate T on $\mathbb{A}_k^{1,\mathrm{an}}$. However, by studying the automorphisms of $\mathbb{A}_k^{1,\mathrm{an}}$, one can prove that any change of coordinate will have the effect of multiplying all the radii by the same constant (in $|k^\times|$), see Proposition 5.5.12 and Remark 5.5.13. In particular, the quotient of the radii of two points is well-defined.

Definition 5.3.17 The field k is said *spherically complete* if every family of discs that is totally ordered by inclusion has a non-empty intersection.

The field k is said *maximally complete* if it has no non-trivial immediate extensions, i.e. extensions with the same value group and residue field.

We refer to the paper [Poo93] by B. Poonen for more on those topics and in particular the construction of spherical completions, i.e. minimal spherically complete extensions. We only quote the following important result.

Theorem 5.3.18 *A valued field is spherically complete if, and only if, it is maximally complete.* $\qquad\square$

Remark 5.3.19 Assume that k is algebraically closed. Then, the completed residue field of a point of type 4 is an immediate extension of k. Using Remark 5.3.13, we can deduce a proof of Theorem 5.3.18 in this case.

To make things more concrete, we would now like to give a rather explicit example of a point of type 4.

Example 5.3.20 Let $r \in (0, 1)$ and consider the field of Laurent series $\mathbb{C}((t))$ endowed with the absolute value defined by $|f| = r^{v_t(f)}$. Recall that the t-adic valuation $v_t(f)$ of f is the infimum of the indices of the non-zero terms of f in its Taylor expansion $f = \sum_{n \in \mathbb{Z}} a_n t^n$. (Note that, for $f = 0$, we have $v_t(0) = +\infty$, hence $|f| = 0$.)

The algebraic closure of $\mathbb{C}((t))$ is the field of Puiseux series:

$$\mathbb{C}((t))^a = \bigcup_{m \in \mathbb{N}_{\geqslant 1}} \mathbb{C}((t^{1/m})).$$

In particular, the exponents of t in the expansion of any given element of $\mathbb{C}((t))^a$ are rational numbers with bounded denominators.

We choose our field k to be the completion of $\mathbb{C}((t))^a$. Its elements may still be written as power series with rational exponents. This time, the exponents may have unbounded denominators but they need to tend to $+\infty$.

Consider a power series of the form $\sum_{n \in \mathbb{N}} t^{q_n}$ where $(q_n)_{n \in \mathbb{N}}$ is a strictly increasing bounded sequence of rational numbers. (For instance, $q_n = 1 - 2^{-n}$ would do.) The associated point of $\mathbb{A}_k^{1,\mathrm{an}}$ is then a point of type 4. In this case, one can explicitly describe an associated family of discs by taking, for each $m \in \mathbb{N}$, the disc with center $\alpha_m := \sum_{n=0}^{m} t^{q_n}$ and radius $r_m := r^{q_{m+1}}$.

One can go even further in this case and describe a spherical completion of k. It is the field of Hahn series $\mathbb{C}((t^{\mathbb{Q}}))$ consisting of the power series $f = \sum_{q \in \mathbb{Q}} a_q t^q$, where the a_q's are rational numbers and the support $\{q \in \mathbb{Q} \mid a_q \neq 0\}$ of f is well-ordered: each of its non-empty subsets has a smallest element.

5.4   Topology

We endow the set $\mathbb{A}_k^{1,\mathrm{an}}$ with the coarsest topology such that, for each $P \in k[T]$, the map

$$x \in \mathbb{A}_k^{1,\mathrm{an}} \longmapsto |P(x)| \in \mathbb{R}$$

is continuous. In more concrete terms, a basis of the topology is given by the sets

$$\{x \in \mathbb{A}_k^{1,\mathrm{an}} : r < |P(x)| < s\},$$

for $P \in k[T]$ and $r, s \in \mathbb{R}$.

Remark 5.4.1 By Example 5.2.2, we can see k as a subset of $\mathbb{A}_k^{1,\mathrm{an}}$. The topology on k induced by that on $\mathbb{A}_k^{1,\mathrm{an}}$ then coincides with that induced by the absolute value $|\cdot|$.

Lemma 5.4.2 *The Berkovich affine line $\mathbb{A}_k^{1,\mathrm{an}}$ is Hausdorff.*

Proof Let $x \neq y \in \mathbb{A}_k^{1,\mathrm{an}}$. Then, there exists $P \in k[T]$ such that $|P(x)| \neq |P(y)|$. We may assume that $|P(x)| < |P(y)|$. Let $r \in (|P(x)|, |P(y)|)$. Set

$$U := \{z \in \mathbb{A}_k^{1,\mathrm{an}} : |P(z)| < r\} \text{ and } V := \{z \in \mathbb{A}_k^{1,\mathrm{an}} : |P(z)| > r\}.$$

The sets U and V are disjoint open subsets of $\mathbb{A}_k^{1,\mathrm{an}}$ containing respectively x and y. The result follows. □

Definition 5.4.3 For $\alpha \in k$ and $r \in \mathbb{R}_{>0}$, the *open disc of center α and radius r* is

$$D^-(\alpha, r) = \{x \in \mathbb{A}_k^{1,\mathrm{an}} : |(T - \alpha)(x)| < r\}.$$

For $\alpha \in k$ and $r \in \mathbb{R}_{>0}$, the *closed disc of center α and radius r* is

$$D^+(\alpha, r) = \{x \in \mathbb{A}_k^{1,\mathrm{an}} : |(T - \alpha)(x)| \leqslant r\}.$$

For $\alpha \in k$ and $r < s \in \mathbb{R}_{>0}$, the *open annulus of center α and radii r and s* is

$$A^-(\alpha, r, s) = \{x \in \mathbb{A}_k^{1,\mathrm{an}} : r < |(T - \alpha)(x)| < s\}.$$

For $\alpha \in k$ and $r \leqslant s \in \mathbb{R}_{>0}$, the *closed annulus of center α and radii r and s* is

$$A^+(\alpha, r, s) = \{x \in \mathbb{A}_k^{1,\mathrm{an}} : r \leqslant |(T - \alpha)(x)| \leqslant s\}.$$

For $\alpha \in k$ and $r \in \mathbb{R}_{>0}$, the *flat closed annulus of center α and radius r* is

$$A^+(\alpha, r, r) = \{x \in \mathbb{A}_k^{1,\mathrm{an}} : |(T - \alpha)(x)| = r\}.$$

These analytic spaces are represented in Figs. 5.3 and 5.4. In the result that follows, we study the topology of discs and annuli as subsets of $\mathbb{A}_k^{1,\mathrm{an}}$.

Lemma 5.4.4 *Let $\alpha \in k$ and $r \in \mathbb{R}_{>0}$. The closed disc $D^+(\alpha, r)$ is compact and has a unique boundary point: $\eta_{\alpha,r}$. The open disc $D^-(\alpha, r)$ is open and its closure is $D^-(\alpha, r) \cup \{\eta_{\alpha,r}\}$.*

Let $\alpha \in k$ and $r < s \in \mathbb{R}_{>0}$. The closed annulus $A^+(\alpha, r, s)$ is compact and has two boundary points: $\eta_{\alpha,r}$ and $\eta_{\alpha,s}$. The open annulus $A^-(\alpha, r, s)$ is open and its closure is $A^-(\alpha, r, s) \cup \{\eta_{\alpha,r}, \eta_{\alpha,s}\}$.

Let $\alpha \in k$ and $r \in \mathbb{R}_{>0}$. The flat closed annulus $A^+(\alpha, r, r)$ is compact and has a unique boundary point: $\eta_{\alpha,r}$.

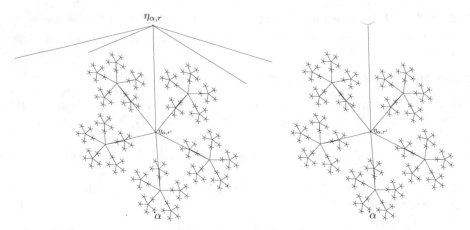

Fig. 5.3 On the left, the closed disc $D^+(\alpha, r)$. On the right, the open disc $D^-(\alpha, r)$, which is a maximal open sub-disc of $D^+(\alpha, r)$, but not the only one

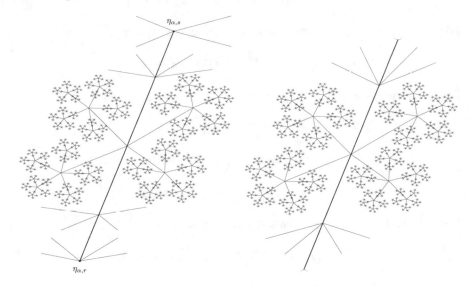

Fig. 5.4 On the left, the closed annulus $A^+(\alpha, r, s)$. On the right, the open annulus $A^-(\alpha, r, s)$, which is the unique maximal open sub-annulus of $A^+(\alpha, r, s)$

Proof Let $x \in D^+(\alpha, r)$. We have $|T - \alpha|_x \leqslant r$, hence it follows from the non-Archimedean triangle inequality that we have $|\cdot|_x \leqslant |\cdot|_{\alpha,r}$, as seminorms on $k[T]$.

Consider the product $\prod_{P \in k[T]}[0, |P|_r]$ endowed with the product topology and its closed subset F consisting of the elements $(x_P)_{P \in k[T]}$ satisfying the conditions

$$\begin{cases} \forall P, Q \in k[T], \ \lambda_{P+Q} \leqslant \max(\lambda_P, \lambda_Q); \\ \forall P, Q \in k[T], \ \lambda_{PQ} = \lambda_P \, \lambda_Q. \end{cases}$$

It follows from the previous argument that the map

$$p\colon D^+(\alpha, r) \longrightarrow \prod_{P \in k[T]} [0, |P|_r]$$
$$x \longmapsto (|P|_x)_{P \in k[T]}$$

induces a bijection between $D^+(\alpha, r)$ and F. (The only non-trivial point is to check that the seminorm on $k[T]$ associated to an element of F induces the given absolute value $|\cdot|$ on k.) Moreover, it follows from the very definition of the topology that p is a homeomorphism onto its image. Since F is closed in $\prod_{P \in k[T]} [0, |P|_r]$, and the latter is compact by Tychonoff's theorem, F is compact, hence $D^+(\alpha, r)$ is compact too.

Let $x \in D^+(\alpha, r) - \{\eta_{\alpha, r}\}$. Then, there exists $P \in k[T]$ such that $|P|_x \neq |P|_{\alpha, r}$, hence $|P|_x < |P|_{\alpha, r}$. In other words, the point x belongs to the open subset $\{y \in \mathbb{A}_k^{1,\mathrm{an}} : |P|_y < |P|_{\alpha, r}\}$ of $\mathbb{A}_k^{1,\mathrm{an}}$, which is contained in $D^+(\alpha, r)$. It follows that x belongs to the interior of $D^+(\alpha, r)$.

Let U be an open subset of $\mathbb{A}_k^{1,\mathrm{an}}$ containing $\eta_{\alpha, r}$. By definition of the topology, there exist $P_1, \ldots, P_n \in k[T]$ and $u_1, v_1, \ldots, u_n, v_n \in \mathbb{R}$ such that

$$\eta_{\alpha, r} \in \{y \in \mathbb{A}_k^{1,\mathrm{an}} : u_i < |P_i|_y < v_i\} \subseteq U.$$

Using the explicit definition of the norms $|\cdot|_{\alpha, s}$, one shows that, for each $s \in \mathbb{R}_{\geqslant 0}$ that is close enough to r, we have $\eta_{\alpha, s} \in U$. We deduce that $\eta_{\alpha, r}$ belongs to the boundary of $D^+(\alpha, r)$ (because we can choose $s > r$) and to the closure of $D^-(\alpha, r)$ (because we can choose $s < r$).

This finishes the proof that the boundary of $D^+(\alpha, r)$ is equal to $\{\eta_{\alpha, r}\}$.

By definition of the topology, the disc $D^-(\alpha, r)$ is open. Since $D^+(\alpha, r)$ is compact, it contains the closure of $D^-(\alpha, r)$. We have already proved that its closure contains $\eta_{\alpha, r}$.

Let $x \in D^+(\alpha, r) - (D^-(\alpha, r) \cup \{\eta_{\alpha, r}\})$. We have $|T - \alpha|_x = r$ and there exists $P \in k[T]$ such that $|P|_x < |P|_{\alpha, r}$. Let us choose such a polynomial P with minimal degree.

Arguing by contradiction, assume that $|P(\alpha)| < |P|_{\alpha, r}$. Write $P = P(\alpha) + (T - \alpha) Q$, with $Q \in k[T]$. We then have

$$|P|_{\alpha, r} = \max(|P(\alpha)|, r|Q|_{\alpha, r}) = r|Q|_{\alpha, r}.$$

If $|P(\alpha)| \neq r|Q|_x$, we have

$$r|Q|_x \leqslant \max(|P(\alpha)|, r|Q|_x) = |P|_x < |P|_{\alpha, r} = r|Q|_{\alpha, r}.$$

If $|P(\alpha)| = r|Q|_x$, the same inequality holds. In any case, we have $|Q|_x < |Q|_{\alpha, r}$, which contradicts the minimality of the degree of P.

We have just proved that $|P(\alpha)| = |P|_{\alpha,r}$. It follows that, for each $y \in D^-(\alpha, r)$, we have $|P|_y = |P|_{\alpha,r}$, hence the open set $\{y \in \mathbb{A}_k^{1,\mathrm{an}} : |P|_y < |P|_{\alpha,r}\}$ contains x and is disjoint from $D^-(\alpha, r)$, so x does not belong to the boundary of $D^-(\alpha, r)$.

We have finally proven that the closure of $D^-(\alpha, r)$ is $D^-(\alpha, r) \cup \{\eta_{\alpha,r}\}$.

The results for the annuli are proven similarly. $\qquad\square$

Since $\mathbb{A}_k^{1,\mathrm{an}}$ may be exhausted by closed discs, we deduce the following result.

Corollary 5.4.5 *The Berkovich affine line $\mathbb{A}_k^{1,\mathrm{an}}$ is countable at infinity and locally compact.* $\qquad\square$

It is possible to give a characterization of the fields k for which the space $\mathbb{A}_k^{1,\mathrm{an}}$ is metrizable.

Corollary 5.4.6 *The following assertions are equivalent:*

(i) *the Berkovich affine line $\mathbb{A}_k^{1,\mathrm{an}}$ is metrizable;*
(ii) *the field k contains a countable dense subset.*

Proof (i) $\implies$ (ii) Assume that $\mathbb{A}_k^{1,\mathrm{an}}$ is metrizable. We fix a metric on $\mathbb{A}_k^{1,\mathrm{an}}$ and will consider balls with respect to it. Let $(\varepsilon_n)_{n \in \mathbb{N}}$ be a sequence of positive real numbers converging to 0.

Let $r \in \mathbb{R}_{>0}$. By Lemma 5.4.4, the closed disc $D^+(0, r)$ is compact. As a consequence, for each $n \in \mathbb{N}$, it is covered by finitely many metric balls of radius ε_n. For each such ball that contains a point of k, pick a point of k in it. The collection of those points is a finite subset $k_{r,n}$ of k. The set $k_r := \bigcup_{n \in \mathbb{N}} k_{r,n}$ is a countable subset of k that is dense in $k \cap D^+(0, r)$.

It follows that the set $k' := \bigcup_{m \in \mathbb{N}_{\geqslant 1}} k_m$ is a countable dense subset of k.

(ii) $\implies$ (i) Assume that the field k contains a countable dense subset k'. Then, the family of sets

$$\{x \in \mathbb{A}_k^{1,\mathrm{an}} : r < |P(x)| < s\}$$

with $P \in k'[T]$ and $r, s \in \mathbb{Q}$ is a countable basis of the topology. The result now follows from Corollary 5.4.5 and Urysohn's metrization theorem. $\qquad\square$

By removing the boundary point of a closed disc, one may obtain either one or infinitely many discs, depending on the radius. We will deal with this question assuming that k is algebraically closed and consider first the case where the radius does not belong to the value group of k.

Lemma 5.4.7 *Assume that k is algebraically closed. For each $\alpha \in k$ and $r \in \mathbb{R}_{>0} - |k^\times|$, we have*

$$D^+(\alpha, r) = \{\eta_{\alpha,r}\} \sqcup D^-(\alpha, r).$$

Proof Let $x \in D^+(\alpha, r) - D^-(\alpha, r)$. We then have $|T - \alpha|_x = r$.

Let $P(T) = a_d T^d + \cdots + a_0 \in k[T]$. Since k is algebraically closed, $|k^\times|$ is divisible and since $r \notin |k^\times|$, all the terms $|a_i| r^i$ are distinct. It follows that

$$|P|_x = \max_{1 \leqslant i \leqslant d} (|a_i| r^i) = |P|_{\alpha,r}.$$

$\square$

We now handle the case of the disc $D^+(0, 1)$. When k is algebraically closed, any disc of the form $D(\alpha, r)$ with $r \in |k^\times|$ may be turned into the latter by a suitable linear change of variable.

Notation 5.4.8 For each $u \in \tilde{k}$, we set $D^-(u, 1) := D^-(\alpha, 1)$, where α is a lift of u in k°.

Since any two lifts α_1 and α_2 satisfy $|\alpha_1 - \alpha_2| < 1$, the definition does not depend on the choice of α.

Lemma 5.4.9 *Assume that k is algebraically closed. We have*

$$D^+(0, 1) = \{\eta_{0,1}\} \sqcup \bigsqcup_{u \in \tilde{k}} D^-(u, 1).$$

Proof Let $u_1 \in \tilde{k}$ and let $\alpha_1 \in k^\circ$ such that $\tilde{\alpha}_1 = u_1$. We have $|T - \alpha_1|_{0,1} = \max(1, |\alpha_1|) = 1$, hence $\eta_{0,1} \notin D^-(\alpha_1, 1)$.

Let $u_2 \neq u_1 \in \tilde{k}$ and let $\alpha_2 \in k^\circ$ such that $\tilde{\alpha}_2 = u_2$. For each $x \in D^-(u_2, 1)$, we have

$$|T - \alpha_1|_x = |(T - \alpha_2) + (\alpha_2 - \alpha_1)|_x = 1,$$

since $|T - \alpha_2|_x < 1$ and $|\alpha_2 - \alpha_1| < 1$. It follows that $x \notin D^-(u_1, 1)$.

To finish, it remains to prove that $D^+(0, 1) - \{\eta_{0,1}\}$ is covered by the discs $D^-(u, 1)$ with $u \in \tilde{k}$. Let $x \in D^+(0, 1) - \{\eta_{0,1}\}$. There exists $P \in k[T]$ such that $|P|_x \neq |P|_{0,1}$, hence $|P|_x < |P|_{0,1}$. Since k is algebraically closed, we may find such a P that is a monomial: $P = T - \alpha$ for some $\alpha \in k$. If $|\alpha| > 1$, then we have $|T - \alpha|_x = |\alpha| = |T - \alpha|_{0,1}$, which contradicts the assumption. We deduce that $|\alpha| \leqslant 1$, hence $|T - \alpha|_x < |T - \alpha|_{0,1} = 1$ and $x \in D^-(\tilde{\alpha}, 1)$. $\square$

We now want to describe bases of neighborhoods of the points of $\mathbb{A}_k^{1,\mathrm{an}}$, at least in the algebraically closed case. To do this, contrary to the usual complex setting, discs are not enough. We will also need annuli and even more complicated subsets.

Definition 5.4.10 An open (resp. closed) *Swiss cheese*[2] over k is a non-empty subset of $\mathbb{A}_k^{1,\mathrm{an}}$ that may be written as the complement of finitely many closed (resp. open) discs over k in an open (resp. a closed) disc over k.

---

[2] This is called a "standard set" in [Ber90, Section 4.2].

Proposition 5.4.11 *Assume that k is algebraically closed. Let $x \in \mathbb{A}_k^{1,\mathrm{an}}$.*
If x has type 1 or 4, it admits a basis of neighborhoods made of discs.
If x has type 2, it admits a basis of neighborhoods made of Swiss cheeses.
If x has type 3, it admits a basis of neighborhoods made of annuli.

Proof By definition of the topology, every neighborhood of x contains a finite intersection of sets of the form $\{u < |P| < v\}$ with $P \in k[T]$ and $u, v \in \mathbb{R}$. Since the sets in the statement are stable under finite intersections, it is enough to prove that each set of the form $\{u < |P| < v\}$ that contains x contains a neighborhood of x as described in the statement.

Let $P \in k[T]$ and $u, v \in \mathbb{R}$ such that $|P(x)| \in (u, v)$. Write $P = c \prod_{j=1}^m (T - \gamma_j)$ with $c \in k^\times$ and $\gamma_1, \ldots, \gamma_m \in k$.

Assume that x has type 1. Since k is algebraically closed, it is a rational point, hence associated to some $\alpha \in k$. One checks that $\{u < |P| < v\}$ then contains a disc of the form $D^-(\alpha, r)$ for $r \in \mathbb{R}_{>0}$.

Assume that x has type 3. By Lemma 5.3.11, there exist $\alpha \in k$ and $r \in \mathbb{R}_{>0}$ such that $x = \eta_{\alpha,r}$. One checks that $\{u < |P| < v\}$ then contains an annulus of the form $A^-(\alpha, r_1, r_2)$ for some $r_1, r_2 \in \mathbb{R}_{>0}$ with $r \in (r_1, r_2)$.

Assume that x has type 4. By Proposition 5.3.12 and Remark 5.3.13, it is associated to a family of closed discs $(D^+(\alpha_i, r_i))_{i \in I}$ whose intersection contains no rational point. Because of this condition, there exists $i \in I$ such that $D^+(\alpha_i, r_i)$ contains none of the γ_j's. Then, for each $j \in \{1, \ldots, m\}$, we have

$$|\alpha_i - \gamma_j| > r_i,$$

hence, for each $y \in D^+(\alpha_i, r_i)$, we have

$$|T(y) - \gamma_j| = |T(y) - \alpha_i + \alpha_i - \gamma_j| = |\alpha_i - \gamma_j|.$$

We deduce that, for each $y \in D^+(\alpha_i, r_i)$, we have $|P(y)| = |P(x)|$ and the result follows.

Assume that x has type 2. By Lemma 5.3.11, there exist $\alpha \in k$ and $r \in \mathbb{R}_{>0}$ such that $x = \eta_{\alpha,r}$. We have

$$|P(x)| = |c| \prod_{j=1}^m \max(r, |\alpha - \gamma_j|) < v,$$

hence there exists $\rho > r$ such that $|c| \prod_{j=1}^m \max(\rho, |\alpha - \gamma_j|) < v$. Then, for each $y \in D^-(\alpha, \rho)$, we have $|P(y)| < v$.

There exists $(\rho_1, \ldots, \rho_m) \in \prod_{j=1}^m (0, |T(x) - \gamma_j|)$ such that $|c| \prod_{j=1}^m \rho_i > u$. Then, for each $y \in \mathbb{A}_k^{1,\mathrm{an}} - \bigcup_{j=1}^m D^+(\gamma_j, \rho_j)$, we have $|P(y)| > u$.

It follows that $D^-(\alpha, \rho) - \bigcup_{j=1}^m D^+(\gamma_j, \rho_j)$ is a neighborhood of x contained in $\{u < |P| < v\}$. $\qquad\square$

Remark 5.4.12 If k is not algebraically closed, bases of neighborhoods of points may be more complicated. Let us give an example. Let p be a prime number that is congruent to 3 modulo 4, so that -1 is not a square in $\mathbb{Q}_p$. Consider the point x of $\mathbb{A}_k^{1,\mathrm{an}}$ associated to a square root of -1 in $\mathbb{C}_p$. Equivalently, the point x is the unique point of $\mathbb{A}_k^{1,\mathrm{an}}$ satisfying $|T^2 + 1|_x = 0$.

The subset U of $\mathbb{A}_k^{1,\mathrm{an}}$ defined by the inequality $|T^2 + 1| < 1$ is an open neighborhood of x. It does not contain 0, so the function T is invertible on it and we may write

$$-1 = T^2 - (T^2 + 1) = T^2 \left(1 - \frac{T^2 + 1}{T^2} \right).$$

At each point y of U, we have $|T^2 + 1|_y < 1$, hence $|T^2|_y = 1$ and we deduce that -1 has a square root on U. In particular, U contains no $\mathbb{Q}_p$-rational points and no discs.

Note that the topology of $\mathbb{A}_k^{1,\mathrm{an}}$ is quite different from the topology of k. We have already seen that $\mathbb{A}_k^{1,\mathrm{an}}$ is always locally compact, whereas k is if, and only if, $|k^\times|$ is discrete and $\tilde{k}$ is finite. In another direction, k is always totally disconnected, but $\mathbb{A}_k^{1,\mathrm{an}}$ contains paths, as the next result shows.

Lemma 5.4.13 *Let $\alpha \in k$. The map*

$$r \in \mathbb{R}_{\geq 0} \longmapsto \eta_{\alpha,r} \in \mathbb{A}_k^{1,\mathrm{an}}$$

is a homeomorphism onto its image I_α.

Proof It is clear that the map is injective and open, so to prove the result, it is enough to prove that it is continuous. By definition, it is enough to prove that, for each $P \in k[T]$, the map $r \in \mathbb{R}_{\geq 0} \mapsto |P|_{\alpha,r} \in \mathbb{R}_{\geq 0}$ is continuous. The result then follows from the explicit description of $|\cdot|_{\alpha,r}$ (see Example 5.2.6). $\qquad\square$

Remark 5.4.14 One may use the paths from the previous lemma to connect the points of k. Let $\alpha, \beta \in k$ and consider the paths I_α and I_β. Example 5.2.6 tells us that they are not disjoint but meet at the point $\eta_{\alpha,|\alpha-\beta|} = \eta_{\beta,|\alpha-\beta|}$ (and actually coincide from this point on). The existence of a path from α to β inside $\mathbb{A}_k^{1,\mathrm{an}}$ follows.

We will use this construction in Sect. 5.7 to show that $\mathbb{A}_k^{1,\mathrm{an}}$ is path-connected.

5.5 Analytic Functions

So far, we have described the Berkovich affine line $\mathbb{A}_k^{1,\mathrm{an}}$ as a topological space. It may actually be endowed with a richer structure, since we may define analytic functions over it.

Definition 5.5.1 Let U be an open subset of $\mathbb{A}_k^{1,\mathrm{an}}$. An *analytic function* on U is a map

$$F: U \to \bigsqcup_{x \in U} \mathscr{H}(x)$$

such that, for each $x \in U$, the following conditions hold:

(i) $F(x) \in \mathscr{H}(x)$;
(ii) there exist a neighborhood V of x and sequences $(P_n)_{n \in \mathbb{N}}$ and $(Q_n)_{n \in \mathbb{N}}$ of elements of $k[T]$ such that the Q_n's do not vanish on V and

$$\lim_{n \to +\infty} \sup_{y \in V} \left(\left| F(y) - \frac{P_n(y)}{Q_n(y)} \right| \right) = 0.$$

Remark 5.5.2 The last condition can be reformulated by saying that F is locally a uniform limit of rational functions without poles, which then makes the definition similar to the usual complex one (where analytic functions are locally uniform limits of polynomials).

The Berkovich affine line $\mathbb{A}_k^{1,\mathrm{an}}$ together with its sheaf of analytic functions $\mathcal{O}$ now has the structure of a locally ringed space. It satisfies properties that are similar to those of the usual complex analytic line $\mathbb{C}$. We state a few of them here without proof.

The set of global analytic functions on some simple open subsets of $\mathbb{A}_k^{1,\mathrm{an}}$ may be described explicitly.

Proposition 5.5.3 *Let* $r \in \mathbb{R}_{>0}$. *Then* $\mathcal{O}(D^-(0,r))$ *is the set of elements*

$$\sum_{i \in \mathbb{N}} a_i T^i \in k[\![T]\!]$$

with radius of convergence greater than or equal to r:

$$\forall s \in (0, r), \quad \lim_{i \to +\infty} |a_i| s^i = 0.$$

$\square$

Corollary 5.5.4 *The local ring* $\mathcal{O}_0$ *at the point* 0 *of* $\mathbb{A}_k^{1,\mathrm{an}}$ *consists of the power series in* $k[\![T]\!]$ *with positive radius of convergence.*

Proof It follows from Proposition 5.5.3 and the fact that the family of discs $D^-(0, r)$, with $r > 0$, forms a basis of neighborhoods of 0 in $\mathbb{A}_k^{1,\mathrm{an}}$ (see the proof of Proposition 5.4.11). $\square$

Corollary 5.5.5 *Assume that* k *is algebraically closed. Let* $x \in \mathbb{A}_k^{1,\mathrm{an}}$ *be a point of type 4, associated to a family of closed discs* $(D^+(\alpha_i, r_i))_{i \in I}$ *as in Proposition 5.3.12. The local ring* $\mathcal{O}_x$ *at the point* x *of* $\mathbb{A}_k^{1,\mathrm{an}}$ *consists of the union over* $i \in I$ *of the sets of power series in* $k[\![T - \alpha_i]\!]$ *with radius of convergence bigger than or equal to* r_i.

Proof It follows from Proposition 5.5.3 and the fact that the family of discs $(D^+(\alpha_i, r_i))_{i \in I}$, with $i \in I$, forms a basis of neighborhoods of x in $\mathbb{A}_k^{1,an}$ (see the proof of Proposition 5.4.11). □

Proposition 5.5.6 *Let* $r, s \in \mathbb{R}_{>0}$ *with* $r < s$. *Then* $\mathcal{O}(A^-(0, r, s))$ *is the set of elements*

$$\sum_{i \in \mathbb{Z}} a_i T^i \in k[[T, T^{-1}]]$$

satisfying the following condition:

$$\forall t \in (r, s), \quad \lim_{i \to \pm\infty} |a_i| t^i = 0.$$

□

Corollary 5.5.7 *Let* $t \in \mathbb{R}_{>0} - |k^\times|$. *The local ring* $\mathcal{O}_{\eta_t}$ *at the point* η_t *of* $\mathbb{A}_k^{1,an}$ *is the set of elements*

$$\sum_{i \in \mathbb{Z}} a_i T^i \in k[[T, T^{-1}]]$$

satisfying the following condition:

$$\exists t_1, t_2 \in \mathbb{R}_{>0} \text{ with } t_1 < t < t_2 \text{ such that } \lim_{i \to -\infty} |a_i| t_1^i = \lim_{i \to +\infty} |a_i| t_2^i = 0.$$

□

Proof It follows from Proposition 5.5.6 and the fact that the family of annuli $A^-(0, t_1, t_2)$, with $t_1 < t < t_2$, forms a basis of neighborhoods of η_t in $\mathbb{A}_k^{1,an}$ (see the proof of Proposition 5.4.11). □

Remark 5.5.8 The local rings at points of type 2 of $\mathbb{A}_k^{1,an}$ do not admit descriptions as simple as that of the other points, due to the fact that they have more complicated bases of neighborhoods (see Proposition 5.4.11). Over an algebraically closed field k, one may still obtain a rather concrete statement as follows. Let C be a set of lifts of the elements of $\tilde{k}$ in k°. Then, the local ring $\mathcal{O}_{\eta_1}$ at the point η_1 of $\mathbb{A}_k^{1,an}$ consists in the functions that may be written as a sum of power series with coefficients in k of the form

$$\sum_{i \in \mathbb{N}} a_i T^i + \sum_{c \in C_0} \sum_{i \in \mathbb{N}_{\geqslant 1}} a_{c,i} (T - c)^{-i},$$

where C_0 is a finite subset of C and the series all have radius of convergence strictly bigger than 1. We refer to [FvdP04, Proposition 2.2.6] for details.

Let us now state some properties of the local rings, i.e. germs of analytic functions at one point. They are easily seen to hold for $\mathcal{O}_0$ using its explicit description as a ring of power series (see Proposition 5.5.3).

Proposition 5.5.9 *Let $x \in \mathbb{A}_k^{1,\mathrm{an}}$. The ring $\mathcal{O}_x$ is a local ring with maximal ideal*

$$\mathfrak{m}_x = \{F \in \mathcal{O}_x : F(x) = 0\}.$$

The quotient $\mathcal{O}_x/\mathfrak{m}_x$ is naturally a dense subfield of $\mathscr{H}(x)$.

If x is a rigid point (see Example 5.2.3), then $\mathcal{O}_x$ is a discrete valuation ring that is excellent and henselian. Otherwise, $\mathcal{O}_x$ is a field.   □

Remark 5.5.10 The existence of the square of -1 in Remark 5.4.12 may be reproved using Henselianity. With the notation of that remark, we have $\mathcal{O}_x/\mathfrak{m}_x = \mathscr{H}(x) = k[T]/(T^2 + 1)$, hence the residue field $\mathcal{O}_x/\mathfrak{m}_x$ contains a root of $T^2 + 1$. Since we are in characteristic 0, this root is simple, hence, by Henselianity, it lifts to a root of $T^2 + 1$ in $\mathcal{O}_x$.

The next step is to define a notion of morphism between open subsets of $\mathbb{A}_k^{1,\mathrm{an}}$. As one should expect, such a morphism $\varphi \colon U \to V$ underlies a morphism of locally ringed spaces (hence a morphism of sheaves $\varphi^\sharp \colon \mathcal{O}_V \to \varphi_* \mathcal{O}_U$), but the precise definition is more involved since we want the seminorms associated to the points of U and V to be compatible. For instance, for each $x \in U$, we want the map $\mathcal{O}_{\varphi(x)}/\mathfrak{m}_{\varphi(x)} \to \mathcal{O}_x/\mathfrak{m}_x$ induced by $\varphi^\sharp$ to be an isometry with respect to $|\cdot|_{\varphi(x)}$ and $|\cdot|_x$ (so that it induces an isometry $\mathscr{H}(\varphi(x)) \to \mathscr{H}(x)$). We will not dwell on those questions, which would lead us too far for this survey. Anyway, in the rest of the text, we will actually make only a very limited use of morphisms.

Let us mention that, as in the classical theories, global sections of the structure sheaf correspond to morphisms to the affine line.

Lemma 5.5.11 *Let U be an open subset of $\mathbb{A}_k^{1,\mathrm{an}}$. Then, the map*

$$\mathrm{Hom}(U, \mathbb{A}_k^{1,\mathrm{an}}) \longrightarrow \mathcal{O}(U)$$
$$\varphi \longmapsto \varphi^\sharp(T)$$

is a bijection.   □

For later use, we record here the description of isomorphisms of discs and annuli. Once the rings of global sections are known (see Propositions 5.5.3 and 5.5.6), those results are easily proven using the theory of Newton polygons (or simple considerations on the behaviour of functions as in Sect. 5.10). We refer to [Duc, 3.6.11 and 3.6.12] for complete proofs.

Proposition 5.5.12 *Let $r_1, r_2 \in \mathbb{R}_{>0}$. Let $\varphi \colon D^-(0, r_1) \to D^-(0, r_2)$ be an isomorphism such that $\varphi(0) = 0$. Write*

$$\varphi^\sharp(T) = \sum_{i \in \mathbb{N}_{\geqslant 1}} a_i T^i \in k[\![T]\!]$$

using Proposition 5.5.3. Then, we have

$$\forall s \in (0, r_1), \ |a_1| \, s > \sup_{i \geqslant 2}(|a_i| \, s^i).$$

In particular, we have $r_2 = |a_1| \, r_1$ and, for each $s \in [0, r_1)$, $\varphi(\eta_s) = \eta_{|a_1| s}$.   □

Remark 5.5.13 The previous result still holds if we allow the radii to be infinite, considering the affine line as the disc of infinite radius.

Proposition 5.5.14 *Let $r_1, s_1, r_2, s_2 \in \mathbb{R}_{>0}$ with $r_1 < s_1$ and $r_2 < s_2$. Let $\varphi \colon A^-(0, r_1, s_1) \to A^-(0, r_2, s_2)$ be an isomorphism. Write*

$$\varphi^\sharp(T) = \sum_{i \in \mathbb{Z}} a_i T^i \in k[\![T, T^{-1}]\!]$$

using Proposition 5.5.6. Then, there exists $i_0 \in \{-1, 1\}$ such that we have

$$\forall t \in (r_1, s_1), \ |a_{i_0}| \, s^{i_0} > \sup_{i \neq i_0}(|a_i| \, s^i).$$

In particular, if $i_0 = 1$, we have $r_2 = |a_1| \, r_1$, $s_2 = |a_1| \, s_1$ and, for each $t \in (r_1, s_1)$, $\varphi(\eta_s) = \eta_{|a_1| s}$. If $i_0 = -1$, we have $r_2 = |a_1|/s_1$, $s_2 = |a_1|/r_1$ and, for each $t \in (r_1, s_1)$, $\varphi(\eta_s) = \eta_{|a_1|/s}$.   □

Remark 5.5.15 The previous result still holds if we allow r_1 or r_2 to be 0 and s_1 or s_2 to be infinite.

Definition 5.5.16 Let $A = A^-(\alpha, r, s)$ be an open annulus. We define the *modulus* of A to be

$$\mathrm{Mod}(A) := \frac{s}{r} \in (1, +\infty).$$

By Proposition 5.5.14, the modulus of an annulus only depends on its isomorphism class and not on the coordinate chosen to describe it.

5.6 Extension of Scalars

Let $(K, |\cdot|)$ be a complete valued extension of $(k, |\cdot|)$. The ring morphism $k[T] \to K[T]$ induces a map

$$\pi_{K/k} \colon \mathbb{A}_K^{1,\mathrm{an}} \longrightarrow \mathbb{A}_k^{1,\mathrm{an}}$$

called the *projection map*. In this section, we study this map.

Proposition 5.6.1 *Let K be a complete valued extension of k. The projection map $\pi_{K/k}$ is continuous, proper and surjective.*

Proof The map $\pi_{K/k}$ is continuous as a consequence of the definitions. To prove that it is proper, note that the preimage of a closed disc in $\mathbb{A}_k^{1,\mathrm{an}}$ is a closed disc in $\mathbb{A}_K^{1,\mathrm{an}}$ with the same center and radius, hence a compact set by Lemma 5.4.4.

It remains to prove that $\pi_{K/k}$ is surjective. Let $x \in \mathbb{A}_k^{1,\mathrm{an}}$ and consider the associated character $\chi_x : k[T] \to \mathscr{H}(x)$. By Lemma 5.3.6, there exists a complete valued field L containing both K and $\mathscr{H}(x)$. The character χ_x over k induces a character over K given by

$$K[T] \xrightarrow{\chi_x \otimes K} \mathscr{H}(x) \otimes_k K \to L.$$

The associated point of $\mathbb{A}_K^{1,\mathrm{an}}$ belongs to $\pi_{K/k}^{-1}(x)$.   □

The following result shows that the fibers of the projection map may be quite big.

Lemma 5.6.2 *Let K be a complete valued extension of k. Assume that k and K are algebraically closed and that K is maximally complete. Let $x \in \mathbb{A}_k^{1,\mathrm{an}}$ be a point of type 4. Then $\pi_{K/k}^{-1}(x)$ is a closed disc of radius $r(x)$.*

Proof Fix notation as in Proposition 5.3.12. We have $r(x) = \inf_{i \in I}(r_i)$. Since K is maximally complete, the intersection of all the discs $D^+(\alpha_i, r_i)$ in $\mathbb{A}_K^{1,\mathrm{an}}$ contains a point $\gamma \in K$.

Let us prove that $\pi_{K/k}^{-1}(x) = D^+(\gamma, r(x))$ in $\mathbb{A}_K^{1,\mathrm{an}}$.

Let $c \in k$. For i big enough, c is not contained in the disc of center α_i and radius r_i, that is to say $|\alpha_i - c| > r_i$. It follows that, for each $y \in D^+(\alpha_i, r_i)$, we have

$$|T - c|_y = |\alpha_i - c| = |T - c|_{\alpha_i, r_i}.$$

We have $D^+(\gamma, r(x)) \subseteq \bigcap_{i \in I} D^+(\alpha_i, r_i)$. It then follows from the previous argument that, for each $z \in D^+(\gamma, r(x))$ and each $c \in k$, we have

$$|T - c|_z = \inf_{i \in I}(|\alpha_i - c|) = |T - c|_x.$$

Since k is algebraically closed, every polynomial is a product of linear terms and we deduce that $D^+(\gamma, r(x)) \subseteq \pi_{K/k}^{-1}(x)$.

Let $y \in \mathbb{A}_K^{1,\mathrm{an}} \setminus D^+(\gamma, r(x))$. Then, there exists $i \in I$ such that

$$|T - \alpha_i|_y > r_i = |T - \alpha_i|_{\alpha_i, r_i} \geqslant |T - \alpha_i|_x.$$

The result follows.   □

Remark 5.6.3 Fix notation as in Proposition 5.3.12. The preceding proof also shows that we have $\bigcap_{i \in I} D^+(\alpha_i, r_i) = \{x\}$ in $\mathbb{A}_k^{1,\mathrm{an}}$.

Remark 5.6.4 The preceding lemma shows that the projection map is not open and does not preserve the types of points in general.

We now deal more specifically with the case of Galois extensions. Let $\sigma \in \mathrm{Aut}(K/k)$ and assume that it preserves the absolute value on K. (This condition is automatic if K/k is algebraic.) For each $x \in \mathbb{A}_K^{1,\mathrm{an}}$, the map

$$P \in K[T] \longmapsto |\sigma(P)|_x \in \mathbb{R}_{\geqslant 0}$$

is a multiplicative seminorm. We denote by $\sigma(x)$ the corresponding point of $\mathbb{A}_K^{1,\mathrm{an}}$.

Proposition 5.6.5 *Let K be a finite Galois extension of k. The map*

$$(\sigma, x) \in \mathrm{Gal}(K/k) \times \mathbb{A}_K^{1,\mathrm{an}} \longmapsto \sigma(x) \in \mathbb{A}_K^{1,\mathrm{an}}$$

is continuous and proper.

The projection map $\pi_{K/k}$ induces a homeomorphism

$$\mathbb{A}_K^{1,\mathrm{an}} / \mathrm{Gal}(K/k) \xrightarrow{\sim} \mathbb{A}_k^{1,\mathrm{an}}.$$

In particular, $\pi_{K/k}$ is continuous, proper, open and surjective.

Proof To prove the first continuity statement, it is enough to prove that, for each $P \in K[T]$, the map

$$(\sigma, x) \in \mathrm{Gal}(K/k) \times \mathbb{A}_K^{1,\mathrm{an}} \longmapsto |P(\sigma(x))| = |\sigma(P)|_x \in \mathbb{R}_{\geqslant 0}$$

is continuous.

Let $P \in K[T]$. We may assume that $P \neq 0$. For each $\sigma, \tau \in \mathrm{Gal}(K/k)$ and each $x, y \in \mathbb{A}_K^{1,\mathrm{an}}$, we have

$$\left| |\sigma(P)|_x - |\tau(P)|_y \right| \leqslant \left| |\sigma(P)|_x - |\sigma(P)|_y \right| + \left| |\sigma(P)|_y - |\tau(P)|_y \right|$$

$$\leqslant \left| |\sigma(P)|_x - |\sigma(P)|_y \right| + \|(\sigma - \tau)(P)\|_\infty \max(1, |T|_y)^{\deg(P)},$$

where, for each $R \in K[T]$, we denote by $\|R\|_\infty$ the maximum of the absolute values of its coefficients. The continuity now follows from the continuity of the maps $z \in \mathbb{A}_K^{1,\mathrm{an}} \mapsto |\sigma(P)|_z$ and $\sigma \in \mathrm{Gal}(K/k) \mapsto \sigma(c)$, for $c \in K$.

Let us now prove properness. Since any compact subset of $\mathbb{A}_K^{1,\mathrm{an}}$ is contained in a disc of the form $D^+(0, r)$ for some $r \in \mathbb{R}_{>0}$, it is enough to prove that the preimage of such a disc is compact. Since this preimage is equal to $\mathrm{Gal}(K/k) \times D^+(0, r)$, the result follows from the compactness of $D^+(0, r)$ (see Lemma 5.4.4).

We now study the map $\pi_{K/k}$. It is continuous and proper, by Proposition 5.6.1.

Let $x \in \mathbb{A}_k^{1,\mathrm{an}}$ and consider the associated character $\chi_x \colon k[T] \to \mathscr{H}(x)$. It induces a morphism of K-algebras $\chi_{K,x} \colon K[T] \to \mathscr{H}(x) \otimes_k K$.

Let $\alpha \in K$ be a primitive element for K/k. Denote by P its minimal polynomial over k and by $P_1, \ldots, P_r$ the irreducible factors of P in $\mathscr{H}(x)[T]$. For each $i \in \{1, \ldots, r\}$, let L_i be an extension of $\mathscr{H}(x)$ generated by a root of P_i. We then have an isomorphism of K-algebras $\varphi \colon \mathscr{H}(x) \otimes_k K \xrightarrow{\sim} \prod_{i=1}^r L_i$.

For each $i \in \{1, \ldots, r\}$, by composing $\varphi \circ \chi_{K,x}$ with the projection on the ith factor, we get a character $\chi_i \colon K[T] \to L_i$. Denote by y_i the associated point of $\mathbb{A}_K^{1,\mathrm{an}}$. We have $\pi_{K/k}(y_i) = x$.

Conversely, let $y \in \mathbb{A}_K^{1,\mathrm{an}}$ such that $\pi_{K/k}(y) = x$. The field $\mathscr{H}(y)$ is an extension of both $\mathscr{H}(x)$ and K, so the universal property of the tensor product yields a natural morphism $\mathscr{H}(x) \otimes_k K \to \mathscr{H}(y)$. It follows that there exists $i \in \{1, \ldots, r\}$ such that $\mathscr{H}(y)$ is an extension of L_i, and we then have $y = y_i$. We have proven that $\pi_{K/k}^{-1}(x) = \{y_1, \ldots, y_r\}$.

Since P is irreducible in $k[T]$, the group $\mathrm{Gal}(K/k)$ acts transitively on its roots, hence on the P_i's. It follows that, for each $i, j \in \{1, \ldots, r\}$, there exists $\sigma \in \mathrm{Gal}(K/k)$ such that $\chi_i \circ \sigma = \chi_j$, hence $\sigma(y_i) = y_j$. We have proven that $\pi_{K/k}^{-1}(x)$ is a single orbit under $\mathrm{Gal}(K/k)$.

The arguments above show that the projection map $\pi_{K/k} \colon \mathbb{A}_K^{1,\mathrm{an}} \to \mathbb{A}_k^{1,\mathrm{an}}$ factors through a map $\pi'_{K/k} \colon \mathbb{A}_K^{1,\mathrm{an}}/\mathrm{Gal}(K/k) \to \mathbb{A}_k^{1,\mathrm{an}}$ and that the latter is continuous and bijective. Since $\pi_{K/k}$ is proper, $\pi'_{K/k}$ is proper too, hence a homeomorphism. $\qquad\square$

Recall that every element of $\mathrm{Gal}(k^s/k)$ preserves the absolute value on k^s. In particular, it extends by continuity to an automorphism of $\widehat{k^s} = \widehat{k^a}$. We endow $\mathrm{Gal}(k^s/k)$ with its usual profinite topology.

Corollary 5.6.6 *The map*

$$(\sigma, x) \in \mathrm{Gal}(k^s/k) \times \mathbb{A}_{\widehat{k^a}}^{1,\mathrm{an}} \mapsto \sigma(x) \in \mathbb{A}_{\widehat{k^a}}^{1,\mathrm{an}}$$

is continuous and proper.

The projection map $\pi_{\widehat{k^a}/k}$ induces a homeomorphism

$$\mathbb{A}_{\widehat{k^a}}^{1,\mathrm{an}}/\mathrm{Gal}(k^s/k) \xrightarrow{\sim} \mathbb{A}_k^{1,\mathrm{an}}.$$

In particular, $\pi_{\widehat{k^a}/k}$ is continuous, proper, open and surjective.

Proof The first part of the statement is proven as in Proposition 5.6.5, using the fact that the group $\mathrm{Gal}(k^s/k)$ is compact.

Let $x, y \in \mathbb{A}_{\widehat{k^a}}^{1,\mathrm{an}}$ such that $\pi_{\widehat{k^a}/k}(x) = \pi_{\widehat{k^a}/k}(y)$. By Proposition 5.6.5, for each finite Galois extension K of k, there exists σ_K in $\mathrm{Gal}(K/k)$ such that $\sigma_K(\pi_{K/k}(x)) = \pi_{K/k}(y)$. By compactness, the family $(\sigma_K)_K$ admits a subfamily converging to some $\sigma \in \text{Gal}(k^s/k)$. We then have

$$\forall P \in k^s[T], \ |\sigma(P)|_x = |P|_y.$$

Let $Q \in \widehat{k^a}[T]$. We want to prove that we still have $|\sigma(Q)|_x = |Q|_y$. We may assume that Q is non-zero. Let d be its degree. By density of k^s into $\widehat{k^a}$, there exists a sequence $(Q_n)_{n \geqslant 0}$ of elements of $k^s[T]$ of degree d that converge to Q for the norm $\|\cdot\|_\infty$ that is the supremum norm of the coefficients. Note that we have

$$|Q - Q_n|_y \leqslant \|Q - Q_n\|_\infty \max(1, |T|_y)^d,$$

so that $(|Q_n|_y)_{n \geqslant 0}$ converges to $|Q|_y$. The same argument shows that $(|\sigma(Q_n)|_x)_{n \geqslant 0}$ converges to $|\sigma(Q)|_x$ and the results follows.

We have just proven that the map $\mathbb{A}^{1,\text{an}}_{\widehat{k^a}} / \text{Gal}(k^s/k) \to \mathbb{A}^{1,\text{an}}_k$ induced by $\pi_{\widehat{k^a}/k}$ is a bijection. The rest of the statement follows as in the proof of Proposition 5.6.5. $\quad\square$

Lemma 5.6.7 *Let $y, z \in \mathbb{A}^{1,\text{an}}_{\widehat{k^a}}$ such that $\pi_{\widehat{k^a}/k}(y) = \pi_{\widehat{k^a}/k}(z)$. Then y and z are of the same type and have the same radius.*

Proof By Corollary 5.6.6, there exists $\sigma \in \text{Gal}(k^s/k)$ such that $z = \sigma(y)$. The result follows easily. $\quad\square$

As a consequence, we may define the type and the radius of a point of the Berkovich affine line over any complete valued field.

Definition 5.6.8 Let $x \in \mathbb{A}^{1,\text{an}}_k$.

We define the *type of the point* x to be the type of the point y, for any $y \in \pi^{-1}_{\widehat{k^a}/k}(x)$.

We define the *radius of the point* x to be the radius of the point y, for any $y \in \pi^{-1}_{\widehat{k^a}/k}(x)$. We denote it by $r(x)$.

We end this section with a finiteness statement that is often useful.

Lemma 5.6.9 *Let X be a subset of $\mathbb{A}^{1,\text{an}}_{\widehat{k^a}}$ that is either a disc, an annulus or a singleton containing a point of type 2 or 3. Then, the orbit of X under $\text{Gal}(k^s/k)$ is finite.*

In particular, for each $x \in \mathbb{A}^{1,\text{an}}_k$ of type 2 or 3, the fiber $\pi^{-1}_{\widehat{k^a}/k}(x)$ is finite.

Proof Let us first assume that X is a closed disc: there exists $\alpha \in \widehat{k^a}$ and $r \in \mathbb{R}_{>0}$ such that $X = D^+(\alpha, r)$. Since k^a is dense in $\widehat{k^a}$, we may assume $\alpha \in k^a$. Then orbit of α is then finite, hence so is the orbit of X. The case of an open disc is dealt with similarly.

Points of type 2 or 3 are boundary points of closed discs by Lemma 5.4.4, hence the orbits of such points are finite too. Since closed and open annuli are determined by their boundary points, which are of type 2 or 3 (see Lemma 5.4.4 again), their orbits are finite too.

Finally, if $x \in \mathbb{A}_k^{1,\mathrm{an}}$ is a point of type 2 (resp. 3), then, by Corollary 5.6.6, the fiber $\pi_{\bar{k}^a/k}^{-1}(x)$ is an orbit of a point of type 2 (resp. 3). The last part of the result follows. $\square$

5.7   Connectedness

In this section, we study the connectedness properties of the Berkovich affine line and its subsets. Our main source here is [Ber90, Section 4.2] for the connectedness properties (see also [Ber90, Section 3.2] for higher-dimensional cases) and [Duc, Section 1.9] for properties of quotients of graphs.

Proposition 5.7.1 *Open and closed discs, annuli and Swiss cheeses are path-connected. The Berkovich affine line $\mathbb{A}_k^{1,\mathrm{an}}$ is path-connected and locally path-connected.*

Proof Let us first handle the case of a closed disc, say $D^+(\alpha, r)$ with $\alpha \in k$ and $r \in \mathbb{R}_{>0}$. By Proposition 5.6.1, the projection map $\pi_{K/k}$ is continuous and surjective for any complete valued extension K of k. As a result, it is enough to prove the result on some extension of k, hence we may assume that k is algebraically closed and maximally complete.

Let $x \in D^+(\alpha, r)$. By Lemma 5.3.11 and Remark 5.3.19, there exist $\beta \in k$ and $s \in \mathbb{R}_{\geqslant 0}$ such that $x = \eta_{\beta,s}$. Since $x \in D^+(\alpha, r)$, we have

$$|T - \alpha|_{\beta,s} = \max(|\alpha - \beta|, s) \leqslant r,$$

hence $s \leqslant r$ and $\eta_{\alpha,r} = \eta_{\beta,r}$. As a consequence, the map

$$\lambda \in [0, 1] \mapsto \eta_{\alpha, s+(r-s)\lambda} \in \mathbb{A}_k^{1,\mathrm{an}}$$

defines a continuous path from x to $\eta_{\alpha,r}$ in $D^+(\alpha, r)$ (see Lemma 5.4.13). It follows that the disc $D^+(\alpha, r)$ is path-connected.

The same argument may be used in order to prove that closed annuli and Swiss cheeses are connected. Indeed, if such a set S is written as the complement of some open discs in $D^+(\alpha, r)$, it is easy to check that, for each $x \in S$, the path joining x to $\eta_{\alpha,r}$ that we have just described actually remains in S.

Since the open figures may be written as increasing unions of the closed ones, the result holds for them too.

The last statement follows from the fact that $\mathbb{A}_k^{1,\mathrm{an}}$ may be written as an increasing union of discs for the global part, and from Proposition 5.4.11 and Corollary 5.6.6 for the local part. $\square$

The Berkovich affine line actually satisfies a stronger connectedness property: it is uniquely path-connected. We will now prove this result, starting with the case of an algebraically closed field.

Definition 5.7.2 We define a partial ordering $\leqslant$ on $\mathbb{A}^{1,\mathrm{an}}_{\widehat{k^a}}$ by setting

$$x \leqslant y \text{ if } \forall P \in \widehat{k^a}[T], \ |P|_x \leqslant |P|_y.$$

For each $x \in \mathbb{A}^{1,\mathrm{an}}_{\widehat{k^a}}$, we set

$$I_x := \{y \in \mathbb{A}^{1,\mathrm{an}}_{\widehat{k^a}} : y \geqslant x\}.$$

Remark 5.7.3 Let $\sigma \in \mathrm{Gal}(k^s/k)$. By definition of the action of σ, for every $x, y \in \mathbb{A}^{1,\mathrm{an}}_{\widehat{k^a}}$, we have $x \leqslant y$ if, and only if, $\sigma(x) \leqslant \sigma(y)$. It follows that, for each $z \in \mathbb{A}^{1,\mathrm{an}}_{\widehat{k^a}}$, we have $\sigma(I_z) = I_{\sigma(z)}$.

The proof of the following lemma is left as an exercise for the reader.

Lemma 5.7.4 *Let $\alpha \in \widehat{k^a}$ and $r \in \mathbb{R}_{>0}$. For each $x \in \mathbb{A}^{1,\mathrm{an}}_{\widehat{k^a}}$, we have $x \leqslant \eta_{\alpha,r}$ if, and only if, $x \in D^+(\alpha, r)$.*
In particular, for $\beta \in \widehat{k^a}$ and $s \in \mathbb{R}_{\geqslant 0}$, we have $\eta_{\alpha,r} \leqslant \eta_{\beta,s}$ if, and only if, $\max(|\alpha - \beta|, r) \leqslant s$. Moreover, when those conditions hold, we have $\eta_{\beta,s} = \eta_{\alpha,s}$.

$\square$

Corollary 5.7.5 *The minimal points for the ordering $\leqslant$ on $\mathbb{A}^{1,\mathrm{an}}_{\widehat{k^a}}$ are exactly the points of type 1 and 4.*

Proof It follows from Lemma 5.7.4 that points of type 2 and 3 are not minimal and that points of type 1 are. To prove the result, it remains to show that points of type 4 are minimal too. This assertion follows from the fact that any such point is the unique point contained in the intersection of a family of closed discs (see Remark 5.6.3) by applying Lemma 5.7.4 again. $\square$

Corollary 5.7.6 *Let $\alpha \in \widehat{k^a}$ and $r \in \mathbb{R}_{\geqslant 0}$. We have*

$$I_{\eta_{\alpha,r}} = \{\eta_{\alpha,s}, s \geqslant r\}.$$

Let $x \in \mathbb{A}^{1,\mathrm{an}}_{\widehat{k^a}}$ be a point of type 4 and fix notation as in Proposition 5.3.12. Then, for each $i, j \in I$, we have $I_{\eta_{\alpha_i,r_i}} \subseteq I_{\eta_{\alpha_j,r_j}}$ or $I_{\eta_{\alpha_j,r_j}} \subseteq I_{\eta_{\alpha_i,r_i}}$ and we have

$$I_x = \{x\} \cup \bigcup_{i \in I} I_{\eta_{\alpha_i,r_i}}.$$

$\square$

The former result shows in particular that our notation is consistent with that of Lemma 5.4.13.

Corollary 5.7.7 *Let* $x \in \mathbb{A}^{1,\mathrm{an}}_{\widehat{k^a}}$. *The radius map* $r\colon \mathbb{A}^{1,\mathrm{an}}_{\widehat{k^a}} \to \mathbb{R}_{\geqslant 0}$ *induces a homeomorphism from* I_x *onto* $[r(x), +\infty)$.

The restriction of the projection map $\pi_{\widehat{k^a}/k}$ *to* I_x *is injective.*

Proof The fact that the radius map induces is a bijection from I_x onto its image follows from Corollary 5.7.6. One can then prove directly that its inverse is continuous and open by arguing as in the proof of Lemma 5.4.13.

Let us now prove the second part of the statement. Let $y, z \in I_x$ such that $x < y < z$. Then, there exist $\alpha \in \widehat{k^a}$ and $r < s \in \mathbb{R}_{>0}$ such that $y = \eta_{\alpha,r}$ and $z = \eta_{\alpha,s}$. Since k^s in dense in $\widehat{k^a}$, we may assume that $\alpha \in k^s$. Let $P_\alpha \in k[T]$ be the minimal polynomial of α over k. Since there exists $Q_\alpha \in \widehat{k^a}[T]$ such that $P_\alpha = (T - \alpha) Q_\alpha$, we have

$$|P_\alpha|_y = r \, |Q_\alpha|_y < s \, |Q_\alpha|_y = |P_\alpha|_z.$$

It follows that $\pi_{\widehat{k^a}/k}(y) \neq \pi_{\widehat{k^a}/k}(z)$.

Finally, assume that there exists $z \in I_x - \{x\}$ such that $\pi_{\widehat{k^a}/k}(z) = \pi_{\widehat{k^a}/k}(x)$. For each $y \in I_x$ such that $x < y < z$ and each $P \in k[T]$, we have

$$|P|_x \leqslant |P|_y \leqslant |P|_z = |P|_x,$$

hence $|P|_y = |P|_z$, which contradicts what we have just proved. $\square$

Corollary 5.7.8 *Let* $x, y \in \mathbb{A}^{1,\mathrm{an}}_{\widehat{k^a}}$. *The set* $\{z \in \mathbb{A}^{1,\mathrm{an}}_{\widehat{k^a}} : z \geqslant x \text{ and } z \geqslant y\}$ *admits a smallest element. We will denote it by* $x \vee y$.

In particular, if x *and* y *are comparable for* $\leqslant$, *then* $I_x \cup I_y$ *is homeomorphic to a half-open interval and, otherwise,* $I_x \cup I_y$ *is homeomorphic to a tripod with one end-point removed, i.e. the union of two closed intervals and one half-open interval glued along a common end-point.*

Proof We have $\{z \in \mathbb{A}^{1,\mathrm{an}}_{\widehat{k^a}} : z \geqslant x \text{ and } z \geqslant y\} = I_x \cap I_y$. By Corollary 5.7.7, I_x and I_y may be sent to intervals of the form $[*, +\infty)$ by order-preserving homeomorphisms. We deduce that, in order to prove the result, it is enough to prove that $I_x \cap I_y$ is non-empty. Setting $R := \max(|T|_x, |T|_y)$, the points x and y belong to $D^+(0, R)$, hence η_R belongs to $I_{x,y}$, by Lemma 5.7.4.

Denoting by $x \vee y$ the smallest element of $I_x \cap I_y$, we have $I_x \cap I_y = I_{x \vee y}$. In particular, by Corollary 5.7.7, $I_x \cap I_y$ is homeomorphic to a half-open interval with end-point $x \vee y$. Set

$$[x, x \vee y] := \{z \in I_x : x \leqslant z \leqslant (x \vee y)\} \text{ and } [y, x \vee y] := \{z \in I_y : y \leqslant z \leqslant (x \vee y)\}.$$

By Corollary 5.7.7 again, $[x, x \vee y]$ and $[y, x \vee y]$ are homeomorphic to closed intervals (possibly singletons if x and y are comparable). Finally, the result follows by writing

$$I_x \cup I_y = [x, x \vee y] \cup [y, x \vee y] \cup I_{x \vee y}.$$

$\square$

Corollary 5.7.9 *Let* $x \in \mathbb{A}^{1,\mathrm{an}}_{\widehat{k^a}}$. *The set* $\mathbb{A}^{1,\mathrm{an}}_{\widehat{k^a}} - I_x$ *is a union of open discs. In particular,* I_x *is closed.*

Proof Let $y \in \mathbb{A}^{1,\mathrm{an}}_{\widehat{k^a}} - I_x$. The point $x \vee y$ defined in Corollary 5.7.8 belongs to I_y and is not equal to y. It follows that there exists $z \in I_y - I_x$ such that $y < z$. By Corollary 5.7.6, y is a point of type 2 or 3, hence, by Lemma 5.7.4, the set $D := \{u \in \mathbb{A}^{1,\mathrm{an}}_{\widehat{k^a}} : u \leqslant z\}$ is a closed disc of positive radius. It is contained in $\mathbb{A}^{1,\mathrm{an}}_{\widehat{k^a}} - I_x$. Since z is not the boundary point of D, y indeed belongs to some open subdisc of D by Lemmas 5.4.7 and 5.4.9. $\square$

Proposition 5.7.10 *Let* Γ *be a subset of* $\mathbb{A}^{1,\mathrm{an}}_{\widehat{k^a}}$ *such that, for each* $x \in \Gamma$, $I_x \subseteq \Gamma$. *Then,* $\mathbb{A}^{1,\mathrm{an}}_{\widehat{k^a}} - \Gamma$ *is a union of discs and points of type 4.*

If, moreover, Γ *is closed, then* $\mathbb{A}^{1,\mathrm{an}}_{\widehat{k^a}} - \Gamma$ *is a union of open discs.*

Proof To prove the first statement, it is enough to show that each point of $\mathbb{A}^{1,\mathrm{an}}_{\widehat{k^a}} - \Gamma$ that is not of type 4 is contained in a closed disc. Let x be such a point. Then, by Lemma 5.7.4, $\{y \in \mathbb{A}^{1,\mathrm{an}}_{\widehat{k^a}} : y \leqslant x\}$ is a closed disc. By assumption, it is contained in $\mathbb{A}^{1,\mathrm{an}}_{\widehat{k^a}} - \Gamma$, and the result follows.

Let us now assume that Γ is closed. Let $x \in \mathbb{A}^{1,\mathrm{an}}_{\widehat{k^a}} - \Gamma$. Since Γ is closed, there exists $y \in I_x - \Gamma$ such that $y > x$. We then conclude as in the proof of Corollary 5.7.9. $\square$

Corollary 5.7.11 *For every* $x, y \in \mathbb{A}^{1,\mathrm{an}}_{\widehat{k^a}}$, *there exists a unique injective path from* x *to* y.

Proof Set $I_{x,y} := I_x \cup I_y$. It follows from Corollaries 5.7.7 and 5.7.8 that it is homeomorphic to a half-open interval or a tripod with one end-point removed. It follows that there exists a unique injective path from x to y inside $I_{x,y}$. In particular, there exists an injective path from x to y in $\mathbb{A}^{1,\mathrm{an}}_{\widehat{k^a}}$.

By Corollary 5.7.9 and Proposition 5.7.10, $\mathbb{A}^{1,\mathrm{an}}_{\widehat{k^a}} - I_{x,y}$ is a union of open discs. By Lemma 5.4.4, every open disc has a unique boundary point. As a consequence, an injective path going from x to y cannot meet any of these open discs, since otherwise it would contain its boundary point twice. It follows that such a path is contained in $I_{x,y}$. $\square$

We want to deduce the result for $\mathbb{A}^{1,\mathrm{an}}_k$ by using the projection map $\pi_{\widehat{k^a}/k} \colon \mathbb{A}^{1,\mathrm{an}}_{\widehat{k^a}} \to \mathbb{A}^{1,\mathrm{an}}_k$. Recall that, by Corollary 5.6.6, it is a quotient map by the group $\mathrm{Gal}(k^s/k)$.

Proposition 5.7.12 *For every $x, y \in \mathbb{A}_k^{1,\text{an}}$, there exists a unique injective path from x to y.*

Proof To ease notation, we will write π instead of $\pi_{\widehat{k^a}/k}$.

Let $x, y \in \mathbb{A}_k^{1,\text{an}}$. Choose $x' \in \pi^{-1}(x)$ and $y' \in \pi^{-1}(y)$. Set $J := I_{x'} \cup I_{y'}$. By Lemma 5.7.7, $\pi(I_{x'})$ and $\pi(I_{y'})$ are half-open intervals and, by Corollary 5.7.8, they meet.

Let $z \in \pi(I_{x'}) \cap \pi(I_{y'})$. There exists $z' \in I_{x'}$ such that $\pi(z') = z$ and $\sigma \in \text{Gal}(k^s/k)$ such that $\sigma(z') \in I_{y'}$. By Remark 5.7.3, we have $\sigma(I_{z'}) = I_{\sigma(z')}$ and we deduce that $\pi(J)$ is a tripod with one end-point removed. In particular, there exists a unique injective path from x to y in $\pi(J)$, and at least one such path in $\mathbb{A}_k^{1,\text{an}}$.

To conclude, let us prove that any injective path from x to y in $\mathbb{A}_k^{1,\text{an}}$ is contained in $\pi(J)$. Set

$$\Gamma := \pi^{-1}(\pi(J)) = \bigcup_{\sigma \in \text{Gal}(k^s/k)} \sigma(J).$$

Since the action of the elements of $\text{Gal}(k^s/k)$ preserves the norm, for each $R \in \mathbb{R}_{>0}$, we have

$$\Gamma \cap D^+(0, R) = \pi^{-1}(\pi(J \cap D^+(0, R))),$$

and it follows that $\Gamma \cap D^+(0, R)$ is compact. We deduce that Γ is closed.

Moreover, we have

$$\Gamma = \bigcup_{z \in \pi^{-1}(x) \cup \pi^{-1}(y)} I_z,$$

hence, by Proposition 5.7.10, $\mathbb{A}_{\widehat{k^a}}^{1,\text{an}} - \Gamma$ is a union of open discs. Using the fact that it is invariant under the action of $\text{Gal}(k^s/k)$, we may now conclude as in the proof of Corollary 5.7.11. $\qquad\square$

Notation 5.7.13 For $x, y \in \mathbb{A}_k^{1,\text{an}}$, we will denote by $[x, y]$ the unique injective path between x and y. We set $(x, y) := [x, y] - \{x, y\}$.

Remark 5.7.14 The fact that $\mathbb{A}_k^{1,\text{an}}$ is uniquely path-connected means that it has the structure of a real tree. The type of the points may be easily read off this structure. Indeed the end-points are the points of type 1 and 4 (see Remark 5.6.3 for points of type 4). Among the others, type 2 points are branch-points (with infinitely many edges meeting there, see Lemma 5.4.9) while type 3 points are not (see Lemma 5.4.7). For a graphical representation of this fact, see Fig. 5.2.

5.8 Virtual Discs and Annuli

In this section, we introduce generalizations of discs and annuli that are more suitable when working over arbitrary fields and study them from the topological point of view. We explain that they retract onto some simple subsets of the real line, namely singletons and intervals respectively. Here we borrow from [Ber90, Section 6.1], which also contains a treatment of more general spaces.

5.8.1 Definitions

Definition 5.8.1 A connected subset U of $\mathbb{A}_k^{1,\mathrm{an}}$ is called a *virtual open (resp. closed) disc* if $\pi_{\widehat{k^a}/k}^{-1}(U)$ is a disjoint union of open (resp. closed) discs. We define similarly virtual open, closed and flat annuli and virtual open and closed Swiss cheeses.

We now introduce particularly interesting subsets of virtual annuli.

Definition 5.8.2 Let A be a virtual open or closed annulus. The *skeleton* of A is the complement of all the virtual open discs contained in A. We denote it by Σ_A.

Example 5.8.3 Consider the open annulus $A := A^-(\gamma, \rho_1, \rho_2)$, with $\gamma \in k$ and $\rho_1 < \rho_2 \in \mathbb{R}_{>0}$. Its skeleton is

$$\Sigma_A = \{\eta_{\gamma,s}, \rho_1 < s < \rho_2\}.$$

It is represented in Fig. 5.5.

Lemma 5.8.4 *Let A be a virtual open (resp. closed) annulus. Let C be a connected component of $\pi_{\widehat{k^a}/k}^{-1}(A)$. Then, C is an open (resp. closed) annulus and $\pi_{\widehat{k^a}/k}$ induces a homeomorphism between Σ_C and Σ_A. In particular, Σ_A is an open (resp. closed) interval.*

Proof Let us handle the case of a virtual open annulus, the other one being dealt with similarly. The connected component C is an open annulus, by the very definition of virtual open annulus. It also follows from the definitions that we have $\Sigma_A = \pi_{\widehat{k^a}/k}(\Sigma_C)$.

Denote by G_C the subgroup of $\mathrm{Gal}(k^s/k)$ consisting of those elements that preserve C. By Corollary 5.6.6, $\pi_{\widehat{k^a}/k}$ induces a homeomorphism $C/G_C \simeq A$, hence a homeomorphism $\Sigma_C/G_C \simeq \Sigma_A$.

Write $C = A^-(\gamma, \rho_1, \rho_2)$, with $\gamma \in \widehat{k^a}$ and $\rho_1 < \rho_2 \in \mathbb{R}_{>0}$. Its complement in $\mathbb{A}_{\widehat{k^a}}^{1,\mathrm{an}}$ has two connected components, namely $D^+(\gamma, \rho_1)$ and

$$D_\infty^+(\gamma, \rho_2) := \{x \in \mathbb{A}_{\widehat{k^a}}^{1,\mathrm{an}} : |(T - \gamma)(x)| \geqslant \rho_2\}.$$

Fig. 5.5 The skeleton of an
open annulus is the open line
segment joining its boundary
points

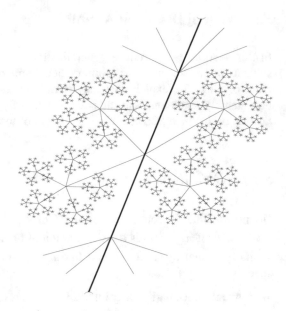

For r big enough, we have $\eta_{\gamma,r} = \eta_r$, which belongs to $D_\infty^+(\gamma, \rho_2)$ and is stable
under $\mathrm{Gal}(k^s/k)$. It follows that $D^+(\gamma, \rho_1)$ and $D_\infty^+(\gamma, \rho_2)$ are stable under G_C.

Let $\sigma \in G_C$. We have

$$\sigma(D^+(\gamma, \rho_1)) = D^+(\sigma(\gamma), \rho_1) = D^+(\gamma, \rho_1),$$

hence $|\sigma(\gamma) - \gamma| \leqslant \rho_1$. It follows that, for each $s \in (\rho_1, \rho_2)$, we have

$$\sigma(\eta_{\gamma,s}) = \eta_{\sigma(\gamma),s} = \eta_{\gamma,s},$$

that is to say σ acts as the identity on Σ_C. The result follows. $\square$

Remark 5.8.5 Virtual discs and annuli are usually defined as arbitrary connected
k-analytic curves (see [PT20, Section 2.2]) whose base change to $\widehat{k^a}$ is a disjoint
union of discs or annuli, without requiring an embedding into $\mathbb{A}_k^{1,\mathrm{an}}$. Our definition
is *a priori* more restrictive.

With this definition of virtual annulus, an additional difficulty appears. If A is
such a virtual annulus and C is a connected component of $\pi_{\widehat{k^a}/k}^{-1}(A)$, then there may
exist elements of $\mathrm{Gal}(k^s/k)$ that preserve C but swap its two ends. In this case, the
skeleton Σ_A is a half-open interval. (For an example of such a behaviour, consider
a Galois orbit in $\widehat{k^a}$ consisting of two points, its image x in $\mathbb{A}_k^{1,\mathrm{an}}$ and let A be the
complement in $\mathbb{P}_k^{1,\mathrm{an}}$ of a small virtual closed disc containing x.) Some authors (for
instance A. Ducros) explicitly rule out this possibility in the definition of virtual
annulus.

We may also extend the notion of modulus (see Definition 5.5.16) from annuli to virtual annuli. Recall that, by Lemma 5.6.9, the set of connected components of the preimage of a virtual open annulus by $\pi_{\widehat{k^a}/k}^{-1}$ is finite and that, by Corollary 5.6.6, $\mathrm{Gal}(k^s/k)$ acts transitively on it.

Definition 5.8.6 Let A be a virtual open annulus. Denote by $\mathcal{C}$ the set of connected components of $\pi_{\widehat{k^a}/k}^{-1}(A)$ and let C_0 be one of them. We define the *modulus* of A to be

$$\mathrm{Mod}(A) := \mathrm{Mod}(C_0)^{1/\#\mathcal{C}} \in (1, +\infty).$$

It is independent of the choice of C_0.

Remark 5.8.7 Beware that other normalizations exist in the literature for the modulus of a virtual open annulus. For instance, A. Ducros sets $\mathrm{Mod}(A) := \mathrm{Mod}(C_0)$ (see [Duc, 3.6.15.11]).

We refer the reader to [Duc, Section 3.6] for a thorough treatment of classical and virtual discs and annuli.

5.8.2 The Case of an Algebraically Closed and Maximally Complete Base Field

In this section, we assume that k is algebraically closed and maximally complete. This will allow us to prove our results through direct computations.

Recall that, by Lemma 5.3.11 and Remark 5.3.19, each point of $\mathbb{A}_k^{1,\mathrm{an}}$ (or a disc or an annulus) is of the form $\eta_{\alpha,r}$, for $\alpha \in k$ and $r \in \mathbb{R}_{\geq 0}$.

Let $\gamma \in k$ and $\rho \in \mathbb{R}_{>0}$. We consider the closed disc $D^+(\gamma, \rho)$.

Lemma 5.8.8 *The map*

$$[0, 1] \times D^+(\gamma, \rho) \ni (t, \eta_{\alpha,r}) \longmapsto \eta_{\alpha,\max(r,t\rho)} \in D^+(\gamma, \rho)$$

is well-defined and continuous. It induces a deformation retraction of $D^+(\gamma, \rho)$ and $D^-(\gamma, \rho) \cup \{\eta_{\gamma,\rho}\}$ onto their unique boundary point $\eta_{\gamma,\rho}$. □

Let $\gamma \in k$ and $\rho_1 < \rho_2 \in \mathbb{R}_{>0}$. We consider the open annulus $A := A^-(\gamma, \rho_1, \rho_2)$. Each rational point of A is contained in an open disc. Indeed, let $\alpha \in k \cap A$ (which implies that $\rho_1 < |\gamma - \alpha| < \rho_2$). Then, the open disc $D^-(\alpha, |\gamma - \alpha|)$ is the maximal open disc containing α that lies in A. This open disc is relatively compact in A and its unique boundary point is $\eta_{\alpha,|\gamma-\alpha|} = \eta_{\gamma,|\gamma-\alpha|}$.

We have

$$A - \Sigma_A = \bigcup_{\alpha \in k \cap A} D^-(\alpha, |\gamma - \alpha|).$$

In particular, each connected component of $A - \Sigma_A$ is an open disc. For each such open disc D, we denote by η_D its boundary point in A.

Lemma 5.8.9 *The map*

$$[0, 1] \times A \ni (t, \eta_{\alpha,r}) \text{ with } r \leqslant |\gamma - \alpha| \longmapsto \eta_{\alpha,\max(r,t|\gamma-\alpha|)} \in A$$

is well-defined and continuous. It induces a deformation retraction of A onto Σ_A.

Its restriction to each connected component D of $A - \Sigma_A$ coincides with the map from Lemma 5.8.8 and induces a deformation retraction of D onto η_D.

For each $\eta \in \Sigma_A$, the set of points that are sent to η by the retraction map is the union of η and all the connected component D of $A - \Sigma_A$ such that $\eta_D = \eta$. It is a flat closed annulus. $\square$

5.8.3 The General Case

We now remove the assumption on k. Let K be an extension of k that is algebraically closed and maximally complete. If X is a virtual disc or a virtual annulus, then each connected component of $\pi_{K/k}^{-1}(X)$ is a true disc or annulus over K and the results of the previous section apply. By continuity and surjectivity of $\pi_{K/k}$, we deduce retraction results in this setting too.

Proposition 5.8.10 *Let D be a virtual open disc in $\mathbb{A}_k^{1,\mathrm{an}}$. Then, D has a unique boundary point η_D in $\mathbb{A}_k^{1,\mathrm{an}}$ and there exists a canonical deformation retraction $\tau_D \colon D \cup \{\eta_D\} \to \{\eta_D\}$.* $\square$

Proposition 5.8.11 *Let A be a virtual open annulus. Each connected component of $A - \Sigma_A$ is a virtual open disc.*

There exists a canonical deformation retraction $\tau_A \colon A \to \Sigma_A$. Its restriction to any connected component D of $A - \Sigma_A$ induces the map τ_D from Proposition 5.8.10.

Moreover, for each open interval (resp. closed interval, resp. singleton) I in Σ_A, the set $\tau_A^{-1}(I)$ is a virtual open annulus (resp. a virtual closed annulus, resp. a virtual flat closed annulus). $\square$

Lemma 5.8.12 *Let A be a virtual open annulus. Let F be a finite subset of Σ_A and denote by $\mathcal{I}$ the set of connected components of $\Sigma_A - F$. The elements of $\mathcal{I}$ are open intervals and we have*

$$\mathrm{Mod}(A) = \prod_{I \in \mathcal{I}} \mathrm{Mod}(\tau_A^{-1}(I)).$$

$\square$

5.9 Lengths of Intervals

In this section, we show that intervals inside $\mathbb{A}_k^{1,\mathrm{an}}$ may be endowed with a canonical (exponential) length. To start with, we define the useful notion of branch at a point.

Notation 5.9.1 For $\alpha \in \widehat{k^a}$ and $r \in \mathbb{R}_{>0}$, we set

$$D_\infty^-(\alpha, r) := \{x \in \mathbb{A}_k^{1,\mathrm{an}} : |(T - \alpha)(x)| > r\}.$$

Lemma 5.9.2 *Let* $x \in \mathbb{A}_{\widehat{k^a}}^{1,\mathrm{an}}$.

If x is of type 1 or 4, then $\mathbb{A}_{\widehat{k^a}}^{1,\mathrm{an}} - \{x\}$ is connected.

If $x = \eta_{\alpha,r}$, with $\alpha \in \widehat{k^a}$ and $r \in \mathbb{R}_{>0} - |k^\times|$, then the connected components of $\mathbb{A}_{\widehat{k^a}}^{1,\mathrm{an}} - \{x\}$ are $D^-(\alpha, r)$ and $D_\infty^-(\alpha, r)$.

If $x = \eta_{\alpha,r}$, with $\alpha \in \widehat{k^a}$ and $r \in |k^\times|$, then the connected components of $\mathbb{A}_{\widehat{k^a}}^{1,\mathrm{an}} - \{x\}$ are $D_\infty^-(\alpha, r)$ and the discs of the form $D^-(\beta, r)$ with $|\beta - \alpha| \leqslant r$.

Proof Assume that x is of type 1 or 4. Let $y, y' \in \mathbb{A}_{\widehat{k^a}}^{1,\mathrm{an}} - \{x\}$. By Corollary 5.7.8, $I_y \cup I_{y'}$ is a connected subset of $\mathbb{A}_{\widehat{k^a}}^{1,\mathrm{an}}$ containing y and y'. By Corollary 5.7.5, it does not contain x. It follows that $\mathbb{A}_{\widehat{k^a}}^{1,\mathrm{an}} - \{x\}$ is connected.

Assume that $x = \eta_{\alpha,r}$, with $\alpha \in \widehat{k^a}$ and $r \in \mathbb{R}_{>0} - |k^\times|$. By Proposition 5.7.1, $D^-(\alpha, r)$ is connected. Since $D_\infty^-(\alpha, r)$ may be written as an increasing union of open annuli, we deduce from Proposition 5.7.1) that it is connected too. The result now follows from Lemma 5.4.7.

Assume that x is of type 2. Up to a change of variables, we may assume that $x = \eta_{0,1}$. As before, we deduce from Proposition 5.7.1 that $D_\infty^-(0, 1)$ and the discs of the form $D^-(\beta, 1)$ with $|\beta| \leqslant 1$ are connected. The result now follows from Lemma 5.4.9.   □

Remark 5.9.3 We have $D^-(\beta, r) = D^-(\beta', r)$ if, and only if, $|\beta - \beta'| < r$.

Definition 5.9.4 Let $x \in \mathbb{A}_k^{1,\mathrm{an}}$. Let $y, z \in \mathbb{A}_k^{1,\mathrm{an}} - \{x\}$. We say that the intervals $(x, y]$ and $(x, z]$ are *x-equivalent* if $(x, y] \cap (x, z] \neq \emptyset$.

An x-equivalence class of intervals $(x, y]$, with $x \neq y$, is called a *branch at x*. We denote by $\mathcal{B}_x$ the set of branches at x.

Remark 5.9.5 If $(x, y]$ and $(x, z]$ are equivalent, if follows from the unique path-connectedness of $\mathbb{A}_k^{1,\mathrm{an}}$ (see Proposition 5.7.12) that there exists $t \in \mathbb{A}_k^{1,\mathrm{an}} - \{x\}$ such that $(x, t] = (x, y] \cap (x, z]$.

Lemma 5.9.6 *Let* $x \in \mathbb{A}_k^{1,\mathrm{an}}$. *Denote by $\mathcal{C}_x$ the set of connected component of $\mathbb{A}_k^{1,\mathrm{an}} - \{x\}$.*

For each $y \in \mathbb{A}_k^{1,\mathrm{an}} - \{x\}$, denote by $C_x(y)$ the connected component of $\mathbb{A}_k^{1,\mathrm{an}} - \{x\}$ containing y. The map

$$C: \quad \mathcal{B}_x \quad \longrightarrow \quad \mathcal{C}_x$$
$$(x, y] \longmapsto C_x(y)$$

is well-defined and bijective.

Proof Let $y \in \mathbb{A}_k^{1,\mathrm{an}} - \{x\}$. For each $t \in (x, y]$, the interval $[t, y]$ is connected and does not contain x, hence $C_x(t) = C_x(y)$. It then follows from Remark 5.9.5 that $C((x, y))$ only depends on the equivalence class of $(x, y]$. In other words, C is well-defined.

Let $y, z \in \mathbb{A}_k^{1,\mathrm{an}} - \{x\}$ such that $(x, y]$ is not equivalent to $(x, z]$. It follows that $[y, x] \cup [x, z]$ is an injective path from y to z. Since, by Proposition 5.7.12, $\mathbb{A}_k^{1,\mathrm{an}}$ is uniquely path-connected, the unique injective path $[y, z]$ from y to z contains x. It follows that y and z belong to different connected components of $\mathbb{A}_k^{1,\mathrm{an}} - \{x\}$. This proves the injectivity of C.

Finally, the surjectivity of C is obvious. $\qquad\qquad\qquad\qquad\qquad\qquad\qquad\square$

Lemma 5.9.7 *Let $x \in \mathbb{A}_k^{1,\mathrm{an}}$ be a point of type 2 or 3. Let $y \in \mathbb{A}_k^{1,\mathrm{an}} - \{x\}$. Then, there exists z in (x, y) such that, for each $t \in (x, z)$, the interval (x, t) is the skeleton of a virtual open annulus.*

Proof Let $x' \in \pi_{\widehat{k^a}/k}^{-1}(x)$. Let $y' \in \pi_{\widehat{k^a}/k}^{-1}(y)$. The image of the path $[x', y']$ by $\pi_{\widehat{k^a}/k}$ is a path between x and y. It follows that, up to changing x' and y', we may assume that $\pi_{\widehat{k^a}/k}$ restricts to a bijection between $[x', y']$ and $[x, y]$.

By Lemma 5.9.6 and the explicit description of the connected components of $\mathbb{A}_{\widehat{k^a}}^{1,\mathrm{an}} - \{x'\}$ from Lemma 5.9.2, there exists z' in (x, y') such that, for each $t' \in (x, z')$, the interval (x, t') is the skeleton of a open annulus.

By Lemma 5.6.9, the orbit of any open annulus under the action of $\mathrm{Gal}(k^s/k)$ is finite. It follows that, up to choosing z' closer to x, we may assume that, for each $t' \in (x, z')$, the interval (x, t') is the skeleton of a open annulus, all of whose conjugates either coincide with it or are disjoint from it. The image of such an open annulus by $\pi_{\widehat{k^a}/k}$ is a virtual open annulus, and the result follows. $\qquad\square$

As a consequence, we obtain the following result, which is the key-point to define lengths of intervals.

Lemma 5.9.8 *Let $x, y \in \mathbb{A}_k^{1,\mathrm{an}}$ be points of type 2 or 3. Then, there exists a finite subset F of (x, y) such that each connected component of $(x, y) - F$ is the skeleton of a virtual open annulus.* $\qquad\qquad\qquad\qquad\qquad\qquad\qquad\qquad\qquad\square$

Definition 5.9.9 Let $x, y \in \mathbb{A}_k^{1,\mathrm{an}}$ be points of type 2 or 3. Let F be a finite subset of (x, y) such that each connected component of $(x, y) - F$ is the skeleton of a virtual open annulus. Let $\mathcal{I}$ be the set of connected components of $(x, y) - F$ and, for each $J \in \mathcal{J}$, denote by A_J the virtual open annulus with skeleton J.

We define the *(exponential) length* of (x, y) to be

$$\ell((x, y)) := \prod_{J \in \mathcal{J}} \text{Mod}(A_J) \in [1, +\infty).$$

It is independent of the choices, by Lemma 5.8.12.

Definition 5.9.10 Let I be an interval inside $\mathbb{A}_k^{1,\text{an}}$ that is not a singleton. We define the *(exponential) length* of I to be

$$\ell(I) := \sup(\{\ell((x, y)) : x, y \in I \text{ of type 2 or 3}\}) \in [1, +\infty].$$

Example 5.9.11 Let $\alpha \in k$ and $r \in \mathbb{R}_{>0}$ with $r \leqslant |\alpha|$. Then, we have

$$\ell([\eta_1, \eta_{\alpha,r}]) = \ell((\eta_1, \eta_{\alpha,r})) = \begin{cases} \frac{1}{r} & \text{if } |\alpha| \leqslant 1; \\ \frac{|\alpha|}{r} & \text{if } |\alpha| \geqslant 1. \end{cases}$$

In particular, we always have $\ell([\eta_1, \eta_{\alpha,r}]) \geqslant 1/r$.

Lemma 5.9.12 *Let I be an interval in $\mathbb{A}_k^{1,\text{an}}$. We have $\ell(I) = +\infty$ if, and only if, the closure of I contains a point of type 1.*

Let I_1, I_2 be intervals in $\mathbb{A}_k^{1,\text{an}}$ such that $I = I_1 \cup I_2$ and $I_1 \cap I_2$ is either empty or a singleton. Then, we have

$$\ell(I) = \ell(I_1)\,\ell(I_2).$$

$\square$

5.10 Variation of Rational Functions

In this section, for every rational function $F \in k(T)$, we study the variation of $|F|$ on $\mathbb{A}_k^{1,\text{an}}$. We will explain that it is controlled by a finite subtree of $\mathbb{A}_k^{1,\text{an}}$ and investigate metric properties.

Notation 5.10.1 Let $x \in \mathbb{A}_k^{1,\text{an}}$. We set

$$I_x := \pi_{\widehat{k^a}/k}(I_{x'}),$$

for $x' \in \pi_{\widehat{k^a}/k}^{-1}(x)$.

By Remark 5.7.3, this does not depend on the choice of x'.

As in the case of an algebraically closed base field, I_x may be thought of as a path from x to ∞.

Proposition 5.10.2 *Let $F \in k(T) - \{0\}$. Let Z be the set rigid points of $\mathbb{A}_k^{1,\mathrm{an}}$ that are zeros or poles of F. Set*

$$I_Z := \bigcup_{z \in Z} I_z.$$

Then $|F|$ is locally constant on $\mathbb{A}_k^{1,\mathrm{an}} - I_Z$.

Proof One immediately reduces to the case where the base field is $\widehat{k^a}$. Since $\widehat{k^a}$ is algebraically closed, F may be written as a quotient of products of linear polynomials. It follows that is is enough to prove the results for linear polynomials.

Let $\alpha \in \widehat{k^a}$ and let us prove the result for $F = T - \alpha$. Let C be a connected component of $\mathbb{A}_k^{1,\mathrm{an}} - I_\alpha$. Let η in the closure of C. It belongs to I_α, hence is of the form $\eta_{\alpha,r}$ for $r \in \mathbb{R}_{\geqslant 0}$.

Recall that discs are connected, by Proposition 5.7.1. By Lemma 5.4.7, the case $r \notin |k^\times|$ leads to a contradiction. It follows that $r \in |k^\times|$. Performing an appropriate change of variables and using Lemma 5.4.9, we deduce that there exists $\beta \in \widehat{k^a}$ with $|\alpha - \beta| = r$ such that $C = D^-(\beta, r)$. For each $x \in C$, we have

$$|(T - \alpha)(x)| = |(T - \beta)(x) + (\beta - \alpha)| = r,$$

because of the non-Archimedean triangle inequality. The results follows. $\qquad\square$

Our next step is to describe the behaviour of $|F|$ on I_Z using metric data. Recall that, for each $x \in \mathbb{A}_k^{1,\mathrm{an}}$, we denote by $\mathcal{B}_x$ the set of branches at x (see Definition 5.9.4).

Definition 5.10.3 Let $x \in \mathbb{A}_k^{1,\mathrm{an}}$ and $b \in \mathcal{B}_x$. Let E be a set. A *map* $f : b \to E$ is the data of a non-empty subset $\mathcal{I}$ of representatives of b and a family of maps $(f_I : I \to E)_{I \in \mathcal{I}}$ such that, for each $I, J \in \mathcal{I}$, f_I and f_J coincide on $I \cap J$.

Let I be a representative of b. We say that f *is defined on* I if I belongs to $\mathcal{I}$. In this case, we usually write $f : I \to E$ instead of $f_I : I \to E$.

Note that a map $f : I \to E$ defined on some representative I of b naturally gives rise to a map $f : b \to E$.

Definition 5.10.4 Let $x \in \mathbb{A}_k^{1,\mathrm{an}}$ and $b \in \mathcal{B}_x$. Let $f : b \to \mathbb{R}_{\geqslant 0}$. Let $N \in \mathbb{Z}$.

We say that f is *monomial along b of exponent N* if there exists a representative $(x, y]$ of b such that f is defined on $(x, y]$ and

$$\forall z \in (x, y], \forall t \in (x, z], \ f(z) = f(t) \, \ell([t, z])^N.$$

We then set

$$\mu_b(f) := N.$$

We say that f is *constant along b* if it is monomial along b of exponent 0.

Remark 5.10.5 Written additively, the last condition becomes

$$\forall z \in (x, y], \forall t \in (x, z], \ \log(f(z)) = \log(f(t)) + N \log(\ell([t, z])).$$

This explains why, in the literature, such maps are often referred to as *log-linear* and N denoted by $\partial_b \log(f)$.

Let $x \in \mathbb{A}_k^{1,\mathrm{an}}$ and $x' \in \pi_{\widehat{k^a}/k}^{-1}(x)$. Let $b' \in \mathcal{B}_{x'}$. It follows from Lemmas 5.9.7 and 5.8.4 that, for each small enough representative $(x', y']$ of b', $\pi_{\widehat{k^a}/k}$ induces a homeomorphism from $(x', y']$ to $(x, \pi_{\widehat{k^a}/k}(y')]$. This property allows to define the image of the branch b'.

Definition 5.10.6 Let $x \in \mathbb{A}_k^{1,\mathrm{an}}$ and $x' \in \pi_{\widehat{k^a}/k}^{-1}(x)$. Let $b' \in \mathcal{B}_{x'}$. The *image of the branch b' by $\pi_{\widehat{k^a}/k}$* is the branch

$$\pi_{\widehat{k^a}/k}(b') := (x, \pi_{\widehat{k^a}/k}(y')] \in \mathcal{B}_x,$$

for a small enough representative $(x', y']$ of b'.

Lemma 5.10.7 *Let $x \in \mathbb{A}_k^{1,\mathrm{an}}$ and $x' \in \pi_{\widehat{k^a}/k}^{-1}(x)$. Let $b \in \mathcal{B}_x$. For each $b \in \mathcal{B}_x$, there exists $b' \in \mathcal{B}_{x'}$ such that $\pi_{\widehat{k^a}/k}(b') = b$. The set of such b''s is finite and $\mathrm{Gal}(k^s/k)$ acts transitively on it.*

Proof The existence of b' is proved as in the beginning of the proof of Lemma 5.9.7. The rest of the statement follows from Lemmas 5.9.7 and 5.6.9 and Corollary 5.6.6. ☐

The following result is a direct consequence of the definitions.

Lemma 5.10.8 *Let $x \in \mathbb{A}_k^{1,\mathrm{an}}$ and $b \in \mathcal{B}_x$. Let $f : b \to \mathbb{R}_{\geqslant 0}$. Assume that there exists $N \in \mathbb{Z}$ such that, for each $b' \in \pi_{\widehat{k^a}/k}^{-1}$, $f \circ \pi_{\widehat{k^a}/k}$ is monomial along b' of exponent N. Then, f is monomial along b of exponent $N \cdot \sharp\pi_{\widehat{k^a}/k}^{-1}(b)$.*

Definition 5.10.9 Let $F \in \widehat{k^a}(T) - \{0\}$. Let $\alpha \in \widehat{k^a}$. The *order of F at α* is the unique integer v such there exists $P \in \widehat{k^a}[T]$ with $P(\alpha) \neq 0$ satisfying

$$F(T) = (T - \alpha)^v P(T).$$

We denote it by $\mathrm{ord}_\alpha(P)$.

Theorem 5.10.10 *Let $F \in \widehat{k^a}(T) - \{0\}$. Let $x \in \mathbb{A}_{\widehat{k^a}}^{1,\mathrm{an}}$ and $b \in \mathcal{B}_x$. Then the map $|F|$ is monomial along b.*
If x is of type 1, then $\mu_b(|F|) = \mathrm{ord}_x(F)$.
If x is of type 2 or 3 and $C(b)$ is bounded, then

$$\mu_b(|F|) = - \sum_{z \in \widehat{k^a} \cap C(b)} \mathrm{ord}_z(F).$$

If x is of type 2 or 3 and $C(b)$ is unbounded, then

$$\mu_b(|F|) = \deg(F) - \sum_{z \in \widehat{k^a} \cap C(b)} \mathrm{ord}_z(F).$$

If x is of type 4, then $\mu_b(|F|) = 0$.

Proof Let us first remark that if the result holds for G and H in $\widehat{k^a}(T) - \{0\}$, then it also holds for GH and G/H. As a result, since $\widehat{k^a}$ is algebraically closed, it is enough to prove the result for a linear polynomial, so we may assume that $F = T - a$, with $a \in k$.

- Assume that x is of type 1.

There exists $\alpha \in k$ such that $x = \alpha$. By Lemmas 5.9.2 and 5.9.6, there is a unique branch at x. It is represented by

$$(\alpha, \eta_{\alpha,s}] = \{\eta_{\alpha,t}, t \in (0, s]\},$$

for any $s \in \mathbb{R}_{>0}$.

If $\alpha = a$, then for each $t \in \mathbb{R}_{>0}$, we have $|(T - a)(\eta_{a,t})| = t$, hence $|T - a|$ is monomial along b of exponent 1. Since we have $\mathrm{ord}_a(T - a) = 1$, the result holds in this case.

If $\alpha \neq a$, then for each $t \in (0, |a - \alpha|)$, we have $|(T - a)(\eta_{\alpha,t})| = |a_\alpha|$, hence $|T - a|$ is monomial along b of exponent 0. Since we have $\mathrm{ord}_\alpha(T - a) = 0$, the result holds in this case too.

- Assume that x is of type 2 or 3 and that $C(b)$ is bounded.

There exist $\alpha \in k$ and $r \in \mathbb{R}_{>0}$ such that $x = \eta_{\alpha,r}$. By Lemma 5.9.2, there exists $\beta \in k$ with $|\beta - \alpha| \leqslant r$ such that $C(b) = D^-(\beta, r)$. Since $\eta_{\alpha,r} = \eta_{\beta,r}$, we may assume that $\alpha = \beta$. The branch b is then represented by

$$(\eta_{\alpha,r}, \eta_{\alpha,s}] = \{\eta_{\alpha,t}, t \in [s, r)\},$$

for any $s \in (0, r]$.

If $a \in C(b)$, then we have $|a - \alpha| < r$, hence, for each $t \in [|a - \alpha|, r)$, we have $|(T - a)(\eta_{\alpha,t})| = t$, hence $|T - a|$ is monomial along b of exponent -1. It follows that the result holds in this case.

If $a \notin C(b)$, then we have $|a - \alpha| = r$, hence, for each $t \in [0, r)$, we have $|(T - a)(\eta_{\alpha,t})| = |a - \alpha|$, hence $|T - a|$ is monomial along b of exponent 0. It follows that the result holds in this case too.

- Assume that x is of type 2 or 3 and that $C(b)$ is unbounded.

There exist $\alpha \in k$ and $r \in \mathbb{R}_{>0}$ such that $x = \eta_{\alpha,r}$. By Lemma 5.9.2, the branch b is then represented by

$$(\eta_{\alpha,r}, \eta_{\alpha,s}] = \{\eta_{\alpha,t}, t \in (r, s]\},$$

for any $s \in (r, +\infty)$.

If $a \in C(b)$, then we have $|a - \alpha| > r$, hence, for each $t \in (r, |a - \alpha|)$, we have $|(T - a)(\eta_{\alpha,t})| = |a - \alpha|$, hence $|T - a|$ is monomial along b of exponent 0. We have

$$\deg(T - a) - \sum_{z \in \widehat{k^a} \cap C(b)} \mathrm{ord}_z(T - a) = \deg(T - a) - \mathrm{ord}_a(T - a) = 1 - 1 = 0,$$

hence the result holds in this case.

If $a \notin C(b)$, then we have $|a - \alpha| \leqslant r$, hence, for each $t \in (r, +\infty)$, we have $|(T - a)(\eta_{\alpha,t})| = t$, hence $|T - a|$ is monomial along b of exponent 1. We have

$$\deg(T - a) - \sum_{z \in \widehat{k^a} \cap C(b)} \mathrm{ord}_z(T - a) = \deg(T - a) = 1,$$

hence the result holds in this case too.

- Assume that x is of type 4.

By Proposition 5.4.11, x admits a basis of neighborhood made of discs. It follows that there exist $\alpha \in k$ and $r \in \mathbb{R}_{>0}$ such that $x \in D^-(\alpha, r)$ and $a \notin D^-(\alpha, r)$. For each $y \in D^-(\alpha, r)$, we have $|(T - a)(y)| = |(T - \alpha)(y) + (\alpha - a)| = |a - \alpha|$, hence $|T - a|$ is constant in the neighborhood of x. The result follows. □

Remark 5.10.11 The term $\deg(R)$ that appears in the formula when $C(b)$ is unbounded may be identified with the opposite of the order of R at ∞. If we had worked on $\mathbb{P}^{1,\mathrm{an}}_{\widehat{k^a}}$ instead of $\mathbb{A}^{1,\mathrm{an}}_{\widehat{k^a}}$, it would not have been necessary to discuss this case separately.

Corollary 5.10.12 *Let $F \in k(T) - \{0\}$. Let $x \in \mathbb{A}^{1,\mathrm{an}}_k$ be a point of type 2 or 3. Then, there exists a finite subset $B_{x,F}$ of B_x such that, for each $b \in B_x \setminus B_{x,F}$, $|F|$ is constant along b and we have*

$$\sum_{b \in B_x} \mu_b(|F|) = \sum_{b \in B_{x,F}} \mu_b(|F|) = 0.$$

Proof Using Lemma 5.10.8, one reduces to the case where the base field is $\widehat{k^a}$. The result then follows from Theorem 5.10.10, since we have $\mathrm{ord}_z(F) = 0$ for almost all $z \in \widehat{k^a}$ and

$$\sum_{z \in \widehat{k^a}} \mathrm{ord}_z(F) = \deg(F).$$

□

Remark 5.10.13 The statement of Corollary 5.10.12 corresponds to a harmonicity property. This is more visible written in the additive form (see Remark 5.10.5):

$$\sum_{b \in B_x} \partial_b \log(|F|) = 0.$$

A full-fledged potential theory actually exists over Berkovich analytic curves. We refer to A. Thuillier's thesis [Thu05] for the details (see also [BR10] for the more explicit case of the Berkovich line over an algebraically closed field).

Since analytic functions are, by definition, locally uniform limits of rational functions, the results on variations of functions extend readily.

Theorem 5.10.14 Let $x \in \mathbb{A}_k^{1,\mathrm{an}}$ be a point of type 2 or 3 and let $F \in \mathcal{O}_x - \{0\}$. Then, for $b \in \mathcal{B}_x$, $|F|$ is monomial along b with integer slope. Moreover, there exists a finite subset $\mathcal{B}_{x,F}$ of $\mathcal{B}_x$ such that, for each $b \in \mathcal{B}_x \setminus \mathcal{B}_{x,F}$, $|F|$ is constant along b and we have

$$\sum_{b \in \mathcal{B}_x} \mu_b(|F|) = 0.$$

$\square$

Corollary 5.10.15 Let $x \in \mathbb{A}_k^{1,\mathrm{an}}$ be a point of type 2 or 3 and let $F \in \mathcal{O}_x - \{0\}$. If $|F|$ has a local maximum at x, then it is locally constant at x. $\square$

Corollary 5.10.16 Let U be a connected open subset of $\mathbb{A}_k^{1,\mathrm{an}}$ and let $F \in \mathcal{O}(U)$. If $|F|$ is not constant on U, then there exists $y \in \partial U$ and $b \in \mathcal{B}_y$ such that

$$\lim_{z \xrightarrow{b} y} |F(z)| = \sup_{t \in U}(|F(t)|),$$

where the limit is taken on points z converging to y along b, and $|F|$ has a negative exponent along b. $\square$

We conclude with a result of a different nature, showing that, if φ is a finite morphism of curves, the relationship between the length of an interval at the source and the length of its image is controlled by the degree of the morphism. We state a simplified version of the result and refer to [Duc, Proposition 3.6.40] for a more general statement.

Theorem 5.10.17 Let A_1 and A_2 be two virtual annuli over k with skeleta Σ_1 and Σ_2. Let $\varphi \colon A_1 \to A_2$ be a finite morphism such that $\varphi(\Sigma_1) = \Sigma_2$. Then, for each $x, y \in \Sigma_1$, we have

$$\ell(\varphi([x, y])) = \ell([x, y])^{\deg(\varphi)}.$$

$\square$

Example 5.10.18 Let $n \in \mathbb{N}_{\geq 1}$ and consider the morphism $\varphi \colon \mathbb{A}_k^{1,\mathrm{an}} \to \mathbb{A}_k^{1,\mathrm{an}}$ given by $T \mapsto T^n$. For each $r \in \mathbb{R}_{>0}$, we have $\varphi(\eta_r) = \eta_{r^n}$. In particular, for $r < s \in \mathbb{R}_{>0}$, we have

$$\ell(\varphi([\eta_r, \eta_s])) = \ell([\eta_{r^n}, \eta_{s^n}]) = \frac{s^n}{r^n} = \ell([\eta_r, \eta_s])^n.$$

Acknowledgments We warmly thank Marco Maculan for his numerous comments on an earlier version of this text, and the anonymous referees for their remarks and corrections.

While writing this text, the authors were supported by the ERC Starting Grant "TOSSIBERG" (grant agreement 637027). The second author was partially funded by a prize of the *Fondation des Treilles*. The *Fondation des Treilles* created by Anne Gruner-Schlumberger aims to foster the dialogue between sciences and arts in order to further research and creation. She welcomes researchers and creators in her domain of Treilles (Var, France).

References

[Ber90] V.G. Berkovich, *Spectral Theory and Analytic Geometry Over Non-Archimedean Fields*. Mathematical Surveys and Monographs, vol. 33 (American Mathematical Society, Providence, RI, 1990)

[Bou06] N. Bourbaki, *Éléments de mathématique. Algèbre commutative. Chapitres 5 à 7* (Springer, Berlin, Heidelberg, 2006)

[BR10] M. Baker, R. Rumely, *Potential Theory and Dynamics on the Berkovich Projective Line*. Mathematical Surveys and Monographs, vol. 159 (American Mathematical Society, Providence, RI, 2010)

[Cas86] J.W.S. Cassels, *Local Fields*. London Mathematical Society Student Texts, vol. 3 (Cambridge University Press, Cambridge, 1986)

[Chr83] G. Christol, *Modules différentiels et équations différentielles p-adiques*. Queen's Papers in Pure and Applied Mathematics, vol. 66 (Queen's University, Kingston, ON, 1983)

[Duc] A. Ducros, La structure des courbes analytiques. Manuscript available at http://www.math.jussieu.fr/~ducros/trirss.pdf

[FvdP04] J. Fresnel, M. van der Put, *Rigid Analytic Geometry and Its Applications*. Progress In Mathematics, vol. 218 (Birkhäuser, Basel, 2004)

[Ked15] K.S. Kedlaya, Reified valuations and adic spectra. Res. Number Theory **1**, 42 (2015). Id/No 20

[Pel20] F. Pellarin, From the Carlitz exponential to Drinfeld modular forms, in *Arithmetic and Geometry Over Local Fields*, ed. by B. Anglès, T. Ngo Dac (Springer International Publishing, Cham, 2020)

[Poo93] B. Poonen, Maximally complete fields. Enseign. Math. (2) **39**(1–2), 87–106 (1993)

[PT20] J. Poineau, D. Turchetti, Berkovich curves and Schottky uniformization. II: Analytic uniformization of Mumford curves, in *Arithmetic and Geometry Over Local Fields*, ed. by B. Anglès, T. Ngo Dac (Springer International Publishing, Cham, 2020)

[Tat71] J. Tate, Rigid analytic spaces. Invent. Math. **12**, 257–289 (1971)

[Thu05] A. Thuillier, Théorie du potentiel sur les courbes en géométrie analytique non-Archimédienne. Applications à la théorie d'Arakelov. PhD thesis, Université de Rennes 1 (2005)

Chapter 6
Berkovich Curves and Schottky Uniformization II: Analytic Uniformization of Mumford Curves

Jérôme Poineau and Daniele Turchetti

Abstract This is the second part of a survey on the theory of non-Archimedean curves and Schottky uniformization from the point of view of Berkovich geometry. It is more advanced than the first part and covers the theory of Mumford curves and Schottky uniformization. We start by briefly reviewing the theory of Berkovich curves, then introduce Mumford curves in a purely analytic way (without using formal geometry). We define Schottky groups acting on the Berkovich projective line, highlighting how geometry and group theory come together to prove that the quotient by the action of a Schottky group is an analytic Mumford curve. Finally, we present an analytic proof of Schottky uniformization, showing that any analytic Mumford curve can be described as a quotient of this kind. The guiding principle of our exposition is to stress notions and fully prove results in the theory of non-Archimedean curves that, to our knowledge, are not fully treated in other texts.

6.1 Introduction

In the first part [PT20] of this survey, we provided a concrete description of the Berkovich affine line over a non-Archimedean complete valued field $(k, |\cdot|)$ and investigated its main properties. It is a remarkable fact that, combining topology, algebra, and combinatorics, one can still get a very satisfactory description of more general analytic curves over k, in the sense of Berkovich theory.

If k is algebraically closed, for instance, one can show that a smooth compact Berkovich curve X can always be decomposed into a finite graph and an infinite number of open discs. If the genus of X is positive, there exists a smallest graph

J. Poineau (✉)
Laboratoire de mathématiques Nicolas Oresme, Université de Caen-Normandie, Normandie University, Caen, France
e-mail: jerome.poineau@unicaen.fr; https://poineau.users.lmno.cnrs.fr/

D. Turchetti
Department of Mathematics and Statistics, Dalhousie University, Halifax, NS, Canada
e-mail: daniele.turchetti@dal.ca; https://www.mathstat.dal.ca/~dturchetti/

© The Author(s), under exclusive license to Springer Nature Switzerland AG 2021
B. Anglès, T. Ngo Dac (eds.), *Arithmetic and Geometry over Local Fields*,
Lecture Notes in Mathematics 2275, https://doi.org/10.1007/978-3-030-66249-3_6

satisfying this property. It is classically called the *skeleton* of X, an invariant that encodes a surprising number of properties of X. As an example, if the Betti number of the skeleton of X is equal to the genus of X and is at least 2, then the curve X can be described analytically as a quotient $\Gamma \backslash O$, where O is an open dense subset of the projective analytic line $\mathbb{P}_k^{1,\mathrm{an}}$ and Γ a suitable subgroup of $\mathrm{PGL}_2(k)$. This phenomenon is known as *Schottky uniformization*, and it is the consequence of a celebrated theorem of D. Mumford, which is the main result of [Mum72a].

Obviously, D. Mumford's proof did not make use of Berkovich spaces, as they were not yet introduced at that time, but rather of formal geometry and the theory of Bruhat-Tits trees. A few years later, L. Gerritzen and M. van der Put recasted the theory purely in the language of rigid analytic geometry (using in a systematic way the notion of reduction of a rigid analytic curve). We refer the reader to the reference manuscript [GvdP80] for a detailed account of the theory and related topics, enriched with examples and applications.

In this text, we develop the whole theory of *Schottky groups* and *Mumford curves* from scratch, in a purely analytic manner, relying in a crucial way on the nice topological properties of Berkovich spaces, and the tools that they enable us to use: the theory of proper actions of groups on topological spaces, of fundamental groups, etc. We are convinced that those features, and Berkovich's point of view in general, will help improve our understanding of Schottky uniformization.

In this second part of the survey, we have allowed ourselves to be sometimes more sketchy than in the first part [PT20], but this should not cause any trouble to anyone familiar enough with the theory of Berkovich curves. We begin by reviewing standard material. In Sect. 6.2, we define the Berkovich projective line $\mathbb{P}_k^{1,\mathrm{an}}$ over k, consider its group of k-linear automorphisms $\mathrm{PGL}_2(k)$ and introduce the Koebe coordinates for the loxodromic transformations. In Sect. 6.3, we give an introduction to the theory of Berkovich analytic curves, starting with those that locally look like the affine line. For the more general curves, we review the theory without proofs, but provide some references. We conclude this section by an original purely analytic definition of Mumford curves. In Sect. 6.4, we propose two definitions of Schottky groups, first using the usual description of their fundamental domains and second, via their group theoretical properties, using their action of $\mathbb{P}_k^{1,\mathrm{an}}$. We show that they coincide by relying on the nice topological properties of Berkovich spaces. In Sect. 6.5, we prove that every Mumford curve may be uniformized by a dense open subset of $\mathbb{P}_k^{1,\mathrm{an}}$ with a group of deck transformations that is a Schottky group. Once again, our proof is purely analytic, relying ultimately on arguments from potential theory. To the best of our knowledge, this is the first complete proof of this result. We conclude the section by investigating automorphisms of Mumford curves and giving explicit examples.

We put a great effort in providing a self-contained presentation of the results above and including details that are often omitted in the literature. Both the theories of Berkovich curves and Schottky uniformization have a great amount of ramifications and interactions with other branches of mathematics. For the interested readers, we provide an appendix with a series of references that will hopefully help them to navigate through this jungle of wonderful mathematical objects.

The idea of writing down these notes came to the first author when he was taking part to the VIASM School on Number Theory in June 2018 in Hanoi. Just as the school was, the material presented here is primarily aimed at graduate students, although we also cover some of the most advanced developments in the field. Moreover, we have included questions that we believe could be interesting topics for young researchers (see Remarks 6.4.20 and 6.5.7). The appendix provides additional material leading to active subjects of research and open problems.

The different chapters in this volume are united by the use of analytic techniques in the study of arithmetic geometry. While they treat different topics, we encourage the reader to try to understand how they are related and may shed light on each other. In particular, the lecture notes of F. Pellarin [Pel20], about Drinfeld modular forms, mention several topics related to ours, although phrased in the language of rigid analytic spaces, such as Schottky groups (Section 5) or quotient spaces (Section 6). It would be interesting to investigate to what extent the viewpoint of Berkovich geometry presented here could provide a useful addition to this theory.

We retain notation as in [PT20]. In particular, $(k, |\cdot|)$ is a non-Archimedean complete valued field, k^a is a fixed algebraic closure of it, and $\widehat{k^a}$ is the completion of the latter.

6.2   The Berkovich Projective Line and Möbius Transformations

6.2.1   Affine Berkovich Spaces

We generalize the constructions of [PT20], replacing $k[T]$ by an arbitrary k-algebra of finite type. Our reference here is [Ber90, Section 1.5].

Definition 6.2.1 Let A be k-algebra of finite type. The *Berkovich spectrum* $\mathrm{Spec}^{\mathrm{an}}(A)$ of A is the set of multiplicative seminorms on A that induce the given absolute value $|\cdot|$ on k.

As in [PT20, Definition 5.3.1], we can associate a completed residue field $\mathscr{H}(x)$ to each point x of $\mathrm{Spec}^{\mathrm{an}}(A)$. As in [PT20, Section 5.4], we endow $\mathrm{Spec}^{\mathrm{an}}(A)$ with the coarsest topology that makes continuous the maps of the form

$$x \in \mathrm{Spec}^{\mathrm{an}}(A) \longmapsto |f(x)| \in \mathbb{R}$$

for $f \in A$. Properties similar to that of the Berkovich affine line still hold in this setting: the space $\mathrm{Spec}^{\mathrm{an}}(A)$ is countable at infinity, locally compact and locally path-connected.

We could also define a sheaf of function on $\mathrm{Spec}^{\mathrm{an}}(A)$ as in [PT20, Definition 5.5.1][1] with properties similar to that of the usual complex analytic spaces.

Lemma 6.2.2 *Each morphism of k-algebras $\varphi \colon A \to B$ induces a continuous map of Berkovich spectra*

$$\mathrm{Spec}^{\mathrm{an}}(\varphi) \colon \mathrm{Spec}^{\mathrm{an}}(B) \longrightarrow \mathrm{Spec}^{\mathrm{an}}(A)$$
$$|\cdot|_x \longmapsto |\varphi(\cdot)|_x \quad .$$

Let us do the example of a localisation morphism.

Notation 6.2.3 Let A be a k-algebra of finite type and let $f \in A$. We set

$$D(f) := \{x \in \mathrm{Spec}^{\mathrm{an}}(A) : f(x) \neq 0\}.$$

It is an open subset of $\mathrm{Spec}^{\mathrm{an}}(A)$.

Lemma 6.2.4 *Let A be a k-algebra of finite type and let $f \in A$. The map $\mathrm{Spec}^{\mathrm{an}}(A[1/f]) \to \mathrm{Spec}^{\mathrm{an}}(A)$ induced by the localisation morphism $A \to A[1/f]$ induces a homeomorphism onto its image $D(f)$.*   □

6.2.2 The Berkovich Projective Line

In this section, we explain how to construct the Berkovich projective line over k. It can be done, as usual, by gluing upside-down two copies of the affine line $\mathbb{A}_k^{1,\mathrm{an}}$ along $\mathbb{A}_k^{1,\mathrm{an}} - \{0\}$. We refer to [BR10, Section 2.2] for a definition in one step reminiscent of the "Proj" construction from algebraic geometry.

To carry out the construction of the Berkovich projective line more precisely, let us introduce some notation. We consider, as before, the Berkovich affine line $X := \mathbb{A}_k^{1,\mathrm{an}}$ with coordinate T, i.e. $\mathrm{Spec}^{\mathrm{an}}(k[T])$. By Lemma 6.2.4, its subset $U := \mathbb{A}_k^{1,\mathrm{an}} - \{0\} = D(T)$ may be identified with $\mathrm{Spec}^{\mathrm{an}}(k[T, 1/T])$.

We also consider another Berkovich affine line X' with coordinate T' and identify its subset $U' := X' - \{0\}$ with $\mathrm{Spec}^{\mathrm{an}}(k[T', 1/T'])$.

By Lemma 6.2.2, the isomorphism $k[T', 1/T'] \xrightarrow{\sim} k[T, 1/T]$ sending T' to $1/T$ induces an isomorphism $\iota \colon U \xrightarrow{\sim} U'$.

---

[1]Note however that the ring of global sections is always reduced, so that we only get the right notion when A is reduced. The proper construction involves defining first the space $\mathbb{A}_k^{n,\mathrm{an}} := \mathrm{Spec}^{\mathrm{an}}(k[T_1, \ldots, T_n])$, then open subsets of it, and then closed analytic subsets of the latter, as we usually proceed for analytifications in the complex setting.

Definition 6.2.5 The Berkovich projective line $\mathbb{P}_k^{1,\mathrm{an}}$ is the space obtained by gluing the Berkovich affine lines X and X' along their open subsets U and U' via the isomorphim ι.

We denote by ∞ the image in $\mathbb{P}_k^{1,\mathrm{an}}$ of the point 0 in X'.

The basic topological properties of $\mathbb{P}_k^{1,\mathrm{an}}$ follow from that of $\mathbb{A}_k^{1,\mathrm{an}}$.

Proposition 6.2.6 *We have* $\mathbb{P}_k^{1,\mathrm{an}} = \mathbb{A}_k^{1,\mathrm{an}} \cup \{\infty\}$.

The space $\mathbb{P}_k^{1,\mathrm{an}}$ *is Hausdorff, compact, uniquely path-connected and locally path-connected.* $\square$

For $x, y \in \mathbb{P}_k^{1,\mathrm{an}}$, we denote by $[x, y]$ the unique injective path between x and y.

6.2.3 Möbius Transformations

Let us recall that, in the complex setting, the group $\mathrm{PGL}_2(\mathbb{C})$ acts on $\mathbb{P}^1(\mathbb{C})$ via Möbius transformations. More precisely, to an invertible matrix $A = \begin{pmatrix} a & b \\ c & d \end{pmatrix}$, one associates the automorphism

$$\gamma_A \colon z \in \mathbb{P}^1(\mathbb{C}) \longmapsto \frac{az+b}{cz+d} \in \mathbb{P}^1(\mathbb{C})$$

with the usual convention that, if $c \neq 0$, then $\gamma_A(\infty) = a/c$ and $\gamma_A(-d/c) = \infty$, and, if $c = 0$, then $\gamma_A(\infty) = \infty$.

We would like to define an action of $\mathrm{PGL}_2(k)$ on $\mathbb{P}_k^{1,\mathrm{an}}$ similar to the complex one. Let $A := \begin{pmatrix} a & b \\ c & d \end{pmatrix} \in \mathrm{GL}_2(k)$.

First note that we can use the same formula as above to associate to A an automorphism γ_A of the set of rational points $\mathbb{P}_k^{1,\mathrm{an}}(k)$.

It is actually possible to deal with all the points this way. Indeed, let x be a point of $\mathbb{P}_k^{1,\mathrm{an}} - \mathbb{P}_k^{1,\mathrm{an}}(k)$. In [PT20, Section 5.3], we have associated to x a character $\chi_x \colon k[T] \to \mathscr{H}(x)$. Since x is not a rational point, $\chi_x(T)$ does not belong to k, hence the quotient $(a\chi_x(T) + b)/(c\chi_x(T) + d)$ makes sense. We can then define $\gamma_A(x)$ as the element of $\mathbb{A}_k^{1,\mathrm{an}}$ associated to the character

$$P(T) \in k[T] \mapsto P\left(\frac{a\chi_x(T) + b}{c\chi_x(T) + d}\right) \in \mathscr{H}(x).$$

This construction can also be made in a more algebraic way. By Lemmas 6.2.2 and 6.2.4, the morphism of k-algebras

$$P(T) \in k[T] \mapsto P\left(\frac{aT + b}{cT + d}\right) \in k\left[T, \frac{1}{cT + d}\right]$$

induces a map $\gamma_{A,1}\colon \mathbb{A}_k^{1,\mathrm{an}} - \{-\frac{d}{c}\} \to \mathbb{A}_k^{1,\mathrm{an}} \subseteq \mathbb{P}_k^{1,\mathrm{an}}$ (with the convention that $-d/c = \infty$ if $c = 0$).

Similarly, the morphism of k-algebras

$$Q(U) \in k[T'] \mapsto Q\left(\frac{c + dT'}{a + bT'}\right) \in k\left[T', \frac{1}{a + bT'}\right]$$

induces a map $\gamma_{A,2}\colon \mathbb{P}_k^{1,\mathrm{an}} - \{0, -\frac{b}{a}\} \to \mathbb{P}_k^{1,\mathrm{an}}$ (with the convention that $-b/a = \infty$ if $a = 0$).

A simple computation shows that the maps $\gamma_{A,1}$ and $\gamma_{A,2}$ are compatible with the isomorphism ι from Sect. 6.2.2. Note that we always have $-\frac{d}{c} \neq -\frac{b}{a}$. If $ad \neq 0$, it follows that we have $\left(\mathbb{A}_k^{1,\mathrm{an}} - \{-\frac{d}{c}\}\right) \cup \left(\mathbb{P}_k^{1,\mathrm{an}} - \{0, -\frac{b}{a}\}\right) = \mathbb{P}_k^{1,\mathrm{an}}$, so the two maps glue to give a global map

$$\gamma_A\colon \mathbb{P}_k^{1,\mathrm{an}} \to \mathbb{P}_k^{1,\mathrm{an}}.$$

We let the reader handle the remaining cases by using appropriate changes of variables.

Notation 6.2.7 For $a, b, c, d \subset k$ with $ad - bc \neq 0$, we denote by $\begin{bmatrix} a & b \\ c & d \end{bmatrix}$ the image in $\mathrm{PGL}_2(k)$ of the matrix $\begin{pmatrix} a & b \\ c & d \end{pmatrix}$.

From now on, we will identify each element A of $\mathrm{PGL}_2(k)$ with the associated automorphism γ_A of $\mathbb{P}_k^{1,\mathrm{an}}$.

Lemma 6.2.8 *The image of a closed (resp. open) disc of $\mathbb{P}_k^{1,\mathrm{an}}$ by a Möbius transformation is a closed (resp. open) disc.*

Proof Let $A \in \mathrm{GL}_2(k)$. We may extend the scalars, hence assume that k is algebraically closed. In this case, A is similar to an upper triangular matrix. In other words, up to changing coordinates of $\mathbb{P}_k^{1,\mathrm{an}}$, we may assume that A is upper triangular. The transformation γ_A is then of the form

$$\gamma_A\colon z \in \mathbb{P}_k^{1,\mathrm{an}} \mapsto \alpha z \in \mathbb{P}_k^{1,\mathrm{an}}$$

or

$$\gamma_A\colon z \in \mathbb{P}_k^{1,\mathrm{an}} \mapsto z + \alpha \in \mathbb{P}_k^{1,\mathrm{an}}$$

for some $\alpha \in k$. In both cases, the result is clear. $\qquad\square$

6.2.4   Loxodromic Transformations and Koebe Coordinates

Definition 6.2.9 A matrix in $\mathrm{GL}_2(k)$ is said to be *loxodromic* if its eigenvalues in k^a have distinct absolute values.

A Möbius transformation is said to be *loxodromic* if some (or equivalently every) representative is.

Lemma 6.2.10 *Let* $a, b, c, d \in k$ *with* $ad - bc \neq 0$ *and set* $A := \begin{pmatrix} a & b \\ c & d \end{pmatrix} \in \mathrm{GL}_2(k)$.
Then A *is loxodromic if, and only if, we have* $|ad - bc| < |a + d|^2$.

Proof Let λ and λ' be the eigenvalues of A in k^a. We may assume that $|\lambda| \leqslant |\lambda'|$.
On the one hand, if $|\lambda| = |\lambda'|$, then we have

$$|a + d|^2 = |\lambda + \lambda'|^2 \leqslant |\lambda'|^2 = |\lambda| \, |\lambda'| = |ad - bc|.$$

On the other hand, if $|\lambda| < |\lambda'|$, then we have

$$|a + d|^2 = |\lambda + \lambda'|^2 = |\lambda'|^2 > |\lambda| \, |\lambda'| = |ad - bc|.$$

$\square$

Let $A \in \mathrm{PGL}_2(k)$ be a loxodromic Möbius transformation.

Fix some representative B of A in $\mathrm{GL}_2(k)$. Denote by λ and λ' its eigenvalues in k^a. We may assume that $|\lambda| < |\lambda'|$. The characteristic polynomial χ_B of B cannot be irreducible over k, since otherwise its roots in k^a would have the same absolute values. It follows that λ and λ' belong to k. Set $\beta := \lambda / \lambda' \in k^{\circ\circ}$.

The eigenspace of B associated to the eigenvalue λ (resp. λ') is a line in k^2. Denote by α (resp. α') the corresponding element in $\mathbb{P}^1(k)$.

Definition 6.2.11 The elements $\alpha, \alpha' \in \mathbb{P}^1(k)$ and $\beta \in k^{\circ\circ}$ depend only on A and not on the chosen representative. They are called the *Koebe coordinates* of A.

There exists a Möbius transformation $\varepsilon \in \mathrm{PGL}_2(k)$ such that $\varepsilon(0) = \alpha$ and $\varepsilon(\infty) = \alpha'$. The Möbius transformation $\varepsilon^{-1} A \varepsilon$ now has eigenspaces corresponding to 0 and ∞ in $\mathbb{P}^1(k)$ and the associated automorphism of $\mathbb{P}^{1,\mathrm{an}}_k$ is

$$\gamma_{\varepsilon^{-1} A \varepsilon} \colon z \in \mathbb{P}^{1,\mathrm{an}}_k \mapsto \beta z \in \mathbb{P}^{1,\mathrm{an}}_k.$$

We deduce that 0 and ∞ are respectively the attracting and repelling fixed points of $\gamma_{\varepsilon^{-1} A \varepsilon}$ in $\mathbb{P}^{1,\mathrm{an}}_k$. It follows that α and α' are respectively the attracting and repelling fixed points of γ_A in $\mathbb{P}^{1,\mathrm{an}}_k$.

The same argument shows that the Koebe coordinates determine uniquely the Möbius transformation A. In fact, given $\alpha, \alpha', \beta \in k$ with $\alpha \neq \alpha'$ and $0 < |\beta| < 1$, the Möbius transformation that has these elements as Koebe coordinates is given explicitly by

$$M(\alpha, \alpha', \beta) = \begin{bmatrix} \alpha - \beta\alpha' & (\beta - 1)\alpha\alpha' \\ 1 - \beta & \beta\alpha - \alpha' \end{bmatrix}, \tag{6.2.1}$$

In an analogous way, whenever $\infty \in \mathbb{P}_k^{1,\mathrm{an}}$ is an attracting or repelling point of a loxodromic Möbius transformation, we can recover the latter as:

$$M(\alpha, \infty, \beta) = \begin{bmatrix} \beta & (1 - \beta)\alpha \\ 0 & 1 \end{bmatrix} \quad \text{or} \quad M(\infty, \alpha', \beta) = \begin{bmatrix} 1 & (\beta - 1)\alpha' \\ 0 & \beta \end{bmatrix}. \tag{6.2.2}$$

Remark 6.2.12 Let $A \in \mathrm{PGL}_2(k)$ be a Möbius transformation that is not loxodromic. Then, extending the scalars to $\widehat{k^a}$ and possibly changing the coordinates, the associated automorphism of $\mathbb{P}_{\widehat{k^a}}^{1,\mathrm{an}}$ is a homothety

$$z \in \mathbb{P}_{\widehat{k^a}}^{1,\mathrm{an}} \mapsto \beta z \in \mathbb{P}_{\widehat{k^a}}^{1,\mathrm{an}} \text{ with } |\beta| = 1$$

or a translation

$$z \in \mathbb{P}_{\widehat{k^a}}^{1,\mathrm{an}} \mapsto z + b \in \mathbb{P}_{\widehat{k^a}}^{1,\mathrm{an}}.$$

Note that those automorphisms have several fixed points in $\mathbb{P}_{\widehat{k^a}}^{1,\mathrm{an}}$ (η_r with $r \geqslant 0$ in the first case and $r \geqslant |b|$ in the second). It follows that A itself also has infinitely many fixed points in $\mathbb{P}_k^{1,\mathrm{an}}$.

6.3 Berkovich k-Analytic Curves

6.3.1 Berkovich $\mathbb{A}^1$-like Curves

In this section we go one step beyond the study of affine and projective lines, by introducing a class of curves that "locally look like the affine line", and see that there are interesting examples of curves belonging to this class.

A much more general theory of k-analytic curves exists but it will be discussed only briefly in this text in Sect. 6.3.2, in the case of smooth curves. For more on this topic, the standard reference is [Ber90, Chapter 4]. The most comprehensive account to-date can be found in A. Ducros' book project [Duc], while deeper discussions of specific aspects are contained in the references in the Appendix A.1 of the present text.

Definition 6.3.1 A *k-analytic $\mathbb{A}^1$-like curve* is a locally ringed space in which every point admits an open neighborhood isomorphic to an open subset of $\mathbb{A}_k^{1,\mathrm{an}}$.

It follows from the explicit description of bases of neighborhoods of points of $\mathbb{A}_k^{1,\mathrm{an}}$ (see [PT20, Proposition 5.4.11]) that each k-analytic $\mathbb{A}^1$-like curve admits a covering by virtual open Swiss cheeses. By local compactness, such a covering can always be found locally finite. It can be refined into a partition (no longer locally finite) consisting of simpler pieces.

Theorem 6.3.2 *Let X be a k-analytic $\mathbb{A}^1$-like curve. Then, there exist*

(i) a locally finite set S of type 2 points of X;
(ii) a locally finite set $\mathcal{A}$ of virtual open annuli of X;
(iii) a set $\mathcal{D}$ of virtual open discs of X

such that $S \cup \mathcal{A} \cup \mathcal{D}$ is a partition of X.

Proof Each virtual open Swiss cheese may be written as a union of finitely many points of type 2, finitely many virtual open annuli and some virtual open discs (as in Example 6.3.5 below). By a combinatorial argument that is not difficult but quite lengthy, the covering so obtained can be turned into a partition.   □

Definition 6.3.3 Let X be a k-analytic $\mathbb{A}^1$-like curve. A partition $\mathcal{T} = (S, \mathcal{A}, \mathcal{D})$ of X satisfying the properties (i), (ii), (iii) of Theorem 6.3.2 is called a *triangulation* of X. The locally finite graph naturally arising from the set

$$\Sigma_{\mathcal{T}} := S \cup \bigcup_{A \in \mathcal{A}} \Sigma_A$$

is called the *skeleton* of $\mathcal{T}$. It is such that $X - \Sigma_{\mathcal{T}}$ is a disjoint union of virtual open discs.

A triangulation $\mathcal{T}$ is said to be *finite* if the associated set S is finite. If this is the case, then $\Sigma_{\mathcal{T}}$ is a finite graph. By the results of [PT20, Section 5.9], for each triangulation $\mathcal{T}$, $\Sigma_{\mathcal{T}}$ may be naturally endowed with a metric structure.

Remark 6.3.4 It is more usual to define a triangulation as the datum of the set S only. Note that S determines uniquely $\mathcal{A}$ and $\mathcal{D}$ since their elements are exactly the connected components of $X - S$, so our change of convention is harmless.

Example 6.3.5 Consider the curve

$$X := D^-(0, 1) - (D^+(a, r) \cup D^+(b, r))$$

for $r \in (0, 1)$ and $a, b \in k$ with $|a|, |b| < 1, |a - b| > r$. Set

$$S := \{\eta_{a,|a-b|}\},$$

$$\mathcal{A} := \{A^-(a, |a - b|, 1), A^-(a, r, |a - b|), A^-(b, r, |a - b|)\}$$

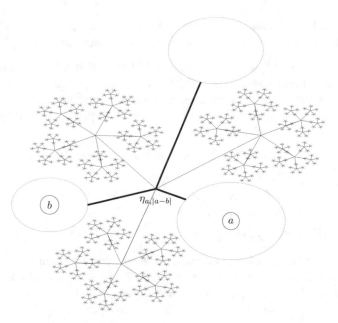

Fig. 6.1 The Swiss cheese X described in Example 6.3.5. Its skeleton Σ_X is the union of the three edges in evidence

and

$$\mathcal{D} := \{D^-(u, |a - b|), u \in k, |u - a| = |u - b| = |a - b|\}.$$

Then, the triple $\mathcal{T} := (S, \mathcal{A}, \mathcal{D})$ is a triangulation of X. The associated skeleton is a finite tree with three (half open) edges (Fig. 6.1).

Proposition 6.3.6 *Let X be a connected $\mathbb{A}^1$-like curve. Let $\mathcal{T} = (S, \mathcal{A}, \mathcal{D})$ be a triangulation of X such that $S \neq \emptyset$ or $\mathcal{A} \neq \emptyset$.*

There exists a canonical deformation retraction $\tau_{\mathcal{T}} \colon X \to \Sigma_{\mathcal{T}}$. Its restriction to any virtual open annulus $A \in \mathcal{A}$ induces the map τ_A from [PT20, Proposition 5.8.11] and its restriction to any connected component D of $A - \Sigma_A$ (which is a virtual open disc) induces the map τ_D from [PT20, Proposition 5.8.10].

In particular, for each $\eta \in \Sigma_A$, the set $\tau_{\mathcal{T}}^{-1}(\eta)$ is a virtual flat closed annulus. $\quad\square$

Definition 6.3.7 Let X be a k-analytic $\mathbb{A}^1$-like curve. The *skeleton* of X is the complement of all the virtual open discs contained in X. We denote it by Σ_X.

Remark 6.3.8 Let X be a k-analytic $\mathbb{A}^1$-like curve. It is not difficult to check that we have

$$\Sigma_X = \bigcap_{\mathcal{T}} \Sigma_{\mathcal{T}},$$

for $\mathcal{T}$ ranging over all triangulations of X. In particular, Σ_X is a locally finite metric graph (possibly empty).

Assume that X is connected and that Σ_X is non-empty. Then there exists a triangulation $\mathcal{T}_0$ of X such that $\Sigma_X = \Sigma_{\mathcal{T}_0}$. In particular, there is a canonical deformation retraction $\tau_X \colon X \to \Sigma_X$.

6.3.2   Arbitrary Smooth Curves

It goes beyond the scope of this survey to develop the full theory of Berkovich analytic curves. We only state a few definitions and general facts, to which we would like to refer later.

Definition 6.3.9 A *smooth k-analytic curve* is a locally ringed space X that is locally isomorphic to an open subset of a Berkovich spectrum of the form $\mathrm{Spec}^{\mathrm{an}}(A)$, where A is the ring of functions on a smooth affine algebraic curve over k.

For each smooth k-analytic curve X and each complete valued extension K of k, one may define the *base-change X_K* of X to K, by replacing each $\mathrm{Spec}^{\mathrm{an}}(A)$ by $\mathrm{Spec}^{\mathrm{an}}(A \otimes_k K)$ in its definition. It is a smooth K-analytic curve and there is a canonical projection morphism $\pi_{K/k} \colon X_K \to X$. The analogues of [PT20, Proposition 5.6.5] and [PT20, Corollary 5.6.6] hold in this more general setting.

Example 6.3.10 For each complete valued extension K of k, the base-change of $\mathbb{A}_k^{1,\mathrm{an}}$ to K is $\mathbb{A}_K^{1,\mathrm{an}}$.

If one starts with a smooth algebraic curve $\mathcal{X}$ over k, one may cover it by curves of the form $\mathrm{Spec}(A)$, with A as in Definition 6.3.9 above, and then glue the corresponding analytic spaces $\mathrm{Spec}^{\mathrm{an}}(A)$ to get a smooth k-analytic curve, called the *analytification* of $\mathcal{X}$, and denoted by $\mathcal{X}^{\mathrm{an}}$.

Example 6.3.11 The analytification of $\mathbb{A}_k^1$ is $\mathbb{A}_k^{1,\mathrm{an}}$.

As in the complex case, smooth compact k-analytic curves are automatically algebraic.

Theorem 6.3.12 *Let X be a smooth compact k-analytic curve. Then, there exists a projective smooth algebraic curve over k such that $X = \mathcal{X}^{\mathrm{an}}$.*

The invariants we have defined so far for the Berkovich affine line $\mathbb{A}_k^{1,\mathrm{an}}$ have natural counterparts for smooth k-analytic curves. Let X be a smooth k-analytic curve. For each point $x \in X$, the completed residue field $\mathcal{H}(x)$ is the completion of a finitely generated extension of k of transcendence degree less than or equal to 1. We may then define integers $s(x)$ and $t(x)$ such that $s(x) + t(x) \leqslant 1$ and the type of x, as we did in the case of $\mathbb{A}_k^{1,\mathrm{an}}$ (see [PT20, Definition 5.3.9]).

If x is of type 2, then, by the equality case in Abhyankar's inequality (see [PT20, Theorem 5.3.8]), the group $|\mathcal{H}(x)^\times|/|k^\times|$ is finitely generated, hence finite, and the field extension $\widetilde{\mathcal{H}(x)}/\tilde{k}$ is finitely generated.

Let us fix the definition of genus of an algebraic curve.

Definition 6.3.13 Let F be a field and let C be a projective curve over F, i.e. a connected normal projective scheme of finite type over F of dimension 1.

If F is algebraically closed, then C is smooth, and we define the *geometric genus of C* to be

$$g(C) := \dim_F H^0(C, \Omega_C).$$

In general, let $\bar{F}$ be an algebraic closure of F. Let C' be the normalization of a connected component of $C \times_F \bar{F}$. It is a projective curve over $\bar{F}$ and we define the *geometric genus of C* to be

$$g(C) := g(C').$$

It does not depend on the choice of C'.

Definition 6.3.14 Let X be a smooth k-analytic curve and let $x \in X$ be a point of type 2.

The *residue curve* at x is the unique (up to isomorphism) projective curve $\mathscr{C}_x$ over $\tilde{k}$ with function field $\widetilde{\mathcal{H}(x)}$. The *genus of x* is the geometric genus of $\mathscr{C}_x$. We denote it by $g(x)$.

The *stable genus of x*, is the genus of any point x' over x in $X_{\widehat{\bar{k}^a}}$. We denote it by $g_{\mathrm{st}}(x)$. It does not depend on the choice of x'.

Example 6.3.15 Let $\alpha \in k$ and $r \in |k^\times|^{\mathbb{Q}}$. By [PT20, Example 5.3.10], the residue curve at the point $\eta_{\alpha,r}$ in $\mathbb{A}_k^{1,\mathrm{an}}$ is the projective line $\mathbb{P}_{\tilde{k}}^1$ over $\tilde{k}$. In particular, we have $g(\eta_{\alpha,r}) = 0$.

By [PT20, Lemma 5.3.11], any point of type 2 in $\mathbb{A}_k^{1,\mathrm{an}}$ (hence in any k-analytic $\mathbb{A}^1$-like curve) has stable genus 0.

The fact that the stable genus does not need to coincide with the genus is what motivates our definition. Let us give an example of this phenomenon.

Remark 6.3.16 Let $p \geqslant 5$ be a prime number. Consider the affine analytic plane $\mathbb{A}_{\mathbb{Q}_p}^{2,\mathrm{an}}$ with coordinates x, y. Let X be the smooth $\mathbb{Q}_p$-analytic curve inside $\mathbb{A}_{\mathbb{Q}_p}^{2,\mathrm{an}}$ given by the equation $y^2 = x^3 + p$ and let $\pi \colon X \to \mathbb{A}_{\mathbb{Q}_p}^{1,\mathrm{an}}$ be the projection onto the first coordinate x.

The fiber $\pi^{-1}(\eta_{0,|p|^{-1/3}})$ contains a unique point, that we will denote by a. One may check that $\widetilde{\mathcal{H}(a)}$ is a purely transcendental extension of $\mathbb{F}_p$ generated by the class u of px^3 (which coincides with the class of py^2):

$$\widetilde{\mathcal{H}(a)} \simeq \mathbb{F}_p(u).$$

In particular, we have $\mathscr{C}_a = \mathbb{P}^1_{\mathbb{F}_p}$ and $g(a) = 0$.

Let us now extend the scalars to the field $\mathbb{C}_p$, whose residue field is an algebraic closure $\overline{\mathbb{F}}_p$ of $\mathbb{F}_p$. Let b be the unique point of $X_{\mathbb{C}_p}$ over a. The field $\widetilde{\mathcal{H}(b)}$ is now generated by the class v of $p^{-1/3}x$ and the class w of $p^{-1/2}y$:

$$\widetilde{\mathcal{H}(b)} \simeq \overline{\mathbb{F}}_p(v)[w]/(v^3 - w^2 + 1).$$

In particular, $\mathscr{C}_b$ is an elliptic curve over $\overline{\mathbb{F}}_p$, and we have $g_{st}(a) = g(b) = 1$.

We always have an inequality between genus and stable genus.

Lemma 6.3.17 *Let X be a smooth k-analytic curve and let $x \in X$ be a point of type 2. Then, we have $g(x) \leqslant g_{st}(x)$.*

Proof Let x' be a point of $X_{\widehat{k^a}}$ over x. By definition, the residue curve $\mathscr{C}_x$ at x is defined over $\tilde{k}$ and the residue curve $\mathscr{C}_{x'}$ at x' is defined over an algebraic closure $\overline{\tilde{k}}$ of $\tilde{k}$.

The projection morphism $\pi_{\widehat{k^a}/k} \colon X_{\widehat{k^a}} \to X$ induces an isometric embedding $\mathcal{H}(x) \to \mathcal{H}(x')$, hence an embedding $\widetilde{\mathcal{H}(x)} \to \widetilde{\mathcal{H}(x')}$. It follows that we have a morphism $\mathscr{C}_{x'} \to \mathscr{C}_x$, hence a morphism $\varphi \colon \mathscr{C}_{x'} \to \mathscr{C}_x \times_{\tilde{k}} \overline{\tilde{k}}$. Its image is a connected component C of $\mathscr{C}_x \times_{\tilde{k}} \overline{\tilde{k}}$. The morphism φ factors through C, and even through the normalization $\widetilde{C}$ of C. By definition, we have $g(\widetilde{C}) = g(x)$ and $g(\mathscr{C}_{x'}) = g_{st}(x)$. The result now follows from the Riemann–Hurwitz formula. $\square$

Proposition 6.3.18 *Let X be a smooth k-analytic curve and let $x \in X$ be a point of type 2. There is a natural bijection between the closed points of the residue curve $\mathscr{C}_x$ at x and the set of directions emanating from x in X.* $\square$

Example 6.3.19 Assume that k is algebraically closed. For $X = \mathbb{A}^{1,\mathrm{an}}_k$ and $x = \eta_1$, the result of Proposition 6.3.18 follows from [PT20, Lemma 5.4.9].

The structure of smooth k-analytic curves is well understood.

Theorem 6.3.20 *Every smooth k-analytic curve admits a triangulation in the sense of Theorem 6.3.2.*

The result of Proposition 6.3.6 also extends. If $\mathcal{T}$ is a non-empty triangulation of a smooth connected k-analytic curve X, then there is a canonical deformation retraction of X onto the skeleton $\Sigma_{\mathcal{T}}$ of $\mathcal{T}$, which is a locally finite metric graph.

We may also define the skeleton of X as in Definition 6.3.7, and it satisfies the properties of Remark 6.3.8.

Remark 6.3.21 With this purely analytic formulation, Theorem 6.3.20 is due to A. Ducros, who provided a purely analytic proof in [Duc]. It is very closely related to the semi-stable reduction theorem of S. Bosch and W. Lütkebohmert (see [BL85]): for each smooth k-analytic curve X, there exists a finite extension ℓ/k such that X_ℓ admits a model over ℓ° whose special fiber is a semi-stable curve over $\tilde{\ell}$, that is, it is reduced and its singularities are at worst double nodes.

If a smooth k-analytic curve X admits a semi-stable model over k°, then we may associate to it a triangulation of X. The points of S, $\mathcal{A}$ and $\mathcal{D}$ then correspond respectively to the irreducible components, the singular points and the smooth points of the special fiber of the model. Moreover, the genus of a point of S (which, in this case, coincides with its stable genus) is equal to the genus of the corresponding component. We refer to [Ber90, Theorem 4.3.1] for more details.

In the other direction, it is always possible to associate a model over k° to a triangulation of X, but it may fail to be semi-stable in general. The reader may consult [Duc, Sections 6.3 and 6.4] for general results.

Definition 6.3.22 Assume that k is algebraically closed. Let X be a smooth connected k-analytic curve. We define the *genus of* X to be

$$g(X) := b_1(X) + \sum_{x \in X^{(2)}} g(x),$$

where $b_1(X)$ is the first Betti number of X and $X^{(2)}$ the set of type 2 points of X.

If k is arbitrary, we define the genus of a smooth geometrically connected k-analytic curve X to be the genus of $X_{\widehat{k^a}}$.

This notion of genus is compatible with the one defined in the algebraic setting.

Theorem 6.3.23 *For each smooth geometrically connected projective algebraic curve $\mathcal{X}$ over k, we have*

$$g(\mathcal{X}) = g(\mathcal{X}^{\mathrm{an}}).$$

Let us finally comment that, among the results that are presented here, Theorem 6.3.20 is deep and difficult, but we will not need to use it since an easier direct proof is available for k-analytic $\mathbb{A}^1$-like curves (see Theorem 6.3.2). The others are rather standard applications of the general theory of curves.

6.3.3 Mumford Curves

Let us now return to $\mathbb{A}^1$-like curves over k. A special kind of such curves is obtained by asking for the existence of an open covering made of actual open Swiss cheeses over k rather than virtual ones. Recall that open Swiss cheeses over k are defined as the complement of closed discs in an open disc over k.

Definition 6.3.24 A connected, compact k-analytic ($\mathbb{A}^1$-like) curve X is called a k-*analytic Mumford curve* if every point $x \in X$ has a neighborhood that is isomorphic to an open Swiss cheese over k.

Remark 6.3.25 Such a curve is automatically projective algebraic by Theorem 6.3.12.

The following proposition relates the definition of a k-analytic Mumford curve with the existence of a triangulation of a certain type, and therefore with the original algebraic definition given by Mumford in [Mum72a]. Its proof uses some technical notions that were not fully presented in the first sections of this text, but we believe that the result of the proposition is important enough to deserve to be fully included for completeness.

Proposition 6.3.26 *Let X be a compact k-analytic curve.*

If $g(X) = 0$, then X is a k-analytic Mumford curve if and only if X is isomorphic to $\mathbb{P}_k^{1,\mathrm{an}}$.

If $g(X) \geqslant 1$, then X is a k-analytic Mumford curve if and only if there exists a triangulation $(S, \mathcal{A}, \mathcal{D})$ of X such that the points of S are of stable genus 0 and the elements of $\mathcal{A}$ are open annuli.

Proof

- Assume that $g(X) = 0$. If X is isomorphic to $\mathbb{P}_k^{1,\mathrm{an}}$, then it is obviously a Mumford curve.

  Conversely, assume that X is a k-analytic Mumford curve. By Theorems 6.3.12 and 6.3.23, it is isomorphic to the analytification of a projective smooth algebraic curve over k. Therefore, to prove that it is isomorphic to $\mathbb{P}_k^{1,\mathrm{an}}$, it is enough to prove that it has a k-rational point.

  By assumption, X contains an open Swiss cheese over k. In particular, it contains an open annulus A over k. Let x be a boundary point of the skeleton of A. By assumption, x has a neighborhood that is isomorphic to an open Swiss cheese over k. It follows that A is contained in a strictly bigger annulus A' whose skeleton strictly contains that of A. Arguing this way (possibly considering the union of all the annuli and applying the argument again), we show that X contains an open annulus over k of infinite modulus. At least one of its boundary points is a k-rational point, and the result follows.

- Assume that $g(X) \geqslant 1$. If X is a k-analytic Mumford curve, then it may be covered by finitely many Swiss cheeses over k. The result then follows from the fact that every Swiss cheese over k admits a triangulation $(S, \mathcal{A}, \mathcal{D})$ such that the points of S are of stable genus 0 and the elements of $\mathcal{A}$ are annuli.

Conversely, assume that there exists a triangulation $(S, \mathcal{A}, \mathcal{D})$ of X satisfying the properties of the statement. Since $g(X) \geqslant 1$, we have $\mathcal{A} \neq \emptyset$. Up to adding a point of S in the skeleton of each element of $\mathcal{A}$, we may assume that all the elements of $\mathcal{A}$ have two distinct endpoints in X.

Let $x \in S$. Denote by $\mathcal{D}_x$ (resp. $\mathcal{A}_x$) the set of elements of $\mathcal{D}$ (resp. $\mathcal{A}$) that have x as an endpoint and set

$$U_x := \{x\} \cup \bigcup_{D \in \mathcal{D}_x} D \cup \bigcup_{A \in \mathcal{A}_x} A.$$

It is an open neighborhood of x in X. Let us now enlarge U_x in the following way: for each $A \in \mathcal{A}_x$, we paste a closed disc at the extremity of A that is different from x. The resulting curve V_x is compact, hence the analytification of a projective smooth algebraic curve over k, by Theorem 6.3.12. Since x is of stable genus 0, the genus of the base-change $(V_x)_{\widehat{k^a}}$ of V_x to $\widehat{k^a}$ is 0. By Theorem 6.3.23, we deduce that $(V_x)_{\widehat{k^a}}$ is isomorphic to $\mathbb{P}^{1,\mathrm{an}}_{\widehat{k^a}}$. Since V_x contains k-rational points (inside the pasted discs, for instance), V_x itself is isomorphic to $\mathbb{P}^{1,\mathrm{an}}_k$. We deduce that U_x is a Swiss cheese over k.

Since any point of X has a neighborhood that is of the form U_x for some $x \in S$, it follows that X is a Mumford curve.

$\square$

Remark 6.3.27 If X is a compact k-analytic curve and k is algebraically closed, then Proposition 6.3.26 shows that the following properties are equivalent:

(i) X is a Mumford curve;
(ii) X is an $\mathbb{A}^1$-like curve;
(iii) the points of type 2 of X are all of genus 0.

Remark 6.3.28 Using the correspondence between triangulations and semi-stable models (see Remark 6.3.21), the result of Proposition 6.3.26 says that k-analytic Mumford curves are exactly those for which there exists a semi-stable model over k° whose special fiber consists of projective lines over $\tilde{k}$, intersecting transversally in $\tilde{k}$-rational points. This is indeed how algebraic Mumford curves are defined in Mumford's paper [Mum72a].

Corollary 6.3.29 *Let X be a k-analytic Mumford curve and $\mathcal{T}$ be a triangulation of X. Then the following quantities are equal:*

(i) the genus of X;
(ii) the cyclomatic number of the skeleton $\Sigma_{\mathcal{T}}$;
(iii) the first Betti number of X.

Proof We may assume that $\mathcal{T} = (S, \mathcal{A}, \mathcal{D})$ satisfies the conclusions of Proposition 6.3.26. We will assume that $\mathcal{A} \neq \emptyset$, the other case being dealt with similarly.

Consider the base-change morphism $\pi_{\widehat{k^a}/k}\colon X_{\widehat{k^a}} \to X$. By assumption, every element A of $\mathcal{A}$ is an annulus over k, hence its preimage $\pi_{\widehat{k^a}/k}^{-1}(A)$ is an annulus over $\widehat{k^a}$. In particular, $\pi_{\widehat{k^a}/k}$ induces a homeomorphism between the skeleton of $\pi_{\widehat{k^a}/k}^{-1}(A)$ and that of A. Since each point of S lies at the boundary of the skeleton of an element of $\mathcal{A}$, we deduce that each point of S has exactly one preimage by $\pi_{\widehat{k^a}}$.

It follows that the set $\mathcal{T}' = (S', \mathcal{A}', \mathcal{D}')$ of $X_{\widehat{k^a}}$, where

- S' is the set of preimages of the elements of S by $\pi_{\widehat{k^a}/k}$;
- $\mathcal{A}'$ is the set of preimages of the elements of $\mathcal{A}$ by $\pi_{\widehat{k^a}/k}$;
- $\mathcal{D}'$ is the set of connected components of the preimages of the elements of $\mathcal{D}$ by $\pi_{\widehat{k^a}/k}$

is a triangulation of $X_{\widehat{k^a}}$ and, moreover, that $\pi_{\widehat{k^a}/k}$ induces a homeomorphism between the skeleta $\Sigma_{\mathcal{T}'}$ and $\Sigma_{\mathcal{T}}$. In particular, their cyclomatic numbers are equal.

Since X is a Mumford curve, all the points of type 2 of the curve $X_{\widehat{k^a}}$ are of genus 0, hence the genus of $X_{\widehat{k^a}}$ coincides with its first Betti number, hence with the cyclomatic number of $\Sigma_{\mathcal{T}'}$, by Proposition 6.3.6. The equality between (i) and (ii) follows.

The equality between (ii) and (iii) follows from Proposition 6.3.6 again. $\qquad\square$

6.4 Schottky Groups

Let $(k, |\cdot|)$ be a complete valued field. Some of the material of this section is adapted from Mumford [Mum72a], Gerritzen and van der Put [GvdP80] and Berkovich [Ber90, Section 4.4].

6.4.1 Schottky Figures

Let $g \in \mathbb{N}_{\geqslant 1}$.

Definition 6.4.1 Let $\gamma_1, \ldots, \gamma_g \in \mathrm{PGL}_2(k)$. Let $\mathcal{B} = (D^+(\gamma_i^\varepsilon), 1 \leqslant i \leqslant g, \varepsilon \in \{\pm 1\})$ be a family of pairwise disjoint closed discs in $\mathbb{P}_k^{1,\mathrm{an}}$. For each $i \in \{1, \ldots, g\}$ and $\varepsilon \in \{-1, 1\}$, set

$$D^-(\gamma_i^\varepsilon) := \gamma_i^\varepsilon(\mathbb{P}_k^1 - D^+(\gamma_i^{-\varepsilon})).$$

We say that $\mathcal{B}$ is a *Schottky figure* adapted to $(\gamma_1, \ldots, \gamma_g)$ if, for each $i \in \{1, \ldots, g\}$ and $\varepsilon \in \{-1, 1\}$, $D^-(\gamma_i^\varepsilon)$ is a maximal open disc inside $D^+(\gamma_i^\varepsilon)$. (See Fig. 6.2 for an illustration.)

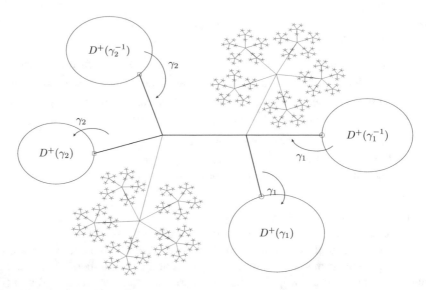

Fig. 6.2 A Schottky figure adapted to a pair (γ_1, γ_2)

Remark 6.4.2 Let $i \in \{1, \ldots, g\}$. It follows from Remark 6.2.12 that γ_i is loxodromic. Moreover, denoting by α_i and α'_i the attracting and repelling fixed points of γ_i respectively, we have

$$\alpha'_i \in D^-(\gamma_i^{-1}) \text{ and } \alpha_i \in D^-(\gamma_i).$$

The result is easily proven for $\gamma = \begin{bmatrix} 1 & 0 \\ 0 & q \end{bmatrix}$ and one may reduce to this case by choosing a suitable coordinate on $\mathbb{P}_k^{1,\mathrm{an}}$.

For the rest of the section, we fix $\gamma_1, \ldots, \gamma_g \in \mathrm{PGL}_2(k)$ and a Schottky figure adapted to $(\gamma_1, \ldots, \gamma_g)$, with the notation of Definition 6.4.1.

Notation 6.4.3 For $\sigma \in \{-, +\}$, we set

$$F^\sigma := \mathbb{P}_k^1 - \bigcup_{\substack{1 \leqslant i \leqslant g \\ \varepsilon = \pm 1}} D^{-\sigma}(\gamma_i^\varepsilon).$$

Note that, for $\gamma \in \{\gamma_1^{\pm 1}, \ldots, \gamma_g^{\pm 1}\}$, $D^+(\gamma)$ is the unique disc that contains $\gamma(F^+)$ among those defining the Schottky figure.

Remark 6.4.4 The sets F^- and F^+ are open and closed Swiss cheeses respectively.

Denote by ∂F^+ the boundary of F^+ in $\mathbb{P}_k^{1,\mathrm{an}}$. It is equal to the set of boundary points of the $D^+(\gamma_i^{\pm 1})$'s, for $i \in \{1, \ldots, g\}$. The skeleton Σ_{F^+} of F^+ is the convex envelope of ∂F^+, that is to say the minimal connected graph containing ∂F^+, or

$$\Sigma_{F^+} = \bigcup_{x,y \in \partial F^+} [x, y].$$

The skeleton Σ_{F^-} of F^- satisfies

$$\Sigma_{F^-} = \Sigma_{F^+} \cap F^- = \Sigma_{F^+} - \partial F^+.$$

Set $\Delta := \{\gamma_1, \ldots, \gamma_g\}$. Denote by F_g the abstract free group with set of generators Δ and by Γ the subgroup of $\mathrm{PGL}_2(k)$ generated by Δ. The existence of a Schottky figure for the g-tuple $(\gamma_1, \ldots, \gamma_g)$ determines important properties of the group Γ. In fact, we have a natural morphism $\varphi \colon F_g \to \Gamma$ inducing an action of F_g on $\mathbb{P}_k^{1,\mathrm{an}}$. We now define a disc in $\mathbb{P}_k^1$ associated with each element of F_g. As usual, we will identify these elements with the words over the alphabet $\Delta^\pm := \{\gamma_1^{\pm 1}, \ldots, \gamma_g^{\pm 1}\}$.

Notation 6.4.5 For a non-empty reduced word $w = w'\gamma$ over Δ and $\sigma \in \{-, +\}$, we set

$$D^\sigma(w) := w' D^\sigma(\gamma).$$

Lemma 6.4.6 *Let u be a non-empty reduced word over $\Delta^\pm$. Then we have $uF^+ \subseteq D^+(u)$.*

Let v be a non-empty reduced word over $\Delta^\pm$. If there exists a word w over $\Delta^\pm$ such that $u = vw$, then we have $uF^+ \subseteq D^+(u) \subseteq D^+(v)$. If, moreover, $u \neq v$, then we have $D^+(u) \subseteq D^-(v)$.

Conversely, if we have $D^+(u) \subseteq D^+(v)$, then there exists a word w over $\Delta^\pm$ such that $u = vw$.

Proof Write in a reduced form $u = u'\gamma$ with $\gamma \in \Delta^\pm$. We have $\gamma F^+ \subseteq D^+(\gamma)$, by definition. Applying u', it follows that $uF^+ \subseteq D^+(u)$.

Assume that there exists a word w such that $u = vw$ and let us prove that $D^+(u) \subseteq D^+(v)$. We first assume that v is a single letter. We will argue by induction on the length $|u|$ of u. If $|u| = 1$, then $u = v$ and the result is trivial. If $|u| \geqslant 2$, denote by δ the first letter of w. By induction, we have $D^+(w) \subseteq D^+(\delta)$. Since $\delta \neq v^{-1}$, we also have $D^+(\delta) \subseteq \mathbb{P}_k^1 - D^+(v^{-1})$. The result follows by applying v.

Let us now handle the general case. Write in a reduced form $v = v'\gamma$ with $\gamma \in \Delta^\pm$. By the former case, we have $D^+(\gamma w) \subseteq D^+(\gamma)$ and $D^+(\gamma w) \subseteq D^-(\gamma)$ if w is non-empty. The result follows by applying v'.

Assume that we have $D^+(u) \subseteq D^+(v)$. We will prove that there exists a word w such that $u = vw$ by induction on $|v|$. Write in reduced forms $u = \gamma u'$ and $v = \delta v'$. By the previous result, we have $D^+(u) \subseteq D^+(\gamma)$ and $D^+(v) \subseteq D^+(\delta)$, hence $\gamma = \delta$. If $|v| = 1$, this proves the result. If $|v| \geqslant 2$, then we deduce that we have

$D^+(u') \subseteq D^+(v')$, hence, by induction, there exists a word w such that $u' = v'w$. It follows that $u = vw$. $\square$

Proposition 6.4.7 *The morphism φ is an isomorphism and the group Γ is free on the generators $\gamma_1, \ldots, \gamma_g$.*

Proof If w is a non-empty word, then the previous lemma ensures that $wF^+ \neq F^+$. The result follows. $\square$

As a consequence, we now identify Γ with F_g and express the elements of Γ as words over the alphabet $\Delta^\pm$. In particular, we allow us to speak of the length of an element γ of Γ, that we denote by $|\gamma|$. Set

$$O_n := \bigcup_{|\gamma| \leqslant n} \gamma F^+.$$

Since the complement of F^+ is the disjoint union of the open disks $D^-(\gamma)$ with $\gamma \in \Delta^\pm$, it follows from the description of the action that, for each $n \geqslant 0$, we have

$$\mathbb{P}_k^{1,\mathrm{an}} - O_n = \bigsqcup_{|w|=n+1} D^-(w).$$

It follows from Lemma 6.4.6 that, for each $n \geqslant 0$, O_n is contained in the interior of O_{n+1}. We set

$$O := \bigcup_{n \geqslant 0} O_n = \bigcup_{\gamma \in \Gamma} \gamma F^+.$$

We now compute the orbits of discs under Möbius transformations $\mathbb{P}_k^{1,\mathrm{an}}$. Set $\iota := \begin{bmatrix} 0 & 1 \\ 1 & 0 \end{bmatrix} \in \mathrm{PGL}_2(k)$. It corresponds to the map $z \mapsto 1/z$ on $\mathbb{P}_k^{1,\mathrm{an}}$. The first result follows from an explicit computation.

Lemma 6.4.8 *Let $\alpha \in k^\times$ and $\rho \in [0, |\alpha|)$. Then, we have $\iota\big(D^+(\alpha, \rho)\big) = D^+\big(\frac{1}{\alpha}, \frac{\rho}{|\alpha|^2}\big)$.* $\square$

Lemma 6.4.9 *Let $r > 0$ and let $\gamma = \begin{bmatrix} a & b \\ c & d \end{bmatrix}$ in $\mathrm{PGL}_2(k)$ such that $\gamma\big(D^+(0, r)\big) \subseteq \mathbb{A}_k^{1,\mathrm{an}}$. Then, we have $|d| > r|c|$ and $\gamma\big(D^+(0, r)\big) = D^+\big(\frac{b}{d}, \frac{|ad-bc|r}{|d|^2}\big)$.*

Proof Let us first assume that $c = 0$. Then, we have $d \neq 0$, so the inequality $|d| > r|c|$ holds, and γ is affine with ratio a/d. The result follows.

Let us now assume that $c \neq 0$. In this case, we have $\gamma^{-1}(\infty) = -\frac{d}{c}$, which does not belong to $D(0, r)$ if, and only if, $|d| > r|c|$. Note that we have the following equality in $k(T)$:

$$\frac{aT + b}{cT + d} = \frac{a}{c} - \frac{ad - bc}{c^2} \frac{1}{T + \frac{d}{c}}.$$

By Lemma 6.4.8, there exist $\beta \in k$ and $\sigma > 0$ such that $\iota\big(D^+(\frac{d}{c}, r)\big) = D^+(\beta, \sigma)$. Then, we have $\gamma\big(D^+(0, r)\big) = D^+(\frac{a}{c} - \frac{ad-bc}{c^2}\beta, \big|\frac{ad-bc}{c^2}\big|\sigma)$ and the result follows from an explicit computation.   □

Lemma 6.4.10 *Let $D' \subseteq D$ be closed discs in $\mathbb{A}_k^{1,\mathrm{an}}$. Let $\gamma \in \mathrm{PGL}_2(k)$ such that $\gamma D' \subseteq \gamma D \subseteq \mathbb{A}_k^{1,\mathrm{an}}$. Then, we have*

$$\frac{\text{radius of } \gamma D'}{\text{radius of } \gamma D} = \frac{\text{radius of } D'}{\text{radius of } D}.$$

Proof Let p be a k-rational point in D' and let τ be the translation sending p to 0. Up to changing D into τD, D' into $\tau D'$, γ into $\gamma\tau^{-1}$ and γ' into $\gamma'\tau^{-1}$, we may assume that D and D' are centered at 0. The result then follows from Lemma 6.4.9.   □

Proposition 6.4.11 *Assume that $\infty \in F^-$. Then, there exist $R > 0$ and $c \in (0, 1)$ such that, for each $\gamma \in \Gamma - \{\mathrm{id}\}$, $D^+(\gamma)$ is a closed disc of radius at most $R\,c^{|\gamma|}$.*

Proof Let $\delta, \delta' \in \Delta^{\pm}$ such that $\delta' \neq \delta^{-1}$. By Lemma 6.4.6, we have $D^+(\delta'\delta) \subset D^-(\delta') \subseteq D^+(\delta')$. Set

$$c_{\delta,\delta'} := \frac{\text{radius of } D^+(\delta'\delta)}{\text{radius of } D^+(\delta')} \in (0, 1).$$

For each $\gamma \in \Gamma$ such that $\gamma\delta'$ is a reduced word, by Lemma 6.4.10, we have

$$\frac{\text{radius of } D^+(\gamma\delta'\delta)}{\text{radius of } D^+(\gamma\delta')} = \frac{\text{radius of } \gamma D^+(\delta'\delta)}{\text{radius of } \gamma D^+(\delta')} = c_{\delta,\delta'}.$$

Set

$$R := \max(\{\text{radius of } D^+(\delta) \mid \delta \in \Delta^{\pm}\})$$

and

$$c := \max(\{c_{\delta,\delta'} \mid \delta, \delta' \in \Delta^{\pm}, \delta' \neq \delta^{-1}\}).$$

By induction, for each $\gamma \in \Gamma - \{\mathrm{id}\}$, we have

$$\text{radius of } D^+(\gamma) \leqslant R\, c^{|\gamma|}.$$

□

Corollary 6.4.12 *Every element of* $\Gamma - \{\mathrm{id}\}$ *is loxodromic.*

Proof In order to prove the result, we may extend the scalars. As a result, we may assume that $F^- \cap \mathbb{P}_k^{1,\mathrm{an}}(k) \neq \emptyset$, hence up to changing coordinates, that $\infty \notin F^-$. Let $\gamma \in \Gamma - \{\mathrm{id}\}$. By Proposition 6.4.11 the radii of the discs $\gamma^n(D^+(\gamma))$ tend to 0 when n tends to ∞, which forces γ to be loxodromic, by Remark 6.2.12.   □

Corollary 6.4.13 *Let* $w = (w_n)_{\neq 0}$ *be a sequence of reduced words over* $\Delta^\pm$ *such that the associated sequence of discs* $(D^+(w_n))_{n \geqslant 0}$ *is strictly decreasing. Then, the intersection* $\bigcap_{n \geqslant 0} D^+(w_n)$ *is a single k-rational point* p_w. *Moreover, the discs* $D^+(w_n)$ *form a basis of neighborhoods of* p_w *in* $\mathbb{P}_k^{1,\mathrm{an}}$.

Proof Let k_0 be a finite extension of k such that $F^- \cap \mathbb{P}^1(k_0) \neq \emptyset$. Consider the projection morphism $\pi_0 : \mathbb{P}_{k_0}^{1,\mathrm{an}} \to \mathbb{P}_k^{1,\mathrm{an}}$. For each $i \in \{1, \ldots, g\}$, γ_i may be identified with an element $\gamma_{i,0}$ in $\mathrm{PGL}_2(k_0)$. The family $(\pi_0^{-1}(D^-(\gamma_i^{\pm 1}), 1 \leqslant i \leqslant g, \varepsilon = \pm 1)$ is a Schottky figure adapted to $(\gamma_{1,0}, \ldots, \gamma_{g,0})$. We will denote with a subscript 0 the associated sets: F_0^-, $D_0^+(w)$, etc. Note that these sets are all equal to the preimages of the corresponding sets by π_0.

Up to changing coordinates on $\mathbb{P}_{k_0}^{1,\mathrm{an}}$, we may assume that $\infty \in F_0^-$. The sequence of discs $(D_0^+(w_n))_{n \geqslant 0}$ is strictly decreasing, so by Lemma 6.4.6, the length of w_n tends to ∞ when n goes to ∞ and, by Proposition 6.4.11, the radius of $D_0^+(w_n)$ tends to 0 when n goes to ∞. It follows that $\bigcap_{n \geqslant 0} D_0^+(w_n)$ is a single point $p_{w,0}$ of type 1 and that the discs $D_0^+(w_n)$ form a basis of neighborhood of $p_{w,0}$ in $\mathbb{P}_{k_0}^{1,\mathrm{an}}$.

Set $p_w := \pi_0(p_{w,0})$. It follows from the results over k_0 that $\bigcap_{n \geqslant 0} D^+(w_n) = \{p_w\}$ and that the discs $D^+(w_n)$ form a basis of neighborhoods of p_w in $\mathbb{P}_k^{1,\mathrm{an}}$.

It remains to show that p_w is k-rational. Note that p_w belongs to the closure of $\mathbb{P}^1(k)$, since it is the limit of the centers of the $D^+(w_n)$'s. Since k is complete, $\mathbb{P}^1(k)$ is closed in $\mathbb{P}^1(\widehat{k^a})$ and the result follows.   □

Corollary 6.4.14 *The set* O *is dense in* $\mathbb{P}_k^{1,\mathrm{an}}$ *and its complement is contained in* $\mathbb{P}^1(k)$.   □

Definition 6.4.15 We say that a point $x \in \mathbb{P}_k^{1,\mathrm{an}}$ is a *limit point* if there exist a point $x_0 \in \mathbb{P}_k^{1,\mathrm{an}}$ and a sequence $(\gamma_n)_{n \geqslant 0}$ of distinct elements of Γ such that $\lim_{n \to \infty} \gamma_n(x_0) = x$.

The *limit set* L of Γ is the set of limit points of Γ.

Let us add a short reminder on proper group actions.

Definition 6.4.16 ([Bou71, III, §4, Définition 1]) We say that the action of a topological group G on a topological space E is *proper* if the map

$$\Gamma \times E \to E \times E$$
$$(\gamma, x) \mapsto (x, \gamma \cdot x)$$

is proper.

Proposition 6.4.17 ([Bou71, III, §4, Propositions 3 and 7]) *Let G be a locally compact topological group and E be a Hausdorff topological space. Then, the action of G on E is proper if, and only if, for every $x, y \in E$, there exist neighborhoods U_x and U_y of x and y respectively such that the set $\{\gamma \in \Gamma \mid \gamma U_x \cap U_y \neq \emptyset\}$ is relatively compact (that is to say finite, if G is discrete).*

In this case, the quotient space $\Gamma \backslash E$ is Hausdorff.   □

We denote by C the set of points $x \in \mathbb{P}_k^{1,\mathrm{an}}$ that admit a neighborhood U_x satisfying $\{\gamma \in \Gamma : \gamma U_x \cap U_x \neq \emptyset\} = \{\mathrm{id}\}$. The set C is an open subset of $\mathbb{P}_k^{1,\mathrm{an}}$ and the quotient map $C \to \Gamma \backslash C$ is a local homeomorphism. In particular, the topological space $\Gamma \backslash C$ is naturally endowed with a structure of analytic space via this map.

Theorem 6.4.18 *We have $O = C = \mathbb{P}_k^{1,\mathrm{an}} - L$. Moreover, the action of Γ on O is free and proper and the quotient $\Gamma \backslash O$ is a Mumford curve of genus g.*

Set $X := \Gamma \backslash O$ and denote by $p \colon O \to X$ the quotient map. Let Σ_O, Σ_{F^+} and Σ_X denote the skeleta of O, F^+ and X respectively. Then, Σ_O is the trace on O of the convex envelope of L:

$$\Sigma_O = O \cap \bigcup_{x,y \in L} [x, y]$$

and we have

$$p^{-1}(\Sigma_X) = \Sigma_O \text{ and } p(\Sigma_O) = p(\Sigma_{F^+}) = \Sigma_X.$$

(See Fig. 6.3 for an illustration.)

Proof Let $x \in L$. By definition, there exist $x_0 \in \mathbb{P}_k^{1,\mathrm{an}}$ and a sequence $(\gamma_n)_{n \geqslant 0}$ of distinct elements of Γ such that $\lim_{n \to \infty} \gamma_n(x_0) = x$. Assume that $x \in F^+$. Since F^+ is contained in the interior of O_1, there exists $N \geqslant 0$ such that $\gamma_N(x_0) \in O_1$, hence we may assume that $x_0 \in O_1$. Lemma 6.4.6 then leads to a contradiction. It follows that L does not meet F^+, hence, by Γ-invariance, L is contained in $\mathbb{P}_k^{1,\mathrm{an}} - O$.

Let $y \in \mathbb{P}_k^{1,\mathrm{an}} - O$. By definition, there exists a sequence $(w_n)_{n \geqslant 0}$ of reduced words over $\Delta^{\pm}$ such that, for each $n \geqslant 0$, $|w_n| \geqslant n$ and $y \in D^-(w_n)$. Let $y_0 \in F^-$. By Lemma 6.4.6, for each $n \geqslant 0$, we have $w_n(y_0) \in D^-(w_n)$ and the sequence of

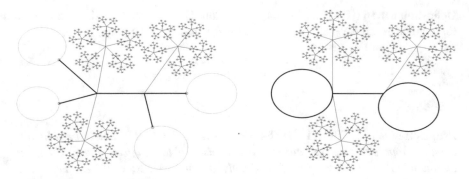

Fig. 6.3 The closed fundamental domain F^+ (on the left) of the Schottky group Γ is a Swiss cheese. The group Γ identifies the ends of the skeleton Σ_{F^+}, so that the corresponding Mumford curve (on the right) contains the finite graph Σ_X

discs $(D^+(w_n))_{n \geqslant 0}$ is strictly decreasing. By Corollary 6.4.13, $(w_n(y_0))_{n \geqslant 0}$ tends to y, hence $y \in L$. It follows that $\mathbb{P}^{1,\mathrm{an}}_k - O = L$.

Set

$$U := F^+ \cup \bigcup_{\gamma \in \Delta^{\pm}} \gamma F^- = \mathbb{P}^{1,\mathrm{an}}_k - \bigsqcup_{|\gamma|=2} D^+(\gamma).$$

It is an open subset of $\mathbb{P}^{1,\mathrm{an}}_k$ and it follows from the properties of the action (see Lemma 6.4.6) that we have $\{\gamma \in \Gamma \mid \gamma U \cap U \neq \emptyset\} = \{\mathrm{id}\} \cup \Delta^{\pm}$. Using the fact that the stabilizers of the points of U are trivial, we deduce that $U \subseteq C$. Letting Γ act, it follows that $O \subseteq C$. Since no limit point may belong to C, we deduce that this is actually an equality.

We have already seen that the action is free on O. Let us prove that it is proper. Let $x, y \in O$. There exists $n \geqslant 0$ such that x and y belong to the interior of O_n. By Lemma 6.4.6, the set $\{\gamma \in \Gamma : \gamma O_n \cap O_n \neq \emptyset\}$ is made of elements of length at most $2n + 1$. In particular, it is finite. We deduce that the action of Γ on O is proper.

The compact subset F^+ of $\mathbb{P}^{1,\mathrm{an}}_k$ contains a point of every orbit of every element of O. It follows that $\Gamma \backslash O$ is compact. The set F^- is an open k-Swiss cheese and the map p is injective on it, which implies that $p_{|F^-}$ induces an isomorphism onto its image. In addition, one may check that each subset of the form $D^+(\gamma) - D^-(\gamma)$ for $\gamma \in \{\gamma_1^{\pm 1}, \ldots, \gamma_g^{\pm 1}\}$ is contained in an open k-annulus on which p is injective. It follows that any element of $\Gamma \backslash O$ has a neighborhood isomorphic to a k-Swiss cheese, hence $\Gamma \backslash O$ is a Mumford curve.

Set $\Sigma := O \cap \bigcup_{x,y \in L} [x, y]$. It is clear that no point of Σ is contained in a virtual open disc inside O, hence $\Sigma \subseteq \Sigma_O$. It follows from [PT20, Proposition 5.7.10] that $\mathbb{P}^{1,\mathrm{an}}_k - \Sigma$ is a union of virtual open discs, hence $\Sigma_O \cap (\mathbb{P}^{1,\mathrm{an}}_k - \Sigma) = \emptyset$. We deduce that $\Sigma_O = \Sigma$. Note that it follows that $\Sigma_{F^+} = \Sigma_O \cap F^+$.

Let $x \in O - \Sigma_O$. Then x is contained in a virtual open disc inside O. Assume that there exists $\gamma \in \Gamma$ such that $x \in \gamma F^-$. Then, the said virtual open disc is contained in γF^-. Since $p_{|\gamma F^-}$ induces an isomorphism onto its image, $p(x)$ is contained in a virtual open disc in X, hence $p(x) \notin \Sigma_X$. As above, the argument may be adapted to handle all the points of $O - \Sigma_O$. It follows that $p^{-1}(\Sigma_X) \subseteq \Sigma_O$.

Let $x \in \Sigma_O$. In order to show that $p(x) \in \Sigma_X$, we may replace x by $\gamma(x)$ for any $\gamma \in \Gamma$, hence assume that $x \in F^+ \cap \Sigma_O = \Sigma_{F^+}$. From the explicit description of the action of Γ on F^+, we may describe precisely the behaviour of p on $\Sigma_{F^+} = \Sigma_{F^-} \cup \partial F^+$: it is injective on Σ_{F^-} and identifies pairs of points in ∂F^+. It follows that $p(x)$ belongs to a injective loop inside X and Remark 6.3.8 then ensures that $p(x) \in \Sigma_X$. The results about the skeleta follow directly.

It remains to prove that the genus of $X = \Gamma \backslash O$ is equal to g. The arguments above show that $\Sigma_X \simeq \Gamma \backslash \Sigma_{F^+}$ is a graph with cyclomatic number g. The result now follows from Corollary 6.3.29.   □

Example 6.4.19 (Tate Curves) If $g = 1$ in the theory above, one starts with the data of an element $\gamma \in \mathrm{PGL}_2(k)$ and of two disjoint closed discs $D^+(\gamma)$ and $D^+(\gamma^{-1})$ in such a way that $\gamma(\mathbb{P}_k^{1,\mathrm{an}} - D^+(\gamma^{-1}))$ is a maximal open disc inside $D^+(\gamma)$. Since γ is loxodromic, up to conjugation, it is represented by a matrix of the form $\begin{bmatrix} q & 0 \\ 0 & 1 \end{bmatrix}$ for some $q \in k$ satisfying $0 < |q| < 1$. In other words, up to a change of coordinate in $\mathbb{P}_k^{1,\mathrm{an}}$, the transformation γ is the multiplication by q and hence the limit set L consists only of the two points 0 and ∞. The quotient curve obtained from applying Theorem 6.4.18 is an elliptic curve, whose set of k-points is isomorphic to the multiplicative group $k^\times / q^{\mathbb{Z}}$.

Remark 6.4.20 It follows from Theorem 6.4.18 and Corollary 6.4.13 that each point in the limit set may be described as the intersection of a nested sequence of discs of the form $\bigcap_{n \geq 0} D^+(w_n)$, for a sequence of words w_n whose lengths tend to infinity. This is a rather concrete description, that could easily be implemented to any precision on a computer. The complex version of this idea gave rise to beautiful pictures in [MSW15].

Actually, we highly recommend the whole book [MSW15] to the reader. It starts with the example of a complex Schottky group with two generators in a very accessible way and then carefully presents a large amount of advanced material, with an original and colorful terminology, enriched with many pictures. Among the subjects covered are the Hausdorff dimension of the limit set ("fractal dust"), the degeneration of the notion of Schottky groups when the discs in the Schottky figures become tangent ("kissing Schottky groups"), etc. We believe that it is worth investigating those questions in the non-Archimedean setting too. In particular, finding a way to draw meaningful non-Archimedean pictures would certainly be very rewarding.

6.4.2 Group-Theoretic Version

We now give the general definition of Schottky group over k and explain how it relates to the geometric situation considered in the previous sections. As regards proper actions, recall Definition 6.4.16 and Proposition 6.4.17.

Definition 6.4.21 A subgroup Γ of $\mathrm{PGL}_2(k)$ is said to be a *Schottky group over k* if

(i)   it is free and finitely generated;
(ii) all its non-trivial elements are loxodromic;
(iii) there exists a non-empty Γ-invariant connected open subset of $\mathbb{P}_k^{1,\mathrm{an}}$ on which the action of Γ is free and proper.

Remark 6.4.22 Schottky groups are discrete subgroups of $\mathrm{PGL}_2(k)$. Indeed any element of $\mathrm{PGL}_2(k)$ that is close enough to the identity has both eigenvalues of absolute value 1, hence cannot be loxodromic.

Remark 6.4.23 There are other definitions of Schottky groups in the literature. L. Gerritzen and M. van der Put use a slightly different version of condition (iii) (see [GvdP80, I (1.6)]). This is due to the fact that they work in the setting of rigid geometry, where the space consists only of our rigid points. We chose to formulate our definition this way in order to take advantage of the nice topological properties of Berkovich spaces and make it look closer to the definition used in complex geometry.

D. Mumford considered a more general setting where k is the fraction field of a complete integrally closed noetherian local ring and he requires only properties (i) and (ii) in his definition of Schottky group (see [Mum72a, Definition 1.3]). The intersection with our setting consists of the complete discretely valued fields k.

When k is a local field, all the definitions coincide (see [GvdP80, I (1.6.4)] and Sect. 6.4.4).

Schottky groups arise naturally when we have Schottky figures as in Sect. 6.4.1. Indeed, the following result follows from Proposition 6.4.7, Corollary 6.4.12 and Theorem 6.4.18.

Proposition 6.4.24 *Let Γ be a subgroup of $\mathrm{PGL}_2(k)$ generated by finitely many elements $\gamma_1, \ldots, \gamma_g$. If there exists a Schottky figure adapted to $(\gamma_1, \ldots, \gamma_g)$, then Γ is a Schottky group.* $\qquad\qquad\Box$

We now turn to the proof of the converse statement.

Lemma 6.4.25 *Let γ be a loxodromic Möbius transformation. Let A and A' be disjoint virtual flat closed annuli. Denote by I the open interval equal to the interior of the path joining their boundary points. Assume that $\gamma A_1 = A_2$ and $\gamma I \cap I = \emptyset$. For $\varepsilon \in \{\emptyset,'\}$, denote by D^ε the connected component of $\mathbb{P}_k^{1,\mathrm{an}} - A^\varepsilon$ that does* *not meet* I. *Then, for* $\varepsilon \in \{\emptyset,'\}$, A^{ε} *is a flat closed annulus*, D^{ε} *is an open disc*, $E^{\varepsilon} := D^{\varepsilon} \cup A^{\varepsilon}$ *is a closed disc and we have*

$$\gamma D = \mathbb{P}_k^{1,\mathrm{an}} - E' \text{ and } \gamma E = \mathbb{P}_k^{1,\mathrm{an}} - D'.$$

Proof For each $\varepsilon \in \{\emptyset,'\}$, D^{ε} and E^{ε} are respectively a virtual open disc and a virtual closed disc. Note that the set $\mathbb{P}_k^{1,\mathrm{an}} - A^{\varepsilon}$ has two connected components, namely D^{ε} and $\mathbb{P}_k^{1,\mathrm{an}} - E^{\varepsilon}$, and that the latter contains I.

Since γ is an automorphism, it sends the connected component $\mathbb{P}_k^{1,\mathrm{an}} - E$ of $\mathbb{P}_k^{1,\mathrm{an}} - A$ to a connected component C of $\mathbb{P}_k^{1,\mathrm{an}} - \gamma A = \mathbb{P}_k^{1,\mathrm{an}} - A'$. Denote by η and η' the boundary points of A and A'. Let $z \in \mathbb{P}_k^{1,\mathrm{an}} - E$. The unique path $[\eta, z]$ between η and z then meets I. Its image is the unique path $[\eta', \gamma(z)]$ between $\gamma(\eta) = \eta'$ and $\gamma(z)$. If $\gamma(z) \notin E'$, then this path meets I, contradicting the assumption $\gamma I \cap I = \emptyset$. We deduce that $\gamma(z) \in E'$, hence that $C = D'$. It follows that we have

$$\gamma D = \mathbb{P}_k^{1,\mathrm{an}} - E' \text{ and } \gamma E = \mathbb{P}_k^{1,\mathrm{an}} - D',$$

as wanted.

In particular, D and D' contain respectively the attracting and repelling fixed point of γ. Since those points are k-rational, we deduce that D and D' are discs. The rest of the result follows. $\square$

Theorem 6.4.26 *Let* Γ *be a Schottky group over* k. *Then, there exists a basis* β *of* Γ *and a Schottky figure* $\mathcal{B}$ *that is adapted to* β.

Proof By assumption, there exists a non-empty Γ-invariant connected open subset U of $\mathbb{P}_k^{1,\mathrm{an}}$ on which the action of Γ is free and proper. The quotient $X := \Gamma \backslash U$ is then an $\mathbb{A}_k^{1,\mathrm{an}}$-like curve in the sense of Sect. 6.3.1. Since U is a connected subset of $\mathbb{P}_k^{1,\mathrm{an}}$, it is simply connected, hence the fundamental group $\pi_1(X)$ of X is isomorphic to Γ. Since X is finitely generated, the topological genus g of X is finite.

Fix a skeleton Σ of X and consider the associated retraction $\tau \colon X \to \Sigma$. Fix g elements $\gamma_1, \ldots, \gamma_g$ of Γ corresponding to disjoint simple loops in Σ. Note that $\gamma_1, \ldots, \gamma_g$ is a basis of Γ.

For each $i \in \{1, \ldots, g\}$, pick a point $x_i \in \alpha_i$ that is not a branch point of Σ. Its preimage by the retraction $A_i := \tau^{-1}(x_i)$ is then a virtual flat closed annulus.

Let Y' be an open subset of U such that the morphism $Y' \to X$ induced by the quotient is an isomorphism onto $X - \bigcup_{1 \leqslant i \leqslant A_i}$. We extend it to a compact lift Y of X in U by adding, for each $i \in \{1, \ldots, g\}$, two virtual flat annuli B_i and B_i' that are isomorphic preimages of A_i. Up to switching the names, we may assume that $\gamma_i B_i = B_i'$.

Let $i \in \{1, \ldots, g\}$. The complement of B_i (resp. B_i') has two connected components. Let us denote by $D^-(\gamma_i)$ (resp. $D^-(\gamma_i^{-1})$) the one that does not meet Y. It is a virtual open disc. We set $D^+(\gamma_i^{-1}) = D^-(\gamma_i^{-1}) \cup B_i$ and $D^+(\gamma_i) = D^-(\gamma_i) \cup B_i'$.

By construction of Y', for each $\gamma \in \Gamma - \{\mathrm{id}\}$, we have $\gamma Y' \cap Y' = \emptyset$. It now follows from Lemma 6.4.25 that the family $(D^+(\gamma_i^\sigma), 1 \leqslant i \leqslant g, \sigma = \pm)$ is a Schottky figure adapted to $(\gamma_1, \ldots, \gamma_g)$.

□

Remark 6.4.27 The fact that Γ is free is actually not used in the proof of Theorem 6.4.26. As a result, Proposition 6.4.24 shows that it is a consequence of the other properties appearing in the definition of a Schottky group. It could also be deduced from the fact that the fundamental group of a Berkovich curve (which is the same as that of its skeleton) is free.

6.4.3   Twisted Ford Discs

We can actually be more precise about the form of the discs in the Schottky figure from Theorem 6.4.26. To do so, we introduce some terminology.

Definition 6.4.28 Let $\gamma = \begin{bmatrix} a & b \\ c & d \end{bmatrix} \in \mathrm{PGL}_2(k)$, with $c \neq 0$, be a loxodromic Möbius transformation and let $\lambda \in \mathbb{R}_{>0}$. We call open and closed *twisted Ford discs* associated to (γ, λ) the sets

$$D^-_{(\gamma,\lambda)} := \left\{ z \in k \ : \ \lambda|\gamma'(z)| = \lambda \frac{|ad - bc|}{|cz + d|^2} > 1 \right\}$$

and

$$D^+_{(\gamma,\lambda)} := \left\{ z \in k \ : \ \lambda|\gamma'(z)| = \lambda \frac{|ad - bc|}{|cz + d|^2} \geqslant 1 \right\}.$$

Lemma 6.4.29 *Let* $\alpha, \alpha', \beta \in k$ *with* $\alpha \neq \alpha'$ *and* $|\beta| < 1$ *and let* $\lambda \in \mathbb{R}_{>0}$. *Set* $\gamma := M(\alpha, \alpha', \beta) = \begin{bmatrix} a & b \\ c & d \end{bmatrix}$. *The twisted Ford discs* $D^-_{(\gamma,\lambda)}$ *and* $D^+_{(\gamma,\lambda)}$ *have center*

$$\frac{\alpha' - \beta\alpha}{1 - \beta} = -\frac{d}{c}$$

and radius

$$\rho = \frac{(\lambda|\beta|)^{1/2}|\alpha - \alpha'|}{|1 - \beta|} = \frac{(\lambda|ad - bc|)^{1/2}}{|c|}.$$

In particular, $\alpha' \in D^-_{(\gamma,\lambda)}$ *if, and only if,* $|\beta| < \lambda$.

The twisted Ford discs $D^-_{(\gamma^{-1},\lambda^{-1})}$ and $D^+_{(\gamma^{-1},\lambda^{-1})}$ have center

$$\frac{\alpha - \beta\alpha'}{1 - \beta} = \frac{a}{c}$$

and radius $\rho' = \rho/\lambda$.

In particular, $\alpha \in D^-_{(\gamma^{-1},\lambda^{-1})}$ if, and only if, $|\beta| < \lambda^{-1}$.   □

Lemma 6.4.30 *Let $\gamma \in \mathrm{PGL}_2(k)$ be a loxodromic Möbius transformation that does not fix ∞ and let $\lambda \in \mathbb{R}_{>0}$. Then, we have $\gamma(D^+_{(\gamma,\lambda)}) = \mathbb{P}^{1,\mathrm{an}}_k - D^-_{(\gamma^{-1},\lambda^{-1})}$.*

Proof Let us write $\gamma = \begin{bmatrix} a & b \\ c & d \end{bmatrix}$. Since γ does not fix ∞, we have $c \neq 0$. Let K be a complete valued extension of k and let $z \in K$. We have $|-c\gamma(z) + a| |cz + d| = |ad - bc|$, hence

$$z \in D_{(\gamma,\lambda)} \iff \lambda \frac{|ad - bc|}{|cz + d|^2} \geq 1 \iff \lambda^{-1} \frac{|ad - bc|}{|-c\gamma(z) + a|^2} \leq 1.$$

Since we have $\gamma^{-1} = \begin{bmatrix} d & -b \\ -c & a \end{bmatrix}$, the latter condition describes precisely the complement of $D^-_{(\gamma^{-1},\lambda^{-1})}$.   □

Lemma 6.4.31 *Let $\gamma \in \mathrm{PGL}_2(k)$ be a loxodromic Möbius transformation. Let $D^+(\gamma)$ and $D^+(\gamma^{-1})$ be disjoint closed discs in $\mathbb{P}^{1,\mathrm{an}}_k$. Set*

$$D^-(\gamma) := \gamma(\mathbb{P}^{1,\mathrm{an}}_k - D^+(\gamma^{-1})) \text{ and } D^-(\gamma^{-1}) := \gamma^{-1}(\mathbb{P}^{1,\mathrm{an}}_k - D^+(\gamma)).$$

Assume that $D^-(\gamma)$ and $D^-(\gamma^{-1})$ are maximal open discs inside $D^+(\gamma)$ and $D^+(\gamma^{-1})$ respectively and that they are contained in $\mathbb{A}^{1,\mathrm{an}}_k$.

Then, there exists $\lambda \in \mathbb{R}_{>0}$ such that, for each $\sigma \in \{-, +\}$, we have

$$D^\sigma(\gamma) = D^\sigma_{\gamma,\lambda} \text{ and } D^\sigma(\gamma^{-1}) = D^\sigma_{\gamma^{-1},\lambda^{-1}}.$$

Proof Denote by α and α' the attracting and repelling fixed points of γ respectively. By the same argument as in Remark 6.4.2, we have $\alpha \in D^-(\gamma^{-1})$ and $\alpha' \in D^-(\gamma)$. Let $r, r' > 0$ such that $D^-(\gamma) = D^-(\alpha', r')$ and $D^-(\gamma^{-1}) = D^-(\alpha, r)$.

Write $\gamma = \begin{bmatrix} a & b \\ c & d \end{bmatrix}$ with $a, b, c, d \in k$. Since $\alpha, \alpha' \in \mathbb{A}^{1,\mathrm{an}}_k$, we have $c \neq 0$. By assumption, $\infty \in \gamma(D^-(\gamma^{-1}))$, hence $-d/c \in D^-(\gamma^{-1})$ and $D^-(\gamma^{-1}) = D^-(-d/c, r)$. Similarly, we have $D^-(\gamma) = D^-(a/c, r')$.

Writing

$$\frac{aT+b}{cT+d} = \frac{a}{c} - \frac{ad-bc}{c^2}\frac{1}{T+\frac{d}{c}},$$

it is not difficult to compute $\gamma(D^-(\gamma^{-1}))$ and prove that we have

$$r = \frac{|ad-bc|}{|c|^2\,r'} = \frac{|\beta|\,|\alpha-\alpha'|^2}{r'}.$$

Set

$$\lambda := \frac{r^2}{|\beta|\,|\alpha-\alpha'|^2} = \frac{r}{r'} = \frac{|\beta|\,|\alpha-\alpha'|^2}{(r')^2}.$$

Since $D^+(\gamma)$ and $D^+(\gamma^{-1})$ are disjoint, we have $\max(r,r') < |\alpha-\alpha'|$, hence $|\beta| < \min(\lambda, \lambda^{-1})$. It follows that $D^-_{\gamma,\lambda}$ and $D^-_{\gamma^{-1},\lambda^{-1}}$ contains respectively α' and α, hence

$$D^-_{(\gamma,\lambda)} = D^-(\alpha',r') = D^-(\gamma) \text{ and } D^-_{(\gamma^{-1},\lambda^{-1})} = D^-(\alpha,r) = D^-(\gamma^{-1}).$$

$\square$

Corollary 6.4.32 *Let Γ be a Schottky group over k whose limit set does not contain ∞. Then, there exists a basis $(\gamma_1,\dots,\gamma_g)$ of Γ and $\lambda_1,\dots,\lambda_g \in \mathbb{R}_{>0}$ such that the family of discs $\left(D^+_{(\gamma_i^\varepsilon,\lambda_i^\varepsilon)}, 1 \leqslant i \leqslant g, \varepsilon \in \{\pm 1\}\right)$ is a Schottky figure that is adapted to $(\gamma_1,\dots,\gamma_g)$.*

Proof By Theorem 6.4.26, there exists a basis $\beta = (\gamma_1,\dots,\gamma_g)$ of Γ and a Schottky figure $\mathcal{B} = (D^+(\gamma_i^\varepsilon), 1 \leqslant i \leqslant g, \varepsilon \in \{\pm 1\})$ that is adapted to β. As in Sect. 6.4.1, define the open discs $D^-(\gamma_i^{\pm 1})$ and set

$$F^+ := \mathbb{P}^1_k - \bigcup_{\substack{1 \leqslant i \leqslant g \\ \varepsilon = \pm 1}} D^-(\gamma_i^\varepsilon).$$

By Theorem 6.4.18, since ∞ is not a limit point of Γ, there exists $\gamma \in \Gamma$ such that $\infty \in \gamma F^+$.

Set $\beta' := (\gamma\gamma_1\gamma^{-1},\dots,\gamma\gamma_g\gamma^{-1})$. It is a basis of Γ and the family of discs $\mathcal{B}' := (\gamma D^+(\gamma_i^\varepsilon), 1 \leqslant i \leqslant g, \varepsilon \in \{\pm 1\})$ is a Schottky figure that is adapted to it. Since all the discs $\gamma D^+(\gamma_i^{\pm 1})$ are contained in $\mathbb{A}^{1,\mathrm{an}}_k$, we may now apply Lemma 6.4.31 to conclude. $\square$

6.4.4   Local Fields

When k is a local field, the definition of a Schottky group can be greatly simplified. Our treatment here borrows from [GvdP80, I (1.6)] (see also [Mar07, Lemma 2.1.1] in the complex setting).

Lemma 6.4.33 *Let* $(\gamma_n)_{n \in \mathbb{N}}$ *be a sequence of loxodromic Möbius transformations such that*

(i) $(\gamma_n)_{n \in \mathbb{N}}$ *has no convergent subsequence in* $\mathrm{PGL}_2(k)$*;*
(ii) the sequence of Koebe coordinates $((\alpha_n, \alpha'_n, \beta_n))_{n \in \mathbb{N}}$ *converges to some* $(\alpha, \alpha', \beta) \in (\mathbb{P}^1(k))^3$.

Then, $(\gamma_n)_{n \in \mathbb{N}}$ *converges to the constant function* α *uniformly on compact subsets of* $\mathbb{P}^{1,\mathrm{an}}_k - \{\alpha'\}$.

Proof By definition, for each $n \in \mathbb{N}$, we have $|\beta_n| < 1$, which implies that $|\beta| < 1$.

Up to changing coordinates, we may assume that $\alpha, \alpha' \in k$. Up to modifying finitely many terms of the sequences, we may assume that, for each $n \in \mathbb{N}$, we have $\alpha_n, \alpha'_n \in k$. In this case, for each $n \in \mathbb{N}$, we have

$$\gamma_n =: \begin{bmatrix} \alpha_n - \beta_n \alpha'_n & (\beta_n - 1)\alpha_n \alpha'_n \\ 1 - \beta_n & \beta_n \alpha_n - \alpha'_n \end{bmatrix} \text{ in } \mathrm{PGL}_2(k).$$

The determinant of the above matrix is $\beta_n(\alpha_n - \alpha'_n)^2$. Since $(\gamma_n)_{n \in \mathbb{N}}$ has no convergent subsequence in $\mathrm{PGL}_2(k)$, we deduce that $\beta(\alpha - \alpha')^2 = 0$. In each of the two cases $\beta = 0$ and $\alpha = \alpha'$, it is not difficult to check that the claimed result holds.

□

The result below shows that the definition of Schottky group may be simplified when k is a local field. Note that, in this case, $\mathbb{P}^1(k)$ is compact, hence closed in $\mathbb{P}^{1,\mathrm{an}}_k$.

Corollary 6.4.34 *Assume that k is a local field. Let Γ be a subgroup of $\mathrm{PGL}_2(k)$ all of whose non-trivial elements are loxodromic.*

Let Λ be the set of fixed points of the elements of $\Gamma - \{\mathrm{id}\}$ and let $\bar{\Lambda}$ be its closure in $\mathbb{P}^{1,\mathrm{an}}_k$. Then, $\bar{\Lambda}$ is a compact subset of $\mathbb{P}^{1,\mathrm{an}}_k$ that is contained in $\mathbb{P}^1(k)$ and the action of Γ on $\mathbb{P}^{1,\mathrm{an}}_k - \bar{\Lambda}$ is free and proper.

Proof Since k is locally compact for the topology given by the absolute value, $\mathbb{P}^1(k)$ is compact. By [PT20, Remark 5.4.1], the topology on k given by the absolute value coincides with that induced by the topology on $\mathbb{A}^{1,\mathrm{an}}_k$. We deduce that $\mathbb{P}^1(k)$ is a compact subset of $\mathbb{P}^{1,\mathrm{an}}_k$. It follows that $\bar{\Lambda}$ is contained in $\mathbb{P}^1(k)$ and that it is compact, as it is closed.

The action of Γ is obviously free on $\mathbb{P}_k^{1,\mathrm{an}} - \bar{\Lambda}$. Assume, by contradiction, that it is not proper. Then, there exist $x, y \notin \bar{\Lambda}$ such that, for every neighborhoods U and V of x and y respectively, the set $\{\gamma \in \Gamma : \gamma U \cap V \neq \emptyset\}$ is infinite.

Since k is a local field, [PT20, Corollary 5.4.6] ensures that the space $\mathbb{A}_k^{1,\mathrm{an}}$ is metrizable. In particular, we may find countable bases of neighborhoods $(U_n)_{n \in \mathbb{N}}$ and $(V_n)_{n \in \mathbb{N}}$ of x and y respectively. By assumption, there exist a sequence $(\gamma_n)_{n \in \mathbb{N}}$ of distinct elements of Γ and a sequence $(x_n)_{n \in \mathbb{N}}$ of elements of $\mathbb{P}_k^{1,\mathrm{an}} - \bar{\Lambda}$ such that, for each $n \in \mathbb{N}$, we have $x_n \in U_n$ and $\gamma_n(x_n) \in V_n$. In particular, $(x_n)_{n \in \mathbb{N}}$ converges to x and $(\gamma_n(x_n))_{n \in \mathbb{N}}$ converges to y.

Since all the non-trivial elements of Γ are loxodromic, by the same argument as in Remark 6.4.22, the group Γ is discrete. As a result, up to passing to a subsequence, we may assume that the assumptions of Lemma 6.4.33 are satisfied. Define α and α' as in this Lemma. Since x does not belong to $\bar{\Lambda}$, it cannot be equal to α'. It follows that the sequences $(\gamma_n(x_n))_{n \in \mathbb{N}}$ and $(\gamma_n(x))_{n \in \mathbb{N}}$ converge to the same limit $y = \alpha$, and we get a contradiction since $\alpha \in \bar{\Lambda}$. $\qquad\square$

Corollary 6.4.35 *Assume that k is a local field. Then, a subgroup Γ of $\mathrm{PGL}_2(k)$ is a Schottky group if, and only if, it is finitely generated and all its non-trivial elements are loxodromic.* $\qquad\square$

6.5   Uniformization of Mumford Curves

The main result of this section, Theorem 6.5.3, states that the procedure described in Sect. 6.4.1 can be reversed: any Mumford curve may be uniformized by an open subset of the Berkovich projective line $\mathbb{P}_k^{1,\mathrm{an}}$ with a Schottky group as group of deck transformations. The consequences of this result are many and far-reaching. Some of them are discussed in Appendix A.3.

This was first proved by D. Mumford in his influential paper [Mum72a], where he introduces this as a non-Archimedean analogue of the uniformization of handlebodies by means of Schottky groups in the complex setting. His arguments make a heavy use of formal models of the curves. Here, we argue directly on the curves themselves, following the strategy of [GvdP80, Chapter IV] and [Lüt16, Proposition 4.6.6]. Note, however that the proof in the first reference is flawed (since it relies on the wrong claim that every k-analytic curve of genus 0 embeds into $\mathbb{P}_k^{1,\mathrm{an}}$, see Remark 6.5.7) and that the second reference assumes that the curve contains at least three rational points.

As an application, we discuss how Theorem 6.5.3 can be used to study the automorphism groups of Mumford curves. This is far from being the sole purpose of uniformization. Other important consequences are mentioned in Appendix A.3.

6.5.1 The Uniformization Theorem

In this section, we prove that any analytic Mumford curve as defined in 6.3.24 can be obtained as the quotient of an open dense subspace of $\mathbb{P}_k^{1,\mathrm{an}}$ by the action of a Schottky group, leading to a purely analytic proof of Mumford's theorem. We begin with a few preparatory results.

Lemma 6.5.1 *Let L be a compact subset of $\mathbb{P}^1(k)$. Set $O := \mathbb{P}_k^{1,\mathrm{an}} - L$.*

(i) Every bounded analytic function on O is constant.
(ii) Every automorphism of O is induced by an element of $\mathrm{PGL}_2(k)$.

Proof

(i) Let $F \in \mathcal{O}(O)$. The function F is constant if, and only if, its pullback to $O_{\widehat{k^a}}$ is, hence we may assume that k is algebraically closed.

Assume, by contradiction, that F is not constant. Then, there exists $x \in O$ and a branch b at x such that $F(x) \neq 0$ and $|F|$ is monomial at x along b with a positive integer exponent. We may assume that x is of type 2 or 3. Then, there exists $y \in O - \{x\}$ and $N \in \mathbb{N}_{\geqslant 1}$ such that, for each $z \in [x, y]$, we have $|F(z)| = |F(x)| \, \ell([x, z])^N$.

Let us now consider a path $[x, y]$, with $y \in \mathbb{P}_k^{1,\mathrm{an}}$, with the following property: for each $z \in (x, y)$, $|F|$ is monomial at z with positive integer slope along the branch in (x, y) going away from x. By Zorn's lemma, we may find a maximal path $[x, y]$ among those.

We claim that y is of type 1. If y is of type 4, then, by [PT20, Theorem 5.10.10], $|F|$ is constant in the neighborhood of y in (x, y), and we get a contradiction. Assume that b is of type 2 or 3. Then, the exponent of $|F|$ at y along the branch corresponding to $[y, x]$ is negative. By [PT20, Corollary 5.10.12], there exists a branch b at y such that $|F|$ is monomial with positive exponent at y along b, which contradicts the maximality. Finally, y is of type 1.

By assumption, $|F|$ has a positive integer exponent everywhere on (x, y). It follows that, for each $z \in (x, y)$, we have $|F(z)| \geqslant |F(x)| \, \ell([x, z])$. Since y is of type 1, by [PT20, Lemma 5.9.12], we have $\ell([x, y]) = \infty$, hence F is unbounded. This is a contradiction.

(ii) Let σ be an automorphism of O.

Let us first assume that O contains at least 2 k-rational points. Up to changing coordinates, we may assume that $0, \infty \in O$. Let us choose an automorphism $\tau \in \mathrm{PGL}_2(k)$ that agrees with σ on 0 and ∞. Then $\tau^{-1} \circ \sigma$ is an automorphism of O that fixes 0 and ∞. In particular, it corresponds to an analytic function with a zero of order 1 at 0 and a pole of order 1 at ∞. Let us consider the quotient analytic function $\varphi := (\tau^{-1} \circ \sigma)/\mathrm{id}$. There exist a neighborhood U of 0 and a neighborhood V of ∞ on which φ is bounded. Since $\tau^{-1} \circ \sigma$ is an automorphism, it sends V to a neighborhood of ∞, hence it is bounded on $O - V$. It follows that φ is bounded on $O - (U \cup V)$, hence on O. By (i), we deduce that φ is constant, and the result follows.

Let us now handle the case where $O \cap \mathbb{P}^1(k) = \emptyset$. There exists a finite extension K of k such that O_K contains a K-rational point. Applying the previous argument after extending the scalars to K, we deduce that σ belongs to $\mathrm{PGL}_2(K)$. Since it preserves $\mathbb{P}^1(k)$, it actually belongs to $\mathrm{PGL}_2(k)$. □

Lemma 6.5.2 *Let Y be a connected k-analytic $\mathbb{A}^1$-like curve of genus 0. Let $\mathcal{T} = (S, \mathcal{A}, \mathcal{D})$ be a triangulation of Y. Assume that $\mathcal{A}$ is non-empty and consists of annuli. Let U be an open relatively compact subset of $\Sigma_{\mathcal{T}}$. Then, there exists an embedding of $\tau_{\mathcal{T}}^{-1}(U)$ into $\mathbb{P}_k^{1,\mathrm{an}}$ such that the complement of $\tau_{\mathcal{T}}^{-1}(U)$ is a disjoint union of finitely many closed discs.*

Proof Recall that $\Sigma_{\mathcal{T}}$ is a locally finite graph (see Theorem 6.3.2). As a consequence, the boundary ∂U of U in $\Sigma_{\mathcal{T}}$ is finite. For each $z \in \partial U$, let I_z be an open interval in $\Sigma_{\mathcal{T}}$ having z as an end-point. Up to shrinking the I_z's, we may assume that they are disjoint.

Let $z \in \partial U$. Set $A_z := \tau_{\mathcal{T}}^{-1}(I_z)$. Since every element of $\mathcal{A}$ is an annulus, up to shrinking I_z (so that it contains no points of S), we may assume that A_z is an annulus. The open annulus A_z may be embedded into an open disc D_z such that the complement is a closed disc.

Let us construct a curve Y' by starting from $\tau_{\mathcal{T}}^{-1}(U)$ an gluing D_z along A_z for each $z \in \partial U$. By construction, the curve Y' is compact and of genus 0. Moreover, it contains rational points, as the D_z do. It follows from Theorems 6.3.12 and 6.3.23 that Y' is isomorphic to $\mathbb{P}_k^{1,\mathrm{an}}$. By construction,

$$\mathbb{P}_k^{1,\mathrm{an}} - \tau_{\mathcal{T}}^{-1}(U) = \bigcup_{z \in \partial U} D_z - A_z$$

is a disjoint union of finitely many closed discs. □

We now state and prove the uniformization theorem.

Theorem 6.5.3 *Let X be a k-analytic Mumford curve. Then the fundamental group Γ of X is a Schottky group. If we denote by L the limit set of Γ, then $O := \mathbb{P}_k^{1,\mathrm{an}} - L$ is a universal cover of X. In particular, we have $X \simeq \Gamma \backslash O$.*

Proof Assume that the genus of X is bigger than or equal to 2.

Let $p: Y \to X$ be the topological universal cover of X. Since p is a local homeomorphism, we may use it to endow Y with a k-analytic structure. The set Y then becomes an $\mathbb{A}^1$-like curve and the map p becomes a local isomorphism of locally ringed spaces. Note that the curve Y has genus 0.

We claim that it is enough to prove that Y is isomorphic to an open subset of $\mathbb{P}_k^{1,\mathrm{an}}$ whose complement lies in $\mathbb{P}^1(k)$. Indeed, in this case, Y is simply connected, hence the fundamental group Γ of X may be identified with the group of deck transformations of p. By Lemma 6.5.1, it embeds into $\mathrm{PGL}_2(k)$. It now follows from the properties of the universal cover and the fundamental group that Γ is a Schottky group (see Remark 6.2.12 for the fact that the non-trivial elements of Γ are loxodromic). Moreover, by Theorem 6.4.18, we have $Y \subseteq \mathbb{P}_k^{1,\mathrm{an}} - L$, where L is the limit set of Γ, hence $X = \Gamma \backslash Y \subset \Gamma \backslash (\mathbb{P}_k^{1,\mathrm{an}} - L)$. Since $\Gamma \backslash Y$ and $\Gamma \backslash (\mathbb{P}_k^{1,\mathrm{an}} - L)$ are both connected proper curves, they have to be equal, hence $Y = \mathbb{P}_k^{1,\mathrm{an}} - L$.

In the rest of the proof, we show that Y embeds into $\mathbb{P}_k^{1,\mathrm{an}}$ with a complement in $\mathbb{P}^1(k)$. Since X is a k-analytic Mumford curve of genus at least 2, it has a minimal skeleton Σ_X and the connected components of Σ_X deprived of its branch points are skeleta of open annuli over k. Its preimage $p^{-1}(\Sigma_X)$ coincides with the minimal skeleton Σ_Y of Y. Similarly, the connected components of Σ_Y deprived of its branch points are skeleta of open annuli over k. We denote by $\tau_Y \colon Y \to \Sigma_Y$ the canonical retraction.

Let $x_0 \in X$ and $y_0 \in p^{-1}(x_0)$. Let ℓ_X be a loop in Σ_X based at x_0 that is not homotopic to 0. It lifts to a path in Σ_Y between y_0 and a point y_1 of $p^{-1}(x_0)$. We may then lift again ℓ_X to a path in Σ_Y between y_1 and a point y_2 of $p^{-1}(x_0)$. Repeating the procedure, we obtain a non-relatively compact path $\lambda(\ell_X)$ in Σ_Y starting at y_0. Note that the length of $\lambda(\ell_X)$ is infinite since it contains infinitely many copies of ℓ_X.

More generally, all the maximal paths starting from y_0 in Σ_Y are of infinite length, since they contain infinitely many lifts of loops from Σ_X.

Since X is of genus at least 2, we may find two loops $\ell_{X,0}$ and $\ell_{X,1}$ based at x_0 in Γ_X that are not homotopic to 0 and not homotopic one to the other. Set $\ell_0 := \lambda(\ell_{X,0})$, $\ell_\infty := \lambda(\ell_{X,0}^{-1})$ and $\ell_1 := \lambda(\ell_{X,1})$. Away from some compact set of Y, the three paths $\ell_0, \ell_\infty, \ell_1$ are disjoint. Up to moving x_0 and y_0, we may assume that

$$\ell_0 \cap \ell_1 = \ell_\infty \cap \ell_1 = \ell_0 \cap \ell_\infty = \{y_0\}.$$

For $i \in \{0, 1, \infty\}$ and $r \in \mathbb{R}_{\geq 1}$, we denote by $\xi_{i,r}$ the unique point of ℓ_i such that $\ell([y_0, \xi_{i,r}]) = r$.

Let $n \in \mathbb{N}_{\geq 1}$. Set

$$U_n := \{z \in \Sigma_Y : \ell([y_0, z]) < 2^n\} \quad \text{and} \quad Y_n := \tau_Y^{-1}(U_n).$$

We already saw that all the maximal paths starting from y_0 in Σ_Y are of infinite length, hence U_n is relatively compact in Σ_Y. Denote by ∂U_n the boundary of U_n in Σ_Y. For each $z \in \partial U_n$, we have $\ell([y_0, z]) = 2^n$.

By Lemma 6.5.2, there exists an open subset O_n of $\mathbb{P}_k^{1,\mathrm{an}}$ and an isomorphism $\varphi_n \colon Y_n \xrightarrow{\sim} O_n$ such that $\mathbb{P}_k^{1,\mathrm{an}} - O_n$ is a disjoint union of closed discs. For each $z \in \partial U_n$, we denote by p_z the end-point of $\varphi_n([y_0, z))$ in $\mathbb{P}_k^{1,\mathrm{an}} - O_n$ and by D_z the connected component of $\mathbb{P}_k^{1,\mathrm{an}} - O_n$ whose boundary point is p_z. To ease the notation, for $i \in \{0, 1, \infty\}$, we set $D_{i,n} := D_{\xi_{i,2^n}}$.

Let us fix a point at infinity on $\mathbb{P}_k^{1,\mathrm{an}}$ and a coordinate T on $\mathbb{A}_k^{1,\mathrm{an}} \subset \mathbb{P}_k^{1,\mathrm{an}}$. We may assume that, for each $i \in \{0, 1, \infty\}$, we have $i \in D_{i,n}$. By pulling back the analytic function T on O_n by φ_n, we get a analytic function on Y_n. We denote it by ψ_n. Recall that it is actually equivalent to give oneself φ_n or ψ_n, see [PT20, Lemma 5.5.11].   □

Lemma 6.5.4 *We have $\varphi_n(y_0) = \eta_1$. For each $r \in [1, 2^n)$, we have*

$$\varphi_n(\xi_{0,r}) = \eta_{1/r}, \ \varphi_n(\xi_{\infty,r}) = \eta_r \ \text{and} \ \varphi_n(\xi_{1,r}) = \eta_{1,1/r}.$$

Let C be a connected component C of $Y - (\ell_0 \cup \ell_\infty)$. For each $y \in C \cap Y_n$, we have

$$|\psi_n(y)| = \begin{cases} 1/r & \text{if the boundary point of } C \text{ is } \xi_{0,r}; \\ r & \text{if the boundary point of } C \text{ is } \xi_{\infty,r}. \end{cases}$$

Let $N \in [\![1, n]\!]$. The image $\varphi_n(Y_N)$ is an open Swiss cheese. More precisely, there exist $d \in \mathbb{N}_{\geqslant 2}$, $\alpha_2, \ldots, \alpha_d \in k^*$ and, for each $j \in [\![2, d]\!]$, $r_j \in [2^{-N}, |\alpha_j|)$ such that $\varphi_n(Y_N)$ is the subset of $\mathbb{A}_k^{1,\mathrm{an}}$ defined by the following conditions:

$$\begin{cases} 2^{-N} < |T| < 2^N; \\ |T - 1| > 2^{-N}; \\ \forall j \in [\![2, d]\!], \ |T - \alpha_j| > r_j. \end{cases}$$

Proof It follows from the construction that, for each $i \in \{0, 1, \infty\}$, $\varphi_n([y_0, \xi_{i,2^n})$ is an injective path joining $\varphi_n(y_0)$ to the boundary point of a disc centered at i. Since those paths only meet at $\varphi_n(y_0)$, the only possibility is that $\varphi_n(y_0) = \{\eta_1\}$.

Let $r \in [1, 2^n)$. Since lengths are preserved by automorphism (see [PT20, Proposition 5.5.14]), for each $i \in \{0, 1, \infty\}$, we have $\ell([\eta_1, \varphi_n(\xi_{i,r})]) = r$. Since $\varphi_n(\xi_{\infty,r})$ belongs to $[\eta_1, \infty]$, it follows that $\varphi_n(\xi_{\infty,r}) = \eta_r$. By a similar argument, we have $\varphi_n(\xi_{0,r}) = \eta_{1/r}$ and $\varphi_n(\xi_{1,r}) = \eta_{1,1/r}$.

Recall that we have $I_0 = \{\eta_r : r \in \mathbb{R}_{\geqslant 0}\} \subset \mathbb{A}_k^{1,\mathrm{an}}$. Let C be a connected component of $\mathbb{A}_k^{1,\mathrm{an}} - I_0$ and let η_r be its boundary point. Then, for each $z \in C$, we have $|T(z)| = r$.

We have $\varphi_n^{-1}(I_0 \cap O_n) = (\ell_0 \cup \ell_\infty) \cap Y_n$. By definition of ψ_n, for each $y \in Y_n$, we have $|\psi_n(y)| = |T(\varphi_n(y))|$. It follows that, for each connected component C of $Y - (\ell_0 \cup \ell_\infty)$ and each $y \in C \cap Y_n$, we have

$$|\psi_n(y)| = \begin{cases} 1/r & \text{if the boundary point of } C \text{ is } \xi_{0,r}; \\ r & \text{if the boundary point of } C \text{ is } \xi_{\infty,r}. \end{cases}$$

The set O_n is an open Swiss cheese. The set $\varphi_n(U_N)$ is a connected open subset of its skeleton and $\varphi_n(Y_N)$ is the preimage of it by the retraction. It follows that $\varphi_n(Y_N)$ is an open Swiss cheese too, hence the complement in $\mathbb{P}_k^{1,\mathrm{an}}$ of finitely many closed discs $E_\infty, E_0, \ldots, E_d$. Let $z_\infty, z_0, \ldots, z_d$ denote the corresponding boundary points. The set $\varphi_n(Y_N)$ contains $\varphi_n(y_0) = \eta_1$ and, by construction of Y_N, for each $i \in \{\infty\} \cup [\![0, d]\!]$, we have $\ell([\eta_1, z_i]) = 2^N$.

Since 0, 1 and ∞ do not belong to O_n, some of those discs E_i contain those points. Since $\varphi_n(Y_N)$ contains η_1, those discs are disjoint. We may assume that, for each $i \in \{0, 1, \infty\}$, we have $i \in E_i$. The length property then implies that we have $z_\infty = \eta_{2^N}$, $z_0 = \eta_{2^{-N}}$ and $z_0 = \eta_{1,2^{-N}}$. In other words,

$$\mathbb{P}_k^{1,\mathrm{an}} - (E_\infty \cup E_0 \cup E_1) = \{x \in \mathbb{A}_k^{1,\mathrm{an}} : 2^{-N} < |T(x)| < 2^N, \ |T - 1| > 2^{-N}\}.$$

For $j \in [\![2, d]\!]$, let α_j be a k-rational point of E_j. The boundary point z_j of E_j is then of the form η_{α_j, r_j} for some $r_j \in \mathbb{R}_{\geq 0}$. Since E_j does not contain 0, we have $r_j < |\alpha_j|$. Moreover, the condition $\ell([\eta_1, \eta_{\alpha_j, r_j}]) = 2^N$ implies that $r_j \geqslant 2^{-N}$ (see [PT20, Example 5.9.11]). The result follows. $\qquad \square$

Let $N, n, m \in \mathbb{N}_{\geqslant 1}$ with $n \geqslant m > N$. The analytic function ψ_m has no zeros on Y_m, hence the quotient $\psi_{n|Y_m}/(\psi_m)$ defines an analytic function on Y_m. Set

$$h_{n,m} := \frac{\psi_{n|Y_m}}{\psi_m} - 1 \in \mathcal{O}(Y_m).$$

Lemma 6.5.5 *For $N, n, m \in \mathbb{N}_{\geqslant 1}$ with $n \geqslant m > N$, we have $\|h_{n,m}\|_{Y_N} \leqslant \max(2^{N-m}, 2^{-m/2})$.*

Proof By Lemma 6.5.4, for each $y \in Y_m$, we have $|\psi_n(y)| = |\psi_m(y)|$. It follows that $\|h_{m,n}\|_{Y_m} \leqslant 1$. We now distinguish two cases.

- Assume that $|h_{n,m}|$ is not constant on Y_N.

By [PT20, Corollary 5.10.16], there exists $y \in \partial Y_N$ such that $\|h_{n,m}\|_{Y_N} = |h_{n,m}(y)|$ and $|h_{n,m}|$ has a negative exponent at y along the branch entering Y_N. By harmonicity (see [PT20, Theorem 5.10.14]), there exist a branch b at y not belonging to Y_N such that the exponent of $|h_{n,m}|$ along b is positive. Repeating the procedure, we construct a path joining y to a boundary point y' of Y_m such that $|h_{n,m}|$ has a positive exponent at each point of $[y, y')$ along the branch pointing towards y'. It follows that we have

$$\|h_{n,m}\|_{Y_n} \geqslant |h_{n,m}(y)| \, \ell([y, y']) \geqslant \|h_{n,m}\|_{Y_N} \, 2^{m-N},$$

hence

$$\|h_{n,m}\|_{Y_N} \leqslant 2^{N-m}.$$

- Assume that $|h_{n,m}|$ is constant on Y_N.

Let N' be the maximum integer smaller than or equal to m such that $|h_{n,m}|$ is constant on $Y_{N'}$. Then, for every $r \in [1, 2^{N'})$, we have $|h_{n,m}(\xi_{1,r})| = \|h_{n,m}\|_{Y_{N'}}$. We also have

$$
\begin{aligned}
|h_{n,m}(\xi_{1,r})| &= \frac{|(\psi_n - \psi_m)(\xi_{1,r})|}{|\psi_m(\xi_{1,r})|} \\
&= \frac{|(\psi_n - \psi_m)(\xi_{1,r})|}{|T(\eta_{1,1/r})|} \\
&\leqslant \max(|(\psi_n - 1)(\xi_{1,r})|, |(\psi_m - 1)(\xi_{1,r})|) \\
&\leqslant |(T - 1)(\eta_{1,1/r})| \\
&\leqslant \frac{1}{r}.
\end{aligned}
$$

We deduce that $\|h_{n,m}\|_{Y_{N'}} \leqslant 2^{-N'}$.

If $N' < m$, it follows from the previous case that we have $\|h_{n,m}\|_{Y_{N'}} \leqslant 2^{N'-m}$. In any case, we have

$$
\|h_{n,m}\|_{Y_N} \leqslant 2^{-m/2}.
$$

$\square$

It follows from Lemma 6.5.5 that the sequence $(\psi_n)_{n>N}$ converges uniformly on Y_N. Let $\psi^{(N)}$ be its limit. It is an analytic function on Y_N.

The functions $\psi^{(N)}$ are compatible, by uniqueness of the limit, which gives rise to an analytic function $\psi \in \mathcal{O}(Y)$. By [PT20, Lemma 5.5.11], there exists a unique analytic morphism $\varphi \colon Y \to \mathbb{A}_k^{1,\mathrm{an}}$ such that the pull-back of T by φ is ψ.

Let $N \in \mathbb{N}_{\geqslant 1}$. By Lemma 6.5.5, there exists $m > N$ such that, for each $n \geqslant m$, we have $\|h_{n,m}\|_{Y_N} \leqslant 2^{-2N}$. (For instance, one could choose $m = 4N$.) By Lemma 6.5.4, we have $\|\psi_m\|_{Y_N} = \|T\|_{\varphi_m(Y_N)} = 2^N$. It follows that $\|\psi_n - \psi_m\|_{Y_N} \leqslant \|\psi_m\|_{Y_N} \|h_{n,m}\|_{Y_N} \leqslant 2^{-N}$. By passing to the limit over n, we deduce that

$$
\|\psi - \psi_m\|_{Y_N} \leqslant 2^{-N}.
$$

Lemma 6.5.6 *We have $\varphi(Y_N) = \varphi_m(Y_N)$ and $\varphi_{|Y_N}$ is an isomorphism onto its image.*

Proof By Lemma 6.5.4, there exist $d \in \mathbb{N}_{\geqslant 2}$, $\alpha_2, \ldots, \alpha_d \in k^*$ and, for each $j \in [\![2, d]\!]$, $r_j \in [2^{-N}, |\alpha_j|)$ such that $\varphi_m(Y_N)$ is the subset of $\mathbb{A}_k^{1,\mathrm{an}}$ defined by

$$
\begin{cases}
2^{-N} < |T| < 2^N; \\
|T - 1| > 2^{-N}; \\
\forall j \in [\![2, d]\!], \ |T - \alpha_j| > r_j.
\end{cases}
$$

For $t \in (1, 2^N)$, let W_t be the subset of $\mathbb{A}_k^{1,\mathrm{an}}$ defined by

$$\begin{cases} 2^{-N} t \leqslant |T| \leqslant 2^N t^{-1}; \\ |T - 1| \geqslant 2^{-N} t; \\ \forall j \in [\![2, d]\!], \ |T - \alpha_j| \geqslant r_j t. \end{cases}$$

Each W_t is compact and the family $(W_t)_{t \in (1,2^N)}$ is an exhaustion of $\varphi_m(Y_N)$.

Let $n \geqslant m$. For $t \in (1, 2^N)$, the set $\varphi^{-1}(W_t) \cap Y_N$ is the subset of points $y \in Y_N$ such that

$$\begin{cases} 2^{-N} t \leqslant |\psi(y)| \leqslant 2^N t^{-1}; \\ |\psi(y) - 1| \geqslant 2^{-N} t; \\ \forall j \in [\![2, d]\!], \ |\psi(y) - \alpha_j| \geqslant r_j t. \end{cases}$$

From the inequality $\|\psi - \psi_m\|_{Y_N} \leqslant 2^{-N}$, we deduce that $\varphi^{-1}(W_t) \cap Y_N = \varphi_m^{-1}(W_t) \cap Y_N$.

It follows that $\varphi(Y_N) = \varphi_m(Y_N)$ and that the morphism $\varphi_{|Y_N} : Y_N \to \varphi(Y_N)$ is proper. Since Y_N is a smooth curve and $\varphi_{|Y_N}$ is not constant, it is actually finite.

To prove that $\varphi_{|Y_N}$ is an isomorphism, it is enough to show that it is of degree 1. We will prove that, for each $r \in [1, 2^N)$, we have $\varphi_{|Y_N}^{-1}(\xi_{\infty,r}) = \{\eta_r\}$. This implies the result, by [PT20, Theorem 5.10.17].

Let $r \in [1, 2^N)$. Let $y \in Y_N$ such that $\varphi(y) = \eta_r$. To prove that $y = \xi_{\infty,r}$, we may extend the scalars to $\widehat{k^a}$. The point η_r of $\mathbb{A}_k^{1,\mathrm{an}}$ is characterized by the following equalities:

$$\begin{cases} |T(\eta_r)| = r; \\ \forall \alpha \in \widehat{k^a} \text{ with } |\alpha| = r, \ |(T - \alpha)(\eta_r)| = r. \end{cases}$$

Since $\varphi(y) = \eta_r$, we have

$$\begin{cases} |\psi(y)| = r; \\ \forall \alpha \in \widehat{k^a} \text{ with } |\alpha| = r, \ |\psi(y) - \alpha| = r. \end{cases}$$

Since $\|\psi - \psi_m\|_{Y_N} \leqslant 2^{-N} < r$, the same equalities hold with ψ_m instead of ψ. It follows that $\psi_m(y) = \eta_r$, hence $y = \xi_{\infty,r}$ since ψ_m is injective. $\qquad\square$

It follows from Lemmas 6.5.4 and 6.5.6 that, for each $N \in \mathbb{N}_{\geqslant 1}$, $\mathbb{P}_k^{1,\mathrm{an}} - \varphi(Y_N)$ is a disjoint union of closed discs with radii smaller than or equal to 2^{-N}. It follows that

$$\mathbb{P}_k^{1,\mathrm{an}} - \varphi(Y) = \bigcap_{N \geqslant 1} \mathbb{P}_k^{1,\mathrm{an}} - \varphi(Y_N)$$

is a compact subset of $\mathbb{P}^1(k)$ (see the proof of Corollary 6.4.13 for details on k-rationality). By Lemma 6.5.6 again, φ induces an isomorphism onto its image.

We briefly sketch how the proof needs to be modified to handle the case of genus 0 and 1. One may use similar arguments but the paths $\ell_0, \ell_\infty, \ell_1$ have to be constructed in a different way. In genus 0, one first proves that X has rational points and consider paths joining y_0 to them. (In this case, one may also argue more directly to prove that X is isomorphic to $\mathbb{P}_k^{1,an}$ by Theorems 6.3.12 and 6.3.23.) In genus 1, the skeleton provides two paths and we can use a rational point to construct the third one. Such a point has to exist, since any annulus over k whose skeleton is of large enough length contains some.

Remark 6.5.7 The most difficult part of the proof of Theorem 6.5.3 consists in proving that the k-analytic curve Y, which is known to be of genus 0, may be embedded into $\mathbb{P}_k^{1,an}$. Contrary to what happens over the field of complex numbers, this is not automatic. This problem was studied extensively by Q. Liu under the assumption that k is algebraically closed. He proved that the answer depends crucially on the maximal completeness of k. If it holds, then any smooth connected k-analytic curve of finite genus may be embedded into the analytification of an algebraic curve of the same genus (hence into $\mathbb{P}_k^{1,an}$ in the genus 0 case), see [Liu87b, Théorème 3] or [Liu87a, Théorème 3.2]. Otherwise, there exists a smooth connected k-analytic curve of genus 0 with no embedding into $\mathbb{P}_k^{1,an}$, see [Liu87b, Proposition 5.5]. Q. Liu also prove several other positive results that hold over any algebraically closed base field.

The results of Q. Liu are stated and proved in the language of rigid analytic geometry. We believe that it is worth adapting them to the setting of Berkovich geometry and that this could lead to a different point of view on the sufficient conditions for algebraizablity. One may also wonder whether it is necessary to assume that the base field is algebraically closed to obtain an unconditional positive result. The case of a discretely valued base field (hence maximally complete but not algebraically closed) is, of course, particularly interesting.

6.5.2   Automorphisms of Mumford Curves

In this section, we use the uniformization of Mumford curves to study their groups of k-linear automorphisms. The fundamental result, proven by Mumford in [Mum72a, Corollary 4.12], is the following theorem. We include a proof of this fact that relies on the topology of Berkovich curves.

Theorem 6.5.8 *Let X be a k-analytic Mumford curve. Let $\Gamma \subset \mathrm{PGL}_2(k)$ be its fundamental group, and let $N := N_{\mathrm{PGL}_2(k)}(\Gamma)$ be the normalizer of Γ in $\mathrm{PGL}_2(k)$. Then, we have*

$$\mathrm{Aut}(X) \cong N/\Gamma.$$

Proof Let $p: O \to X$ be the universal cover of X provided by Theorem 6.5.3, and let $\sigma \in \mathrm{Aut}(X)$. Since p is locally an isomorphism of k-analytic curves, the automorphism σ can be lifted to an analytic automorphism $\tilde{\sigma} \in \mathrm{Aut}(O)$ such that $p \circ \tilde{\sigma} = \sigma \circ p$. By Lemma 6.5.1, $\tilde{\sigma}$ extends uniquely to an automorphism of $\mathbb{P}_k^{1,\mathrm{an}}$, that is, an element $\tau \in \mathrm{PGL}_2(k)$. The automorphism τ has to normalize Γ: in fact, for any $\gamma \in \Gamma$, the element $\tau \gamma \tau^{-1} \in \mathrm{Aut}(O)$ induces the automorphism $\sigma \sigma^{-1} = \mathrm{id}$ on X. It follows that $\tau \gamma \tau^{-1} \in \Gamma$, so that $\tau \in N$.

Conversely, let $\tau \in N$. By definition, the limit set L of Γ is preserved by τ. It follows that τ induces an automorphism of $O = \mathbb{P}_k^{1,\mathrm{an}} - L$. Moreover, for each $\gamma \in \Gamma$ and each $x \in \mathbb{P}_k^{1,\mathrm{an}}$, we have

$$\tau(\gamma(x)) = (\tau \gamma \tau^{-1})(\tau(x)) \in \Gamma \cdot \tau(x).$$

It follows that τ descends to an automorphism of $X \simeq \Gamma \backslash O$. □

As was the case for the uniformization, Mumford's proof relies on non-trivial results in formal geometry. The Berkovich analytic proof turns out to be shorter and much less technical due to the fact that the uniformization of a Mumford curve can be interpreted as a universal cover of analytic spaces.

Recall from Remark 6.3.8 that the skeleton Σ_X of the Mumford curve X is a finite metric graph. We will denote by $\mathrm{Aut}(\Sigma_X)$ the group of isometric automorphisms of Σ_X. An interesting feature of the automorphism group of an analytic curve, which is immediate in the Berkovich setting, is the existence of a *restriction homomorphism*

$$\rho : \mathrm{Aut}(X) \longrightarrow \mathrm{Aut}(\Sigma_X)$$
$$\sigma \longmapsto \sigma_{|\Sigma_X}.$$

Proposition 6.5.9 *Let X be a Mumford curve of genus at least 2. Then, the restriction homomorphism $\rho : \mathrm{Aut}(X) \to \mathrm{Aut}(\Sigma_X)$ is injective.*

Proof Let $\sigma \in \mathrm{Aut}(X)$ such that $\rho(\sigma) = \mathrm{id}$, that is, σ acts trivially on the skeleton Σ_X. Then, as in the proof of Theorem 6.5.8, one can lift σ to an automorphism of the universal cover $p : O \longrightarrow X$. By possibly composing this lifting with an element of the Schottky group, we can find a lifting $\tilde{\sigma}$ that fixes a point x in the preimage $p^{-1}(\Sigma_X) \subset O$. Since σ fixes Σ_X pointwise, then $\tilde{\sigma}$ fixes the fundamental domain in $p^{-1}(\Sigma_X)$ by the action of the Schottky group Γ_X containing x. By continuity of the action of Γ_X on $p^{-1}(\Sigma_X)$, the automorphism $\tilde{\sigma}$ has to fix the whole $p^{-1}(\Sigma_X)$ pointwise. But then the corresponding element $\tau \in \mathrm{PGL}_2(k)$ obtained by extending $\tilde{\sigma}$ thanks to Lemma 6.5.1(ii) has to fix the limit set of Γ_X, which is infinite when $g(X) \geq 2$. It follows that τ is the identity of $\mathrm{PGL}_2(k)$, hence that σ is the identify automorphism. □

Remark 6.5.10 The previous proposition can be proved also using algebraic methods as follows. The fact that $g(X) \geq 2$ implies that $\mathrm{Aut}(X)$ is a finite group. Then, for every $\sigma \in \mathrm{Aut}(X)$, $Y := X/\langle \sigma \rangle$ makes sense as a k-analytic curve, and the quotient map $f_\sigma : X \to Y$ is a ramified covering. Let us now suppose that $\rho(\sigma) = \mathrm{id}$. Then Y contains an isometric image of the graph Σ_X, whose cyclomatic number is $g(X)$, by Corollary 6.3.29. It follows from the definition of the genus that $g(Y) \geqslant g(X)$. We can now apply Riemann–Hurwitz formula to find that

$$2g(X) - 2 = \deg(f_\sigma)(2g(Y) - 2) + R,$$

where R is a positive quantity. Since $g(Y) \geqslant g(X) \geq 2$, we deduce that $\deg(f_\sigma) = 1$, hence $\sigma = \mathrm{id}$.

The proposition shows that $\mathrm{Aut}(\Sigma_X)$ controls $\mathrm{Aut}(X)$, but it is a very coarse bound when the genus is high. Much better bounds are known, as one can see in the examples below and in the first part of Appendix A.3, containing an outline of further results about automorphisms of Mumford curves, including the case of positive characteristic.

Example 6.5.11 Let X be a Mumford curve such that $\mathrm{Aut}(\Sigma_X) = \{1\}$. Then Proposition 6.5.9 ensures that X has no non-trivial automorphisms as well. Since, up to replacing k with a suitable field extension, every stable metric graph can be realized as the skeleton of a Mumford curve, one can build in this way plenty of examples of Mumford curves without automorphisms. For example, the graph of genus 3 in Fig. 6.4 below has a trivial automorphism group, as long as the edge lengths are generic enough, for example when all lengths are different.

This graph can be obtained by pairwise identifying the ends of a tree as in Fig. 6.5.

One can realize this tree inside $\mathbb{P}_k^{1,\mathrm{an}}$ as the skeleton of a fundamental domain under the action of a Schottky group in many ways. As an example, if $k = \mathbb{Q}_p$ with $p \geq 5$, a suitable Schottky group is obtained by carefully choosing the

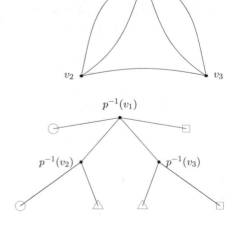

Fig. 6.4 The metric graph Σ_X has trivial group of automorphisms if the edge lengths are all different

Fig. 6.5 The graph in the previous figure is obtained from its universal covering tree by pairwise gluing the ends of the finite sub-tree Σ_F. The gluing is made by identifying the ends that are marked with the same shape

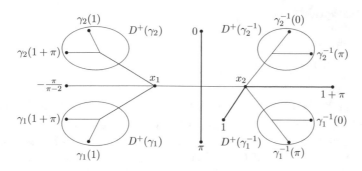

Fig. 6.6 The Schottky figure associated with (γ_1, γ_2)

Koebe coordinates that give rise to the desired skeleton. One can for instance pick $\Gamma = \langle M(0, \infty, p^3), M(1, 2, p^4), M(p, p - 2, p^3) \rangle$ and verify that it gives rise to a fundamental domain whose skeleton is the tree in Fig. 6.5. As a consequence of Theorem 6.5.8, the normalizer of Γ in $PGL_2(k)$ is the group Γ itself.

Example 6.5.12 Assume that k is algebraically closed and that its residue characteristic is different from 2 and 3. Let π, ρ be elements of k satisfying $|\pi| < 1, \rho^3 = 1$ and $\rho \neq 1$. Fix the following elements of $PGL_2(k)$:

$$a = \begin{bmatrix} -\pi & 0 \\ -2 & \pi \end{bmatrix}, \ b = \begin{bmatrix} 1 + \pi - \rho & (1+\pi)(\rho - 1) \\ 1 - \rho & (1+\pi)\rho - 1 \end{bmatrix}.$$

These elements are of finite order, respectively two and three. The fixed rigid points of a are 0 and π, while the fixed rigid points of b are 1 and $1 + \pi$.

Thanks to our assumption that $\mathrm{char}(\widetilde{k}) \neq 2$, the transformation a acting on $\mathbb{P}_k^{1,\mathrm{an}}$ fixes the path joining 0 and π, and sends every open disc whose boundary point lies on this path to a disjoint open disc with the same boundary point. For example, the image by a of the disc $D^-(-\frac{\pi}{\pi-2}, 1)$ is the disc[2] $\mathbb{P}_k^{1,\mathrm{an}} - D^+(0, |\pi|)$, and vice versa.

The same happens for the action of b: the path joining 1 and $1 + \pi$ is fixed, while any open disc with its boundary point on this path is sent to a disjoint open disc with the same boundary point. Since b is of order three, the orbit of such a disc consists of three disjoint discs. For example, the orbit of $D^-(0, 1)$ contains $b(D^-(0, 1)) = D^-(1 - \frac{\pi}{(1+\pi)\rho - 1}, 1)$ and $b^2(D^-(0, 1)) = D^-(1 - \frac{\pi}{(1+\pi)\rho^2 - 1}, 1)$.

Let us consider the elements $\gamma_1 := abab^2$ and $\gamma_2 := ab^2ab$. Using the geometry of a and b described above, one can check that the 4-tuple $(D^+(\gamma_1), D^+(\gamma_1^{-1}), D^+(\gamma_2), D^+(\gamma_2^{-1}))$ represented in Fig. 6.6 provides a Schottky figure adapted to (γ_1, γ_2).

Thanks to Proposition 6.4.24, the existence of a Schottky figure ensures that $\Gamma = \langle \gamma_1, \gamma_2 \rangle$ is a Schottky group of rank 2. Denote its limit set by L. By Theorem 6.4.18,

---

[2]Recall that on the projective line we consider also discs "centered in ∞" such as this one.

Fig. 6.7 The skeleton Σ_X of the Mumford curve uniformized by Γ

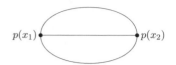

the quotient $X := \Gamma \backslash (\mathbb{P}_k^{1,\text{an}} - L)$ makes sense as a k-analytic space and it is a Mumford curve of genus 2. Let $p \colon (\mathbb{P}_k^{1,\text{an}} - L) \to X$ denote the universal cover. It also follows from Theorem 6.4.18 that the topology of X may be described quite explicitly from the action of Γ. We deduce in this way that the skeleton Σ_X of X is the metric graph represented in Fig. 6.7. By measuring the lengths of the paths joining the boundaries of the discs in the Schottky figure, one can verify that the three edges of Σ_X have equal lengths.

Let us now compute the automorphism group $\text{Aut}(X)$. By Theorem 6.5.8, this can be done by computing the normalizer N of Γ in $\text{PGL}_2(k)$. The elements a and b lie in N, since $\gamma_i a = a \gamma_i^{-1}$ for $i = 1, 2$ and $\gamma_1 b = b \gamma_2^{-1}$, but we can also find elements in N that do not belong to the subgroup generated by a and b. Let

$$c := \begin{bmatrix} 1 + \pi & -\pi(1 + \pi) \\ 2 & -(1 + \pi) \end{bmatrix} \in \text{PGL}_2(k).$$

A direct computation shows that the transformation c is such that $c^2 = \text{id}$, $cac = a$ and $cbc = b^2$, so that c belongs to N. The group $N' = \langle a, b, c \rangle \subset \text{PGL}_2(k)$ is then contained in N, and the quotient N'/Γ is isomorphic to the dihedral group D_6 of order 12. In fact, if we call α, β, γ the respective classes of a, b, c in N'/Γ, we have that $\alpha\beta = \beta\alpha$, and then $\langle \alpha, \beta \rangle$ is a cyclic group of order 6. However, the same computation above shows that γ does not commute with β. The group D_6 is also the automorphism group of the skeleton Σ_X, and so, by Proposition 6.5.9, we have $N = N'$ and the restriction homomorphism $\text{Aut}(X) \to \text{Aut}(\Sigma_X)$ is an isomorphism.

Note that one can extract quite a lot of information from the study of the action of N on $\mathbb{P}_k^{1,\text{an}}$. In this example, $\alpha \in \text{Aut}(X)$ is an order 2 automorphism known as the *hyperelliptic involution*, since it induces a degree 2 cover of the projective line $\varphi : X \to \mathbb{P}_k^{1,\text{an}}$. This last fact can be checked on the skeleton Σ_X by noting that $\alpha(p(x_1)) = p(x_2)$, and hence α has to switch the ends of every edge of Σ_X. As a result, the quotient $X/\langle\alpha\rangle$ is a contractible Mumford curve, and hence it is isomorphic to $\mathbb{P}_k^{1,\text{an}}$.

This description of X as a cover of $\mathbb{P}_k^{1,\text{an}}$ is helpful to compute an explicit equation for the smooth projective curve whose analytification is X. In fact, a genus 2 curve that is a double cover of the projective line can be realized as the smooth compactification of a plane curve of equation

$$y^2 = \prod_{i=1}^{6} (x - a_i),$$

where the $a_i \in k$ are the ramification points of the cover, and the involution defining the cover sends y to $-y$.

In order to find the a_i, we shall first compute the branch locus $B \subset X$ of the hyperelliptic cover. The fixed points of a are 0 and π, so the corresponding points $p(0), p(\pi)$ are in B. The other branch points can be obtained by finding those $x \in \mathbb{P}_k^{1,\mathrm{an}}(k)$ satisfying the condition $\gamma_i(x) = a(x)$ for $i = 1, 2$. We have $\gamma_1(b(0)) = abab^2b(0) = aba(0) = a(b(0))$, and the same applies to $b(\pi)$, so the images by p of these two points are also in B. In the same way, we find that $\gamma_2(b^2(0)) = a(b^2(0))$ and $\gamma_2(b^2(\pi)) = a(b^2(\pi))$. We have found in this way that $B = \{p(0), p(b(0)), p(b^2(0)), p(\pi), p(b(\pi)), p(b^2(\pi))\}$.

To find the ramification locus, we have to compute $\varphi(B)$. Since $\langle \alpha \rangle$ is a normal subgroup of $\mathrm{Aut}(X)$, the element β acts as an automorphism of order 3 of $X/\langle \alpha \rangle \cong \mathbb{P}_k^{1,\mathrm{an}}$. Up to a change of coordinate of this projective line, we can suppose that the fixed points of β are 0 and ∞, so that β is the multiplication by a primititve third root of unity, and that the first ramification point is $a_1 = \varphi(p(0)) = 1$. Then, after possibly reordering them, the remaining ramification points are $a_2 = \rho$, $a_3 = \rho^2$ and a_4, ρa_4, $\rho^2 a_4$, with $|a_4 - 1| < 1$.

With a bit more effort, we can actually compute the value of a_4. To do this, notice that the function p is injective when restricted to the open fundamental domain

$$ F^- = \mathbb{P}_k^{1,\mathrm{an}} - \left(D^+(\gamma_1) \cup D^+(\gamma_1^{-1}) \cup D^+(\gamma_2) \cup D^+(\gamma_2^{-1}) \right). $$

If we set $F' = \varphi \circ p(F^-)$ we then have a two-fold cover $F^- \longrightarrow F' \subset \mathbb{P}_k^{1,\mathrm{an}}$ induced by $\varphi \circ p$, which can be explicitly written as a rational function $z \mapsto \frac{z^2}{(z-\pi)^2}$ (this function can be found by looking at the action of a on F^- explicitly). Note that F^- contains both the fixed rigid points of b, i.e. $1 + \pi$ and 1, and those of a, i.e. 0 and π. When we reparametrize the projective line on the target of φ to get the wanted equation, we are imposing the conditions $\alpha \circ p(1+\pi) \mapsto \infty$, $\alpha \circ p(1) \mapsto 0$ and $\alpha \circ p(0) \mapsto 1 = a_1$. These choices leave only one possibility for the ramification point a_4: it is $\left(\frac{1-\pi}{1+\pi} \right)^2$. We have now found the equation of the plane section of our Mumford curve: it is

$$ y^2 = (x^3 - 1) \cdot \left(x^3 - \frac{(1 - \pi)^6}{(1 + \pi)^6} \right). $$

Note that $|a_4 - a_1| = \left| \frac{(1-\pi)^2}{(1+\pi)^2} - 1 \right| = |\pi|$.

A different example of a hyperelliptic Mumford curve with a similar flavour is discussed in the expository paper [CK05], accompanied with figures and other applications of automorphisms of Mumford curves.

Example 6.5.13 The curve in Example 6.5.12 has the same automorphism group in every characteristic (different from 2 and 3). However, Mumford curves in positive characteristic have in general more automorphism than in characteristic 0.

An interesting class of examples are the so-called *Artin-Schreier-Mumford curves*, first introduced by Subrao in [Sub75]. We sketch here the main results and refer to [CKK10] for more detailed proofs of these facts. Let p be a prime, $q = p^e$ be a power of p, and $k = \mathbb{F}_q((t))$. Let X be the analytification of the curve defined inside $\mathbb{P}_k^1 \times \mathbb{P}_k^1$ by the equation

$$(y^q - y)(x^q - x) = f(t) \text{ with } f \in t\mathbb{F}_q[[t]].$$

This is an ordinary curve in characteristic $p > 0$ with many automorphisms, and for this reason has caught the attention of cryptographers and positive characteristic algebraic geometers alike. One way to study its automorphisms is to observe that X is a Mumford curve. A Schottky group attached to it can be constructed by fixing an element $v \in k$ and looking at the automorphisms of $\mathbb{P}_k^{1,\mathrm{an}}$ of the form

$$a_u = \begin{bmatrix} 1 & u \\ 0 & 1 \end{bmatrix}, b_u = \begin{bmatrix} v & 0 \\ u & v \end{bmatrix} \in \mathrm{PGL}_2(k), \ u \in \mathbb{F}_q^\times.$$

These transformations are all of order p, a_u represent translations by elements of $\mathbb{F}_q^\times$ and b_u their conjugates under the inversion $z \mapsto \frac{v}{z}$. The subgroup $\Gamma_v = \langle a_u^{-1} b_{u'}^{-1} a_u b_{u'} : (u, u') \in \mathbb{F}_q^{\times 2} \rangle$ of $\mathrm{PGL}_2(k)$ is a Schottky group of rank $(q-1)^2$, and for a certain value of v^3 it gives rise to the curve X by Schottky uniformization. The immediate consequence of this fact, is that X is a Mumford curve of genus $(q-1)^2$.

The group of automorphisms $\mathrm{Aut}(X)$ is isomorphic to a semi-direct product $(\mathbb{Z}/p\mathbb{Z})^{2e} \rtimes D_{q-1}$, and its action is easy to describe using the equation of the curve: the elementary abelian subgroup $(\mathbb{Z}/p\mathbb{Z})^{2e}$ consists of those automorphisms of the form $(x, y) \mapsto (x + \alpha, y + \beta)$ with $(\alpha, \beta) \in (\mathbb{F}_q)^2$, while the dihedral subgroup D_{q-1} is generated by $(x, y) \mapsto (y, x)$ and $(x, y) \mapsto (\gamma x, \gamma^{-1} y)$ for $\gamma \in \mathbb{F}_q^\times$. We deduce that the order of $\mathrm{Aut}(X)$ is $2(q-1)q^2$. In characteristic 0, it is not possible to have these many automorphisms, thanks to bounds by Hurwitz and Herrlich that would give rise to a contradiction (see Appendix A.3 for the precise statement of these bounds).

Acknowledgments We warmly thank Marco Maculan for his numerous comments on an earlier version of this text, and the anonymous referees for their remarks and corrections. The Appendix A.3 greatly benefited from insights and remarks by Jan Vonk and Henri Darmon.

While writing this text, the authors were supported by the ERC Starting Grant "TOSSIBERG" (grant agreement 637027). The second author was partially funded by a prize of the *Fondation des Treilles*. The *Fondation des Treilles* created by Anne Gruner-Schlumberger aims to foster the dialogue between sciences and arts in order to further research and creation. She welcomes researchers and creators in her domain of Treilles (Var, France).

---

[3]The relationship between v and $f(t)$ is not immediate, and it is the object of the paper [CKK10].

Appendix: Further Reading

The theory of Berkovich curves has several applications to numerous fields of mathematics, and uniformization plays a role in many of these. A complete description of these applications goes far beyond the scope of the present text, but we would like to provide the interested reader with some hints about the state of the art and where to find more details in the existing literature, as well as point out which simplifications adopted in this text are actually instances of a much richer theory.

A.1 Berkovich Spaces and their Skeleta

We provided a short introduction to the theory of Berkovich curves and their skeleta in Sect. 6.3.2 of this text.

The first discussion of this topic appears already in Chapter 4 of Berkovich's foundational book [Ber90]. In this context, the definition of the skeleton of a Berkovich curve X makes use of formal models and the semi-stable reduction theorem, that states that for the analytification of a smooth proper and geometrically irreducible algebraic curve over k, there exists a finite Galois extension K of k such that the base change X_K has a semi-stable formal model. Berkovich showed that the dual graph of the special fiber of any semi-stable formal model embeds in the curve X_K and that it is invariant by the action of the Galois group $\mathrm{Gal}(K/k)$ over X_K, which allows to define skeleta of X as quotients of skeleta of X_K. This construction is again found in A. Thuillier's thesis [Thu05], where it is exploited to define a theory of harmonic functions on Berkovich curves.

In Definition 6.3.3, we adopted another approach to the study of skeleta, via the use of triangulations. This was first introduced by Ducros in [Duc08] to study étale cohomology groups of Berkovich curves. In the case where k is algebraically closed, a comprehensive exposition of skeletons, retractions, and harmonic functions on non-Archimedean curves can be found in the paper [BPR14]. There, the authors are motivated by connections with tropical geometry, as, for a given algebraic variety over k, the skeletons of its analytification are tightly related to its tropicalization maps. Other than in the aforementioned paper, these connections are exposed in [Wer16], where the higher-dimensional cases are highlighted as well.

As for higher-dimensional spaces, Berkovich introduced skeleta in [Ber99]. They are simplicial sets onto which the spaces retract by deformation. They are constructed using semi-stable formal models and generalizations of them, so they are not known to exist in full generality, but Berkovich nonetheless managed to use them to prove that smooth spaces are locally contractible (hence admits universal covers).

The connections with tropical geometry have proven fruitful, among other things, to study finite covers of Berkovich curves $Y \to X$ over k. The general pattern is that these covers are controlled by combinatorial objects that are enhanced versions of compatible pairs (Σ_Y, Σ_X) of skeletons of the curves Y and X. Assume that k is algebraically closed. Whenever the degree of such a cover is coprime with the residue characteristic of k, the papers [ABBR15a] and [ABBR15b] give conditions on a pair (Σ_Y, Σ_X) to lift to a finite morphism of curves $Y \to X$. In the case of covers of degree divisible by the residue characteristic of k, the situation is still far from understood, but progress has been made thanks to the work of M. Temkin and his collaborators in the papers [CTT16, Tem17, BT20]. The main tool used in these works is the *different function*.

With regard to higher-dimensional varieties, a new approach to skeletons was proposed by E. Hrushosvksi and F. Loeser in [HL16] using techniques coming from model theory. They are able to define skeleta of analytifications of quasi-projective varieties and deduce the remarkable result that any such space has the homotopy type of a CW-complex.

In the specific case of curves over an algebraically closed base field, the paper [CKP18] uses triangulations in order to give a more concrete model-theoretic version of Berkovich curves (and morphisms between them). In particular, the authors manage to give an explicit description of definable subsets of curves and prove some tameness properties.

Without the assumption that k is algebraically closed, or rather that X has a semi-stable formal model over the valuation ring of k, the structure of analytic curves is much harder to grasp, due among other things to the difficulty of classifying virtual discs and virtual annuli. The curious reader will find much food for thought in the book by A. Ducros [Duc], which can nevertheless be of difficult reading for a first approach. If k is a discrete valuation field, a generalization of potential theory on Berkovich curves is provided in [BN16] thanks to a careful study of regular models, and the introduction of the notion of *weight function*. In regard to the problem of determining a minimal extension necessary for the existence of a semi-stable model, an approach via triangulations has been recently proposed in [FT19].

Finally, let us mention that we chose to introduce Berkovich curves as $\mathbb{A}^1$-like curves because we are convinced that this is a natural framework for studying uniformization, but the general theory is much richer, and contains many examples of Berkovich curves that are not $\mathbb{A}^1$-like.

A.2 Non-Archimedean Uniformization in Arithmetic Geometry

In the case of curves over the field of complex numbers, Schottky uniformization can be seen in the context of the classical uniformization theorem for Riemann surfaces, proven independently by P. Koebe and H. Poincaré in 1907. It states that every simply connected complex Riemann surface is conformally equivalent to the complex projective line, the complex affine line, or the Poincaré upper-half plane. As a consequence, the universal covering space of any Riemann surface X is one of these, and when X is compact, Koebe-Poincaré uniformization factors through the Schottky uniformization $\left(\mathbb{P}_{\mathbb{C}}^{1,\mathrm{an}} - L\right) \to X$. A remarkable book on complex uniformization [dSG10] has been written by the group of mathematicians known under the collective name of Henri Paul de Saint-Gervais. It constitutes an excellent reference both on the historical and mathematical aspects of the subject.

In the non-Archimedean case, the history of uniformization is much more recent. The uniformization theory of elliptic curves over a non-Archimedean field $(k, |\cdot|)$ was the main motivation underlying J. Tate's introduction of rigid analytic geometry in the 1960s. Using his novel approach, Tate proved that every elliptic curve with split multiplicative reduction over k is analytically isomorphic to the multiplicative group $k^{\times}/q^{\mathbb{Z}}$ for some q in k with $0 < |q| < 1$. Tate's computations were known to experts, but remained unpublished until 1995, when they were presented in [Tat95] together with a discussion on further aspects of this theory, including automorphic functions, a classification of isogenies of Tate curves, and a brief mention of how to construct "universal" Tate curves over the ring $\mathbb{Z}[[q]][q^{-1}]$ using formal geometry. These formal curves appeared for the first time in the paper [DR73] by P. Deligne and M. Rapoport, who attributed it to M. Raynaud and called them *generalized elliptic curves*. In *loc. cit.* the authors exploited them to give a moduli-theoretic interpretation at the cusps of the modular curves $X_0(Np)$ with $p \nmid N$. Further reading in this direction include the foundational paper [KM85], that concerns the case of modular curves $X(Np^n)$ and [Con07], that provides a more contemporary perspective on generalized elliptic curves.

Interpreting the Schottky uniformization of Mumford curves of [Mum72a] as a higher genus generalization of Tate's theory inspired several novel arithmetic discoveries. One of the most important is the uniformization of Shimura curves, fundamental objects in arithmetic geometry that vastly generalize modular curves. In [Che76], I. Cherednik considered a Shimura curve C associated with a quaternion algebra B over $\mathbb{Q}$. For a prime p where B is ramified, he proved that the p-adic analytic curve $(C \times_{\mathbb{Q}} \mathbb{Q}_p)^{\mathrm{an}}$ can be obtained as a quotient of *Drinfeld p-adic halfplane* $\mathbb{P}_{\mathbb{Q}_p}^{1,\mathrm{an}} - \mathbb{P}_k^{1,\mathrm{an}}(\mathbb{Q}_p)$, by the action of a Schottky group. This Schottky group can be as a subgroup of a different quaternion algebra B' over $\mathbb{Q}$, constructed explicitly from B via a procedure known as *interchange of invariants*. The theory obtained in this way is classically referred to as *Cherednik-Drinfeld uniformization*, since V. Drinfeld gave a different proof of this result in [Dri76], building on a description of C as a moduli space of certain abelian varieties. The excellent paper [BC91] provides a detailed account of these constructions.

By generalizing Drinfeld's modular interpretation, the approach can be extended to some higher dimensional Shimura varieties, resulting in their description as quotients of the Drinfeld upper-half space via a uniformization map introduced independently by G. Mustafin [Mus78] and A. Kurihara [Kur80]. For a firsthand account of the development of this uniformization, we refer the reader to the book [RZ96] by M. Rapoport and T. Zink.

Non-Archimedean uniformization of Shimura varieties has remarkable consequences. First of all, it makes possible to find and describe integral models of Shimura varieties, since the property of being uniformizable imposes restrictions on the special fibers of such models. Furthermore, it gives a way to compute étale and ℓ-adic cohomology groups, as well as the action of the absolute Galois group $\mathrm{Gal}(\overline{\mathbb{Q}_p}/\mathbb{Q}_p)$ on these, making it a powerful tool for studying Galois representations. All the aforementioned results were shown in the framework of formal and rigid geometry. However, more contemporary approaches to uniformization of Shimura varieties and Rapoport-Zink spaces make use of Berkovich spaces (see [Var98, JLV03]), or Huber adic geometry in the form of perfectoid spaces (see [SW13] and [Car19]). In particular, the perfectoid approach can be used to vastly generalize the uniformization of Shimura varieties and establish a theory of *local Shimura varieties*. This construction is exposed in the lecture notes [SW20] by P. Scholze and J. Weinstein.

Local and global uniformization of Shimura varieties are investigated in relation to period mappings, Gauss-Manin connections, and uniformizing differential equations in the book by Y. André [And03], where striking similarities between the complex and p-adic cases are highlighted. For more results about the relevance of Shimura varieties, not necessarily with regard to uniformization, we refer to [Mil05].

Finally, let us mention that Tate's uniformization of elliptic curves with split multiplicative reduction generalizes to abelian varieties. This is also a result of Mumford, contained in the paper [Mum72b], that can be regarded as a sequel to [Mum72a], since the underlying ideas are very similar. In this case, the uniformization theorem is formulated by stating that a totally degenerate abelian variety of dimension g over k is isomorphic to the quotient of the analytic torus $(\mathbb{G}_{m,k}^g)^{\mathrm{an}}$ by the action of a torsion-free subgroup of $(k^\times)^g$. This applies in particular to Jacobians of Mumford curves, a case surveyed in detail in the monograph [Lüt16]. We shall remark that Mumford's constructions are more general than their presentation in this text: they work not only over non-Archimedean fields, but more generally over fields of fractions of complete integrally closed noetherian rings of any dimension.

A.3 The Relevance of Mumford Curves

The uniformization theorem in the complex setting is a very powerful tool, and one of the main sources of analytic methods applied to the study of algebraic curves. This leads to the expectation that, in the non-Archimedean setting, Mumford curves can be more easily studied, turning out to be a good source of examples for testing certain conjectures. This is indeed the case for several topics in algebraic curves and their applications, as we could already sample in Sect. 6.5.2 on the subject of computing the group of automorphisms of curves.

This appendix is a good place to remark that Examples 6.5.12 and 6.5.13 in that section are instances of a much deeper theory. For a smooth projective algebraic curve C of genus $g \geq 2$ over a field of characteristic zero, the Hurwitz bound ensures that the finite group of automorphisms $\text{Aut}(C)$ is of order at most $84(g - 1)$. This bound is sharp: there exist curves of arbitrarily high genus whose automorphism groups attain it, the so-called *Hurwitz curves*. However, if we know that C is (the algebraization of) a Mumford curve, F. Herrlich proved a better bound in [Her80]. Namely, if we denote by p the residue characteristic of K, he showed that:

$$|\text{Aut}(C)| \leq \begin{cases} 48(g - 1) & p = 2 \\ 24(g - 1) & p = 3 \\ 30(g - 1) & p = 5 \\ 12(g - 1) & \text{otherwise.} \end{cases}$$

This result relies on the characterization of automorphism groups of Mumford curves as quotients N/Γ, where Γ is a Schottky group associated with C and N its normalizer in $\text{PGL}_2(K)$ (see Theorem 6.5.8). One can show that the group N acts discontinuously on an infinite tree that contains the universal covering tree of the skeleton $\Sigma_{C^{\text{an}}}$, and use Serre's theory of groups acting on trees to prove that N is an amalgam of finite groups. In his paper, Herrlich achieves the bounds above by classifying those amalgams that contain a Schottky group as a normal subgroup of finite index.

Over a field of characteristic $p > 0$, the Hurwitz bound is replaced by the Stichtenoth bound, stating that $|\text{Aut}(C)| \leq 16g^4$, unless C is isomorphic to a Hermitian curve. When C is a Mumford curve, this bound can be improved in principle using Herrlich's strategy. However, this is not an easy task, as one has to overcome the much bigger difficulties that arise in positive characteristic. This has been achieved recently by M. Van der Put and H. Voskuil, who prove in [VvdP19, Theorem 8.7] that $|\text{Aut}(C)| < \max\{12(g - 1), g\sqrt{8g + 1} + 3\}$ except for three occurrences of (isomorphism classes of) X, which happen when $p = 3$ and $g = 6$. Moreover, in [VvdP19, Theorem 7.1] they show that the bound is achieved for any choice of the characteristic $p > 0$. The bound corrects and extends a bound given by G. Cornelissen, F. Kato and A. Kontogeorgis in [CKK01].

Another application of uniformization of Mumford curves is the *resolution of non-singularities* for hyperbolic curves[1] over $\overline{\mathbb{Q}}_p$. Given such a curve X, and a smooth point P of the special fiber of a semi-stable model of X, it is an open problem to find a finite étale cover $Y \longrightarrow X$ such that a whole irreducible component of the special fiber of the stable model of Y lies above P. Earlier versions of this problem were introduced and proved by S. Mochizuki [Moc96] and A. Tamagawa [Tam04], that showed connections with important problems in

---

[1] A hyperbolic curve in this context is a genus g curve with n marked points satisfying the inequality $2g - 2 + n > 0$.

anabelian geometry. The interest of the version proposed here is also motivated by anabelian geometry: F. Pop and J. Stix proved in [PS17] that any curve for which resolution of non-singularities holds satisfies also a valuative version of Grothendieck's section conjecture. In the paper [Lep13], E. Lepage uses Schottky uniformization in a Berkovich setting to show that resolution of non-singularities holds when X is a hyperbolic Mumford curve. His approach consists in studying μ_{p^n}-torsors of the universal cover of X, which are better understood since they can be studied using logarithmic differentials of rational functions. With this technique, he can show that there is a dense subset of type 2 points $\mathcal{V} \in X$, with the following property: every $x \in \mathcal{V}$ can be associated with a μ_{p^n}-torsor $\tau : Y \to X$ such that $\tau^{-1}(x)$ is a point of positive genus. This last condition ensures that the corresponding residue curve is an irreducible component of the stable model of Y.

Mumford curves have been also proven useful in purely analytic contexts, for instance to study potential theory and differential forms. Using the fact that all type 2 points in a Mumford curve are of genus 0, P. Jell and V. Wanner [JW18] are able to establish a result of Poincaré duality and compute the Betti numbers of the tropical Dolbeaut cohomology arising from the theory of bi-graded real valued differential forms developed in [CLD12].

Finally, let us mention that archimedean and non-archimedean Schottky uniformizations can be studied in a unified framework thanks to work of the authors [PT], where a moduli space $\mathcal{S}_g$ parametrizing Schottky groups of fixed rank g over all possible valued fields is constructed for every $g \geq 2$. This construction is performed in the framework of *Berkovich spaces over* $\mathbb{Z}$ developed in [Poi10, Poi13, LP]. More precisely, the space $\mathcal{S}_g$ is realized as an open, path-connected subspace of $\mathbb{A}_{\mathbb{Z}}^{3g-3,\mathrm{an}}$, it is endowed with a natural action of the group $\mathrm{Out}(F_g)$ of outer automorphisms of the free group, and exhibits interesting connections with other constructions of moduli spaces, in the frameworks of tropical geometry and geometric group theory. The space $\mathcal{S}_g$ seems to be ideal to study phenomena of degeneration of Schottky groups from archimedean to non-archimedean.

A different take on the interplay between archimedean and non-archimedean Schottky uniformizations is provided by Y. Manin's approach to Arakelov geometry. In the paper [Man91] several formulas for computing the Green function on a Riemann surface using Schottky uniformization and are explicitly inspired by Mumford's construction. These formulas involve the geodesics lengths in the hyperbolic handlebody uniformized by the Schottky group associated with such a surface, suggesting connections between hyperbolic geometry and non-archimedean analytic geometry. This result has been reinterpreted in term of noncommutative geometry by C. Consani and M. Marcolli [CM04] by replacing the Riemann surface with a noncommutative space that encodes certain properties of the archimedean Schottky uniformization. This noncommutative formalism has led to applications both in the non-archimedean world (see for example [CM03]) and in the archimedean one, for instance to Riemannian geometry in [CM08]. We think that the theory of Berkovich spaces could fit nicely in this picture, and it would be an interesting project to investigate the relations between noncommutative geometric objects related to Schottky uniformization (e.g. graph C^*-algebras) and Mumford curves in the Berkovich setting.

References

[ABBR15a] O. Amini, M. Baker, E. Brugallé, J. Rabinoff, Lifting harmonic morphisms I: metrized complexes and Berkovich skeleta. Res. Math. Sci. **2**, Art. 7, 67 (2015)

[ABBR15b] O. Amini, M. Baker, E. Brugallé, J. Rabinoff, Lifting harmonic morphisms II: Tropical curves and metrized complexes. Algebra Number Theory **9**(2), 267–315 (2015)

[And03] Y. André, Period mappings and differential equations : from $\mathbb{C}$ to $\mathbb{C}_p$, in *MSJ Memoirs*, vol. 12 (Mathematical Society of Japan, Tokyo, 2003)

[BC91] J.-F. Boutot, H. Carayol, Uniformisation p-adique des courbes de Shimura: Les théorèmes de Čerednik et de Drinfeld, in *Courbes Modulaires et Courbes de Shimura. C. R. Sémin., Orsay/Fr. 1987-88*(196-197) in Astérisque (SMF, New York, 1991), pp. 45–158

[Ber90] V.G. Berkovich, Spectral theory and analytic geometry over non-Archimedean fields, in *Mathematical Surveys and Monographs*, vol. 33 (American Mathematical Society, Providence, 1990)

[Ber99] V.G. Berkovich, Smooth p-adic analytic spaces are locally contractible. Invent. Math. **137**(1), 1–84 (1999)

[BL85] S. Bosch, W. Lütkebohmert, Stable reduction and uniformization of abelian varieties. I. Math. Ann. **270**(3), 349–379 (1985)

[BN16] M. Baker, J. Nicaise, Weight functions on Berkovich curves. Algebra Number Theory **10**(10), 2053–2079 (2016)

[Bou71] N. Bourbaki, *Éléments de mathématique. Topologie générale. Chapitres 1 à 4* (Hermann, Paris, 1971)

[BPR14] M. Baker, S. Payne, J. Rabinoff, On the structure of non-archimedean analytic curves. Trop. Non-Archimedean Geom. **605**, 93 (2014)

[BR10] M. Baker, R. Rumely, Potential theory and dynamics on the Berkovich projective line, in *Mathematical Surveys and Monographs*, vol. 159 (American Mathematical Society, Providence, 2010)

[BT20] U. Brezner, M. Temkin, Lifting problem for minimally wild covers of Berkovich curves. J. Algebraic Geom **29**, 123–166 (2020)

[Car19] A. Caraiani, *Perfectoid Spaces: Lectures from the 2017 Arizona Winter School (Mathematical Surveys and Monographs)*, chapter Perfectoid Shimura varieties (American Mathematical Society, New York, 2019), pp. 258–307

[Che76] I.V. Cherednik, Uniformization of algebraic curves by discrete arithmetic subgroups of $PGL_2(k_w)$ with compact quotient spaces. (Russian) Mat. Sb. (N.S.) **100**(1), 59–88 (1976)

[CK05] G.L.M. Cornelissen, F. Kato, The p-adic icosahedron. Not. Am. Math. Soc. **52**, 08 (2005)

[CKK01] G. Cornelissen, F. Kato, A. Kontogeorgis, Discontinuous groups in positive characteristic and automorphisms of Mumford curves. Math. Ann. **320**(1), 55–85 (2001)

[CKK10] G. Cornelissen, F. Kato, A. Kontogeorgis, The relation between rigid-analytic and algebraic deformation parameters for Artin-Schreier-Mumford curves. Isr. J. Math. **180**, 345–370 (2010)

[CKP18] P. Cubides Kovacsics, J. Poineau, Definable sets of Berkovich curves. J. Inst. Math. Jussieu, 1–65 (2019).

[CLD12] A. Chambert-Loir, A. Ducros, Formes différentielles réelles et courants sur les espaces de Berkovich (2012). https://arxiv.org/abs/1204.6277

[CM03] C. Consani, M. Marcolli, Spectral triples from Mumford curves. Int. Math. Res. Not. **2003**, 11 (2003)

[CM04] C. Consani, M. Marcolli, Noncommutative geometry, dynamics, and ∞-adic Arakelov geometry. Sel. Math. **10**(2), 167 (2004)

[CM08] G. Cornelissen, M. Marcolli, Zeta functions that hear the shape of a Riemann surface. J. Geom. Phys. **58**(5), 619–632 (2008)

[Con07] B. Conrad, Arithmetic moduli of generalized elliptic curves. J. Inst. Math. Jussieu **6**, 04 (2007)

[CTT16] A. Cohen, M. Temkin, D. Trushin, Morphisms of Berkovich curves and the different function. Adv. Math. **303**, 800–858 (2016)

[DR73] P. Deligne, M. Rapoport, Les schémas de modules des courbes elliptiques, in *Modular Functions of One Variable II. Lecture Notes in Mathematics*, vol. 349 (Springer, Berlin, 1973), pp. 143–316

[Dri76] V.G. Drinfeld, Coverings of p-adic symmetric regions. Funct. Anal. and Appl. **10**, 107–115 (1976)

[dSG10] H.P. de Saint Gervais. *Uniformisation des surfaces de Riemann. Retour sur un théorème centenaire*, ENS Éditions (2010), 544 pp.

[Duc] A. Ducros, *La Structure des Courbes Analytiques*. http://www.math.jussieu.fr/~ducros/trirss.pdf.

[Duc08] A. Ducros, Triangulations et cohomologie étale sur une courbe analytique p-adique. J. Algebraic Geom. **17**(3), 503–575 (2008)

[FT19] L. Fantini, D. Turchetti, *Triangulations of Non-Archimedean Curves, Semi-stable Reduction, and Ramification* (2019). https://arxiv.org/abs/1911.04407

[GvdP80] L. Gerritzen, M. van der Put, *Schottky Groups and Mumford Curves. Lecture Notes in Mathematics*, vol. 817 (Springer, Berlin, 1980)

[Her80] F. Herrlich, Die Ordnung der Automorphismengruppe einer p-adischen Schottkykurve. Math. Ann. **246**, 125–130 (1980)

[HL16] E. Hrushovski, F. Loeser, *Non-Archimedean Tame Topology and Stably Dominated Types. Annals of Mathematics Studies*, vol. 192 (Princeton University, Princeton, 2016), vol. 192

[JLV03] B.W. Jordan, R. Livné, Y. Varshavsky, Local points of twisted Mumford quotients and Shimura curves. Math. Ann. **327**(3), 409–428 (2003)

[JW18] P. Jell, V. Wanner, Poincaré duality for the tropical Dolbeault cohomology of non-archimedean Mumford curves. J. Number Theory **187**, 344–371 (2018)

[KM85] N. Katz, B. Mazur, *Arithmetic Moduli of Elliptic Curves (AM-108)* (Princeton University, Princeton, 1985)

[Kur80] A. Kurihara, Construction of p-adic unit balls and the Hirzebruch proportionality. Am. J. Math. **102**(3), 565–648 (1980)

[Lep13] E. Lepage, Resolution of nonsingularities for Mumford curves. Publ. Res. Inst. Math. Sci. **49**(4), 861–891 (2013)

[Liu87a] Q. Liu, Ouverts analytiques d'une courbe algébrique en géométrie rigide. Ann. de l'Institut Fourier **37**, 01 (1987)

[Liu87b] Q. Liu, *Ouverts Analytiques d'une courbe projective sur un corps valué complet, ultramétrique, algébriquement clos*. PhD thesis (Université de Bordeaux, France 1987). https://www.math.u-bordeaux.fr/~qliu/articles/thesis_QL.pdf

[LP] T. Lemanissier, J. Poineau, Espaces de Berkovich sur **Z** : catégorie, topologie, cohomologie (2020). https://arxiv.org/abs/2010.08858

[Lüt16] W. Lütkebohmert, *Rigid geometry of curves and their Jacobians. Ergebnisse der Mathematik und ihrer Grenzgebiete*, vol. 61 (Springer, Berlin, 2016)

[Man91] Y.I. Manin, Three-dimensional hyperbolic geometry as ∞-adic Arakelov geometry. Invent. Math. **104**(1), 223–243 (1991)

[Mar07] A. Marden, *Outer Circles: An Introduction to Hyperbolic 3-manifolds* (Cambridge University, Cambridge, 2007)

[Mil05] J.S. Milne, *Harmonic Analysis, the Trace Formula, and Shimura Varieties* (Clay Mathematical Proceedings of the American Mathematical Society, Providence, 2005), pp. 265–378. Chapter Introduction to Shimura varieties

[Moc96] S. Mochizuki, The profinite Grothendieck conjecture for closed hyperbolic curves over number fields. J. Math. Sci.-Univ. Tokyo **3**(3), 571–628 (1996)

[MSW15] D. Mumford, C. Series, D. Wright, *Indra's pearls. The vision of Felix Klein. With cartoons by Larry Gonick* (Cambridge University, Cambridge, 2015). reprint of the 2002 hardback edition edition

[Mum72a] D. Mumford, An analytic construction of degenerating curves over complete local rings. Compos. Math. **24**(2), 129–174 (1972)

[Mum72b] D. Mumford, An analytic construction of degenerating abelian varieties over complete rings. Compos. Math. **24**(3), 239–272 (1972)

[Mus78] G.A. Mustafin, Nonarchimedean uniformization. Math. USSR, Sb. **34**, 187–214 (1978)

[Pel20] F. Pellarin, From the Carlitz exponential to Drinfeld modular forms, in *Arithmetic and Geometry Over Local Fields*, ed. by B. Anglès, T. Ngo Dac (Springer International Publishing, Cham, 2020)

[Poi10] J. Poineau, La droite de Berkovich sur **Z**. Astérisque **334**, xii+284 (2010)

[Poi13] J. Poineau, Espaces de Berkovich sur **Z**: étude locale. Invent. Math. **194**(3), 535–590 (2013)

[PS17] F. Pop, J. Stix, Arithmetic in the fundamental group of a p-adic curve. On the p-adic section conjecture for curves. Journal für die reine und angewandte Mathematik **2017**(725), 1–40 (2017)

[PT] J. Poineau, D. Turchetti, Schottky spaces and universal Mumford curves over **Z** (in preparation). https://poineau.users.lmno.cnrs.fr/Textes/MumfordZ.pdf

[PT20] J. Poineau, D. Turchetti, Berkovich curves and Schottky uniformization I: The Berkovich affine line, in *Arithmetic and Geometry Over Local Fields*, ed. by B. Anglès, T. Ngo Dac (Springer International Publishing, Cham, 2020)

[RZ96] M. Rapoport, T. Zink, *Period Spaces for p-divisible Groups. Annals of Mathematics Studies* (Princeton University, Princeton, 1996)

[Sub75] D. Subrao, The p-rank of Artin-Schreier curves. Manuscr. Math. **16**, 169–193 (1975)

[SW13] P. Scholze, J. Weinstein, Moduli of p-divisible groups. Cambridge J. Math. **1**(2), 145–237 (2013)

[SW20] P. Scholze, J. Weinstein, *Berkeley Lectures on p-adic Geometry (Annals of Mathematics Studies Book 389)* (Princeton University, Princeton, 2020)

[Tam04] A. Tamagawa, Resolution of nonsingularities of families of curves. Publ. Res. Inst. Math. Sci. **40**(4), 1291–1336 (2004)

[Tat95] J. Tate, A review of non-Archimedean elliptic functions, in *Elliptic Curves, Modular Forms, and Fermat's Last Theorem (Hong Kong, 1993)*. Series Number Theory, I (International Press, Cambridge, 1995), pp. 162–184

[Tem17] M. Temkin, Metric uniformization of morphisms of Berkovich curves. Adv. Math. **317**, 438–472 (2017)

[Thu05] A. Thuillier, *Théorie du potentiel sur les courbes en géométrie analytique non-Archimédienne. Applications à la théorie d'Arakelov.* PhD thesis (Université de Rennes, France, 2005)

[Var98] Y. Varshavsky, p-adic uniformization of unitary shimura varieties. Publ. Mathématiques de l'IHÉS **87**, 57–119 (1998)

[VvdP19] H. Voskuil, M. van der Put, Mumford curves and Mumford groups in positive characteristic. J. Algebra **517**, 119–166 (2019)

[Wer16] A. Werner, Analytification and tropicalization over non-archimedean fields, in *Nonarchimedean and Tropical Geometry*, ed. by M. Baker, S. Payne (Springer, Berlin, 2016), pp. 145–171

Chapter 7
On the Stark Units of Drinfeld Modules

Floric Tavares Ribeiro

Abstract We present the notion of Stark units and various techniques involving it. The Stark units constitute a useful tool to study the unit and class modules of a Drinfeld module as defined by Taelman. We review some recent results on Drinfeld $\mathbb{F}_q[\theta]$-modules which make use of this notion. In particular, we present the "discrete Greenberg conjectures" which explain the structure of the class module of the canonical multi-variable deformations of the Carlitz module, and a result on the non vanishing modulo a given prime of a class of Bernoulli-Carlitz numbers.

7.1 Introduction

This text aims to constitute an introduction, largely accessible to non specialist readers, to the notion of *Stark units* of Drinfeld modules. The germs of the concept of Stark units can be found in [APTR16, APTR18]. The notion has been conceptualized in [ATR17] for Drinfeld modules over $\mathbb{F}_q[\theta]$ and then further developed in the general context of Drinfeld modules in [ANDTR17] and in [ANDTR20a] for t-modules.

Let $\mathbb{F}_q$ be a finite field with q elements, θ be an indeterminate over $\mathbb{F}_q$, $A = \mathbb{F}_q[\theta]$, and B be a finite integral extension of A, and denote by τ the map $x \mapsto x^q$. A Drinfeld A-module defined over B is a ring homomorphism $\phi : A \to B[\tau], a \mapsto \phi_a$ where $\phi_a \equiv a \pmod{\tau}$. We first define the z-deformation of ϕ which consists in twisting the Frobenius τ by a new variable z which commutes with τ. This can be obtained simply by the formula $\tilde{\phi} : A \to B[z][\tau], a \mapsto \tilde{\phi}_a$ where, if $\phi_a = \sum_{i=0}^{r} a_i \tau^i$, then $\tilde{\phi}_a = \sum_{i=0}^{r} a_i z^i \tau^i$.

F. Tavares Ribeiro (✉)
Laboratoire de Mathématiques Nicolas Oresme, Université de Caen Normandie,
Normandie University, Caen, France
e-mail: floric.tavares-ribeiro@unicaen.fr

© The Author(s), under exclusive license to Springer Nature Switzerland AG 2021
B. Anglès, T. Ngo Dac (eds.), *Arithmetic and Geometry over Local Fields*,
Lecture Notes in Mathematics 2275, https://doi.org/10.1007/978-3-030-66249-3_7

This naive construction reveals its interest when one computes the unit module $U(\phi)$ of the Drinfeld module ϕ. This unit module, introduced by L. Taelman along with the class module (see [Tae12]), is, roughly speaking, the A-module of the elements which map to integral elements via the exponential map associated to ϕ. One can obtain a submodule of finite index of $U(\phi)$ by computing $U(\widetilde{\phi})$ and then evaluating at $z = 1$. This is the module of Stark units of ϕ.

The terminology of Stark units comes from the remark of Anderson from [And96] that the elements that he constructed play a role similar to the circular units, which generalize in the classical case to Stark units. The idea of considering Stark units indeed arose from investigations on log-algebraicity. A log-algebraicity result consists in the construction of a specific unit from the L-series of a Drinfeld module. The concept of log-algebraicity is due to D. Thakur and has been notably developed by G. Anderson in [And94, And96]. It has become a very lively topic in the current research. In a log-algebraicity statement, one in fact builds an element in $U(\widetilde{\phi})$, its evaluation at $z = 1$ is then always a Stark unit.

We can track this analogy in particular in Theorem 7.4.6 which states that the Fitting ideal of the quotient of $U(\phi)$ by the module of Stark units is equal to the Fitting ideal of the class module of ϕ.

The chapter is organized as follows. We start defining the basic notions involved in the theory of Drinfeld A-modules and introduce the tools which are necessary to state Taelman's class formula. The first three sections are meant to be self contained and present the general machinery of Stark units in the case of Drinfeld $\mathbb{F}_q[\theta]$-modules. This machinery has been generalized for Anderson A-modules with general A without difficulty. We invite the interested reader to [ANDTR17, ANDTR20a] for more details.

We present in Sect. 7.5 several class formulas and explain how Stark units appear in these formulas or can be computed from them.

We then turn to a slightly more general kind of objects, which are deformations of Drinfeld modules, in particular the multi-variable "canonical" deformations of the Carlitz module, which is canonical in the sense that the Carlitz module is deformed by its own shtuka function. This is a key object for arithmetic applications that we then review. First we show that the class module of the canonical deformation of the Carlitz module is, depending on the case, pseudo cyclic or pseudo null, which reminds of the classical Greenberg conjectures. Then we prove that, given a prime P, almost all Bernoulli-Carlitz numbers of a certain form do not vanish modulo P.

We finish with some words on Stark units in more general settings.

Some new proofs are given when possible and references are provided along the way. For the general references on Drinfeld modules, we refer the reader to [Gos96, Ros02, Tha04]. There are also obvious links between this survey and F. Pellarin's contribution [Pel20] to this volume, although the settings and notation might sometimes differ.

7.2 Background

After some notation, we present in this section the notions of Fitting ideals and ratios of covolumes which will be needed later, in particular in Sect. 7.5 to state class formulas.

7.2.1 Notation

We will use the following notation:

- $\mathbb{F}_q$: a finite field with q elements, of characteristic p,
- θ: an indeterminate over $\mathbb{F}_q$,
- $A = \mathbb{F}_q[\theta]$, $K = \mathbb{F}_q(\theta)$, $K_\infty = \mathbb{F}_q((\frac{1}{\theta}))$,
- v_∞: the valuation at the place ∞ such that $v_\infty(\theta) = -1$,
- $\mathbb{C}_\infty$: the completion of a fixed algebraic closure of K_∞,
- $\tau : \mathbb{C}_\infty \to \mathbb{C}_\infty, x \mapsto x^q$ the Frobenius endomorphism.

Note that K_∞ is the completion of K with respect to v_∞.

If k is a field containing $\mathbb{F}_q$, we set $(kK)_\infty = k \widehat{\otimes}_{\mathbb{F}_q} K_\infty = k((\frac{1}{\theta}))$. This is a field. If $x \in (kK)_\infty^\times$, we can write x uniquely as $x = \sum_{n \geq N} x_n \frac{1}{\theta^n}$, $x_n \in k$ with $x_N \neq 0$. Then we call $x_N \in k^\times$ the sign of x and write $\mathrm{sgn}(x) = x_N$. We say that such an $x \in (kK)_\infty$ is *monic* if $\mathrm{sgn}(x) = 1$. The valuation v_∞ extends naturally to $(kK)_\infty$ which is complete with respect to this valuation.

If L is a finite extension of K we denote by O_L the integral closure of A in L. We write $L_\infty = L \otimes_K K_\infty$ and if k is a field containing $\mathbb{F}_q$, $(kL)_\infty = L \otimes_K (kK)_\infty$. Note that $(kL)_\infty \simeq L_\infty$ when $k = \mathbb{F}_q$. As a finite dimensional $(kK)_\infty$-vector space, $(kL)_\infty$ is endowed with a natural topology. Moreover, O_{kL} or kO_L will denote the sub-k-vector space of $(kL)_\infty$ spanned by O_L. This is isomorphic to $k \otimes_{\mathbb{F}_q} O_L$.

The Frobenius homomorphism τ extends uniquely to a continuous homomorphism on $(kL)_\infty$ by putting $\tau(x) = x$ for all $x \in k$. We then have $\tau(O_{kL}) \subset O_{kL}$.

A case of particular interest in this text will be $k = \mathbb{F}_q(z)$ where z is a new indeterminate over $\mathbb{F}_q$. In this case, we will consider the Tate algebra

$$\mathbb{T}_z(L_\infty) := \left\{ \sum_{n \geq 0} a_n z^n \; ; \; a_n \in L_\infty, \lim_{n \to \infty} a_n = 0 \right\} \subset (\mathbb{F}_q(z)L)_\infty.$$

We have also the description $\mathbb{T}_z(K_\infty) \simeq \mathbb{F}_q[z][[\frac{1}{\theta}]]$ and more generally

$$\mathbb{T}_z(L_\infty) \simeq \mathbb{F}_q[z][[\frac{1}{\theta}]] \otimes_K L.$$

Remark that $\tau(\mathbb{T}_z(L_\infty)) \subset \mathbb{T}_z(L_\infty)$, and $\mathbb{T}_z(L_\infty) \cap O_{\mathbb{F}_q(z)L} = O_L[z]$.

It will be useful to also use the notation $(\mathbb{F}_q[z]L)_\infty = (L[z])_\infty = \mathbb{T}_z(L_\infty)$, and $O_{\mathbb{F}_q[z]L} = O_{L[z]} = O_L[z]$ so that if k denotes either $\mathbb{F}_q$, $\mathbb{F}_q(z)$ or $\mathbb{F}_q[z]$, then $(kL)_\infty$ stands respectively for L_∞, $(\mathbb{F}_q(z)L)_\infty$ or $\mathbb{T}_z(L_\infty)$, and O_{kL} for O_L, $O_{\mathbb{F}_q(z)L}$ or $O_L[z]$.

7.2.2 Fitting Ideals

In this section, we review basic facts on the theory of Fitting ideals. The standard references are the appendix to [MW84] and [Nor76, Eis95, Lan02].

We fix a commutative ring $\mathcal{R}$ and consider a finitely presented $\mathcal{R}$-module M. If for $a, b \in \mathbb{N}$,

$$\mathcal{R}^a \longrightarrow \mathcal{R}^b \longrightarrow M \longrightarrow 0$$

is a presentation of M, and if X is the matrix of the map $\mathcal{R}^a \to \mathcal{R}^b$ then one defines $\mathrm{Fitt}_M(\mathcal{R})$ to be the ideal of $\mathcal{R}$ generated by all the $b \times b$ minors of X if $b \le a$, and $\mathrm{Fitt}_{\mathcal{R}}(M) = 0$ if $b > a$. This is independent from the presentation chosen for M. Note that if M is torsion, one has $b \le a$.

In the case where $\mathcal{R}$ is a principal ideal domain (or more generally a Dedekind domain), the structure theorem asserts that if M is a torsion $\mathcal{R}$-module, then there exist ideals $I_1, \ldots, I_n$ of $\mathcal{R}$ such that M is isomorphic to the product $\mathcal{R}/I_1 \times \cdots \times \mathcal{R}/I_n$. This implies that $\mathrm{Fitt}_{\mathcal{R}}(M) = \prod_{i=1}^n I_i$. Fitting ideals are also multiplicative in exact sequences. That is, if $0 \to M_1 \to M \to M_2 \to 0$ is exact, then

$$\mathrm{Fitt}_{\mathcal{R}}(M_1) \cdot \mathrm{Fitt}_{\mathcal{R}}(M_2) = \mathrm{Fitt}_{\mathcal{R}}(M). \tag{7.1}$$

This can be deduced, for instance, from [Bou65, VII. §4 n.5 Proposition 10].

In the case where k is a field and $\mathcal{R} = k[\theta]$ we will denote by $[M]_{k[\theta]}$ the monic generator of $\mathrm{Fitt}_{k[\theta]}(M)$. Remark that in this case, there is a simple way to compute this quantity:

$$[M]_{k[\theta]} = \det_{k[Z]}(Z - \theta \mid M)\mid_{Z=\theta}. \tag{7.2}$$

We fix a field $k \supset \mathbb{F}_q$ such that $\mathbb{F}_q$ is algebraically closed in k. As an example, one can choose $k = \mathbb{F}_q(z)$. Let us write $\mathcal{R} = k[\theta]$. Let G be a finite abelian group whose order is prime to p. Let us denote by $\widehat{G} = \mathrm{Hom}(G, \overline{\mathbb{F}_q}^\times)$ the set of characters on G. For $\chi \in \widehat{G}$, we denote by $\mathbb{F}_q(\chi)$ the (finite, Galois) extension of $\mathbb{F}_q$ generated by the values of χ:

$$\mathbb{F}_q(\chi) := \mathbb{F}_q[\chi(g), g \in G].$$

And similarly,

$$k(\chi) := k[\chi(g), g \in G]$$

is the compositum of k and $\mathbb{F}_q[\chi]$ and is just isomorphic to $k \otimes_{\mathbb{F}_q} \mathbb{F}_q[\chi]$.

For $\chi \in \widehat{G}$, we define the idempotent

$$e_\chi := \frac{1}{|G|} \sum_{g \in G} \chi(g) g^{-1} \in \mathbb{F}_q(\chi)[G].$$

If $\chi \in \widehat{G}$, we also define:

$$[\chi] := \{\sigma \circ \chi, \ \sigma \in \mathrm{Gal}(\mathbb{F}_q(\chi)/\mathbb{F}_q)\} \subset \widehat{G}$$

and the corresponding idempotent:

$$e_{[\chi]} = \sum_{\psi \in [\chi]} e_\psi \in \mathbb{F}_q[G].$$

We define a map $e_{[\chi]}\mathbb{F}_q[G] \rightarrow \mathbb{F}_q(\chi)$ by associating, for $x \in \mathbb{F}_q[G]$, to $e_{[\chi]}x$ the unique $\lambda \in \mathbb{F}_q(\chi)$ such that $e_\chi x = \lambda e_\chi$ in $\mathbb{F}_q(\chi)[G]$. It is not hard to check that this is a well defined isomorphism, and thus it induces isomorphisms $e_{[\chi]}k[G] \rightarrow k(\chi)$ and $e_{[\chi]}\mathcal{R}[G] \rightarrow k(\chi)[\theta] = \mathcal{R}(\chi)$. Remark that the notion of a *monic element in* $e_{[\chi]}\mathcal{R}[G]$ is then well defined and does not depend on the choice of the representative χ of $[\chi]$.

Then, $\mathcal{R}[G]$ is the direct sum of its $[\chi]$-components $e_{[\chi]}\mathcal{R}[G]$. It is thus a principal ideal ring, and the notion of monic elements on each component leads to a natural notion of monic elements on $\mathcal{R}[G]$. Thus, if M is an $\mathcal{R}[G]$-module which is finite dimensional over k, then we can define $[e_{[\chi]}M]_{e_{[\chi]}\mathcal{R}[G]}$ for all character χ, and

$$[M]_{\mathcal{R}[G]} = \sum_\chi [e_{[\chi]}M]_{e_{[\chi]}\mathcal{R}[G]} \in e_{[\chi]}\mathcal{R}[G].$$

If M is now an $\mathcal{R}(\chi)[G]$ module which is finite dimensional over k, then we can define in a similar way $[e_\chi M]_{e_\chi \mathcal{R}(\chi)[G]} \in e_\chi \mathcal{R}(\chi)[G] = \mathcal{R}(\chi)e_\chi$. So if M is an $\mathcal{R}[G]$-module which is finite dimensional over k, we can set $M(\chi) := M \otimes_\mathcal{R} \mathcal{R}(\chi)$ and then we remark that:

$$\left[e_{[\chi]}M\right]_{e_{[\chi]}\mathcal{R}[G]} = \sum_{\psi \in [\chi]} \left[e_\psi M(\psi)\right]_{e_\psi \mathcal{R}[G]} \in \mathcal{R}[G].$$

If now M is a free $\mathcal{R}[G]$-module, then we also have the equality:

$$[M]_{\mathcal{R}[G]} = \det_{k[G][Z]} (Z - \theta \mid M)|_{Z=\theta}.$$

7.2.3   Ratio of Covolumes

We define here $k[\theta]$-lattices and the notion of ratio of covolumes which will be used to compare two lattices.

We fix k a field containing $\mathbb{F}_q$ and recall that $(kK)_\infty = k\widehat{\otimes}_{\mathbb{F}_q} K_\infty = k((\frac{1}{\theta}))$. In what follows, we fix V to be a finite dimensional $(kK)_\infty$-vector space endowed with the natural topology coming from $(kK)_\infty$.

Definition 7.2.1 A sub-$k[\theta]$-module M of V is a $k[\theta]$-*lattice* in V if M is discrete in V and if M generates V over $(kK)_\infty$.

Lemma 7.2.2 *Let M be a sub-$k[\theta]$-module of V. If M is discrete in V, then M is finitely generated over $k[\theta]$ and its rank is lower or equal to the dimension of V over $(kK)_\infty$. Equality holds if, and only if, M is a $k[\theta]$-lattice in V.*

Proof We choose a norm of $(kK)_\infty$-vector space on V. Let $e_1 \in M$ be an element of minimal norm among the non zero elements of M. Let d be the dimension of the $(kK)_\infty$-vector space generated by M. We build by induction a family $(e_1, \ldots, e_d)$ of elements of M such that for $1 \le i \le d$, e_i is an element of minimal norm among the non zero elements of $M \setminus ((kK)_\infty e_1 \oplus \cdots \oplus (kK)_\infty e_{i-1})$. If $x \in M$, then there are $\lambda_1, \ldots, \lambda_d \in (kK)_\infty$ such that $x = \sum_{i=1}^d \lambda_i e_i$. For $1 \le i \le d$, write $\lambda_i = \lambda_{i,0} + \lambda_{i,1}$ with $\lambda_{i,0} \in k[\theta]$ and $\lambda_{i,1} \in \frac{1}{\theta}k[\frac{1}{\theta}]$. Then

$$x - \sum_{i=1}^d \lambda_{i,0} e_i = \sum_{i=1}^d \lambda_{i,1} e_i \in M. \tag{7.3}$$

Let j be the maximal index, if it exists, for which $\lambda_{j,1} \ne 0$. Then (7.3) contradicts the minimality of e_j. We therefore must have $\lambda_{i,1} = 0$ for all i, and thus, $M = \bigoplus_{i=1}^d k[\theta]e_i$. We get the desired inequality.

This also proves that the dimension of the $(kK)_\infty$-vector space generated by M is the rank of M, whence the case of equality. $\square$

As an immediate consequence, we can state:

Proposition 7.2.3 *Let M be a sub-$k[\theta]$-module of V. The following are equivalent:*

  (i) M is a $k[\theta]$-lattice in V,
  (ii) There exists a $(kK)_\infty$-basis $(e_1, \ldots, e_n)$ of V such that M is the free $k[\theta]$-module of basis $(e_1, \ldots, e_n)$,
  (iii) M is discrete in V and its $k[\theta]$-rank is equal to the dimension of V over $(kK)_\infty$.

We can now proceed with the definition of ratio of co-volumes of lattices.

Let M and M' be two $k[\theta]$-lattices in V. Let $\mathcal{B}$ and $\mathcal{B}'$ be $k[\theta]$-bases of M and M', respectively. The ratio of co-volumes of M in M' is then defined as

$$\left[M' : M\right]_{k[\theta]} = \frac{\det_{\mathcal{B}'} \mathcal{B}}{\text{sgn}(\det_{\mathcal{B}'} \mathcal{B})} \in (kK)_\infty.$$

Note that this is independent of the choices of $\mathcal{B}$ and $\mathcal{B}'$.

Remark 7.2.4

- The definition immediately implies that if M_0, M_1 and M_2 are lattices in V, then

$$[M_0 : M_1]_{k[\theta]} [M_1 : M_2]_{k[\theta]} = [M_0 : M_2]_{k[\theta]}.$$

- We also see that for two lattices M, M' in V, $\left[M' : M\right]_{k[\theta]} = \left[M : M'\right]_{k[\theta]}^{-1}$.

  The two following results are also immediate:

Proposition 7.2.5 *Let M be a $k[\theta]$-lattice of V and u be a $(kK)_\infty$-automorphism of V. Then $u(M)$ is a lattice of V and*

$$[M : u(M)]_{k[\theta]} = \frac{\det u}{\text{sgn}(\det u)}.$$

Proposition 7.2.6 *If M and M' are two $k[\theta]$-lattices of V and $M' \subset M$, then M/M' is a torsion $k[\theta]$-module and*

$$\left[M : M'\right]_{k[\theta]} = \left[M/M'\right]_{k[\theta]}.$$

Now let G be a finite abelian group whose order is prime to p. We suppose further that V is a free $(kK)_\infty[G]$-module. Write $\mathcal{R} = k[\theta]$. An $\mathcal{R}[G]$-lattice M in V is an $\mathcal{R}$-lattice in the $(kK)_\infty$-vector space V which is an $\mathcal{R}[G]$-submodule of V.

Let us fix a character $\chi \in \widehat{G}$. Then $e_{[\chi]}M$ is an $e_{[\chi]}\mathcal{R}[G]$ lattice in $e_{[\chi]}V$. Thus it makes sense to define for two $\mathcal{R}[G]$-lattices M and M' in V the ratio $\left[e_{[\chi]}M : e_{[\chi]}M'\right]_{e_{[\chi]}\mathcal{R}[G]}$. We then set

$$\left[M : M'\right]_{\mathcal{R}[G]} = \sum \left[e_{[\chi]}M : e_{[\chi]}M'\right]_{e_{[\chi]}\mathcal{R}[G]}$$

where the sum runs over the classes of characters $[\chi]$.

7.3 Drinfeld Modules

We review in this section the definition of Drinfeld modules and of the two fundamental associated maps: the exponential and the logarithm maps. We finish with the simplest example of a Drinfeld module, the Carlitz module, which allows some explicit computations. We also refer the reader to [Pel20, §3] where Drinfeld modules are presented for a general ring A.

7.3.1 Drinfeld Modules

In what follows, we fix $k = \mathbb{F}_q$ or $k = \mathbb{F}_q(z)$ and $R = kA$, that is, $R = A = \mathbb{F}_q[\theta]$ or $R = \mathbb{F}_q(z)[\theta]$. Let L be a finite extension of K. We write $S = O_{kL}$, that is $S = O_L$ if $R = A$ and $S = O_{\mathbb{F}_q(z)L}$ otherwise. We recall that S is endowed with the Frobenius homomorphism τ.

Definition 7.3.1 A *Drinfeld R-module* defined over S is a k-algebra homomorphism $\phi : R \to S[\tau]; a \mapsto \phi_a$ such that $\phi_a \equiv a \pmod{S[\tau]\tau}$ for all $a \in A$.

We remark that the data of ϕ_θ is sufficient to define the Drinfeld module ϕ. In particular, a Drinfeld A-module over O_L extends naturally to a Drinfeld $\mathbb{F}_q(z)[\theta]$-module over $O_{\mathbb{F}_q(z)L}$.

The degree $\deg_\tau \phi_\theta$ is called the *rank* of ϕ.

Example 7.3.2 We do not exclude the rank 0 case. In this case the Drinfeld module is the trivial map $\phi : a \mapsto \phi_a = a$.

Example 7.3.3 The *Carlitz module* is the Drinfeld A-module C over A defined by $C_\theta = \theta + \tau$. It is of rank 1. See Sect. 7.3.3 below.

Definition 7.3.4 Let ϕ be a Drinfeld A-module over O_L given by $\phi_\theta = \sum_{i=0}^n a_i \tau^i$ with $a_i \in O_L$. The *z-twist* of ϕ is the Drinfeld $\mathbb{F}_q(z)[\theta]$-module $\widetilde{\phi}$ over $O_{\mathbb{F}_q(z)L}$ given by $\widetilde{\phi}_\theta = \sum_{i=0}^n a_i z^i \tau^i$ and extended by $\mathbb{F}_q(z)$-linearity for any $a \in \mathbb{F}_q(z)[\theta]$.

If M is an $S[\tau]$-module and ϕ is a Drinfeld R-module over S, then ϕ induces a structure of R-module on M via $(a, m) \in R \times M \mapsto \phi_a(m)$. We then write $\phi(M)$ for the R-module M considered with this structure of R-module.

7.3.2 Exponential and Logarithm

We keep the notation of the previous section. Let ϕ be a Drinfeld kA-module over O_{kL}.

Let M be a finitely generated and free $(kL)_\infty$-module equipped with a semi-linear map τ, that is:

$$\forall a \in (kL)_\infty, \ \forall m \in M, \ \tau(a.m) = \tau(a).\tau(m).$$

We call such a module a τ-*module over* $(kL)_\infty$. It is in particular a finite dimensional $(kK)_\infty$-vector space, and all norms of $(kK)_\infty$-vector space on M are equivalent.

Proposition 7.3.5 *There exists a unique series* $\exp_\phi = \sum_{i \geq 0} e_i \tau^i \in kL[[\tau]]$ *such that:*

(i) $e_0 = 1$,
(ii) $\exp_\phi a = \phi_a \exp_\phi$ *holds in* $kL[[\tau]]$ *for all* $a \in A$.

Moreover, if $\| \cdot \|$ *is a norm of* $(kK)_\infty$*-vector spaces over* $(kL)_\infty$, *then*

$$\lim_{n \to \infty} \|e_n\|^{q^{-n}} = 0.$$

As a consequence, if M *is a* τ*-module over* $(kL)_\infty$, *then* $\exp_\phi$ *defines a function which converges everywhere on* M.

Proof We refer the reader to [And86, Proposition 2.1.4] for a proof of this classical result. Since this will be useful later on, we give a short proof of the last assertion: $\exp_\phi$ converges on the whole M.

We fix a norm $\| \cdot \|$ of $(kK)_\infty$-vector spaces on M. From the identification $(kK)_\infty \simeq k((\frac{1}{\theta}))$, we see that for all $x \in (kK)_\infty$, we have $|\tau(x)| \leq |x|^q$. Thus, since M is finite dimensional over $(kK)_\infty$, there exists some constant $\alpha \geq 1$ such that for all $x \in M$, $\|\tau(x)\| \leq \alpha \|x\|^q$. Thus for all $x \in M$ and all $n \geq 1$, we have: $\|\tau^n(x)\| \leq \alpha^{\frac{q^n-1}{q-1}} \|x\|^{q^n} \leq (\alpha \|x\|)^{q^n}$. Thus for all n,

$$\|e_n \tau^n(x)\| \leq \left(\|e_n\|^{q^{-n}} \alpha \|x\| \right)^{q^n} \tag{7.4}$$

which concludes the proof. $\square$

We call $\exp_\phi$ the *exponential map* associated to the Drinfeld module ϕ.

Corollary 7.3.6 *If* M *is a* τ*-module over* $(kL)_\infty$, *then the exponential map* $\exp_\phi :$ $M \to M$ *is locally an isometry.*

Proof We use the same notation as in the previous proof. Let us write $m = \max_n \|e_n\|^{q^{-n}}$. From Inequality (7.4), we get that for all $n \geq 1$, and for all $x \in M$ such that $\|x\| \leq (m\alpha)^{-1}$,

$$\|e_n \tau^n(x)\| \leq (m\alpha \|x\|)^q .$$

Thus, if $\|x\| < \min \left((m\alpha)^{-1}, (m\alpha)^{\frac{q}{1-q}} \right)$, and for all $n \geq 1$, $\|e_n \tau^n(x)\| < \|x\|$. It implies that $\| \exp_\phi(x)\| = \|x\|$. The proof is finished. $\square$

Proposition 7.3.7 *There exists a unique series* $\log_\phi = \sum_{i \geq 0} l_i \tau^i \in kL[[\tau]]$ *such that:*

(i) $l_0 = 1$,
(ii) $\log_\phi \phi_a = a \log_\phi$ *holds in* $kL[[\tau]]$ *for all* $a \in A$.

Moreover, we have $\exp_\phi \log_\phi = \log_\phi \exp_\phi = 1$ *in* $kL[[\tau]]$ *and if* $\| \cdot \|$ *is a norm of* $(kK)_\infty$-*modules over* $(kL)_\infty$, *then* $\|l_n\|^{q^{-n}}$ *is bounded. As a consequence, if* M *is a* τ-*module over* $(kL)_\infty$, *then* $\log_\phi$ *converges on a neighborhood of* 0 *in* M.

Proof The construction of $\log_\phi$ is standard: if it exists, then we must have $\exp_\phi \log_\phi = \log_\phi \exp_\phi = 1$. So it can be obtained as the inverse series (in τ) of the exponential map, and this gives both (i) and (ii). Note that it can also be constructed directly by solving the equation $\log_\phi \phi_\theta = \theta \log_\phi$.

Let $m = \max(1, \max_n \|e_n\|)$. We prove by induction that for all n, $\|l_n\| \le m^{q^n}$. The case $n = 0$ is trivial. The inequality

$$\|l_n\| = \| - \sum_{i=0}^{n-1} l_i e_{n-i}^{q^i}\| \le \max_i \|l_i e_{n-i}^{q^i}\| \le \max_{i \le n-1} m^{2q^i} \le m^{q^n}$$

concludes the proof. $\square$

We call $\log_\phi$ the *logarithm map* associated to the Drinfeld module ϕ.

Corollary 7.3.8 *The logarithm map* $\log_\phi$ *is an isometry on a neighborhood of* 0.

Proof The proof can be done along the same lines as that of Corollary 7.3.6. It is also a consequence of the fact that the logarithm map is formally an inverse map of the exponential map, that it converges on a neighborhood of 0 and that the exponential map is locally an isometry. $\square$

If ϕ is a Drinfeld A-module over O_L, and $\tilde{\phi}$ denotes its z-twist, then we have $\exp_{\tilde{\phi}}(\mathbb{T}_z(L_\infty)) \subset \mathbb{T}_z(L_\infty)$, and if $x \in \mathbb{T}_z(L_\infty)$ and $\log_{\tilde{\phi}}(x)$ converges in $(\mathbb{F}_q(z)L)_\infty$, then it converges in $\mathbb{T}_z(L_\infty)$. Thus Corollary 7.3.6 and 7.3.8 remain true on $\mathbb{T}_z(L_\infty)$.

7.3.3   The Carlitz Module

The Carlitz module is often considered as the first case of a Drinfeld module, and we can make a lot of the constructions completely explicit here. We give a short overview of these explicit constructions and refer the reader to [Pel20, §4] or, for instance, to [Gos96, §3] for more details.

Let us recall that the Carlitz module is the Drinfeld A-module C over A defined by $C_\theta = \theta + \tau$. We define $D_0 = 1$, and for $i \ge 1$, $D_{i+1} = D_i^q (\theta^{q^{i+1}} - \theta)$, so that $v_\infty(D_i) = -iq^i$. Then the exponential map associated to C is $\exp_C = \sum_{i \ge 0} \frac{1}{D_i} \tau^i$. Similarly, if $l_0 = 1$, and for $i \ge 1$, $l_{i+1} = l_i(\theta - \theta^{q^i})$, then $\log_C = \sum_{i \ge 0} \frac{1}{l_i} \tau^i$. The kernel of $\exp_C : \mathbb{C}_\infty \to \mathbb{C}_\infty$ is a rank one A-module. One can give an explicit description of a generator of this kernel as

$$\widetilde{\pi} := (-\theta)^{\frac{1}{q-1}} \theta \prod_{j \geq 1} \frac{1}{1 - \theta^{1-q^j}} \tag{7.5}$$

where $(-\theta)^{\frac{1}{q-1}}$ is a fixed $q - 1$-st root of $-\theta$ in $\mathbb{C}_\infty$. We call $\widetilde{\pi}$ "the" period of the Carlitz module (uniquely determined up to $\mathbb{F}_q^\times$).

7.4 Stark Units

We come to the definition of the Stark units. We first review Taelman's class and unit modules. Then we will be able to define the module of Stark units which is a submodule of the unit module. The section ends with some words on Anderson's [And94] which inspired the notion of Stark units.

7.4.1 Taelman Modules

We define here the class module and the unit module of a Drinfeld module as introduced by L. Taelman in [Tae12].

Let L/K be a finite extension and let ϕ denote a Drinfeld A-module over O_L. We define the *unit module* of ϕ to be

$$U(\phi; O_L) = \left\{ x \in L_\infty, \exp_\phi(x) \in O_L \right\}$$

and the *class module* of ϕ to be

$$H(\phi; O_L) = \frac{\phi(L_\infty)}{\phi(O_L) + \exp_\phi(L_\infty)}.$$

Since $\exp_\phi$ is a homomorphism of A-modules, those are naturally A-modules.

We also write $\widetilde{\phi}$ for the z-twist of ϕ and define the corresponding Taelman modules:

$$U(\widetilde{\phi}; O_{\mathbb{F}_q(z)L}) = \left\{ x \in (\mathbb{F}_q(z)L)_\infty, \exp_{\widetilde{\phi}}(x) \in O_{\mathbb{F}_q(z)L} \right\}$$

and

$$H(\widetilde{\phi}; O_{\mathbb{F}_q(z)L}) = \frac{\widetilde{\phi}((\mathbb{F}_q(z)L)_\infty)}{\widetilde{\phi}(O_{\mathbb{F}_q(z)L}) + \exp_{\widetilde{\phi}}((\mathbb{F}_q(z)L)_\infty)}.$$

And finally, at the "integral" level, we define:

$$U(\widetilde{\phi}; O_L[z]) = \left\{ x \in \mathbb{T}_z(L_\infty), \exp_{\widetilde{\phi}}(x) \in O_L[z] \right\}$$

and

$$H(\widetilde{\phi}; O_L[z]) = \frac{\widetilde{\phi}(\mathbb{T}_z(L_\infty))}{\widetilde{\phi}(O_L[z]) + \exp_{\widetilde{\phi}}(\mathbb{T}_z(L_\infty))}.$$

We fix from now on a Drinfeld A-module ϕ over O_L and write $k = \mathbb{F}_q$, $\mathbb{F}_q(z)$ or $\mathbb{F}_q[z]$ and $\varphi = \phi$ in the first case and $\varphi = \widetilde{\phi}$ otherwise.

Proposition 7.4.1

1. *The class module $H(\varphi; O_{kL})$ is finitely generated over k, thus a finitely generated and torsion kA-module.*
2. *Suppose that $k = \mathbb{F}_q$ or $k = \mathbb{F}_q(z)$. The unit module $U(\varphi; O_{kL})$ is a kA-lattice in $(kL)_\infty$.*

Proof We use the proof of [Dem14, Proposition 2.6].

For Part 1, since $\exp_\varphi$ is locally an isometry on $(kL)_\infty$, we can find a neighborhood V of 0 such that $\exp_\varphi$ is an isometry on V, $\exp_\varphi(V) = V$ and $V \cap O_{kL} = \{0\}$. We remark that $\frac{(kL)_\infty}{O_{kL}+V}$ is finitely generated over k. But we have a surjection $\frac{(kL)_\infty}{O_{kL}+V} \twoheadrightarrow H(\varphi; O_{kL})$ so that $H(\varphi; O_{kL})$ is also finitely generated.

For Part 2, since $\exp_\varphi$ is locally an isometry, we get that $U(\varphi; O_{kL})$ is discrete in $(kL)_\infty$. The exponential map induces a short exact sequence of kA-modules:

$$0 \longrightarrow \frac{(kL)_\infty}{U(\varphi; O_{kL}) + V} \longrightarrow \frac{\varphi((kL)_\infty)}{\varphi(O_{kL}) + V} \longrightarrow H(\varphi; O_{kL}) \longrightarrow 0.$$

Since the vector space in the middle is finite dimensional over k, then so is the first one. If $U(\varphi; O_{kL})$ did not generate $(kL)_\infty$ over $(kK)_\infty$, we could find $x \in (kL)_\infty$ such that $(kK)_\infty U(\varphi; O_{kL}) \cap (kK)_\infty x = \{0\}$. But, there is an injection $O_{kL} \hookrightarrow \frac{(}{k}L)_\infty V$, and $\frac{(kL)_\infty}{U(\varphi; O_{kL})+V}$ is the cokernel of the natural map $U(\varphi; O_{kL}) \to \frac{(}{k}L)_\infty V$. We deduce that the kA-ranks of O_{kL} and $U(\varphi; O_{kL})$ must coincide. Thus $U(\varphi; O_{kL})$ is a lattice in $(kL)_\infty$. $\square$

Proposition 7.4.2 *We have:*

1. $U(\widetilde{\phi}; O_{\mathbb{F}_q(z)L}) = \mathbb{F}_q(z)U(\widetilde{\phi}; O_L[z]) \subset (\mathbb{F}_q(z)L)_\infty$,
2. $H(\widetilde{\phi}; O_{\mathbb{F}_q(z)L}) \simeq \mathbb{F}_q(z) \otimes_{\mathbb{F}_q[z]} H(\widetilde{\phi}; O_L[z])$.

Proof For Part 1, we mimic the proof of [APTR16, Proposition 5.4].

The inclusion $\mathbb{F}_q(z)U(\widetilde{\phi}; O_L[z]) \subset U(\widetilde{\phi}; O_{\mathbb{F}_q(z)L})$ is clear.

We have that $\mathbb{F}_q(z)\mathbb{T}_z(L_\infty)$ is dense in $(\mathbb{F}_q(z)L)_\infty$. We fix a neighborhood V of 0 in $\mathbb{T}_z(L_\infty)$ such that $\exp_{\widetilde{\phi}}(V) = V$. We write V' for the closure of

$\mathbb{F}_q(z)V$ in $(\mathbb{F}_q(z)L)_\infty$. We still have $\exp_{\widetilde{\phi}}(V') = V'$. We then have $(\mathbb{F}_q(z)L)_\infty = \mathbb{F}_q(z)\mathbb{T}_z(L_\infty) + V'$. Let $f \in U(\widetilde{\phi}; O_{\mathbb{F}_q(z)L})$. We can write $f = g + h$ with $g \in \mathbb{F}_q(z)\mathbb{T}_z(L_\infty)$ and $h \in V'$. We get:

$$\exp_{\widetilde{\phi}}(h) = \exp_{\widetilde{\phi}}(f) - \exp_{\widetilde{\phi}}(g) \in \big(O_{\mathbb{F}_q(z)L} + \mathbb{F}_q(z)\mathbb{T}_z(L_\infty)\big) \cap V'.$$

But

$$\big(O_{\mathbb{F}_q(z)L} + \mathbb{F}_q(z)\mathbb{T}_z(L_\infty)\big) \cap V' = \mathbb{F}_q(z)\mathbb{T}_z(L_\infty) \cap V' = \mathbb{F}_q(z)V.$$

Thus, $h \in \mathbb{F}_q(z)V$ and $f \in \mathbb{F}_q(z)\mathbb{T}_z(L_\infty)$. This proves Part 1.

Part 2 is a consequence of the fact that $\mathbb{F}_q(z)\mathbb{T}_z(L_\infty)$ is dense in $(\mathbb{F}_q(z)L)_\infty$ and $\exp_{\widetilde{\phi}}$ is locally an isometry.   □

Proposition 7.4.3 *The $A[z]$-module $H(\widetilde{\phi}; O_L[z])$ is a finitely generated and torsion $\mathbb{F}_q[z]$-module, with no z-torsion.*

Proof We copy the proof of [ATR17, Proposition 2].

By Proposition 7.4.1, $H(\phi; O_L[z])$ is finitely generated over $\mathbb{F}_q[z]$. Since $\exp_{\widetilde{\phi}} \equiv 1 \pmod{L[z][[\tau]]z\tau}$, we get:

$$\mathbb{T}_z(L_\infty) = z\mathbb{T}_z(L_\infty) + \exp_{\widetilde{\phi}}(\mathbb{T}_z(L_\infty)).$$

We deduce that the multiplication by z is surjective on $H(\widetilde{\phi}; O_L[z])$. Thus, if we denote by $H(\widetilde{\phi}; O_L[z])[z]$ the z-torsion of $H(\widetilde{\phi}; O_L[z])$, the multiplication by z induces an exact sequence of finitely generated $\mathbb{F}_q[z]$-modules:

$$0 \longrightarrow H(\widetilde{\phi}; O_L[z])[z] \longrightarrow H(\widetilde{\phi}; O_L[z]) \longrightarrow H(\widetilde{\phi}; O_L[z]) \longrightarrow 0.$$

By the structure theorem for finitely generated modules over $\mathbb{F}_q[z]$, this implies that $H(\widetilde{\phi}; O_L[z])[z] = 0$ and that $H(\widetilde{\phi}; O_L[z])$ is a torsion $\mathbb{F}_q[z]$-module.   □

Corollary 7.4.4 *The class module $H(\widetilde{\phi}; O_{\mathbb{F}_q(z)L})$ vanishes.*

Proof This is a consequence of the previous proposition and Proposition 7.4.2.   □

7.4.2   The Module of Stark Units

We define here the module of Stark units, and compute its covolume in the unit module.

We keep the notation of Sect. 7.4.1. The evaluation $z \mapsto 1$ induces a map ev : $\mathbb{T}_z(L_\infty) \to L_\infty$.

Definition 7.4.5 The *module of Stark units* is defined as:

$$U_{St}(\phi; O_L) = \text{ev}\left(U(\widetilde{\phi}, O_L[z])\right).$$

We observe that $U_{St}(\phi; O_L) \subset U(\phi; O_L)$. We will now prove the following theorem by using the proof of [ATR17, Theorem 1] or [ANDTR17, Proposition 2.7].

Theorem 7.4.6 *The A-module $U_{St}(\phi; O_L)$ is an A-lattice in L_∞ and*

$$\left[\frac{U(\phi; O_L)}{U_{St}(\phi; O_L)}\right]_A = [H(\phi; O_L)]_A.$$

We introduce a map on L_∞:

$$\alpha : \begin{cases} L_\infty \to & \mathbb{T}_z(L_\infty) \\ x \mapsto & \frac{\exp_{\widetilde{\phi}}(x) - \exp_\phi(x)}{z - 1}. \end{cases}$$

The map is well defined since $\text{ev}(\exp_{\widetilde{\phi}}(x)) = \exp_\phi(x)$ so that $z - 1$ divides $\exp_{\widetilde{\phi}}(x) - \exp_\phi(x)$ in $\mathbb{T}_z(L_\infty)$.

Proposition 7.4.7 *The map α induces an isomorphism of A-modules:*

$$\overline{\alpha} : \frac{U(\phi, O_L)}{U_{St}(\phi; O_L)} \xrightarrow{\sim} H(\widetilde{\phi}; O_L[z])[z - 1]$$

where $H(\widetilde{\phi}; O_L[z])[z - 1]$ is the $(z - 1)$-torsion of $H(\widetilde{\phi}; O_L[z])$.

Proof Let us first show that $\alpha : U(\phi, O_L) \to H(\phi; \mathbb{T}_z(L_\infty))$ is a homomorphism of A-modules. Let $x \in U(\phi, O_L)$ and $a \in A$. Write $\phi_a = \sum_{i=0}^n a_i \tau^i$ with $a_i \in O_L$. Thus,

$$\begin{aligned} \alpha(ax) &= \frac{\exp_{\widetilde{\phi}}(ax) - \exp_\phi(ax)}{z - 1} \\ &= \frac{\widetilde{\phi}_a(\exp_{\widetilde{\phi}}(x)) - \phi_a(\exp_\phi(x))}{z - 1} \\ &= \widetilde{\phi}_a(\alpha(x)) + \sum_{i=0}^n a_i \frac{z^i - 1}{z - 1} \tau^i (\exp_\phi(x)) \end{aligned}$$

and this equals $\widetilde{\phi}_a(\alpha(x))$ in $H(\widetilde{\phi}; O_L[z])$ since $\exp_\phi(x) \in O_L$.

We now prove that the image of $U(\phi, O_L)$ in $H(\widetilde{\phi}; O_L[z])$ through α lies in $H(\widetilde{\phi}; O_L[z])[z - 1]$.

Let $x \in U(\phi; O_L)$. We then have

$$(z-1)\alpha(x) = \exp_{\widetilde{\phi}}(x) - \exp_{\phi}(x) \in \exp_{\phi}(\mathbb{T}_z(L_\infty)) + O_L[z]$$

so that it vanishes in $H(\widetilde{\phi}; O_L[z])$.

We now show that α is surjective on $H(\widetilde{\phi}; O_L[z])[z-1]$.

Let $x \in \mathbb{T}_z(L_\infty)$ be such that its image in $H(\widetilde{\phi}; O_L[z])$ lies in $H(\widetilde{\phi}; O_L[z])[z-1]$. Thus, $(z-1)x \in \exp_{\widetilde{\phi}}(\mathbb{T}_z(L_\infty)) + O_L[z]$. Write $(z-1)x = \exp_{\widetilde{\phi}}(u) + v$ with $u \in \mathbb{T}_z(L_\infty)$ and $v \in O_L[z]$. Write $u = u_1 + (z-1)u_2$ with $u_1 \in L_\infty$ and $u_2 \in \mathbb{T}_z(L_\infty)$ and $v = v_1 + (z-1)v_2$ with $v_1 \in O_L$ and $v_2 \in O_L[z]$. Then we have

$$(z-1)x = \exp_{\widetilde{\phi}}(u_1) + v_1 + (z-1)(\exp_{\widetilde{\phi}}(u_2) + v_2)$$

so that, by evaluating at $z = 1$, we get $\exp_{\phi}(u_1) + v_1 = 0$. Thus $u_1 \in U(\phi; O_L)$. Moreover, we get:

$$\alpha(u_1) = \frac{\exp_{\widetilde{\phi}}(u_1) - \exp_{\phi}(u_1)}{z-1}$$

$$= \frac{\exp_{\widetilde{\phi}}(u_1) + v_1}{z-1}$$

$$= x - \exp_{\widetilde{\phi}}(u_2) + v_2$$

so that the images of $\alpha(u_1)$ and x in $H(\widetilde{\phi}; O_L[z])$ coincide.

We claim that the kernel κ of $\alpha : U(\phi; O_L) \to H(\widetilde{\phi}; O_L[z])$ equals $U_{St}(\phi; O_L)$. We start with the inclusion $U_{St}(\phi; O_L) \subset \kappa$.

Let $x \in U_{St}(\phi; O_L)$, it is the evaluation at $z = 1$ of some $u \in U(\widetilde{\phi}; O_L[z])$, so there exists $v \in \mathbb{T}_z(L_\infty)$ such that $x = u + (z-1)v$. Thus

$$\alpha(x) = \frac{\exp_{\widetilde{\phi}}(u) - \exp_{\phi}(x)}{z-1} + \exp_{\widetilde{\phi}}(v)$$

but $\exp_{\phi}(x)$ is the evaluation at $z = 1$ of $\exp_{\widetilde{\phi}}(u) \in O_L[z]$. Thus $\alpha(x) \in O_L[z] + \exp_{\phi}(\mathbb{T}_z(L_\infty))$.

Lastly, we show the other inclusion: $\kappa \subset U_{St}(\phi; O_L)$. Let $x \in U(\phi; O_L)$ be such that $\alpha(x)$ vanishes in $H(\widetilde{\phi}; O_L[z])$, that is, $\alpha(x) \in O_L[z] + \exp_{\widetilde{\phi}}(\mathbb{T}_z(L_\infty))$. Thus $(z-1)\alpha(x) = \exp_{\widetilde{\phi}}(x) - \exp_{\phi}(x) = (z-1)u + \exp_{\widetilde{\phi}}((z-1)v)$ for some $u \in O_L[z]$ and $v \in \mathbb{T}_z(L_\infty)$. Thus $x - (z-1)v \in U(\widetilde{\phi}; O_L[z])$ and its evaluation at $z = 1$ is x, that is, $x \in U_{St}(\phi; O_L)$. $\qquad\square$

Proposition 7.4.8 *We have:*

$$\left[H(\widetilde{\phi}; O_L[z])[z-1]\right]_A = [H(\phi; O_L)]_A .$$

Proof The evaluation map ev induces an exact sequence of A-modules:

$$0 \longrightarrow (z-1)H(\widetilde{\phi}; O_L[z]) \longrightarrow H(\widetilde{\phi}; O_L[z]) \longrightarrow H(\phi; O_L) \longrightarrow 0$$

from which we get the exact sequence of finitely generated k-vector spaces

$$0 \to H(\widetilde{\phi}; O_L[z])[z-1] \to H(\widetilde{\phi}; O_L[z]) \xrightarrow{z-1} H(\widetilde{\phi}; O_L[z]) \to H(\phi; O_L) \to 0.$$

By (7.1), the multiplicativity of the Fitting ideal in exact sequences, we obtain

$$\left[H(\widetilde{\phi}; O_L[z])[z-1]\right]_A = [H(\phi; O_L)]_A.$$

$\square$

Proof of Theorem 7.4.6 It only remains to show that $U_{\mathrm{St}}(\phi; O_L)$ is an A-lattice. It is a direct consequence of the fact that $\frac{U(\phi; O_L)}{U_{\mathrm{St}}(\phi; O_L)}$ is a finite dimensional $\mathbb{F}_q$-vector space. $\square$

Let now E/L be a finite abelian extension of degree prime to p and let $G = \mathrm{Gal}(E/L)$. Then $U(\phi; O_E)$ and $U_{\mathrm{St}}(\phi; O_E)$ are both $A[G]$-lattices in $E_\infty = E \otimes_K K_\infty$ and $H(\phi; O_E)$ is naturally an $A[G]$-module. We remark that the map $\overline{\alpha}$ of Proposition 7.4.7 is G-equivariant, so that the equivalent of Theorem 7.4.6 remains true in the equivariant setting:

Proposition 7.4.9 *We have*

$$\left[\frac{U(\phi; O_E)}{U_{\mathrm{St}}(\phi; O_E)}\right]_{A[G]} = [H(\phi; O_E)]_{A[G]}.$$

An example will be given in Theorem 7.5.10 below in the context of the equivariant class formula.

7.4.3 Link with Anderson's Special Points

Let us finish this section with a few words on the origin of the notion of Stark Units. This notion grew up from attempts to understand the fundamental work [And94] of Anderson. Following Thakur, Anderson considers the formal power series for integers $m \geq 0$:

$$l_m(X, Z) := \sum_{a \in A \text{ monic}} \frac{C_a(X)^m}{a} Z^{q^{\deg a}} \in K[X][[Z]]$$

where τ acts on X and Z via $\tau(X) = X^q$ and $\tau(Z) = Z^q$. He shows [And94, Theorem 3] the following log-*algebraicity* result:

$$S_m(X, Z) := \exp_C(l_m(X, Z)) \in A[X, Z].$$

Let us fix now a monic irreducible polynomial $P \in A$ of degree d and define $\lambda := \exp_C(\frac{\tilde{\pi}}{P})$. Then $L = K(\lambda)$ is the "cyclotomic" extension associated with P. We refer the reader to [Ros02, Chapter 12] for more details on this extension. Anderson considers the A-submodule S of $C(O_L)$ generated by $S_m(\lambda, 1)$ for all $m \geq 0$. He (see [And94, §4.5]) calls S the *module of special points* and remarks that the special points play a role analogue to the circular units in the classical setting of cyclotomic fields.

It turns out that those special elements are just the images under the exponential map of what we called Stark units. More precisely (see [AT15, §7, in particular Theorem 7.5]):

$$S = \exp_C(U_{\mathrm{St}}(C; O_L)).$$

Stark units are therefore a generalization of the analogue of circular units for the Carlitz module, which explains their name.

7.5   Class Formulas

This section is devoted to class formulas: the original Taelman class formula from [Tae12] and some generalizations, in particular in the equivariant setting. We also give some explicit examples.

In what follows, we keep considering a finite extension L of K and a Drinfeld $\mathbb{F}_q[\theta]$-module ϕ defined over O_L.

7.5.1   Taelman's Class Formula

We present Taelman's class formula and how it can be expressed in terms of the regulator of Stark units.

Let I be a non-zero ideal of O_L. Then O_L/IO_L is a finite dimensional $\mathbb{F}_q$-vector space. Since $\tau(I) \subset I$, it makes sense to define both $[O_L/IO_L]_A$ and $[\phi(O_L/IO_L)]_A$.

Remark that the first one is easy to compute:

Lemma 7.5.1 *Let I be a non-zero ideal of O_L and denote by $N_{L/K}$ the norm map from the ideals of O_L to the ones of A. Then $[O_L/IO_L]_A$ is the monic generator of $N_{L/K}(I)$.*

Proof The equality $\text{Fitt}_A (O_L/IO_L) = N_{L/K}(I)$ is immediate from the definitions. □

If $\mathfrak{P}$ is a prime ideal of O_L, the Euler factor at $\mathfrak{P}$ is then the quotient $[O_L/\mathfrak{P}O_L]_A / [\phi(O_L/\mathfrak{P}O_L)]_A$. By putting together all these local factors, we obtain the L-series:

$$L(\phi/O_L) := \prod_{\mathfrak{P}} \frac{[O_L/\mathfrak{P}O_L]_A}{[\phi(O_L/\mathfrak{P}O_L)]_A} \tag{7.6}$$

where the product runs over all the non-zero prime ideals of O_L.

Lemma 7.5.2 *Let I be a non-zero ideal of O_L. Let $n \geq 1$. Then:*

$$\begin{aligned}[O_L/I^n O_L]_A \cdot [\phi(O_L/IO_L)]_A \\ = [\phi(O_L/I^n O_L)]_A \cdot [O_L/IO_L]_A \,.\end{aligned}$$

Proof We prove this equality by induction on n. The case $n = 1$ is clear. The short exact sequence

$$0 \to I^n O_L/I^{n+1} O_L \to O_L/I^{n+1} O_L \to O_L/I^n O_L \to 0$$

gives

$$\left[O_L/I^{n+1} O_L\right]_A = [O_L/I^n O_L]_A \cdot \left[I^n O_L/I^{n+1} O_L\right]_A \,.$$

Similarly, we have the short exact sequence

$$0 \to \phi(I^n O_L/I^{n+1} O_L) \to \phi(O_L/I^{n+1} O_L) \to \phi(O_L/I^n O_L) \to 0$$

but for any $x \in I^n O_L, a \in A, \phi_a(x) \equiv ax \pmod{I}^{qn} O_L$, thus

$$\phi(I^n O_L/I^{n+1} O_L) \simeq I^n O_L/I^{n+1} O_L,$$

so that

$$\left[\phi(O_L/I^{n+1} O_L)\right]_A = [\phi(O_L/I^n O_L)]_A \cdot \left[I^n O_L/I^{n+1} O_L\right]_A \,.$$

Putting altogether we get the desired result. □

The previous lemma, together with the Chinese Remainder Theorem allows to write the L-series as:

$$L(\phi/O_L) := \prod_{P} \frac{[O_L/PO_L]_A}{[\phi(O_L/PO_L)]_A} \tag{7.7}$$

where the product runs over all the monic irreducible polynomials P of A. In this form, the numerator is also very easy to compute:

$$[O_L/PO_L]_A = P^{[L:K]}.$$

The main result of [Tae12] is the following class formula:

Theorem 7.5.3 (Taelman) *The product defining $L(\phi/O_L)$ converges in K_∞, and the following equality holds:*

$$L(\phi/O_L) = [O_L : U(\phi; O_L)]_A [H(\phi; O_L)]_A.$$

Corollary 7.5.4 *We have:*

$$L(\phi/O_L) = [O_L : U_{St}(\phi; O_L)]_A.$$

Proof This is immediate from Taelman's class formula and Theorem 7.4.6.   □

The co-volume of the Taelman units or the Stark units in O_L is very similar to the classical notion of a regulator, so that the previous corollary can nicely translate as: the L-value attached to ϕ is the regulator of its module of Stark units.

Remark that, as in (7.6), we can also define the z-twisted version of the L-series:

$$L(\widetilde{\phi}/O_{\mathbb{F}_q(z)L}) := \prod_{\mathfrak{P}} \frac{\left[O_{\mathbb{F}_q(z)L}/\mathfrak{P}O_{\mathbb{F}_q(z)L}\right]_{\mathbb{F}_q(z)A}}{\left[\widetilde{\phi}(O_{\mathbb{F}_q(z)L}/\mathfrak{P}O_{\mathbb{F}_q(z)L})\right]_{\mathbb{F}_q(z)A}}$$

where the product runs over all the non-zero prime ideals of O_L. Here again, the numerator of the local factor at $\mathfrak{P}$ is

$$\left[O_{\mathbb{F}_q(z)L}/\mathfrak{P}O_{\mathbb{F}_q(z)L}\right]_{\mathbb{F}_q(z)A} = N_{L/K}(\mathfrak{P}).$$

And, similarly to (7.7), we have the alternative expression:

$$L(\widetilde{\phi}/O_{\mathbb{F}_q(z)L}) := \prod_{P} \frac{\left[O_{\mathbb{F}_q(z)L}/PO_{\mathbb{F}_q(z)L}\right]_{\mathbb{F}_q(z)A}}{\left[\widetilde{\phi}(O_{\mathbb{F}_q(z)L}/PO_{\mathbb{F}_q(z)L})\right]_{\mathbb{F}_q(z)A}}$$

where the product runs over all the monic irreducible polynomials P of A. And again:

$$\left[O_{\mathbb{F}_q(z)L}/PO_{\mathbb{F}_q(z)L}\right]_{\mathbb{F}_q(z)A} = P^{[L:K]}.$$

By Demeslay's adaptation of the work of Taelman, [Dem14, Theorem 2.7], we also have the class formula:

Theorem 7.5.5 (Demeslay) *The product defining* $L(\widetilde{\phi}/O_{\mathbb{F}_q(z)L})$ *converges in* $(\mathbb{F}_q(z)K)_\infty$, *and the following equality holds:*

$$L(\widetilde{\phi}/O_{\mathbb{F}_q(z)L}) = \left[O_{\mathbb{F}_q(z)L} : U(\widetilde{\phi}; O_{\mathbb{F}_q(z)L})\right]_{\mathbb{F}_q(z)A} \left[H(\widetilde{\phi}; O_{\mathbb{F}_q(z)L})\right]_{\mathbb{F}_q(z)A}.$$

Remark that, because of Corollary 7.4.4, this result can simply be stated as

$$L(\widetilde{\phi}/O_{\mathbb{F}_q(z)L}) = \left[O_{\mathbb{F}_q(z)L} : U(\widetilde{\phi}; O_{\mathbb{F}_q(z)L})\right]_{\mathbb{F}_q(z)A}.$$

Corollary 7.5.6 *The L-series* $L(\widetilde{\phi}/O_{\mathbb{F}_q(z)L})$ *converges in* $\mathbb{T}_z(K_\infty)$.

Proof For any monic irreducible polynomial $P \in A$, we have:

$$\left[\widetilde{\phi}(O_L/PO_L)\right]_{\mathbb{F}_q(z)A} = \det_{\mathbb{F}_q(z)[Z]} \left(Z - \theta \mid \widetilde{\phi}(O_L/PO_L)\right)|_{Z=\theta}$$

which is a polynomial in z which evaluates to $P^{[L:K]}$ at $z = 0$. But

$$\deg_\theta \left(\left[\widetilde{\phi}(O_L/PO_L)\right]_{\mathbb{F}_q(z)A}\right) = \dim_{\mathbb{F}_q} O_L/PO_L = \deg_\theta P^{[L:K]}.$$

We deduce that the local factor at P belongs to $\mathbb{T}_z(K_\infty)$. The convergence of $L(\widetilde{\phi}/O_{\mathbb{F}_q(z)L})$ in $(\mathbb{F}_q(z)K)_\infty$ then implies its convergence in $\mathbb{T}_z(K_\infty)$. □

7.5.2 The Equivariant Class Formula

We present now the class formula in the equivariant setting.

We consider as previously a Drinfeld A-module ϕ defined over O_L, and E/L a finite abelian extension of degree prime to p and we let $G = \mathrm{Gal}(E/L)$.

In this context, we can define an equivariant L-series via:

$$L(\phi/(O_E/O_L), G) := \prod_{\mathfrak{P}} \frac{[O_E/\mathfrak{P}O_E]_{A[G]}}{[\phi(O_E/\mathfrak{P}O_E)]_{A[G]}}$$

where the product runs over the non-zero prime ideals of O_E. As in (7.7), it is equivalent to taking the product over the non-zero prime ideals of O_L or of A. And we have the z-twisted version:

$$L(\widetilde{\phi}/(O_{\mathbb{F}_q(z)E}/O_{\mathbb{F}_q(z)L}), G) := \prod_{\mathfrak{P}} \frac{\left[O_{\mathbb{F}_q(z)E}/\mathfrak{P}O_{\mathbb{F}_q(z)E}\right]_{\mathbb{F}_q(z)A[G]}}{\left[\widetilde{\phi}(O_{\mathbb{F}_q(z)E}/\mathfrak{P}O_{\mathbb{F}_q(z)E})\right]_{\mathbb{F}_q(z)A[G]}}.$$

The convergence of the L-series $L(\phi/(O_E/O_L), G)$, and an equivariant class formula involving it was proved, in an even more general setting, by Fang in [Fan18, Theorem 1.12]:

Theorem 7.5.7 (Fang) *We have:*

$$L(\phi/(O_E/O_L), G) = [O_E : U(\phi; O_E)]_{A[G]} [H(\phi; O_E)]_{A[G]}.$$

The equivariant class formula has its origin in [AT15, Theorem A] for the Carlitz module. We also signal to the reader the recent work [FGHP20] of Ferrara, Green, Higgins and Popescu where an equivariant class formula is proved without the restrictions that G is abelian and of order prime to p.

Following the proof of [AT15, Theorem A] (the details can be found in [ATR17, Proposition 4]), one can show the z-twisted version:

Theorem 7.5.8 *The L-series $L(\widetilde{\phi}/(O_{\mathbb{F}_q(z)E}/O_{\mathbb{F}_q(z)L}), G)$ converges in $(\mathbb{F}_q(z) K)_\infty[G]$ and we have:*

$$L(\widetilde{\phi}/(O_{\mathbb{F}_q(z)E}/O_{\mathbb{F}_q(z)L}), G) = \left[O_{\mathbb{F}_q(z)E} : U(\widetilde{\phi}; O_{\mathbb{F}_q(z)E})\right]_{\mathbb{F}_q(z)A[G]}.$$

As for $L(\widetilde{\phi}/O_{\mathbb{F}_q(z)L})$, the convergence of $L(\widetilde{\phi}/(O_{\mathbb{F}_q(z)E}/O_{\mathbb{F}_q(z)L}), G)$ in $(\mathbb{F}_q(z)K)_\infty[G]$ implies that it actually converges in $\mathbb{T}_z(K_\infty)[G]$. We can then evaluate it at $z = 1$, and we see that the result is just $L(\phi/(O_E/O_L), G)$.

Combining Theorem 7.5.7 with Proposition 7.4.9, we also get:

Theorem 7.5.9

$$L(\phi/(O_E/O_L), G) = [O_E : U_{\mathrm{St}}(\phi; O_E)]_{A[G]}.$$

In the case where $L = K$, we have a simple description of the Stark units in terms of the equivariant L-series (see [ATR17, Theorem 2]):

Theorem 7.5.10 *Let ϕ be a Drinfeld A-module defined over A and E/K be a finite abelian extension of degree prime to p, and $G = \mathrm{Gal}(E/K)$. We have:*

$$U(\widetilde{\phi}; O_E[z]) = L(\widetilde{\phi}/(O_{\mathbb{F}_q(z)E}/\mathbb{F}_q(z)A), G)O_E[z]$$

and

$$U_{\mathrm{St}}(\phi; O_E) = L(\phi/(O_E/A), G)O_E.$$

Proof Since $A[G]$ and $\mathbb{F}_q(z)A[G]$ are principal ideal rings, we see that O_E is a rank 1 free $A[G]$-module, and that $O_{\mathbb{F}_q(z)E}$ and $U(\widetilde{\phi}; O_{\mathbb{F}_q(z)E})$ are free $\mathbb{F}_q(z)A[G]$-modules of rank 1. By Theorem 7.5.8, we then have:

$$L(\widetilde{\phi}/(O_{\mathbb{F}_q(z)E}/\mathbb{F}_q(z)A), G)O_{\mathbb{F}_q(z)E} = U(\widetilde{\phi}; O_{\mathbb{F}_q(z)E}).$$

And since $L(\widetilde{\phi}/(O_{\mathbb{F}_q(z)E}/\mathbb{F}_q(z)A), G)$ converges in $\mathbb{T}_z(K_\infty)[G]$, we get:

$$L(\widetilde{\phi}/(O_{\mathbb{F}_q(z)E}/\mathbb{F}_q(z)A), G)O_E[z] \subset U(\widetilde{\phi}; O_E[z]).$$

If conversely $x \in U(\widetilde{\phi}; O_E[z]) \subset \mathbb{T}_z(E_\infty)$, there is $y \in O_{\mathbb{F}_q(z)E}$ such that

$$x = L(\widetilde{\phi}/(O_{\mathbb{F}_q(z)E}/\mathbb{F}_q(z)A), G)y.$$

Since $L(\widetilde{\phi}/(O_{\mathbb{F}_q(z)E}/\mathbb{F}_q(z)A), G)$ has sign 1, this implies that $y \in O_E[z]$. Thus

$$U(\widetilde{\phi}; O_E[z]) = L(\widetilde{\phi}/(O_{\mathbb{F}_q(z)E}/\mathbb{F}_q(z)A), G)O_E[z].$$

The second assertion comes now from the evaluation at $z = 1$.   □

7.5.3 Examples

Let us now work out some examples of the class formula. We first treat the Carlitz module C with $L = K$. We refer to Sect. 7.3.3 for the basic facts and notation on the Carlitz module. The L-series associated to C is easily computed. Let $P \in A$ be monic and irreducible. Then obviously $[A/PA]_A = P$. Moreover, as $C_P \equiv \tau^{\deg P}$ (mod $PA[\tau]$), we get $C(A/PA) \simeq A/(P-1)A$ so that the local factor at P is just $(1 - \frac{1}{P})^{-1}$ and

$$L(C/A) = \prod_P (1 - \frac{1}{P})^{-1} = \sum_{a \in A_+} \frac{1}{a}$$

where A_+ stands for the subset of monic polynomials in A. This is also the zeta value at 1 as defined by Carlitz. The other values are, if $n \geq 0$:

$$\zeta_A(n) = \sum_{a \in A_+} \frac{1}{a^n}.$$

Note that at a negative integer, the zeta value is also defined as the (finite!) sum, for $n \geq 0$:

$$\zeta_A(-n) = \sum_{d \geq 0} \sum_{a \in A_+, \deg a = d} a^n.$$

Let us define

$$\mathcal{N} = \{x \in K_\infty, v_\infty(x) > -1\}.$$

Because $v_\infty(D_i) = -iq^i$, we can make Corollary 7.3.6 explicit: $\exp_C$ is isometric on $\mathcal{N}$, so that $\exp_C(\mathcal{N}) = \mathcal{N}$. Consequently, $\exp_C(K_\infty) + A = K_\infty$ so that $H(C; A) = \{0\}$. Hence, by Theorem 7.4.6, $U(C; A) = U_{St}(C; A)$. This is a rank one A-module, and since $1 \in \mathcal{N}$, we see that $U(C; A) = A \log_C(1)$. The class formula for C can then be written as:

$$\zeta_A(1) = L(C/A) = [A : U(C; A)]_A = \log_C(1).$$

We thus recover this well-known equality which is a consequence of a result of Carlitz [Gos96, Theorem 3.1.5].

Let us now fix an integer $d \geq 0$ and consider the Drinfeld A-module ϕ over A defined by $\phi_\theta = \theta + (-\theta)^d \tau$. We see that if $a \in A$ and $C_a = \sum_{i=0}^k a_i \tau^i$ then $\phi_a = \sum_{i=0}^k a_i (-\theta)^{d \frac{q^i-1}{q-1}} \tau^i$. Let $P \in A$ be monic and irreducible. We thus get that $\phi_P \equiv (-\theta)^{d \frac{q^{\deg P}-1}{q-1}} \tau^{\deg P} \pmod{PA[\tau]}$. But

$$\theta^{\frac{q^{\deg P}-1}{q-1}} = \theta^{1+q+\cdots+q^{\deg P-1}} \equiv (-1)^{\deg P} P(0) \mod P.$$

We deduce that $\phi_{P-P(0)^d}$ is identically zero on A/PA and since for any $Q \in A$, ϕ_Q is a polynomial of $A[\tau]$ of degree $\deg Q$ in τ, $P(X) - P(0)^d$ is the minimal polynomial of ϕ_θ, that is $\phi(A/PA) \simeq A/(P - P(0)^d)A$. Thus $[\phi(A/PA)]_A = P - P(0)^d$. We get:

$$L(\phi/A) = \prod_P \left(1 - \frac{P(0)^d}{P}\right)^{-1} = \sum_{a \in A_+} \frac{a(0)^d}{a}.$$

These computations are also consequences of Sect. 7.6.2 below. See in particular Eq. (7.9). Let us now describe the units and Stark units of ϕ. For that purpose, we use results that will be proved later on. We have by Proposition 7.6.5:

$$U_{St}(\phi; A) = L(\phi/A)A.$$

There are now two different cases, whether $n \equiv 1 \pmod{q-1}$ or not. This difference is linked to the fact that the kernel of $\exp_\phi : K_\infty \to K_\infty$ is non trivial if and only if $n \equiv 1 \pmod{q-1}$.

In the case $n \not\equiv 1 \pmod{q-1}$, by the proof of Theorem 7.7.1 we have $H(\phi; A) = \{0\}$ and thus

$$U(\phi; A) = U_{St}(\phi; A) = L(\phi/A)A.$$

In the case $n \equiv 1 \pmod{q-1}$, the unit module is the kernel of $\exp_\phi$ if $n \neq 1$ and more generally the inverse image of the A-torsion submodule of $\phi(K_\infty)$ if $n = 1$.

More explicitly, if $n = 1$:

$$U(\phi; A) = \frac{\tilde{\pi}}{(-\theta)^{\frac{1}{q-1}}\theta} A$$

and if $n > 1$:

$$U(\phi; A) = \frac{\tilde{\pi}}{(-\theta)^{\frac{1}{q-1}}\theta^{\frac{n-1}{q-1}}} A$$

where $(-\theta)^{\frac{1}{q-1}}$ is the fixed $(q-1)$-st root of $-\theta$ (see Eq. (7.5)). Thus, if $n > 1$, there is $B_n \in A$ of degree $\frac{n-q}{q-1}$ such that

$$(-\theta)^{\frac{1}{q-1}}\theta^{\frac{n-1}{q-1}} L(\phi/A) = \tilde{\pi} B_n.$$

Taelman's class formula (Theorem 7.5.3) tells us that $[H(\phi; O_L)]_A = B_n$. Moreover, $[H(\phi; O_L)]_A$ just vanishes when $n = 1$.

7.6 The Multi-Variable Deformation of a Drinfeld A-Module

7.6.1 The Multi-Variable Setting

We have presented in the previous section the z-deformation of a Drinfeld module ϕ, which, roughly speaking, "evaluates" at $z = 1$ to ϕ. It turns out that there are other natural ways to twist a Drinfeld module using multiple variables. The idea here is still to twist the Frobenius τ by a polynomial in the new variables. It is also of interest to combine those two deformations and define Stark units for the multiple variable deformation of our Drinfeld module. Let us now give more precise statements:

Let $t_1, \ldots, t_n$ be new variables, with $n \geq 1$. We will denote by $\mathbf{t}$ the set of variables $t_1, \ldots, t_n$. We fix some additional notation:

- $k = \mathbb{F}_q(\mathbf{t}) = \mathbb{F}_q(t_1, \ldots, t_n)$,
- $A = k[\theta]$, $\mathbb{K} = k(\theta)$, $\mathbb{K}_\infty = k((\frac{1}{\theta}))$,
- v_∞ the valuation at the place ∞ such that $v_\infty(\theta) = -1$, extending the valuation on K_∞.

We fix a complete algebraically closed extension of $\mathbb{K}$ and we identify $\mathbb{C}_\infty$ with the completion of the algebraic closure of K in this extension. For L a fixed finite extension of K, $\mathbb{L}$ will denote the compositum of L and $\mathbb{K}$, and $O_\mathbb{L}$ the integral closure of A in $\mathbb{L}$. We set $\mathbb{L}_\infty = \mathbb{L} \otimes_\mathbb{K} \mathbb{K}_\infty$. We extend τ to $\mathbb{L}$ by k-linearity and thus to $\mathbb{L}_\infty$.

Then, the theory developed in the previous sections remain valid by replacing $\mathbb{F}_q$ by k. We leave to the reader as an exercice to check that the arguments carry over. We will then be interested in Drinfeld $\mathbb{A}$-modules ϕ defined over $O_{\mathbb{L}}$ with an obvious definition. The existence of the exponential and logarithmic maps and their properties described in Sect. 7.3.2 remain valid and we can define the $\mathbb{A}$-modules $U(\phi; O_{\mathbb{L}})$ and $H(\phi; O_{\mathbb{L}})$. By Demeslay's work [Dem14], we have in particular:

Theorem 7.6.1 (Demeslay) *Let ϕ be a Drinfeld $\mathbb{A}$-module defined over $O_{\mathbb{L}}$. Then:*

1. the unit module

$$U(\phi; O_{\mathbb{L}}) = \left\{ x \in \mathbb{L}_\infty, \exp_\phi(x) \in O_{\mathbb{L}} \right\}$$

is an $\mathbb{A}$-lattice in $\mathbb{K}_\infty$,
2. the class module

$$H(\phi; O_{\mathbb{L}}) = \frac{\phi(\mathbb{L}_\infty)}{\phi(O_{\mathbb{L}}) + \exp_\phi(\mathbb{L}_\infty)}$$

is a finite dimensional k-vector space and an $\mathbb{A}$-module via ϕ,
3. the infinite product

$$L(\phi/O_{\mathbb{L}}) := \prod_P \frac{[O_{\mathbb{L}}/PO_{\mathbb{L}}]_{\mathbb{A}}}{[\phi(O_{\mathbb{L}}/PO_{\mathbb{L}})]_{\mathbb{A}}},$$

where the product runs over the monic irreducible polynomials $P \in A$, converges in $\mathbb{L}_\infty^\times$ and we have the class formula:

$$L(\phi/O_{\mathbb{L}}) = [O_{\mathbb{L}} : U(\phi; O_{\mathbb{L}})]_{\mathbb{A}} [H(\phi; O_{\mathbb{L}})]_{\mathbb{A}}.$$

Proof Part 1 and Part 2 follow from [Dem14, Proposition 2.6] and Part 3 from [Dem14, Theorem 2.7] □

As previously, we can define the z-twist $\widetilde{\phi}$ of a Drinfeld $\mathbb{A}$-module ϕ defined over $O_{\mathbb{L}}$ by twisting the frobenius τ by z. It is thus a Drinfeld $k(z)\mathbb{A}$-module over $O_{k(z)\mathbb{L}}$. Demeslay's work also applies to this case and we have similarly:

Theorem 7.6.2 (Demeslay) *Let ϕ be a Drinfeld $\mathbb{A}$-module defined over $O_{\mathbb{L}}$ and $\widetilde{\phi}$ be its z-twist. Then:*

1. the unit module

$$U(\widetilde{\phi}; O_{k(z)\mathbb{L}}) = \left\{ x \in (k(z)\mathbb{L})_\infty, \exp_{\widetilde{\phi}}(x) \in O_{k(z)\mathbb{L}} \right\}$$

is a $k(z)\mathbb{A}$-lattice in $(k(z)\mathbb{K})_\infty$,

2. *the class module*

$$H(\widetilde{\phi}; O_{k(z)\mathbb{L}}) = \frac{\widetilde{\phi}((k(z)\mathbb{L})_\infty)}{\widetilde{\phi}(O_{k(z)\mathbb{L}}) + \exp_{\widetilde{\phi}}((k(z)\mathbb{L})_\infty)}$$

is a finite dimensional $k(z)$-vector space and a $k(z)\mathbb{A}$-module via $\widetilde{\phi}$,
3. *the infinite product*

$$L(\widetilde{\phi}/O_{k(z)\mathbb{L}}) := \prod_P \frac{\left[O_{k(z)\mathbb{L}}/PO_{k(z)\mathbb{L}}\right]_{k(z)\mathbb{A}}}{\left[\widetilde{\phi}(O_{k(z)\mathbb{L}}/PO_{k(z)\mathbb{L}})\right]_{k(z)\mathbb{A}}},$$

where the product runs over the monic irreducible polynomials $P \in A$, converges in $(k(z)\mathbb{L})_\infty^\times$ and we have the class formula:

$$L(\widetilde{\phi}/O_{k(z)\mathbb{L}}) = [O_{k(z)\mathbb{L}} : U(\widetilde{\phi}; O_{k(z)\mathbb{L}})]_{k(z)\mathbb{A}}[H(\widetilde{\phi}; O_{k(z)\mathbb{L}})]_{k(z)\mathbb{A}}.$$

Remark 7.6.3 As in Proposition 7.4.3, $H(\widetilde{\phi}; O_{\mathbb{L}}[z])$ is a finitely generated torsion $k[z]$-module, so that the class module $H(\widetilde{\phi}; O_{k(z)\mathbb{L}})$ vanishes, which simplifies the class formula.

We now want to work at the integral level in $\mathbb{A}$ or $\mathbb{K}_\infty$. We then suppose that $\phi_\theta \in O_L[\mathbf{t}][\tau]$. We can thus consider ϕ either as a Drinfeld $\mathbb{A}$-module defined over $\mathbb{L}$ or as a Drinfeld $A[\mathbf{t}]$-module defined over $O_L[\mathbf{t}]$. We denote by $\mathbb{T}_n(L_\infty)$ the Tate algebra in variables $t_1, \ldots, t_n$ and coefficients in L_∞ and we define the Taelman modules:

$$U(\phi; O_L[\mathbf{t}]) = \{x \in \mathbb{T}_n(L_\infty), \exp_\phi(x) \in O_L[\mathbf{t}]\} \subset U(\phi; O_{\mathbb{L}})$$

and

$$H(\phi; O_L[\mathbf{t}]) = \frac{\phi(\mathbb{T}_n(L_\infty))}{\phi(O_L[\mathbf{t}]) + \exp_\phi(\mathbb{T}_n(L_\infty))}.$$

Since ϕ is defined over $O_L[\mathbf{t}]$, by using the functional equation $\phi_\theta \exp_\phi = \exp_\phi \theta$, one shows that $\exp_\phi$ has coefficients in $L[\mathbf{t}]$, so that $\exp_\phi(\mathbb{T}_n(L_\infty)) \subset \mathbb{T}_n(L_\infty)$. We deduce that:

$$U(\phi; O_L[\mathbf{t}]) = U(\phi; O_{\mathbb{L}}) \cap \mathbb{T}_n(L_\infty).$$

By the same argument as in Proposition 7.4.2, we also have

$$U(\phi; O_{\mathbb{L}}) = kU(\phi; O_L[\mathbf{t}])$$

and

$$H(\phi; O_L[\mathbf{t}]) \otimes_{\mathbb{F}_q[\mathbf{t}]} k \simeq H(\phi; O_{\mathbb{L}}).$$

By evaluation at $z = 1$ of the unit module, we have a well defined notion of the module of Stark units $U_{\mathrm{St}}(\phi; O_{\mathbb{L}})$. Let us be more explicit for the construction at the integral level. We denote by $\mathbb{T}_{n,z}(L_\infty)$ the Tate algebra in variables $t_1, \ldots, t_n, z$ and coefficients in L_∞. Then we define

$$U(\widetilde{\phi}; O_L[\mathbf{t}, z]) = \left\{ x \in \mathbb{T}_{n,z}(L_\infty), \exp_{\widetilde{\phi}}(x) \in O_L[\mathbf{t}, z] \right\}$$

and

$$H(\widetilde{\phi}; O_L[\mathbf{t}, z]) = \frac{\widetilde{\phi}(\mathbb{T}_{n,z}(L_\infty))}{\widetilde{\phi}(O_L[\mathbf{t}, z]) + \exp_{\widetilde{\phi}}(\mathbb{T}_{n,z}(L_\infty))}.$$

The evaluation at $z = 1$ of $U(\widetilde{\phi}; O_L[\mathbf{t}, z])$ is our module of Stark units $U_{\mathrm{St}}(\phi; O_L[\mathbf{t}]) \subset U(\phi; O_L[\mathbf{t}])$.

Theorem 7.4.6 remains true here, in particular we have the following version (see [ATR17, Proposition 6]):

Proposition 7.6.4 *The map*

$$\alpha : \begin{cases} \mathbb{T}_n(L_\infty) \to & \mathbb{T}_{n,z}(L_\infty) \\ x \mapsto & \frac{\exp_{\widetilde{\phi}}(x) - \exp_\phi(x)}{z-1} \end{cases}$$

induces an isomorphism of $A[\mathbf{t}]$-modules:

$$\frac{U(\phi; O_L[\mathbf{t}])}{U_{\mathrm{St}}(\phi; O_L[\mathbf{t}])} \simeq H(\widetilde{\phi}; O_L[\mathbf{t}, z])[z - 1].$$

7.6.2 The Canonical Deformation of the Carlitz Module

We focus here on a natural multi-variable deformation of the Carlitz module built by means of its shtuka function.

Let ϕ be a Drinfeld A-module defined over O_L and $f(\mathbf{t}) = f(t_1, \ldots, t_n) \in O_L[\mathbf{t}]$. Then we can use f to twist ϕ: if $a \in A$ and $\phi_a = \sum_{i=0}^m a_i \tau^i$, then

$$\widehat{\phi}_a = \sum_{i=0}^m a_i (f(\mathbf{t})\tau)^i = \sum_{i=0}^m a_i \left(\prod_{j=0}^i \tau^j(f)(\mathbf{t}) \right) \tau^i.$$

Remark that, as for the z-twist, we in fact twist here the action of the Frobenius τ by $f(\mathbf{t})$, which induces the deformation of ϕ. We get a Drinfeld $A[\mathbf{t}]$-module $\widehat{\phi}$ defined over $O_L[\mathbf{t}]$.

From now on, we will be only interested in the case of the Carlitz module C. Let us recall (see Sect. 7.3.3) that C is the Drinfeld A-module defined over A by $C_\theta = \theta + \tau$. To such a Drinfeld module one can associate a so-called *shtuka function* (see e.g. [Gos96, §7.11], or [Tha93]), from which one recovers the Drinfeld module, and which encodes its arithmetic properties. In the case of the Carlitz module, the shtuka function is simply $t - \theta$. There is therefore a natural n variable twist of the Carlitz module, which we call the *canonical deformation of the Carlitz module*, given by

$$f(\mathbf{t}) = \prod_{i=1}^{n} (t_i - \theta).$$

We thus consider the Drinfeld $A[\mathbf{t}]$-module $\varphi = \widehat{C}$ defined over $A[\mathbf{t}]$ by

$$\varphi_\theta = \theta + f(\mathbf{t})\tau = \theta + \prod_{i=1}^{n} (t_i - \theta)\tau.$$

We will denote for $k \geq 0$, by $f_k(\mathbf{t})$ the polynomial appearing in the formula $(f(\mathbf{t})\tau)^k = f_k(\mathbf{t})\tau^k$, that is:

$$f_k(\mathbf{t}) = \prod_{i=1}^{n} \prod_{j=0}^{k} (t_i - \theta^{q^j}).$$

We get the exponential map $\exp_\varphi = \sum_{i\geq 0} \frac{1}{D_i} f_i(\mathbf{t})\tau^i$ and the logarithm map $\log_\varphi = \sum_{i\geq 0} \frac{1}{l_i} f_i(\mathbf{t})\tau^i$.

We also introduce the Anderson-Thakur ω function:

$$\omega(t) := (-\theta)^{\frac{1}{q-1}} \prod_{j\geq 0} \left(1 - \frac{t}{\theta^{q^j}}\right)^{-1} \in \mathbb{T}_1(K_\infty)^\times.$$

We see from (7.5) that $-\widetilde{\pi}$ is the residue of ω at $t = \theta$ and that ω enjoys the functional equation:

$$\tau(\omega(t)) = (t - \theta)\omega(t).$$

Thus, we get

$$\exp_\varphi = \left(\prod_{i=1}^{n} \omega(t_i)\right)^{-1} \exp_C \left(\prod_{i=1}^{n} \omega(t_i)\right).$$

In particular, we obtain:

$$\ker(\exp_\varphi : \mathbb{T}_n(\mathbb{C}_\infty) \to \mathbb{T}_n(\mathbb{C}_\infty)) = \frac{\widetilde{\pi}}{\prod_{i=1}^n \omega(t_i)} A[\mathbf{t}]. \tag{7.8}$$

And we remark that this kernel is included in $\mathbb{T}_n(K_\infty)$ if, and only if, $n \equiv 1$ (mod $q - 1$).

The L-series associated to φ can be computed similarly to the one of C (see Sect. 7.5.3). We have

$$\varphi_P \equiv f_{\deg P}(\mathbf{t})\tau^{\deg P} \pmod{PA[\mathbf{t}][\tau]}$$

but

$$(t - \theta)(t - \theta^q)\cdots(t - \theta^{q^{\deg P - 1}}) \equiv P(t) \pmod{PA[t]}$$

so that

$$\varphi_P \equiv P(t_1)\cdots P(t_n)\tau^{\deg P} \pmod{PA[\mathbf{t}][\tau]}.$$

We deduce that $P(X) - P(t_1)\cdots P(t_n)$ is an annihilating polynomial of ϕ_θ acting on $A/PA(\mathbf{t})$. Since it is also a monic irreducible polynomial in $\mathbb{F}_q(\mathbf{t})[X]$, of degree $\deg P$, it is its characteristic polynomial and we get by (7.2):

$$\left[\frac{A}{PA}(\mathbf{t})\right]_{\mathbb{A}} = P - P(t_1)\cdots P(t_n).$$

Putting all the local factors together, we obtain

$$L(\varphi/\mathbb{A}) = \prod_P \left(1 - \frac{P(t_1)\cdots P(t_n)}{P}\right)^{-1} = \sum_{a \in A_+} \frac{a(t_1)\cdots a(t_n)}{a} \in \mathbb{T}_n(K_\infty)^\times. \tag{7.9}$$

Similar calculations for the z-twist of φ lead to:

$$L(\widetilde{\varphi}/k(z)\mathbb{A}) = \prod_P \left(1 - \frac{z^{\deg P}P(t_1)\cdots P(t_n)}{P}\right)^{-1}$$

$$= \sum_{a \in A_+} z^{\deg P} \frac{a(t_1)\cdots a(t_n)}{a} \in \mathbb{T}_{n,z}(K_\infty)^\times.$$

Let us compute the units:

Proposition 7.6.5 *We have*

$$U(\widetilde{\varphi}; A[\mathbf{t}, z]) = L(\widetilde{\varphi}/k(z)\mathbb{A})A[\mathbf{t}, z]$$

so that

$$U_{\mathrm{St}}(\varphi; A[\mathbf{t}]) = L(\varphi/\mathbb{A})A[\mathbf{t}].$$

Moreover, $[H(\varphi; \mathbb{A})]_{\mathbb{A}} \in A[\mathbf{t}] \cap \mathbb{T}_n(K_\infty)^\times$ *and*

$$[H(\varphi; \mathbb{A})]_{\mathbb{A}} U(\varphi; A[\mathbf{t}]) = L(\varphi/\mathbb{A})A[\mathbf{t}].$$

Proof We give the proof for $U(\varphi; A[\mathbf{t}])$. The other assertion can be proved in a similar way, since, by Remark 7.6.3, $H(\widetilde{\varphi}; k(z)\mathbb{A})$ vanishes.

First, since φ has coefficients in $A[\mathbf{t}]$ and because we can compute $[H(\varphi; \mathbb{A})]_{\mathbb{A}}$ as a determinant by Eq. (7.2), we see that $[H(\varphi; \mathbb{A})]_{\mathbb{A}} \in A[\mathbf{t}]$.

Since the unit module has rank 1, by the class formula (Theorem 7.5.5), we get $[H(\varphi; \mathbb{A})]_{\mathbb{A}} U(\varphi; \mathbb{A}) = L(\varphi/\mathbb{A})\mathbb{A}$. Since $U(\varphi; \mathbb{A}) = kU(\varphi; A[\mathbf{t}])$, we can find $\eta \in U(\varphi; A[\mathbf{t}])$ such that $U(\varphi; \mathbb{A}) = \mathbb{A}\eta$. We can, and will, also assume that η is primitive in $\mathbb{T}_n(K_\infty)$, that is, not divisible by a non constant polynomial $\delta \in \mathbb{F}_q[\mathbf{t}]$. We get $[H(\varphi; \mathbb{A})]_{\mathbb{A}} \eta\mathbb{A} = L(\varphi/\mathbb{A})\mathbb{A}$, so that

$$L(\varphi/\mathbb{A}) = \lambda\, [H(\varphi; \mathbb{A})]_{\mathbb{A}}\, \eta$$

for some $\lambda \in \mathbb{F}_q^\times$. In particular, $[H(\varphi; \mathbb{A})]_{\mathbb{A}} \in \mathbb{T}_n(K_\infty)^\times$. We get:

$$U(\varphi; A[\mathbf{t}]) = U(\varphi; \mathbb{A}) \cap \mathbb{T}_n(K_\infty) = ([H(\varphi; \mathbb{A})]_{\mathbb{A}}^{-1} L(\varphi/\mathbb{A})\mathbb{A}) \cap \mathbb{T}_n(K_\infty)$$

$$= [H(\varphi; \mathbb{A})]_{\mathbb{A}}^{-1} L(\varphi/\mathbb{A})A[\mathbf{t}]$$

whence the result. $\square$

We set

$$\mathcal{N} = \left\{ x \in \mathbb{T}_n(K_\infty), v_\infty(x) \geq \frac{n}{q-1} - 1 \right\}.$$

Lemma 7.6.6 *If* $x \in \mathcal{N}$, $v_\infty(\exp_\varphi(x) - x) > v_\infty(x)$ *and* $v_\infty(\log_\varphi(x) - x) > v_\infty(x)$. *In particular, both* $\exp_\varphi$ *and* $\log_\varphi$ *define isometries* $\mathcal{N} \to \mathcal{N}$.

Proof For $k \geq 0$, we compute: $v_\infty(D_k) = -kq^k$, $v_\infty(l_k) = -q\frac{q^k-1}{q-1}$ and $v_\infty(f_k(\mathbf{t})) = -n\frac{q^k-1}{q-1}$. Thus, if $x \in \mathcal{N}$, and $k > 0$,

$$v_\infty\left(\frac{f_k(\mathbf{t})}{D_k}\tau^k(x)\right) = v_\infty(x) + (q^k - 1)\left(v_\infty(x) + k - \frac{n}{q-1}\right) + k > v_\infty(x)$$

and

$$v_\infty\left(\frac{f_k(\mathbf{t})}{l_k}\tau^k(x)\right) = v_\infty(x) + (q^k - 1)\left(\frac{q-n}{q-1} + v_\infty(x)\right) > v_\infty(x)$$

whence the result. □

Remark 7.6.7 If $n \leq 2q-2$, we have $\mathbb{T}_n(K_\infty) = \mathcal{N}+A[\mathbf{t}] \subset \exp_\varphi(\mathbb{T}_n(K_\infty))+A[\mathbf{t}]$ so that $H(\varphi; A[\mathbf{t}]) = \{0\}$ and

$$U(\varphi; A[\mathbf{t}]) = U_{St}(\varphi; A[\mathbf{t}]) = L(\varphi/\mathbb{A})A[\mathbf{t}].$$

7.7 Applications

7.7.1 Discrete Greenberg Conjectures

As a first application of the notion of Stark Units, we present a pseudo-nullity and a pseudo-cyclicity result from [ATR17] for the class module of the canonical deformation of the Carlitz module. These theorems are reminiscent of the Greenberg conjectures, in particular after evaluation at characters.

We keep the notation of all the previous sections. In particular, we recall that:

$$\mathcal{N} = \left\{x \in \mathbb{T}_n(K_\infty), v_\infty(x) \geq \frac{n}{q-1} - 1\right\}.$$

We denote now

$$\mathbb{B}_n(\mathbf{t}) = [H(\varphi; \mathbb{A})]_\mathbb{A} \in A[\mathbf{t}] \cap \mathbb{T}_n(K_\infty)^\times.$$

By Remark 7.6.7, we have $\mathbb{B}_n(\mathbf{t}) = 1$ if $1 \leq n \leq 2q - 2$. We also introduce the special elements:

$$u_n(\mathbf{t}, z) = \exp_{\widetilde{\varphi}}(L(\widetilde{\varphi}/k(z)\mathbb{A})) \in A[\mathbf{t}, z]$$

and

$$u_n(\mathbf{t}) = u_n(\mathbf{t}, 1) = \exp_\varphi(L(\varphi/\mathbb{A})) \in A[\mathbf{t}].$$

By Proposition 7.6.5, those elements generate the $A[\mathbf{t}, z]$-module (via $\widetilde{\varphi}$) $U(\widetilde{\varphi}; A[\mathbf{t}, z])$ and the $A[\mathbf{t}]$-module (via φ) of Stark units $U_{\mathrm{St}}(\varphi; A[\mathbf{t}])$.

If $1 \leq n \leq q-1$, $L(\varphi/\mathbb{A}) \in \mathcal{N}$; by Lemma 7.6.6, we have $u_n(\mathbf{t}) \in \mathcal{N} \cap A[\mathbf{t}] = \mathbb{F}_q$ and u_n has the same sign as $L(\varphi/\mathbb{A})$. Thus in this case, $u_n(\mathbf{t}) = 1$.

As we have seen in (7.8), $\exp_\varphi$ is injective on $\mathbb{T}_n(K_\infty)$ if and only if $n \not\equiv 1$ (mod $q - 1$). This leads us to distinguish the two cases, where different phenomena occur.

7.7.1.1   Case $n \not\equiv 1$ (mod $q - 1$)

We prove in this case the following pseudo-nullity result (see [ATR17, Theorem 3]):

Theorem 7.7.1 *We have* $\mathbb{B}_n(\mathbf{t}) = 1$, *that is,* $H(\varphi; A[\mathbf{t}])$ *is a finitely generated and torsion* $\mathbb{F}_q[\mathbf{t}]$-*module.*

Proof Let $r \in \{2, \ldots, q - 1\}$ be such that $n \equiv r$ (mod $q - 1$). We denote by ψ the r-variable twist of the Carlitz module:

$$\psi_\theta = (t_1 - \theta) \cdots (t_r - \theta)\tau + \theta.$$

We set:

$$\Xi := \frac{L(\psi/\mathbb{F}_q(t_1, \ldots, t_r))}{\omega(t_{r+1}) \cdots \omega(t_n)} \in \mathbb{T}_n(K_\infty)^\times.$$

We get for $a \in A[\mathbf{t}]$:

$$\exp_\varphi(a\,\Xi) = \frac{\exp_\psi(a L(\psi/\mathbb{F}_q(t_1, \ldots, t_r)))}{\omega(t_{r+1}) \cdots \omega(t_n)}$$

$$= \frac{\psi_a(u_r(t_1, \ldots, t_r))}{\omega(t_{r+1}) \cdots \omega(t_n)} = \frac{\psi_a(1)}{\omega(t_{r+1}) \cdots \omega(t_n)}.$$

Remark now that $\mathcal{N} = \left\{ x \in \mathbb{T}_n(K_\infty), v_\infty(x) \geq \frac{n-r}{q-1} \right\}$ so that

$$\mathbb{T}_n(K_\infty) = A[\mathbf{t}] \oplus \mathcal{N} \oplus \bigoplus_{k=1}^{\frac{n-r}{q-1}-1} \theta^{k - \frac{n-r}{q-1}} \mathbb{F}_q[\mathbf{t}].$$

We then define for $1 \leq i, j \leq \frac{n-r}{q-1} - 1$, $\beta_{ij} \in \mathbb{F}_q[\mathbf{t}]$ by the formula:

$$\exp_\varphi\left(\theta^i\,\Xi\right) - \sum_{j=1}^{\frac{n-r}{q-1}-1} \theta^{j - \frac{n-r}{q-1}} \beta_{ij} \in A[\mathbf{t}] \oplus \mathcal{N}.$$

Our theorem is now equivalent to $\det(\beta_{ij}) \neq 0$. Since $\det(\beta_{ij}) \in \mathbb{F}_q[\mathbf{t}]$, it will be enough to show that its evaluation at $t_1 = \cdots t_n = 0$ does not vanish. Let us denote by $\mathrm{ev}_0 : \mathbb{T}_n(K_\infty) \to K_\infty$ this evaluation. We have:

$$\mathrm{ev}_0(\exp_\varphi\left(\theta^i \,\Xi\right)) = \frac{\psi'_{\theta^i}(1)}{(-\theta)^{\frac{n-r}{q-1}}} \in \sum_{j=1}^{\frac{n-r}{q-1}-1} \theta^{j-\frac{n-r}{q-1}}\,\mathrm{ev}_0(\beta_{ij}) + A + \mathrm{ev}_0(\mathcal{N})$$

where $\psi'_\theta = (-\theta)^r \tau + \theta$. An immediate induction now shows that for $i \geq 1$,

$$\psi'_{\theta^i}(1) - \theta^i \in \theta^{i+1}A.$$

Thus $\mathrm{ev}_0(\det(\beta_{ij})) \neq 0$ and $\det(\beta_{ij}) \neq 0$. □

7.7.1.2 Case $n \equiv 1 \pmod{q-1}$

Let us first describe the unit module in this case:

Proposition 7.7.2 *If $n = 1$ then*

$$U(\varphi; A[\mathbf{t}]) = \frac{\widetilde{\pi}}{(t_1 - \theta)\omega(t_1)} A[\mathbf{t}].$$

and if $n > 1$, then

$$U(\varphi; A[\mathbf{t}]) = \frac{\widetilde{\pi}}{\prod_{i=1}^n \omega(t_i)} A[\mathbf{t}].$$

Proof Since $\frac{\widetilde{\pi}}{\prod_{i=1}^n \omega(t_i)} A[\mathbf{t}] = \ker \exp_\varphi$, it is clearly included in $U(\varphi; A[\mathbf{t}])$. As the unit module has rank 1, we deduce that if $x \in U(\varphi; A[\mathbf{t}])$, then $y = \exp_\varphi(x)$ is a torsion point for φ, that is there is $a \in A[\mathbf{t}]$ such that $\varphi_a(y) = 0$. But, if $v_\infty(x) \leq 0$, we see that

$$v_\infty((t_1 - \theta) \cdots (t_n - \theta)(\tau(x))) = q v_\infty(x) - n \quad \text{and} \quad v_\infty(\theta x) = v_\infty(x) - 1.$$

If $n > 1$, the first quantity is strictly lower than the second, this easily implies that no non trivial torsion point can exist: if $a \in A[\mathbf{t}]$, $\varphi_a(x)$ has the same (negative, and even explicitly computable) valuation as $\varphi_{\theta^{\deg_\theta(a)}}(x)$. With the same argument in the case $n = 1$ we see that if x is a torsion point, it must have valuation 0, so $x \in \mathbb{F}_q(t)$. Conversely, for $x \in \mathbb{F}_q(t)$, we have $\varphi_{(\theta-t)}(x) = 0$. □

Remark that in both cases we have the decomposition of $\mathbb{F}_q[t]$-modules: $\mathbb{T}_n(K_\infty) = \mathcal{N} \oplus U(\varphi; A[\mathbf{t}])$. In particular, if $n > 1$:

$$\exp_\varphi(\mathbb{T}_n(K_\infty)) = \mathcal{N}. \tag{7.10}$$

In the case $n = 1$, we know that $\mathbb{B}_n(\mathbf{t}) = 1$, so that, units and Stark units coincide, we deduce that $L(\varphi/\mathbb{A})$ equals, up to the sign, $\frac{\tilde{\pi}}{(\theta - t_1)\omega(t_1)}$. But both have sign 1, so that we recover Pellarin's formula (see [Pel12]):

$$L(\varphi/\mathbb{A}) = \frac{\tilde{\pi}}{(\theta - t_1)\omega(t_1)}.$$

If $n > 1$, we obtain another description of $\mathbb{B}_n(\mathbf{t})$:

$$\mathbb{B}_n(\mathbf{t}) = [H(\varphi; \mathbb{A})]_{\mathbb{A}} = (-1)^{\frac{n-1}{q-1}} L(\varphi/\mathbb{A}) \frac{\prod_{i=1}^n \omega(t_i)}{\tilde{\pi}}.$$

We deduce in particular that $\mathbb{B}_n(\mathbf{t})$ has degree in θ equal to $\frac{n-q}{q-1}$. In particular, when $n = q$, we recover that $\mathbb{B}_q(\mathbf{t}) = 1$ so that

$$L(\varphi/\mathbb{A}) = \frac{\tilde{\pi}}{\prod_{i=1}^q \omega(t_i)}.$$

More generally, if one can explicitly compute $\mathbb{B}_n(\mathbf{t})$, this gives us an explicit formula for $L(\varphi/\mathbb{A})$. We also stress that $L(\varphi/\mathbb{A}) \frac{\prod_{i=1}^n \omega(t_i)}{\tilde{\pi}} \in A[\mathbf{t}]$ is one of the main results of [AP15] where it is obtained without using the class formula.

Recall from Proposition 7.4.7 that we can build a map $\bar{\alpha} : \frac{U(\varphi; A[\mathbf{t}])}{U_{\mathrm{St}}(\varphi; A[\mathbf{t}])} \to H(\tilde{\varphi}; A[\mathbf{t}, z])[z - 1]$. We can compose it with the evaluation at $z = 1$ and obtain a map $\beta : \frac{U(\varphi; A[\mathbf{t}])}{U_{\mathrm{St}}(\varphi; A[\mathbf{t}])} \to H(\varphi; A[\mathbf{t}])$. Let us remark that β is induced by:

$$\exp_\varphi^{(1)} \begin{cases} U(\varphi; A[\mathbf{t}]) \to & \mathbb{T}_n(K_\infty) \\ x \mapsto & \sum_{k \geq 1} k \frac{f_k(\mathbf{t})}{D_k} \tau^k(x) \end{cases}$$

since we essentially differentiate $\exp_{\tilde{\varphi}}$ at 1 with respect to z.

Let us denote by $H^{(1)}(\varphi; A[\mathbf{t}]) \subset H(\varphi; A[\mathbf{t}])$ the image of β.

We devote the rest of this section to the proof of the following pseudo-cyclicity result (see [ATR17, Theorem 4]):

Theorem 7.7.3 *Let $n \geq q$. There is an isomorphism of $A[\mathbf{t}]$-modules:*

$$H^{(1)}(\varphi; A[\mathbf{t}]) \simeq \frac{A[\mathbf{t}]}{\mathbb{B}_n(\mathbf{t})A[\mathbf{t}]}$$

and the quotient $\frac{H(\varphi; A[\mathbf{t}])}{H^{(1)}(\varphi; A[\mathbf{t}])}$ is a finitely generated and torsion $\mathbb{F}_q[\mathbf{t}]$-module.

Proof Since $\frac{U(\varphi; A[\mathbf{t}])}{U_{\mathrm{St}}(\varphi; A[\mathbf{t}])}$ is an $A[\mathbf{t}]$-module isomorphic to $\frac{A[\mathbf{t}]}{\mathbb{B}_n(\mathbf{t})A[\mathbf{t}]}$ generated by the image of $\frac{\tilde{\pi}}{\prod_{i=1}^n \omega(t_i)}$, we are led to compute $\exp_\varphi^{(1)}(\frac{\tilde{\pi}}{\prod_{i=1}^n \omega(t_i)})$. But we have once again:

$$\exp_\varphi^{(1)}\left(\frac{\tilde{\pi}}{\prod_{i=1}^n \omega(t_i)}\right) = \frac{1}{\prod_{i=1}^n \omega(t_i)}\exp_C^{(1)}(\tilde{\pi})$$

where $\exp_C^{(1)} = \sum_{k\geq 1}k\frac{1}{D_k}\tau^k$. The proof that can be found in [ATR17, Theorem 3] relies on computations involving $\exp_C^{(1)}(\tilde{\pi})$. We will give here a slightly different proof, more similar to that of Theorem 7.7.1 above.

We denote by ψ the q-variable twist of the Carlitz module:

$$\psi_\theta = (t_1 - \theta)\cdots(t_q - \theta)\tau + \theta.$$

We first compute $u' = \exp_\psi^{(1)}\left(\frac{\tilde{\pi}}{\prod_{i=1}^q \omega(t_i)}\right)$. Since $\mathbb{B}_q(\mathbf{t}) = 1$, we have $u' \in A[t_1, \ldots, t_q] \oplus \mathcal{N}_q$ where $\mathcal{N}_q = \{x \in \mathbb{T}_q(K_\infty), v_\infty(x) \geq 1\}$. But $v_\infty(\frac{\tilde{\pi}}{\prod_{i=1}^q \omega(t_i)}) = 0$ and for $k \geq 1$, $v_\infty(D_k) = -kq^k$, $v_\infty(l_k) = -q\frac{q^k-1}{q-1}$ and $v_\infty(f_k(\mathbf{t})) = -n\frac{q^k-1}{q-1}$.

$$v_\infty\left(\frac{f_k(t_1,\cdots,t_q)}{D_k}\right) = kq^k - q\frac{q^k-1}{q-1} = q^k(k - \frac{q}{q-1}) + \frac{q}{q-1}$$

which is positive if $k > 1$ and equals 0 if $k = 1$. Thus $\frac{(t_1-\theta)\cdots(t_q-\theta)}{\theta^q-\theta}\frac{\tilde{\pi}}{\prod_{i=1}^q \omega(t_i)}$ has sign 1, we obtain that $u' \in 1 + \mathcal{N}_q$.

We get for $a \in A[\mathbf{t}]$:

$$\exp_\varphi^{(1)}\left(a\frac{\tilde{\pi}}{\prod_{i=1}^n \omega(t_i)}\right) = \frac{\exp_\psi^{(1)}(a\frac{\tilde{\pi}}{\prod_{i=1}^q \omega(t_i)})}{\omega(t_{q+1})\cdots\omega(t_n)}$$

$$\equiv \frac{\psi_a(u')}{\omega(t_{q+1})\cdots\omega(t_n)} \pmod{\mathcal{N} + A[\mathbf{t}]}$$

$$\equiv \frac{\psi_a(1)}{\omega(t_{q+1})\cdots\omega(t_n)} \pmod{\mathcal{N} + A[\mathbf{t}]}.$$

Remark now that

$$\mathbb{T}_n(K_\infty) = A[\mathbf{t}] \oplus \mathcal{N} \oplus \bigoplus_{k=1}^{\frac{n-q}{q-1}}\theta^{k-\frac{n-1}{q-1}}\mathbb{F}_q[\mathbf{t}].$$

We then define for $1 \leq i, j \leq \frac{n-q}{q-1}$, $\beta_{ij} \in \mathbb{F}_q[\mathbf{t}]$ by the formula:

$$\exp_\varphi^{(1)}\left(\theta^i\frac{\tilde{\pi}}{\prod_{i=1}^n \omega(t_i)}\right) - \sum_{j=1}^{\frac{n-q}{q-1}}\theta^{j-\frac{n-1}{q-1}}\beta_{ij} \in A[\mathbf{t}] \oplus \mathcal{N}.$$

The injectivity of β is now equivalent to $\det(\beta_{ij}) \neq 0$. It is again enough to show that its evaluation at $t_1 = \cdots t_n = 0$ does not vanish. Let us denote by $\mathrm{ev}_0 : \mathbb{T}_n(K_\infty) \to K_\infty$ this evaluation. We have:

$$\mathrm{ev}_0\left(\exp_\varphi^{(1)}\left(\theta^i \frac{\widetilde{\pi}}{\prod_{i=1}^n \omega(t_i)}\right)\right) = \frac{\psi'_{\theta^i}(1)}{(-\theta)^{\frac{n-q}{q-1}}} \in \sum_{j=1}^{\frac{n-q}{q-1}} \theta^{j-\frac{n-1}{q-1}} \mathrm{ev}_0(\beta_{ij}) + A + \mathrm{ev}_0(\mathcal{N})$$

where $\psi'_\theta = (-\theta)^q \tau + \theta$. But again, for $i \geq 1$,

$$\psi'_{\theta^i}(1) - \theta^i \in \theta^{i+1} A.$$

Thus $\mathrm{ev}_0(\det(\beta_{ij})) \neq 0$ and $\det(\beta_{ij}) \neq 0$.

Finally, $H^{(1)}(\varphi; A[\mathbf{t}])$ is a sub-$\mathbb{F}_q[\mathbf{t}]$-module of $H(\varphi; A[\mathbf{t}])$ with same rank, which gives the last assertion. $\square$

7.7.1.3 Evaluation at Characters

Let us now very briefly explain some consequences of Theorems 7.7.1 and 7.7.3 above. We refer the reader for instance to [APTR16, §9] for more details. Let a be a non constant and square free element in A and $\chi : A/aA \to \overline{\mathbb{F}}_q$ be a Dirichlet character mod a. Let us denote by k_a the extension of $\mathbb{F}_q$ generated by the roots of a. Then one can find $\zeta_1, \ldots, \zeta_n \in k_a$ (in fact all of the ζ_i's are roots of a) such that for all $b \in A$, $\chi(b) = b(\zeta_1) \cdots b(\zeta_n)$. We then have a natural homomorphism of $\mathbb{F}_q$-vector spaces $\mathrm{ev}_\chi : \mathbb{T}_n(K_\infty) \to (k_a K)_\infty$ which evaluates t_i to ζ_i for all $1 \leq i \leq n$.

We get for instance:

$$\mathrm{ev}_\chi(L(\varphi/\mathbb{A})) = L(C/A, \chi) := \sum_{b \in A+} \frac{\chi(b)}{b}.$$

In order to define the class module associated to χ, we define $\tau_a : K_\infty \otimes_{\mathbb{F}_q} k_a K_\infty \otimes_{\mathbb{F}_q} k_a$ by $\tau_a = \tau \otimes \mathrm{id}$. We use it to define the Drinfeld A-module C' over $A \otimes_{\mathbb{F}_q} k_a$ by $C'_\theta = \theta + \prod_{i=1}^n (1 \otimes \zeta_i - \theta \otimes 1)\tau_a$. Then:

$$H_\chi := \frac{C'(K_\infty \otimes_{\mathbb{F}_q} k_a)}{\exp_{C'}(K_\infty \otimes_{\mathbb{F}_q} k_a) + C'(A \otimes_{\mathbb{F}_q} k_a)}.$$

In fact, ev_χ also induces a surjection $H(\varphi; A[\mathbf{t}]) \to H_\chi$. Moreover, although the number n of variables involved in this construction is not unique, it is unique modulo $q - 1$. The minimal number n that can be used is called the *type* of χ. There is a well defined notion of "almost all characters of type n" which is, roughly speaking, all but a Zariski closed non trivial subset.

Then, Theorems 7.7.1 and 7.7.3 imply:

Theorem 7.7.4

1. *If $n \not\equiv 1 \pmod{q - 1}$, then for almost all Dirichlet character χ of type n, we have $H_\chi = \{0\}$.*
2. *If $n \equiv 1 \pmod{q - 1}$, then for almost all Dirichlet character χ of type n, H_χ is a cyclic $A \otimes k_a$-module.*

These two results remind of the celebrated Greenberg conjectures. For details on the analogy between the two contexts we refer the reader to [ATR17, Introduction].

7.7.2 On the Bernoulli-Carlitz Numbers

As a second application, we show the non vanishing of families of Bernoulli-Carlitz numbers modulo monic irreducible polynomials P for almost all P. This is a striking result as it is a stronger function field version of an open conjecture on Bernoulli numbers.

The classical Bernoulli numbers have been discovered and studied by Jacob Bernoulli during the late seventeenth century. They can be defined as the coefficients B_m, $m \geq 0$ which appear in the power series equality

$$\frac{t}{e^t - 1} = \sum_{m \geq 0} B_m \frac{t^m}{m!}. \tag{7.11}$$

Euler computed the zeta values $\zeta(n) = \sum_{k \geq 1} k^{-n}$ for even positive integers n with the help of the Bernoulli numbers: if $n > 0$ is even then

$$\zeta(n) = \frac{-1}{2} \frac{(2i\pi)^n}{n!} B_n. \tag{7.12}$$

For more background on Bernoulli numbers, we refer the reader for instance to [IR90, Chapter 15 §1].

In 1935, Carlitz introduced analogues of the Bernoulli numbers. Those *Bernoulli-Carlitz* numbers are linked with the polynomials $\mathbb{B}_n(\mathbf{t})$. We prove in this section a quite surprising result on the Bernoulli-Carlitz numbers with the help of $\mathbb{B}_n(\mathbf{t})$. Let $N > 1$ be an integer and $N = \sum_{i=0}^r n_i q^i$ be its q-expansion. Then we define the Carlitz factorial as:

$$\Pi(N) = \prod_{i=0}^r D_i^{n_i} \in A$$

where we recall (see Sect. 7.3.3) that $D_0 = 1$, and for $i \geq 1$, $D_{i+1} = D_i^q (\theta^{q^{i+1}} - \theta)$. The Bernoulli-Carlitz numbers are defined as the coefficients BC_N, $N \geq 0$ which appear in the power series equality (similar to (7.11)):

$$\frac{t}{\exp_C(t)} = \sum_{m \geq 0} BC_N \frac{t^N}{\Pi(N)}.$$

We also recall that for $N \geq 1$, we have the Carlitz zeta value:

$$\zeta_A(N) = \sum_{a \in A_+} \frac{1}{a^N}.$$

Then the N-th Bernoulli-Carlitz number is $BC_N = 0$ if $N \not\equiv 0 \pmod{q-1}$ and, if $N \equiv 0 \pmod{q-1}$,

$$\zeta_A(N) = \frac{\widetilde{\pi}^N}{\Pi(N)} BC_N$$

reminding of Euler's formula (7.12). (Remark that the role of 2 is played here by $q - 1$.)

If we have the q-expansion $N = \sum_{i=0}^r n_i q^i$, then we denote $\ell_q(N) = \sum_{i=0}^r n_i$ and define the evaluation map $\mathrm{ev}_N : \mathbb{T}_{\ell_q(N)}(K_\infty) \to K_\infty$ by $\mathrm{ev}_N(t_j) = \theta^{q^k}$ if $\sum_{i=0}^{k-1} n_i < j \leq \sum_{i=0}^k n_i$, so that

$$\mathrm{ev}_N(a(t_1) \cdots a(t_{\ell_q(N)})) = a^N.$$

We recall the link between Bernoulli-Carlitz numbers and the polynomials $\mathbb{B}_n(\mathbf{t})$:

Proposition 7.7.5 *Let $N \geq 2$, $N \equiv 1 \pmod{q-1}$. Let $P \in A$ be a monic irreducible polynomial of degree $d > 1$, such that $q^d > N$. Then $BC_{q^d-N} \equiv 0 \pmod{P}$ if and only if $\mathrm{ev}_N(\mathbb{B}_{\ell_q(N)}(\mathbf{t})) \equiv 0 \pmod{P}$.*

We do not give the proof, which can be found in [ANDTR19, Proposition 4.3] or in [AP14, Theorem 2]. Let us just sketch the main ideas: starting with the identity in $\ell_q(N)$ variables:

$$(-1)^{\frac{\ell_q(N)-1}{q-1}} \frac{L(\varphi/\mathbb{A})}{\widetilde{\pi}} \prod_{i=1}^{\ell_q(N)} \omega(t_i) = \mathbb{B}_{\ell_q(N)}(\mathbf{t}).$$

We then apply τ^d and evaluate with ev_N so that, up to some terms, the left hand side becomes $\frac{\zeta_A(q^d-N)}{\widetilde{\pi}^{q^d-N}} = \frac{BC_{q^d-N}}{\Pi_{q^d-N}}$, and the right hand side is congruent to $\mathrm{ev}_N(\mathbb{B}_{\ell_q(N)}(\mathbf{t})) \bmod P$ since for all $a \in A$, $a^{q^d} \equiv a \pmod{P}$.

As a consequence of Proposition 7.7.5, we see that if $\mathrm{ev}_N(\mathbb{B}_{\ell_q(N)}(\mathbf{t})) \neq 0$, then for all P not dividing $\mathrm{ev}_N(\mathbb{B}_{\ell_q(N)}(\mathbf{t}))$ and such that $q^{\deg P} > N$, that is, for almost all P, we have $BC_{q^d-N} \equiv 0 \pmod{P}$. In fact, we have the more precise result:

Theorem 7.7.6 *Let $N \geq 2$, $N \equiv 1 \pmod{q-1}$. Let $P \in A$ be a monic irreducible polynomial of degree $d > 1$, such that $q^d > N$. If $d \geq \frac{\ell_q(N)-1}{q-1} N$, then $BC_{q^d-N} \not\equiv 0 \pmod{P}$.*

This is a strong version of the following open conjecture on classical Bernoulli numbers:

Conjecture 7.7.7 *Let $N \geq 3$ be an odd integer, then there exist infinitely many prime numbers p such that $B_{p-N} \not\equiv 0 \pmod{p}$.*

It seems however reasonable to expect that the equivalent of Theorem 7.7.6 does not hold for Bernoulli numbers. Namely, if $N \geq 3$ is an odd integer, then there should exist infinitely many prime numbers p such that $B_{p-N} \equiv 0 \pmod{p}$. This would be an example where number fields and function fields lead to different results.

Theorem 7.7.6 is the main theorem of [ANDTR19]. The key result is that $\mathrm{ev}_N(\mathbb{B}_{\ell_q(N)}(\mathbf{t}))$ is not zero. We actually prove more generally:

Theorem 7.7.8 *Let $n \geq 2$, $n \equiv 1 \pmod{q-1}$. Then for any evaluation homomorphism $\mathrm{ev} : A[\mathbf{t}] \to A$ such that $\mathrm{ev}(t_i)$ is non constant for all i, we have*

$$\mathrm{ev}(\mathbb{B}_n(\mathbf{t})) \neq 0.$$

Proof We give a proof different from the one of [ANDTR19]. Recall:

$$H(\varphi; \mathbb{A}) = \frac{\varphi(\mathbb{K})}{\exp_\varphi(\mathbb{K}) + \varphi(\mathbb{A})}.$$

And $\mathbb{B}_n(\mathbf{t}) = [H(\varphi; \mathbb{A})]_\mathbb{A}$, in particular:

$$\mathbb{B}_n(\mathbf{t}) = \det(Z - \varphi_\theta \mid H(\varphi; \mathbb{A}))_{|Z=\theta}.$$

We set $r = \frac{n-q}{q-1}$. As for (7.10), we have

$$\exp_\varphi(\mathbb{K}) = \left\{ x \in \mathbb{K}, v_\infty(x) \geq \frac{n}{q-1} - 1 \right\}.$$

Since $\frac{n}{q-1} - 1 = r + \frac{1}{q-1}$, a basis of $H(\varphi; \mathbb{A})$ is given by $\frac{1}{\theta^r}, \cdots, \frac{1}{\theta}$. We compute the matrix of φ_θ in this basis. It is the sum of a matrix M_n that we must determine and of a nilpotent matrix $N_n = (\delta_{i,j+1})_{1 \leq i,j \leq r}$ where $\delta_{i,j}$ is the Kronecker symbol. That is, the coefficients of N_n immediately above the diagonal are 1, and 0 elsewhere. Note that M_n is the matrix of $(t_1 - \theta) \cdots (t_n - \theta)\tau$. Since $q(r-k) = r + n - q(k+1)$, we get in $H(\varphi; \mathbb{A})$:

$$(t_1 - \theta) \cdots (t_n - \theta) \tau \left(\frac{1}{\theta^{r-k}} \right) = \sum_{j=0}^{r-1} \frac{\sigma(q(k+1) - j)}{\theta^{r-j}}$$

where

$$\sigma(j) = (-1)^{j-1} \sum_{i_1 < i_2 < \cdots < i_j} t_{i_1} \cdots t_{i_j}$$

if $0 \le j \le n$, and $\sigma(j) = 0$ otherwise. (Note that $\sigma(0) = -1$.) Thus,

$$M_n = (\sigma(jq - (i-1)))_{1 \le i, j \le r}.$$

We will replace the polynomials $\sigma(j)$ by symbols independent of the number of variables in order to proceed by induction on n. We define on $\mathbb{F}_q$ variables Σ_j, $j > 0$ and a valuation val on $\mathbb{F}_q[\Sigma_j, j > 0]$ such that $val(\Sigma_j) = j$ by stating that if

$$f = \sum_{k_1, \dots, k_n \ge 0} \alpha_{k_1, \dots, k_n} \prod_{j=1}^{n} \Sigma_j^{k_j}$$

then $val(f) = -\infty$ if $f = 0$ and $val(f) = \inf\{\sum_{j=1}^{n} jk_j; \alpha_{k_1, \dots, k_n} \ne 0\}$ otherwise. We moreover set $\Sigma_0 = -1$ and $\Sigma_j = 0$ if $j < 0$. Let

$$\mathbb{M}_n = (\Sigma_{jq-(i-1)})_{1 \le i, j \le r}.$$

We have the evaluation map $ev_n : \mathbb{F}_q[\Sigma_j, j > 0] \to \mathbb{F}_q[\mathbf{t}]$ defined by $ev_n(\Sigma_j) = \sigma(j)$ (recall that $\sigma(j) = 0$ if $n < j$). Then $val(f)$ equals the valuation of $ev_n(f)$ with respect to the ideal $(t_1, \dots, t_n)$, and

$$M_n = ev_n(\mathbb{M}_n).$$

Developing now $\det(ZI_r - \mathbb{M}_n - N_n)$ with respect to the last column, we find

$$\det(ZI_r - \mathbb{M}_n - N_n) = Z \det(ZI_{r-1} - \mathbb{M}_{n-(q-1)} - N_{n-(q-1)}) + \epsilon$$

where ϵ is a sum of terms which are multiples of elements in the last column of $\mathbb{M}_n$, that is, $\Sigma_{rq-(i-1)}, 0 \le i \le r$ all of them of valuation at least $rq - (r-1) = r(q-1) + 1$.

Thus, by induction, $\det(ZI_r - \mathbb{M}_n - N_n) = Z^r + \sum_{i=1}^{r} \beta_i Z^{r-i}$ with $val(\beta_i) \ge i(q-1) + 1$, and thus

$$\mathbb{B}_n = \theta^r + \sum_{i=1}^{r} B_i(\mathbf{t}) \theta^{r-i}$$

where the valuation of $B_i(\mathbf{t}) \in \mathbb{F}_q[\mathbf{t}]$ with respect to $(t_1, \ldots, t_n)$ is at least $i(q - 1) + 1$. Thus for every evaluation homomorphism ev, $\mathrm{ev}(\mathbb{B}_n(\mathbf{t}))$ has valuation r at the place θ of A. $\qquad\square$

7.8 Stark Units in More General Settings

In this final short section, we want to stress out that the machinery of Stark units carries over to more general settings than Drinfeld $\mathbb{F}_q[\theta]$-modules. The results presented in Sect. 7.4 have indeed been developed in [ANDTR17] for Drinfeld modules over a general A. More precisely, we replace K with a function field in which $\mathbb{F}_q$ is algebraically closed, fix a place ∞ of K and write A for the ring of functions regular outside ∞ (see [Pel20, §2.2]). If L/K is a finite extension, a Drinfeld A-module over O_L is an $\mathbb{F}_q$-algebra homomorphism

$$\phi : \begin{cases} A \to O_L[\tau] \\ a \mapsto \phi_a \end{cases}$$

such that $\phi_a \equiv a \pmod{\tau}$ for all $a \in A$. We refer the reader to [Pel20, §3] for a presentation of the Drinfeld modules in this general setting. We can define units in this setting, and follow the constructions presented in this text, that is, twist the Frobenius by a new variable z, define z-units and evaluate them at $z = 1$ to obtain Stark units.

Let K_∞ denote the completion of K at ∞ and $\mathbb{F}_\infty$ its residue field. We choose a *sign function* sgn : $K_\infty^\times \to \mathbb{F}_\infty^\times$, that is, a group homomorphism which is the identity on $\mathbb{F}_\infty^\times$. A rank one Drinfeld module ϕ is *sign-normalized* if there is an $i \in \mathbb{N}$ such that

$$\forall a \in A \backslash \{0\}, \quad \phi_a = a + a_1 \tau + \cdots + \mathrm{sgn}(a)^{q^i} \tau^{\deg a}.$$

Stark units are used in [ANDTR17] to obtain various results for sign normalized rank one Drinfeld modules: explicitly computing the Taelman units, obtaining a class formula and some log-algebraicity results, that is, constructing explicit units by the mean of the L-series. As in Sect. 7.6.2, canonical deformations of these Drinfeld modules are also introduced by means of their shtuka functions.

In [ANDTR20a], Stark units have been extended to Anderson t-modules (for $A = \mathbb{F}_q[\theta]$) which are defined as $\mathbb{F}_q$-algebra homomorphisms

$$E : \begin{cases} \mathbb{F}_q[\theta] \to & M_n(O_L)[\tau] \\ a \mapsto E_a = E_{a,0} + E_{a,1} \tau + \cdots + E_{a,r \deg a} \tau^{r \deg a} \end{cases}$$

such that $(E_a - E_{a,0})^n = 0$ for all $a \in \mathbb{F}_q[\theta]$. For instance, the n-th tensor power of the Carlitz module is the Anderson t-module defined by

$$
E_\theta := \begin{pmatrix} \theta & 1 & & \\ & \ddots & \ddots & \\ & & \ddots & 1 \\ & & & \theta \end{pmatrix} + \begin{pmatrix} 0 & 0 & \dots & 0 \\ \vdots & \vdots & & \vdots \\ 0 & 0 & \dots & 0 \\ 1 & 0 & \dots & 0 \end{pmatrix} \tau.
$$

We refer the reader to [AT90] for more details about these Anderson t-modules.

Once again, Stark units play a key role in [ANDTR20a] to determine the Taelman's units of t-modules which allows to prove that a large class of t-modules satisfy a conjecture of Taelman stated in [Tae09]. They are also used to establish log-algebraicity identities for the tensor powers of the Carlitz module.

One can finally extend the definition of t-module to a general A and define Stark units in this context where the machinery of Sect. 7.4 still works.

We also signal to the reader two very recent works involving Stark units: in [GND20] Green and Ngo Dac use Stark units to obtain log-algebraic identities for Anderson t-modules. They derive from it some logarithmic identities on multiple zeta values. In [ANDTR20b], the authors prove a class formula generalizing Theorem 7.5.3 to a large class of Anderson modules over a general A, which includes in particular all Drinfeld modules.

We will end this survey with a remark on the level of generality to which one can extend the notion of Stark units. At the beginning of this work, we had an exponential map, that is a power series in the Frobenius τ which satisfies a certain functional identity involving τ, and we wanted to study the Taelman units, that is the inverse image of the integral elements through the exponential map. We then introduced the Stark units by twisting the Frobenius τ with a new variable z and proceeded to the study of the z-units before evaluating at 1 to get a natural submodule of the Taelman units. If we now consider a difference field (K, τ) (see [DV20, §2]), then the above construction should carry over if we have a suitable exponential map. It would be interesting to work out Stark units in this general setting (which involves a definition of a *suitable* exponential map). Due to the formal nature of the construction, one would expect applications mainly in the case of non archimedean fields. L. Di Vizio's contribution [DV20] to this volume gives many examples of difference fields for which one could try to see what comes out from a construction of Stark units.

Acknowledgments This text follows a stay at the VIASM in Hanoi in July 2018. I am very grateful to the institution and all the people at the VIASM for the hospitality and the excellent working conditions of the stay. It is a pleasure to thank B. Anglès and T. Ngo Dac for very helpful and instructive conversations. I also thank the referees for their corrections and suggestions that helped to improve the original text.

References

[And86] G.W. Anderson, t-motives. Duke Math. J. **53**(2), 457–502 (1986)

[And94] G.W. Anderson, Rank one elliptic A-modules and A-harmonic series. Duke Math. J. **73**(3), 491–542 (1994)

[And96] G.W. Anderson, Log-algebraicity of twisted A-harmonic series and special values of L-series in characteristic p. J. Number Theory **60**(1), 165–209 (1996)

[ANDTR17] B. Anglès, T. Ngo Dac, F. Tavares Ribeiro, Stark units in positive characteristic. Proc. Lond. Math. Soc. (3) **115**(4), 763–812 (2017)

[ANDTR19] B. Anglès, T. Ngo Dac, F. Tavares Ribeiro, Exceptional zeros of L-series and Bernoulli-Carlitz numbers. Ann. Sc. Norm. Super. Pisa Cl. Sci. (5) **19**(3), 981–1024 (2019)

[ANDTR20a] B. Anglès, T. Ngo Dac, F. Tavares Ribeiro, On special L-values of t-modules. Adv. Math. **372**, Art. 107313 (2020)

[ANDTR20b] B. Anglès, T. Ngo Dac, F. Tavares Ribeiro, A class formula for admissible Anderson modules (2020). Preprint https://hal.archives-ouvertes.fr/hal-02490566/document

[AP14] B. Anglès, F. Pellarin, Functional identities for L-series values in positive characteristic. J. Number Theory **142**, 223–251 (2014)

[AP15] B. Anglès, F. Pellarin, Universal Gauss-Thakur sums and L-series. Invent. Math. **200**(2), 653–669 (2015)

[APTR16] B.Anglès, F. Pellarin, F. Tavares Ribeiro, Arithmetic of positive characteristic L-series values in Tate algebras. Compos. Math. **152**(1), 1–61 (2016). With an appendix by F. Demeslay

[APTR18] B. Anglès, F. Pellarin, F. Tavares Ribeiro, Anderson-Stark units for $\mathbb{F}_q[\theta]$. Trans. Am. Math. Soc. **370**(3), 1603–1627 (2018)

[AT90] G.W. Anderson, D.S. Thakur, Tensor powers of the Carlitz module and zeta values. Ann. Math. (2) **132**(1), 159–191 (1990)

[AT15] B. Anglès, L. Taelman, Arithmetic of characteristic p special L-values. Proc. Lond. Math. Soc. (3) **110**(4), 1000–1032 (2015). With an appendix by Vincent Bosser

[ATR17] B. Anglès, F. Tavares Ribeiro, Arithmetic of function field units. Math. Ann. **367**(1–2), 501–579 (2017)

[Bou65] N. Bourbaki, *Éléments de mathématique. Fasc. XXXI. Algèbre commutative. Chapitre 7: Diviseurs* (Actualités Scientifiques et Industrielles, No. 1314. Hermann, Paris, 1965)

[Dem14] F. Demeslay, A class formula for L-series in positive characteristic (2014). arXiv:1412.3704

[DV20] L. Di Vizio, Difference Galois theory for the "applied" mathematician, in *Arithmetic and Geometry Over Local Fields*, ed. by B. Anglès, T. Ngo Dac (Springer International Publishing, Cham, 2020)

[Eis95] D. Eisenbud, Commutative algebra, in *Graduate Texts in Mathematics*, vol. 150 (Springer, New York, 1995). With a view toward algebraic geometry

[Fan18] J. Fang, Equivariant special L-values of abelian t-modules. J. Number Theory (2018)

[FGHP20] J. Ferrara, N. Green, Z. Higgins, C.D. Popescu, An equivariant Tamagawa number formula for Drinfeld modules and applications (2020). https://arxiv.org/abs/2004.05144

[Gos96] D. Goss, Basic structures of function field arithmetic, in *Ergebnisse der Mathematik und ihrer Grenzgebiete (3) [Results in Mathematics and Related Areas (3)]*, vol. 35 (Springer, Berlin, 1996)

[GND20] N. Green, T. Ngo Dac, On log-algebraic identities for Anderson t-modules and characteristic p multiple zeta values (2020). https://arxiv.org/abs/2007.11060

[IR90] K. Ireland, M. Rosen, A classical introduction to modern number theory, in *Graduate Texts in Mathematics*, vol. 84, 2nd edn. (Springer, New York, 1990)

[Lan02] S. Lang, Algebra, in *Graduate Texts in Mathematics*, vol. 211, 3rd edn. (Springer, New York, 2002)

[MW84] B. Mazur, A. Wiles, Class fields of abelian extensions of $\mathbf{Q}$. Invent. Math. **76**(2), 179–330 (1984)

[Nor76] D.G. Northcott, *Finite Free Resolutions* (Cambridge University, Cambridge, 1976). Cambridge Tracts in Mathematics, No. 71

[Pel12] F. Pellarin, Values of certain L-series in positive characteristic. Ann. of Math. (2) **176**(3), 2055–2093 (2012)

[Pel20] F. Pellarin, From the Carlitz exponential to Drinfeld modular forms, in *Arithmetic and Geometry Over Local Fields*, ed. by B. Anglès, T. Ngo Dac (Springer International Publishing, Cham, 2020)

[Ros02] M. Rosen, Number theory in function fields, in *Graduate Texts in Mathematics*, vol. 210 (Springer, New York, 2002)

[Tae09] L. Taelman, Special L-values of t-motives: a conjecture. Int. Math. Res. Not. **2009**(16), 2957–2977 (2009)

[Tae12] L. Taelman, Special L-values of Drinfeld modules. Ann. of Math. (2) **175**(1), 369–391 (2012)

[Tha93] D.S. Thakur, Shtukas and Jacobi sums. Invent. Math. **111**(3), 557–570 (1993)

[Tha04] D.S. Thakur, *Function Field Arithmetic* (World Scientific Publishing Co. Inc., River Edge, 2004)

LECTURE NOTES IN MATHEMATICS

Editors in Chief: J.-M. Morel, B. Teissier;

Editorial Policy

1. Lecture Notes aim to report new developments in all areas of mathematics and their applications – quickly, informally and at a high level. Mathematical texts analysing new developments in modelling and numerical simulation are welcome.

   Manuscripts should be reasonably self-contained and rounded off. Thus they may, and often will, present not only results of the author but also related work by other people. They may be based on specialised lecture courses. Furthermore, the manuscripts should provide sufficient motivation, examples and applications. This clearly distinguishes Lecture Notes from journal articles or technical reports which normally are very concise. Articles intended for a journal but too long to be accepted by most journals, usually do not have this "lecture notes" character. For similar reasons it is unusual for doctoral theses to be accepted for the Lecture Notes series, though habilitation theses may be appropriate.

2. Besides monographs, multi-author manuscripts resulting from SUMMER SCHOOLS or similar INTENSIVE COURSES are welcome, provided their objective was held to present an active mathematical topic to an audience at the beginning or intermediate graduate level (a list of participants should be provided).

   The resulting manuscript should not be just a collection of course notes, but should require advance planning and coordination among the main lecturers. The subject matter should dictate the structure of the book. This structure should be motivated and explained in a scientific introduction, and the notation, references, index and formulation of results should be, if possible, unified by the editors. Each contribution should have an abstract and an introduction referring to the other contributions. In other words, more preparatory work must go into a multi-authored volume than simply assembling a disparate collection of papers, communicated at the event.

3. Manuscripts should be submitted either online at www.editorialmanager.com/lnm to Springer's mathematics editorial in Heidelberg, or electronically to one of the series editors. Authors should be aware that incomplete or insufficiently close-to-final manuscripts almost always result in longer refereeing times and nevertheless unclear referees' recommendations, making further refereeing of a final draft necessary. The strict minimum amount of material that will be considered should include a detailed outline describing the planned contents of each chapter, a bibliography and several sample chapters. Parallel submission of a manuscript to another publisher while under consideration for LNM is not acceptable and can lead to rejection.

4. In general, **monographs** will be sent out to at least 2 external referees for evaluation.

   A final decision to publish can be made only on the basis of the complete manuscript, however a refereeing process leading to a preliminary decision can be based on a pre-final or incomplete manuscript.

   Volume Editors of **multi-author works** are expected to arrange for the refereeing, to the usual scientific standards, of the individual contributions. If the resulting reports can be forwarded to the LNM Editorial Board, this is very helpful. If no reports are forwarded or if other questions remain unclear in respect of homogeneity etc, the series editors may wish to consult external referees for an overall evaluation of the volume.

5. Manuscripts should in general be submitted in English. Final manuscripts should contain at least 100 pages of mathematical text and should always include

   – a table of contents;
   – an informative introduction, with adequate motivation and perhaps some historical remarks: it should be accessible to a reader not intimately familiar with the topic treated;
   – a subject index: as a rule this is genuinely helpful for the reader.
   – For evaluation purposes, manuscripts should be submitted as pdf files.

6. Careful preparation of the manuscripts will help keep production time short besides ensuring satisfactory appearance of the finished book in print and online. After acceptance of the manuscript authors will be asked to prepare the final LaTeX source files (see LaTeX templates online: https://www.springer.com/gb/authors-editors/book-authors-editors/manuscriptpreparation/5636) plus the corresponding pdf- or zipped ps-file. The LaTeX source files are essential for producing the full-text online version of the book, see http://link.springer.com/bookseries/304 for the existing online volumes of LNM). The technical production of a Lecture Notes volume takes approximately 12 weeks. Additional instructions, if necessary, are available on request from lnm@springer.com.

7. Authors receive a total of 30 free copies of their volume and free access to their book on SpringerLink, but no royalties. They are entitled to a discount of 33.3 % on the price of Springer books purchased for their personal use, if ordering directly from Springer.

8. Commitment to publish is made by a *Publishing Agreement*; contributing authors of multiauthor books are requested to sign a *Consent to Publish form*. Springer-Verlag registers the copyright for each volume. Authors are free to reuse material contained in their LNM volumes in later publications: a brief written (or e-mail) request for formal permission is sufficient.

Addresses:
Professor Jean-Michel Morel, CMLA, École Normale Supérieure de Cachan, France
E-mail: moreljeanmichel@gmail.com

Professor Bernard Teissier, Equipe Géométrie et Dynamique,
Institut de Mathématiques de Jussieu – Paris Rive Gauche, Paris, France
E-mail: bernard.teissier@imj-prg.fr

Springer: Ute McCrory, Mathematics, Heidelberg, Germany,
E-mail: lnm@springer.com

Printed in the United States
By Bookmasters